THE OF THE COMPUTER

A TECHNICAL AND BUSINESS HISTORY

STEPHEN J MARSHALL

Copyright © 2015 Stephen J Marshall

All rights reserved. No part of this publication may be reproduced, stored in a retrieval system, or transmitted, in any form or in any means – by electronic, mechanical, photocopying, recording or otherwise – without the prior written permission of the copyright owner.

Print edition May 2017

ISBN 978-1-5468-4907-0

Typeset in Georgia and Arial using Adobe InDesign

Cover design by Eleanor M Marshall

www.storyofthecomputer.com

Contents

Introduction   1

PART ONE
MECHANISATION

1. Computer Prehistory – Calculating Machines   7
2. The Analogue Alternative   57
3. The Electrification of Calculating Machinery   85
4. The Dawn of Electronic Computation   123

PART TWO
INDUSTRIALISATION

5. The Metamorphosis of the Calculator – Stored-Program Machines   155
6. Data Processing and the Birth of the Computer Industry   189

PART THREE
SEGMENTATION

7. Revving Up for Higher Performance   237
8. Solid Progress – Transistors and Unified Architectures   261
9. Small is Beautiful – The Minicomputer   313

PART FOUR
PERSONALISATION

10. The Big Picture – Computer Graphics   353
11. The Microcomputer Revolution   397
12. Bringing it all Together – The Graphics Workstation   447
13. Getting Personal – The World According to Wintel   505

Index   555

Introduction

"I have found that the possession of an understanding of technology, just as with an understanding of music, literature, or the arts, brings with it great personal satisfaction and pleasure", Lord Broers, 2005.

The digital computers which have become such an indispensable part of our everyday lives are the culmination of centuries of innovation and development. Their story begins with the earliest efforts to mechanise calculation in the 17th century, steadily moving from mechanical to electromechanical then electronic technologies before accelerating rapidly with the arrival of stored-program machines and the establishment of a nascent computer industry in the 1950s. From this point on, the rate of technological progress has been astounding. In his classic 1986 paper, 'No Silver Bullet – Essence and Accident in Software Engineering', Fred Brooks writes that, "No other technology since civilisation began has seen six orders of magnitude price-performance gain in 30 years". It is this incredible rate of progress which has led to the development of the sophisticated and inexpensive electronic computers that almost every person in the developed world now takes for granted.

Computers are the enablers for many of the technologies that our modern world relies upon and their impact on society cannot be overestimated. Computer technology has facilitated the exploration of space and has helped put a man on the moon. It has allowed scientists to map the human genome and probe the innermost secrets of the atom. Computers are used by engineers to design better, safer, more affordable products. They control the machine tools and robots that manufacture most of the goods in our shops. In their tiniest microprocessor form, they are embedded within our mobile telephones and domestic appliances, and the demand for microprocessors is now so high that billions of them are made each year.

The business aspects of the story are equally impressive. The computer industry has spawned some of the most successful and highest valued companies the world has ever seen. Several of the founders of these companies have become multi-billionaires. A few, such as Bill Gates and the late Steve Jobs, have even become celebrities, their names held up as the epitome of success for every aspiring entrepreneur or business leader. The computer is also the engine of the knowledge economy and has changed the way we do business. The global financial markets which exert such a large influence over the wealth of nations have become so complex that they could not exist without computer technology. Computer systems have also revolutionised personal finance through credit cards, Automated Teller Machines and online banking, a trend which looks likely to continue unabated with recent developments in virtual currency.

The Story of the Computer

The convergence of portable computing and mobile telecommunications technologies has enabled new methods of social interaction. This personalisation of computer technology has made most people much more aware of, and informed about, computers but the complexity of the underlying technology has also increased to the point where few users understand how it works. As our reliance on computer technology increases, it is vitally important that the history of this key technology is properly chronicled. The old cliché that we can all learn from history is usually just that, a cliché. However, by including the concepts and technologies which underpin computers and describing them in non-specialist terms, this story may also contribute to improving the understanding of computer technology for the general reader.

Complex technologies are seldom developed in isolation. They are built upon earlier discoveries and inventions in related disciplines. The networking pioneer Paul Baran likened this process to building a cathedral, with each contributor laying down blocks on top of the old foundations, making it very difficult for historians to give credit where it is due. The computer is perhaps the ultimate example of this evolutionary development process, its story tightly interwoven with many of the great advances in engineering, mathematics and the physical sciences which have taken place over the past 400 years. Related inventions have also contributed enormously and computers would not exist in their present form without such innovations as the telephone, radar, television and the transistor.

The development of the computer has also been shaped by market forces and the changing needs of the customer. The first calculating machines were created by scientists and mathematicians in the 17th century to aid the process of mental calculation. The transition to an industry came as a result of the huge increase in commercial activity during the Industrial Revolution, with banks and insurance companies clamouring for machinery that could help reduce their vast computational workloads. By the 20th century military requirements had come to the fore and early developments in electronic computers were driven by the need to provide ballistic tables and fire control for artillery, cryptanalysis for codebreaking, and inertial guidance for rockets and missiles. In the 1970s a sharp fall in manufacturing costs following the introduction of the microprocessor helped to create a viable market for hobby and home computers which kick-started the process of achieving a mass market for computers. The computer industry has evolved and adapted to reflect these changing needs. As the industry has grown, innovation has become increasingly driven by competition amongst rival firms.

A story must have its principal characters and the story of the computer is no different in this respect. Our cast includes gifted mathematicians such as John Napier, Blaise Pascal, Gottfried Wilhelm Leibniz and George Boole plus many of history's greatest scientists including James Clerk Maxwell, Lord Kelvin and Nikola Tesla. Computing also has its own heroes in Charles Babbage, Alan Turing and Seymour Cray. Like many good stories, there are elements of mystery. These include the discovery of a mysterious object in an ancient shipwreck off the Greek island of Antikythera in 1900, which changed our perception of mechanical technology in the ancient world, and the role of Renaissance artist Leonardo da Vinci, who may

Introduction

or may not have been responsible for the first design for a calculating machine. There are also disputed claims fuelled by professional rivalries and the prestige that accompanies being associated with the development of a landmark invention.

This book sets out to chart the complex evolutionary process that has resulted in the creation of today's computers, picking out those innovations and discoveries which contributed most to the pool of knowledge through their influence on later advances and taking into consideration the business drivers as well as the specific technical breakthroughs. To put developments into context and provide a more rounded picture, it also covers the advances in engineering technology, or 'building blocks', which facilitated them. It is divided into four parts:-

- In Part 1: Mechanisation, we see humanity's earliest efforts to automate the process of calculation, first through mechanical means, then electromechanical and finally electronic.
- Part 2: Industrialisation describes the transformation from sequence-controlled calculators to stored-program computers and the birth of the computer industry.
- In Part 3: Segmentation, we see the industry maturing and new market segments beginning to emerge for faster or smaller computers, facilitated by the introduction of solid-state components.
- Part 4: Personalisation covers the technologies which have led to the development of mass-produced personal computers, how the demand for hobby and home computers reshaped the industry, and how computer technology changed from being hardware driven to software driven.

I began using computers professionally in 1975 when they were room-sized mainframes which were programmed via decks of punched cards. In 1979, I had my first hands-on experience of microcomputers and was amazed by the incredible power and freedom that interactive personal computing offered. Since then, I have followed the development of computers with a fascination which remains undimmed with the passing years. For those readers who share this fascination, I hope my book may help to feed it. For those who don't, perhaps this book will help to nurture a similar fascination.

PART ONE
MECHANISATION

1. Computer Prehistory – Calculating Machines

"For it is unworthy of excellent men to lose hours like slaves in the labour of calculation which could safely be regulated to anyone else if machines were used", Gottfried Wilhelm Leibniz, 1685.

The Origins of the Computer

For centuries, scholars have striven to develop tools to simplify complex calculations. As demand increased and the technology advanced, these became increasingly sophisticated, eventually evolving into the machines that are the direct ancestors of modern computers. Therefore, the origins of the electronic digital computer lie not in the electronic age but with the development of mechanical calculating aids and mathematical instruments. In order to fully understand the amazing developments that took place in the 20th century, it is necessary to go further back in time and examine humanity's earliest efforts to mechanise calculation.

Calculating Aids

The need for tools to aid the process of mental calculation first arose with the adoption of trading by ancient civilisations and the subsequent development of numerical and monetary systems. In the absence of written numerals, and as trading became more widespread and the arithmetic calculations increasingly complex, merchants began using readily available objects such as beads or pebbles to help keep count. These evolved into the earliest calculating aids.

The first calculating aid was probably the counting board, a flat rectangular board made from wood or stone and marked with parallel lines to provide placeholders for the counters. Each line denoted a different amount, such as tens, hundreds, thousands and so on, and numbers were represented by the combined positions of a set of counters on the board.

Many ancient civilisations also employed the abacus as a calculating aid. An abacus is essentially a more convenient form of counting board in which the counters are held in slots or threaded on wires to keep them in place. Numbers are represented by moving the appropriate counters in each line from one side to the other. Its origins are obscure but the earliest known example is the Sumerian abacus from southern Mesopotamia which dates from around 2500 BC. The simplicity of the abacus belies a powerful capability in the hands of a trained user and modern versions can still be seen in use in some parts of Asia.

The counting board and abacus were useful aids for addition and subtraction calculations but less so for multiplication and division. Following a reawakening of interest in mathematics amongst European scholars that began in Italy during the Renaissance, mathematicians began to develop more sophisticated aids to calculation. The first person to make a major contribution in this field was the influential Scottish mathematician and theologian John Napier.

Born in Merchiston, Edinburgh in 1550, John Napier came from a wealthy family of Scottish noblemen and his scholarly pursuits were conducted purely as a hobby rather than in any professional capacity. Nevertheless, he made several important contributions to mathematics, not least of which were his calculating aids. Napier's first major achievement in this field was the invention of logarithms in 1614. Logarithms reduce a tricky multiplication or division calculation to a simple addition or subtraction by representing numbers by the power or exponent to which the base must be raised in order to produce the result. Logarithm values were pre-calculated for a range of numbers and distributed in tabular form. However, the production of such tables was a lengthy and exacting task. Automating this work would be a recurring theme in the history of calculating machinery but for now this process would remain a manual one.

In his book 'Rabdologiae', which was published shortly after his death in 1617, Napier describes another invention also aimed at simplifying the process of multiplication. It comprised a set of ten rods of square section inscribed with numbers grouped in pairs corresponding to the digits 0 to 9 inside a square with a diagonal line separating them, plus an additional rod inscribed with a column of the digits 1 to 9 arranged vertically downward which was used to represent the multiplier. To multiply one number by another, the rod representing the multiplier was placed alongside a set of rods in the correct order necessary to represent the multiplicand, thereby creating what was essentially a multiplication table. For single digit multipliers, the product was then read off from the appropriate multiplier row by simply adding the pairs of numbers within each diagonal strip from right to left, performing a tens carry where necessary. For larger multipliers, the partial products were read off in turn and then added together.

By reducing the task of multiplication to a series of simple additions, Napier's invention significantly increased the speed and reliability with which such calculations could be performed. The rods could also be used to help simplify division calculations and later versions included two additional rods for the calculation of square roots. Napier's rods became more commonly known as Napier's Bones due to the frequent choice of ivory or bone as the material for their manufacture. Being easy to use and relatively cheap to make, they became very popular, their use spreading rapidly throughout Europe and even

1. Computer Prehistory – Calculating Machines

reaching China through the work of Jesuit missionaries.

Napier's Bones

Napier's earlier invention, logarithms, led to the development of the second important early calculating aid, the slide rule. The first practical slide rule was invented by the English mathematician and clergyman William Oughtred in 1622. Oughtred's device was based on the work of Edmund Gunter, an instrument maker and professor of astronomy at Gresham College, London, who two years earlier had created his 'Line of Numbers', a logarithmic scale in the form of a long wooden bar which was used to multiply numbers together by marking off lengths corresponding to the numbers using a pair of dividers. Oughtred realised that the dividers could be eliminated by adding a second scale placed alongside which could be repositioned relative to the first. Furthermore, if these scales were circular rather than linear, a more compact instrument would result.

Oughtred designed his instrument with two concentric rings made from brass and engraved with a series of circular logarithmic scales. Two arms which pivoted about the centre were also fitted to aid reading. He named it his 'Circles of Proportion'. The basic design of the slide rule was gradually refined over the years until it finally reached its familiar (linear) form in 1850. Although requiring more care to use than Napier's Bones, it was a more versatile instrument and remained a popular calculating aid until superseded by electronic pocket calculators in the 1970s.

The Story of the Computer

John Napier's work is the starting point for our story. Calculating aids such as Napier's Bones and the slide rule were a welcome improvement on manual methods but they were far from automatic. Fortunately, advances in mechanical technology were also taking place throughout this period and these would allow mathematicians to create much more capable devices for calculation, calculating machines.

Building Blocks – Geared Mechanisms

Geared mechanisms, in which toothed wheels operate together to impart some form of precision movement, were first introduced by the ancient Greeks. Gears were originally developed to transmit power for heavy lifting applications but were gradually refined for use in more sophisticated applications such as water-powered clocks and mechanised representations of the heavens. They may also have been employed in mechanical automata, animated figures devised for entertainment or ceremonial purposes.

Descriptions of such devices first appear in the writings of Philon of Byzantium who lived during the 3rd century BC. The celebrated Greek mathematician and scientist Archimedes of Syracuse, who also lived during this period, is known to have contributed greatly to the development of geared mechanisms, although his own writings on the subject have been lost. Hero of Alexandria, a Greek inventor who lived in the 1st century AD, also wrote extensively on the subject. However, despite the written evidence supporting their existence, very little was known about these early mechanisms and no examples were thought to have survived until the discovery of a mysterious object in an ancient shipwreck off the Greek island of Antikythera in 1900.

The Antikythera Mechanism

The Antikythera Mechanism consists of seven main fragments plus seventy or so smaller pieces. It was discovered by sponge divers in the wreckage of a Roman ship that had foundered during the 1st century BC. A large number of archeologically significant artefacts such as bronze and marble statues were also recovered from the shipwreck and the mechanism fragments were initially ignored, their true nature obscured by a covering of petrified debris. However, in May 1902 archaeologists noticed an inscription on one of the fragments, prompting further investigation which revealed them to be pieces of the earliest surviving geared mechanism.

Early investigations of the Antikythera Mechanism were hampered by the calcified condition of the fragments and it was not until the 1970s that advances in x-ray imaging techniques allowed researchers to examine the internal structure in sufficient detail to attempt a reconstruction. These studies showed that the complete mechanism would have contained at least 31 bronze gears in

1. Computer Prehistory – Calculating Machines

a wooden-framed case about the size of a shoebox which was fitted with bronze plates on the front and back faces. A large dial dominated the front face of the case and two smaller dials were engraved on the back face. The gears ranged in size from around 9 to 130 millimetres in diameter and feature precision crafted triangular teeth with a pitch of approximately 1.5 millimetres. They were arranged in a complex configuration which included compound gear trains and the first known example of differential gearing, where a large gear is meshed with two smaller gears of different size that rotate at different rates.

The Antikythera Mechanism is likely to have been operated by turning a handle of some description through the desired number of revolutions until a specific date was reached then reading the resulting values from pointers on each dial. However, opinions differ over its intended purpose. The presence of astronomical symbols on the dials and inscriptions referring to the planets Mars and Venus suggest an astronomical or astrological use. The cycles of the sun and moon are accurately represented, rendering the device capable of predicting eclipses but it may also have been designed to model the motions of the five known planets.

The Antikythera Mechanism was a much more sophisticated and higher precision geared mechanism than was previously thought possible in the ancient world, leading to speculation that it was a fake or from a much later period than the other objects from the shipwreck would suggest. However, descriptions of similar devices appear in the writings of the Roman politician and philosopher Marcus Tullius Cicero who lived during the 1st century BC and most experts who have studied the Mechanism now agree that it dates from around 150-100 BC.

Early Astronomical and Calendrical Instruments

The advanced mechanical technology of the Hellenistic period so impressively exemplified by the Antikythera Mechanism is of a level of sophistication not seen again until the Middle Ages. Only one other geared mechanism from antiquity has been found, a portable sundial with associated calendrical gearing from the early Byzantine period which dates from between 400 and 600 AD. However, with only four known gears, this device is much simpler in design and noticeably cruder in construction. The knowledge required to create such mechanisms then seems to have been lost for several centuries, only to re-emerge in the Islamic world at the end of the first millennium.

The great Persian scientist, Al-Bīrūnī, describes a mechanical calendar instrument very similar to the Byzantine device in a book devoted to the construction of various types of astronomical instruments which was written in 996 AD. Al-Bīrūnī's version, which he named 'The Box of the Moon', featured eight brass gearwheels with triangular teeth housed in a circular case. Dials on the front of the case indicated the positions of the sun and moon in the zodiac and also the phase of the moon for any given day of the month. Unlike the other instruments described in his book, Al-Bīrūnī provides no information on the originator of the calendar, suggesting that such devices were already well known to Islamic scholars, having been handed down from ancient times, and that the Byzantine sundial was perhaps a relatively poor example. The calendar would have been fitted to the back of an astrolabe, an early astronomical instrument used to determine the positions and movements of heavenly bodies.

Mechanical calendars and astrolabes became more closely integrated in later designs, resulting in the instrument known as a geared astrolabe. The earliest known example of such an instrument was constructed by the Persian metalworker Muhammad ibn Abi Bakr al-Isfahani in 1221/22. This used an arrangement of seven brass gears to model the movements of the sun and moon in a similar manner to that described by Al-Bīrūnī. Now in the collections of the Museum of the History of Science at Oxford University, it is the oldest surviving complete geared mechanism.

Geared Astrolabe (Photo © Museum of the History of Science)

As Islamic scientific knowledge began to permeate the West via Islamic Spain and the Crusades, geared astrolabes were soon being constructed in France.

1. Computer Prehistory – Calculating Machines

The earliest example dates from around 1300 and features a sophisticated epicyclic gearing arrangement. The design of geared astrolabes may also have influenced the development of mechanical clocks which began to appear in Europe around the late 13th century and it is interesting to note that the earliest applications of mechanical clock technology were also astronomy related.

Astronomical Clocks

The design and manufacture of geared mechanisms advanced rapidly in the Middle Ages through the development of the mechanical clock. The initial focus of this effort was the construction of astronomical clocks; elaborate timepieces that also displayed astronomical information, such as the positions of the sun and moon, and which were often accompanied by automata.

The first astronomical clocks originated not in Europe but in China. Although there is evidence of mechanical clock development in China as early as the 8th century AD, the first fully documented astronomical clock was built by Su Sung, a Chinese scholar and engineer who lived during the Song Dynasty at the end of the 11th century. Named the 'Cosmic Engine', the elaborate design featured a time display surmounted by a large mechanised celestial globe which was coupled to an armillary sphere mounted on the top. Like a giant cuckoo clock, the time of day was announced at regular intervals by mechanical automata emerging from a series of windows in the base of the clock accompanied by the sounding of bells, gongs and cymbals.

A critical component of mechanical clocks is the mechanism that regulates the speed of rotation of the gearing, which is known as the escapement. In the Cosmic Engine, Su Sung employed an escapement originally devised by the Buddhist monk and mathematician Yi Xing and the military engineer Liang Lingzan over three hundred years earlier for a mechanised celestial globe built in the year 725. This comprised a large wheel with 36 spokes, each of which was fitted with a water bucket on the end. When a bucket filled with water, it became heavy enough to overcome the resistance of a ratchet-like arrangement of levers, rotating the wheel through one increment and positioning the next bucket to receive water. A system of bronze gears and a chain drive transferred this movement to the other parts of the clock. Two water tanks were also incorporated into the design to ensure that water was supplied to this mechanism at a constant rate of flow.

The Cosmic Engine was completed in 1094 and installed atop a 12 metre high clock tower which stood in the grounds of the Imperial Palace in the city of Kaifeng. This imposing device operated until 1126 when the city was captured by Jurchen invaders. It was then dismantled and moved to the Jurchen capital city of Beijing where it is known to have remained in operation for several more years.

Water-powered mechanisms were superseded from around 1270 by clockwork mechanisms which employed falling weights as a source of power. These first appeared in Europe but their origins are obscure. Weight-driven mechanisms freed up clockmakers to install clocks in places where water power might not be suitable, such as inside buildings. These clocks continued to display astronomical information but their primary function was now to keep time in places of religious worship such as cathedrals and monasteries.

There are references to the construction of several astronomical clocks in England during the late 13th century, the earliest at Dunstable Priory in 1283, but none of these devices have survived and there are no reliable contemporaneous descriptions of them. The earliest recorded example for which we have detailed information is one begun in 1327 by Richard of Wallingford, a mathematician and astronomer who also served for a number of years as the Abbot of St Albans Abbey.

Richard's design incorporated a large disk showing a map of the constellations which rotated to indicate their positions in the night sky. The position of the sun was indicated by a smaller disk mounted on the end of an arm which rotated around the outside of the star map. The phases of the moon were also represented by a rotating ball with white and black hemispheres that could be viewed through a window on the front of the clock. A separate dial displayed the ebb and flow of the tide at London Bridge. The time of day was indicated by the striking of a bell every hour, with the number of strokes automatically varying according to the number of the hour.

Richard of Wallingford died in 1336 as a result of complications from leprosy and his clock was completed some time after his death. It is known to have survived intact until at least 1534 when it was described by John Leland, antiquarian to King Henry VIII, but is likely to have been dismantled following the dissolution of St Albans Abbey five years later in 1539.

The 14th century saw the rapid spread of mechanical clock technology throughout Europe. One of the most impressive and influential examples of the clockmaker's art was an astronomical clock completed in Italy in 1364 by Giovanni de' Dondi, a professor of medicine at the University of Padua. De' Dondi's clock, which he called an 'Astrarium', comprised a heptagonal-shaped frame approximately 1.3 metres high with dials mounted on each side of the upper section that displayed the positions of the sun, moon and five known planets. The dial representing the sun also incorporated a rotating star map. An eighth dial mounted on one side of the lower section displayed the time of day in 24-hour format and a revolving drum mounted horizontally beneath the upper section indicated important dates in the Christian calendar.

The weight-driven mechanism contained 107 brass gears in a highly sophisticated arrangement which included complex epicyclic gearing and oval gearwheels with irregularly spaced teeth. De' Dondi is reputed to have carried out all of the construction work himself, which took 16 years to complete, but it is possible that his father Jacopo, who also built clocks, may have assisted him.

Following completion, the Astrarium was acquired from de' Dondi by Duke Gian Galeazzo Visconti and installed in the ducal library of the Castello Visconteo in the city of Pavia. Records show that it was still in existence around 1530 but its ultimate fate is unknown. Fortunately, a set of manuscripts by de' Dondi describing the design and operation of the Astrarium did survive and have provided sufficient information for several modern replicas to be built.

Gear Manufacture

Early gears were crafted by hand using only basic hand tools and a pair of dividers to mark out the spacing of the teeth. This was a painfully slow process which required considerable care on the part of the craftsman and did not lend itself to the production of gears with an odd number of teeth. The invention of the spring-driven clockwork mechanism around the year 1450 facilitated the construction of portable clocks that were better suited to domestic use. As mechanical clocks became more commonplace and a clockmaking industry began to take hold in several European countries, a special-purpose tool known as a wheel cutting engine was developed to simplify the manufacture of gears.

A wheel cutting engine is basically a combination of a dividing plate and a cutter in the form of a rotary file. The dividing plate comprises a circular plate drilled with concentric rings of regularly spaced holes and a spring-actuated pin which locks the plate in the desired position. This ensures the correct pitch between teeth by precisely controlling the rotation of the gear blank when moving to the next tooth position. The rotary file is shaped to incorporate the desired profile of a gear tooth and is used to form the gear teeth individually.

The earliest known use of a wheel cutting engine was by Gianello Torriano, an Italian clockmaker and mathematician. In 1530, Torriano was asked to restore the Astrarium of Giovanni de' Dondi for the Spanish king Charles V who had recently been crowned Holy Roman Emperor. Torriano declined the commission on the basis that the Astrarium was beyond repair but offered instead to construct a new astronomical clock which would surpass it. Torriano relocated to Spain and began work on the design of the clock which would become known as El Cristalino due to the rock crystal windows in the case through which the mechanism could be seen. Completed in 1565, it is reputed to have contained over 1,800 gearwheels but was constructed in less than 4 years due to Torriano's use of a wheel cutting engine.

Torriano's wheel cutting engine predates a similar tool for cutting the teeth of clock gearwheels which was invented in 1673 by the prolific English experimental scientist Robert Hooke. Torriano's engine seems to have been unknown outside of Italy and Spain, and it was Hooke's machine that would become the model for the wheel cutting engines which were later used throughout the clockmaking industry. The advent of wheel cutting engines not only reduced the time required to manufacture gears but also significantly increased their quality.

The Earliest Calculating Machines

The precise rotational relationship between two meshed gearwheels also lends itself particularly well to the counting of numbers. Therefore, the introduction of precision geared mechanisms of the type found in mechanical clocks made possible the development of machines which could represent large numbers and perform simple arithmetical operations on them. The need for such machines was stimulated by the Scientific Revolution, the quest for scientific knowledge that took place in Europe during the 17th and 18th Centuries, the initial focus of which was the study of the planets and planetary motion.

Leonardo's 'Device for Calculation'

In February 1967, US newspapers announced the discovery in Spain's national library, the Biblioteca Nacional de España, of two lost collections of manuscripts by the great Renaissance artist, military engineer and inventor Leonardo da Vinci. The discovery was made by Jules Piccus, a professor of Romance languages at the University of Massachusetts Amherst. Piccus had been searching the library for medieval ballads and had stumbled upon the long lost manuscripts by chance, much to the embarrassment of library officials. The manuscripts had been brought to Spain in the 1590s and were acquired by the library some time later but had been subsequently misplaced. Now renamed the Codices Madrid, the first collection, the Codex Madrid I, comprised 191 pages of studies in mechanics produced by Leonardo between 1490 and 1496.

News of Piccus's discovery was picked up by Roberto A Guatelli, a New York based Italian engineer who specialised in creating working replicas of Leonardo's designs. On examining a facsimile copy of the Codex Madrid I which had been placed on display at the University of Massachusetts Amherst, Guatelli's attention was drawn to a sketch showing a gear train of 13 parallel vertical shafts, each fitted with a wheel and pinion gear, plus a horizontal shaft at each end of the train to which were attached weights. Annotations suggested a gear ratio of 10:1 between each vertical shaft. He recalled seeing a similar sketch in the Codex Atlanticus, a huge collection of Leonardo manuscripts dating from 1478 to 1519. Convinced that the two sketches represented Leonardo's designs for a calculating machine, Guatelli set out to build a replica based on them.

Following completion in 1968, the replica was exhibited as a 'Device for Calculation' by the US computer firm IBM, which had commissioned models of Leonardo inventions from Guatelli in the early 1950s. However, the exhibit

1. Computer Prehistory – Calculating Machines

created controversy in academic circles, prompting an inquiry to be held using a panel of academic experts to ascertain the reliability of the replica. The objectors argued that the Codex Madrid sketch represented a simple device for conferring mechanical advantage, hence the weights at each end, and that Guatelli had used his own intuition and imagination to go beyond what Leonardo had intended. This view was supported by a note written below the sketch in which Leonardo compares the arrangement of gears with one of levers on the opposite page. The findings of the panel of experts were ultimately inconclusive, nevertheless, IBM decided to withdraw the exhibit from display, presumably to avoid any damage to its reputation.

Guatelli's reasoning that the Leonardo sketches were of a calculating device probably stemmed from a strong physical resemblance between the Codex Madrid sketch and later calculating machine designs but this appears to have been purely coincidental. It seems that he and the IBM curators had succumbed to the romantic notion that the creative genius who first conceived such inventions as the helicopter and the parachute should also be responsible for inventing the calculator. There is little doubt that Leonardo possessed the capability to invent such a device. He was an accomplished mathematician and his many designs include a sophisticated automaton in the form of a knight in armour plus a number of clock mechanisms, several of which also appear in the Codices Madrid. He is also known to have studied in the ducal library in Pavia which housed the Astrarium of Giovanni de' Dondi. However, there is nothing in the Codex Madrid sketch to indicate its use for calculation. The Codex Atlanticus sketch is of a more complex mechanism and is much more intriguing but even if it was intended as a calculating device, it is unlikely to have been built as Leonardo seldom bothered to construct his inventions and the Codex Atlanticus manuscripts were not widely seen until relatively recently.

The Calculating Clock

Although no examples have survived, there is compelling documentary evidence to support the case that the first calculating machine was invented by the German astronomer, mathematician, cartographer and theologian Wilhelm Schickard. Schickard was born in the town of Herrenberg near Stuttgart in 1592. Academically inclined from an early age, the first part of his career was devoted to studies in theology and oriental languages at the University of Tübingen and in 1613 he became a Lutheran minister in the nearby town of Nürtingen. However, it was an association with the emerging science of astronomy that led to his work on calculating machinery.

In 1617, Schickard met the mathematician and astronomer Johannes Kepler. Kepler was the leading German astronomer in the early 17th century and a key figure in the Scientific Revolution which swept through Europe during this period. Both men had been educated at the University of Tübingen and shared similar backgrounds and interests. Schickard was also a skilled craftsman so Kepler asked him to produce some engravings to illustrate his new book 'Harmonices Mundi' which was published in 1619. As their friendship developed, Schickard's enthusiasm for academic studies was reawakened and in 1619 he was appointed to the position of Professor of Hebrew at Tübingen. Kepler also encouraged Schickard's interest in mathematics and astronomy, two subjects that he would have already been familiar with through his undergraduate studies, and the two men began to correspond on a regular basis. It is through these letters that we are aware of Schickard's invention of a calculating machine.

In 1935, the German historian Franz Hammer discovered a letter written by Schickard to Kepler in September 1623 which referred to a machine constructed by Schickard that could automatically perform all four basic arithmetical operations (addition, subtraction, multiplication and division) by means of a geared mechanism. A second letter written in February 1624 provides further details on the design of the machine, which Schickard called a 'Calculating Clock', but also reports that a copy intended for Kepler had been destroyed in a fire at the workshop of the craftsman Schickard had employed to help build it. Frustratingly, two sketches of the machine which Schickard refers to in the second letter were missing. Without these, it was impossible to determine what the machine looked like or how it functioned.

In 1956, while engaged on research aimed at identifying and cataloguing a complete collection of Kepler's works, Hammer made a chance discovery in the library of the Pulkovo Astronomical Observatory in Russia. He was searching through an early copy of Kepler's Rudolphine Tables when he found a slip of paper that had seemingly been used as a bookmark. It was this slip of paper which contained the missing sketches of Schickard's calculating machine.

Hammer announced his findings at an academic conference on the history of mathematics in April 1957. In the audience was Bruno von Freytag-Löringhoff, a Professor of Philosophy at the University of Tübingen and a specialist in the history of logic machines, a type of mechanical device for solving logical expressions. Von Freytag-Löringhoff was inspired by Hammer's revelation that a fellow Tübingen professor may have invented the first calculating machine and he set out to recreate Schickard's design. However, Hammer was a historian and his focus had been on the historical implications of his discovery rather than the mechanical details. Consequently, it was left for von Freytag-Löringhoff to decipher what little source material was available in order to determine how the machine actually operated.

Working over the next three years, von Freytag-Löringhoff, assisted by mechanic Erwin Epple, was able to piece together sufficient information to construct a working replica of Schickard's machine which he publicly demonstrated in January 1960. The Calculating Clock comprised three separate elements; a geared mechanism for the addition and subtraction of six-digit numbers, a set of Napier's Bones for multiplication and division, and a set of six dials for recording intermediate results. These were all housed one above the other within a single tall box. To perform a multiplication or division calculation, the user first set up the numbers on the Napier's Bones then added or subtracted the partial products using the geared mechanism, utilising the intermediate results dials at the foot of the machine to store the result at each stage in the process.

The addition and subtraction mechanism contained six parallel horizontal shafts arranged in a line to represent the machine's six decimal digits. Each shaft was fitted with a 10-toothed gearwheel and a cylinder with numbers inscribed on its circumference which, when viewed through a small window above, indicated the value of that digit. A dial attached to the front of each digit wheel shaft allowed the numbers which were to be added or subtracted to be set using a stylus. Addition was performed by turning the appropriate digit wheel shaft for the number that you wished to add by the necessary amount using large knobs attached to the front of each shaft, beginning at the rightmost (lowest) digit to be added. Subtraction was accomplished by turning the digit wheel shafts in the opposite direction.

A key feature of Schickard's machine was the method by which it automatically performed a tens carry between digits, a necessary feature for any successful calculating machine design. Schickard could have employed a simple gear train with a gear ratio of 10:1, as had been assumed for the Leonardo da Vinci device, but the cumulative friction from this mechanism would have made the machine very difficult to operate. Instead, he devised a simple carry mechanism which used an additional single-toothed gear mounted on each digit wheel shaft that operated in combination with an intermediate gear to increment the gearwheel representing the next digit to the left by one tenth of a revolution on completion of a full revolution of a digit wheel. However, it is not clear how this mechanism would have coped with the possibility of miscounting due to over-rotation of the intermediate gear. On von Freytag-Löringhoff's replica, this was addressed by the inclusion of a spring-loaded detent that only allowed the intermediate gear to increment by one complete tooth at a time.

We know from a letter written by Kepler to Schickard in 1617 that Schickard was familiar with Napier's Bones through Kepler. In the Calculating Clock, they were implemented using a set of six vertical cylinders with knobs at the top of each to rotate them. In front of these were nine horizontal slides that when correctly positioned revealed windows on the front panel through which the resulting numbers could be viewed. The device for recording intermediate results comprised a set of six disks with numbers inscribed on them, each of which could be rotated via knobs until the desired number appeared in a window.

Replica of Schickard's Calculating Clock (Photo © Herbert Klaeren)

Although astronomical calculations generally require numbers considerably larger than six digits, Schickard may have restricted his machine to only six due to the mechanical limitations of the carry mechanism. The action of adding 1 to the number 999999 would have resulted in a large amount of mechanical stress being placed on the single-toothed gear at the units position due to the cumulative friction from performing a carry at every position simultaneously, possibly causing damage. Therefore, for larger numbers, Schickard provided a set of brass rings for placing over the fingers to remind the user how many carries had been performed beyond the sixth digit. An audible prompt was provided by a bell that rang each time such a carry, or overflow, occurred.

Schickard went on to become Professor of Astronomy at the University of Tübingen in 1631 but there is no evidence that he continued his work on calculating machines. It is possible that he gave up, having been unable to overcome the mechanical limitations of his design but it is also possible that he was satisfied with the performance of the machine and had no reason to develop it further. Sadly, Schickard died of the bubonic plague, aged 43, in October 1635 along with his only remaining child, the disease having also taken the lives of his wife and other children some months earlier. His original machine has never been found.

The Pascaline

Until the discovery of Schickard's letters to Kepler, it was generally accepted that the first calculating machine was invented by the celebrated French mathematician, scientist and philosopher Blaise Pascal. Pascal was born in the city of Clermont (now Clermont-Ferrand) in 1623, the same year that Schickard invented his Calculating Clock. Remarkably, the man who would become one of France's greatest scholars had no formal education, having been tutored at home under the direction of his father Étienne, a senior civil servant who had himself dabbled in scientific study. The family moved to Paris in 1631 following the death of Pascal's mother. In Paris, he was introduced by his father to many of the leading French intellectuals of the day. The young Pascal flourished in the stimulating intellectual atmosphere of the Paris salons and he soon acquired a reputation as an exceptional mathematical talent.

In 1639, Pascal's father was appointed royal tax collector for the Upper Normandy region and the family moved to the city of Rouen. There Pascal continued his mathematical studies, publishing his first academic paper, an essay on conic sections, in 1640. In 1642, Pascal began work on the design of a calculating machine in an effort to help his father with the lengthy tax calculations necessary for France's complicated non-decimal system of currency. However, initial attempts to employ local tradesmen to undertake the construction work were unsuccessful. Undeterred, Pascal decided to teach himself the skills necessary for building mechanical instruments and over the next three years he completed the design and construction of his first calculating machine, which he called the Numerical Calculating Machine or 'Pascaline'.

Without access to Kepler's private correspondence, it is highly unlikely that Pascal would have been aware of Schickard's earlier work and this is evident in the design of the Pascaline which differs considerably from that of the Calculating Clock. Pascal's machine was designed to perform addition and subtraction of five-digit non-decimal numbers. It comprised a rectangular brass box about the size of a shoe box, on the lid of which was a row of five spoked wheels for entering numbers. Above each wheel was a small window through which the resulting numbers, which were displayed on numbered cylinders, could be read. The rims of the spoked wheels, which were fixed, were engraved with numerals at each spoke position to represent the value of the digit entered. The number of spokes in each wheel corresponded to the French system of currency at that time, with 12 on the wheel at the right hand end to represent deniers, 20 on the next wheel to represent sols, and 10 at each of the other three positions to represent livres.

Numbers were entered into the Pascaline by means of a stylus which was placed between the appropriate spokes in each spoked wheel and used to rotate the wheel clockwise until a stop fixed to the lid of the box was reached. Successive numbers would then be dialled in to the machine in the same way until the calculation was completed. Multiplication and division were not possible directly but could be performed by successive addition or subtraction until the desired result was achieved.

In order to overcome the problem of undue mechanical stress on the gearwheels when performing multiple carries between digits, Pascal devised an ingenious carry mechanism which employed falling weights. A weighted lever connected to each digit gearwheel was raised incrementally by the action of the rotating gearwheel. On reaching the correct value for a carry, the lever then dropped under gravity, driving a linkage that rotated the next digit gearwheel by one. However, the complex design of this mechanism rendered it unidirectional, so subtraction could not be accomplished by simply rotating the gearwheels in the opposite direction. Pascal solved this problem by employing the 'nine's complement' method, where the nine's complement of the number that you wish to subtract (the number that must be added to each digit to produce nine) is added to the number to achieve the same result. To facilitate this, an additional row of results windows positioned above displayed the nines complement of each number in the window below and a sliding cover was provided to select which row would be visible during a calculation.

The Pascaline (Photo © David Monniaux)

In an effort to refine his design, Pascal constructed numerous versions of the Pascaline over the following seven years, experimenting with alternative materials and changing the specification according to the intended application. Some of these variants were fully decimal and featured up to eight digits but they all employed the same basic mechanism. During this period, Pascal discovered that a local clockmaker in Rouen had constructed a replica of his machine based on a description that Pascal himself had provided. To make matters worse, the clockmaker had attempted to improve the mechanism and seemed intent on selling the machine commercially. Pascal, who was known to be prone to occasional bouts of pomposity, viewed this as an affront to his

1. Computer Prehistory – Calculating Machines

reputation. It spurred him into seeking an early form of patent protection for his invention, a royal decree granting him a monopoly on the manufacture and distribution, which he obtained in 1649 after sending one of his machines to the French Chancellor.

Enthused by the prospect of commercialising his invention, Pascal then embarked on a series of promotional activities, including a public demonstration of the machine in Paris. These efforts led to the purchase of two machines by Queen Ludwika Maria Gonzaga of Poland, who was a prominent patron of the sciences. He also gifted one machine to Queen Christina of Sweden, another keen supporter of the sciences, in 1650. Pascal's designs made extensive use of lantern wheel gears, a type of gearwheel which uses cylindrical bars instead of teeth, as these were easier to manufacture than toothed gears. However, the machines remained incredibly time consuming to construct due to the intricacy of the mechanism and the lack of suitable tools for manufacturing precision parts. This necessitated a selling price of 100 livres, the equivalent of a year's salary for a middle income worker, so few examples were sold despite the interest shown in the machine by the scientific literati. Reliability was also questionable, as the gravity-driven carry mechanism was prone to errors if the machine was knocked or moved suddenly during a calculation.

Following a decade of development, Pascal appears to have tired of working on calculating machines, probably through frustration with unsuccessful attempts to make money from his invention. For the next few years he devoted his energies more fully to his scientific and mathematical work, later abandoning these in favour of philosophical and theological studies as a result of a profound religious experience in 1654.

Pascal died at the age of 39 in August 1662 after a lifetime of ill health, his place in history nevertheless assured by his remarkable achievements in mathematics, philosophy and the natural sciences. Fortunately Pascal's work on calculating machines is reasonably well documented. In 1644, he wrote a 16-page pamphlet describing the Pascaline, a copy of which was appended to his letter to the French Chancellor requesting patent protection, and a written account of the machine can also be found in Diderot's Encyclopédie of 1751. Nine of the 50 or so original machines known to have been made by Pascal have also survived.

Although no longer credited as the inventor of the first mechanical calculating machine, Blaise Pascal deserves credit for being first to recognise the commercial potential of calculating machinery. His work also inspired others to create calculators, most notably the English diplomat and inventor Samuel Morland, but none were able to improve significantly on Pascal's design until the invention of the first machine that could automatically perform all four basic arithmetical operations in 1674.

The Stepped Reckoner

The next significant advance in calculating machine technology was made by another prominent figure in 17th century intellectual circles, the German mathematician, philosopher and historian, Gottfried Wilhelm Leibniz. Leibniz was born in 1646 in the city of Leipzig. His father, who died when he was six, had been Professor of Moral Philosophy at the University of Leipzig and this was where Leibniz went at the age of fourteen to study law. A gifted scholar, Leibniz graduated with a Bachelor's degree in 1663, swiftly followed by a Master's in philosophy later that same year. He then began working towards obtaining a doctorate, however, his doctoral thesis was rejected, possibly as a result of his youth, and he transferred to the University of Altdorf where he received a doctorate in law in 1667.

Having turned down the offer of an academic post at Altdorf, his first job upon leaving university was as secretary to the Nuremberg Alchemical Society. However, an interest in politics and a desire to help with his country's recovery from the ravages of the Thirty Years War led to a succession of diplomatic roles, the first of which was in the service of the German statesman Baron Johann Christian von Boineburg. During this period Leibniz also pursued a growing interest in science and this appears to have been the trigger for his work on calculating machines.

Leibniz began work on the design of a calculating machine in 1671. In 1672, he was sent to Paris as diplomatic advisor to the Archbishop of Mainz. Paris had become the intellectual capital of Europe and Leibniz was fortunate to be allowed to remain there for four years. Making the most of the situation, he took the opportunity to improve his knowledge of mathematics and physics by studying under the prominent Dutch scientist Christiaan Huygens, who was also based in Paris at this time. As part of these studies, Leibniz familiarised himself with the works of Blaise Pascal and it is very likely that he also encountered Pascal's calculator designs during this period.

Leibniz soon began to encounter the same problems with construction and materials that his predecessors had faced. He exhibited a wooden model of his calculating machine to the Royal Society in London in January 1673 which, although incomplete, was sufficiently impressive that it helped secure his election as a Fellow in April of that year. However, it was only after he engaged the services of a Paris clockmaker by the name of Olivier that he was able to make further progress and his machine, which he described as a Staffelwalze, or Stepped Reckoner in English, was finally completed in the summer of 1674.

The Stepped Reckoner was designed to perform addition, subtraction, multiplication and division on decimal numbers of up to eight digits with a product of up to twelve. It took the form of a long box-shaped instrument, slightly larger than the Pascaline, which was constructed in brass and steel and mounted in a wooden case. The machine was divided lengthwise into two parts, an upper section (when viewed from above) containing the calculating mechanism and a lower section containing the input mechanism. On the top face of the lower section was a row of eight small digit setting dials for entering numbers and a larger multiplier dial at the right hand end for setting the multiplicand or divisor required for a multiplication or division calculation. The top face of the upper section featured twelve windows for displaying the results, each of which could also be set to a specific value. A crank handle mounted on the front of the machine allowed the lower section to be shifted lengthwise relative to the upper in order that the eight digit positions at which the input mechanism engaged with the calculating mechanism could be altered by up to four places. A second crank handle at the left hand end operated the calculating mechanism.

Addition and subtraction calculations were performed by entering the first number directly into the results windows, setting the second number using the digit setting dials then turning the crank handle once in a clockwise or anticlockwise direction depending on whether the user wanted to add or subtract. Negative results were displayed as red numbers in the results windows. Multiplication and division were performed by repeated addition or subtraction. For a single digit multiplicand or divisor, this was done by setting the operand using the digit setting dials, then using a stylus to set the multiplicand or divisor on the multiplier dial and turning the crank handle continually until the number set on the multiplier dial had been reached. For multiplicands or divisors with more than one digit, this sequence was repeated for each additional digit with the input mechanism repositioned each time so that it corresponded to the position of the digit in the multiplicand or divisor.

The key component of the Stepped Reckoner was the 'stepped wheel', a cylindrical drum with 9 gear teeth of varying length, arranged so that their length increased as the drum was rotated. This operated in conjunction with a small pinion gear which could be positioned at any one of ten locations along a shaft mounted parallel to the drum so that it meshed with a different number of teeth on the drum depending on its position, ranging from 0 teeth at one end to 9 at the other. The input mechanism contained eight of these stepped wheels, one for each digit, and the positions of the pinion gears for each were controlled by the digit setting dials on the lower section of the machine.

The Stepped Reckoner incorporated a simple tens carry mechanism that employed a series of connecting rods and star wheels to increment the next digit position by one at the appropriate point. Unlike Pascal's design, this mechanism was bidirectional, however, it was incapable of propagating a carry across more than one digit position simultaneously. To solve this problem, Leibniz devised a clever workaround in the form of an indicator device which employed pentagon-shaped disks whose edges were visible through slots in the top of the machine. These were arranged so that they would come to rest with their edges flush when a carry was correctly performed but with a corner protruding through the slot to indicate when a carry was pending and required user intervention.

Leibniz submitted his completed machine to the Académie des Sciences in Paris in 1675. Two further examples, one of which had an increased product capacity of 16 digits, are known to have been constructed under the supervision of Rudolf Christian Wagner, an associate of Leibniz who was Professor of Mathematics at the University of Helmstedt. Leibniz had also planned to build a machine for the Russian Tzar Peter the Great but there are no records confirming that this was ever produced.

Leibniz died aged 70 in November 1716, having made major contributions in both mathematics and philosophy over a long and highly productive career. However, his later years were overshadowed by a bitter row with arch rival Sir Isaac Newton over the invention of infinitesimal calculus. Although it is now accepted that both men arrived at their versions of calculus independently, the mathematical community sided with Newton and Leibniz's reputation suffered greatly. Unlike other intellectuals of similar stature, Leibniz was not honoured at the time of his death. Only his personal secretary attended his funeral and he was buried in an unmarked grave.

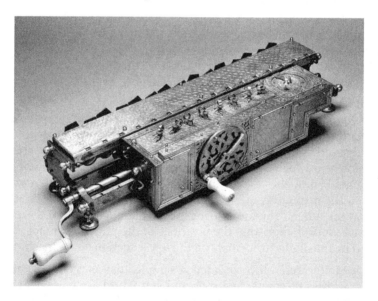

Leibniz's Stepped Reckoner (Photo © Historisches Museum Hannover)

Leibniz's calculating machines were also thought to have been lost until one of them, the 16 digit machine, was discovered in an attic of Göttingen University in 1876 by workmen repairing the roof. It had been sent there for repair in 1764 and subsequently forgotten. This machine was placed in the collections of the State Library in Hannover but was found to be inoperable due to corrosion and missing parts. In 1894, it was sent for restoration to Arthur Burkhardt, the proprietor of a successful calculating machine firm based in the German town of Glashütte. Burkhardt worked over the next two years to restore the machine to working order and wrote this up in a detailed account which was published in 1897.

With its ability to perform all four basic arithmetical operations, the Stepped Reckoner was a major leap forward in mechanical calculating technology and the stepped wheel would become a standard component of later calculator designs. It next appeared in a machine constructed by the German clergyman and inventor, Philipp Matthäus Hahn, in 1774. Hahn built several calculating machines before his death in 1790, all of which incorporated the stepped wheel. However, although Leibniz gave a brief description of the operation of the Stepped Reckoner in a 1710 paper, no details of the mechanism were available until the publication of Burkhardt's article almost two centuries later. Therefore, unless Hahn had access to one of the Leibniz machines (and there is no evidence to support this), it is unlikely that his machines were based directly on Leibniz's work. A more plausible scenario is that Hahn had somehow reinvented the stepped wheel and that later designers had borrowed the concept from Hahn.

Building Blocks – Machine Tools and Mass Production

Despite the pioneering efforts of Schickard, Pascal and Leibniz, it would take more than a century before calculating machines would begin to permeate society. The main reasons for this lengthy hiatus were the difficulty in manufacturing the precision components necessary for such machines and the resulting high cost of manufacture. Both were alleviated by the development of machine tools and mass production techniques during the Industrial Revolution.

The Evolution of Machine Tools

Tools for cutting or shaping material have been used for many thousands of years and the origins of the machine tool can be traced back to the woodworking lathes developed by the ancient Egyptians around 1250 BC. In its simplest form, the lathe is a device that rotates a block of material between centres, allowing cutting tools to 'machine' the material to form cylindrical shapes in a process known as turning. Early lathes were hand operated and required an additional person to rotate the workpiece back and forth by means of a rope or a bow. Foot treadle lathes that required only a single operator began to appear in the 13th century. The earliest versions retained the use of the rope drive which was supported from above by a flexible pole and operated by a foot treadle. This was later replaced by a flywheel and crank mechanism to rotate the workpiece in a unidirectional motion.

The addition of a heavy flywheel and the faster speeds achievable through continuous rotation made lathes much more powerful, thus extending their capability, and the first lathes for cutting screw threads in wood were developed in Germany in the late 15th century. The earliest depiction of a screw-cutting lathe appears in the Mittelalterliche Hausbuch, an important medieval manuscript which was completed around 1480. This incorporated a slide rest to support the cutting tool, the position of which could be accurately controlled by means of two screws; the lead screw to move the tool longitudinally along the length of the workpiece, and the cross feed screw to move the tool towards or away from the workpiece in a transverse motion. Leonardo da Vinci is also known to have produced designs for various machine tools during this period and his sketches for a foot treadle lathe, a screw-cutting lathe and a pipe boring machine appear in the Codex Atlanticus.

The next significant advance in machine tool technology also took place during the 15th century when water power was first applied to machine tools. This further increased the power of the tools to the point where they could be successfully applied to metalworking. The driving force for these applications was the European armaments industry, particularly the manufacture of cannon, the demand for which grew rapidly as nations began to increase the size of their navies.

Early cannon had been manufactured by forging strips of wrought iron into a cylindrical shape in a process similar to barrel making but these crude weapons were unreliable and the technique was limited to the manufacture of smaller calibre weapons. The trend towards larger and more powerful weapons led to the introduction in the early 16th century of cannon which were cast from iron or bronze using a process adapted from one used to cast bells. However, the casting process produced a bore which was insufficiently smooth and cylindrical. This prompted the development of a new type of machine tool known as a boring mill.

A boring mill is essentially a lathe which has had one of its centres replaced by a long bar with a cutting tool on the end. In early designs, the workpiece would be held stationary while the rotating boring bar was fed slowly into the bore of the cannon to allow the cutting tool to machine it to a smooth finish. One of the earliest depictions of a cannon boring mill is given in the Pirotecnia, a treatise on metals and metallurgy written by the Italian engineer and metallurgist Vannoccio Biringuccio, which was published in 1540. Biringuccio's boring mill was driven by a type of treadmill powered by the action of two men walking inside it but the similarity to a water wheel is striking and it is often described as being water powered.

Early boring mills were designed to finish a pre-existing bore that had been formed by casting and were incapable of correcting common defects, such as where the bore was badly misshapen or had been cast eccentric to the body of the cannon. This deficiency was rectified in 1713 by the Swiss inventor Johann Maritz. While employed as Commissioner of the King's Foundry in Strasbourg, France, Maritz invented a new type of vertical boring mill that was capable of machining cannon from solid castings. In Maritz's design, the workpiece was suspended vertically over the boring bar mechanism which was driven by two horses. Maritz continued to refine his design after moving to the Netherlands where the technique of boring from solid castings was eventually adopted for all cannon manufacture.

Development of the cannon boring mill continued throughout Europe and in 1774 the English industrialist John Wilkinson patented a horizontal boring mill that offered improved accuracy. In Wilkinson's water-powered design, the workpiece was positioned horizontally and rotated between bearings while the stationary boring bar was fed into it by means of a rack and pinion mechanism. This arrangement helped to increase rigidity and reduce vibration, resulting in a more precise machining action. Wilkinson's patent was revoked in 1779 due to an embarrassing similarity to an earlier machine invented by the Dutch foundryman and artist Jan Verbruggen in 1758. Undaunted, Wilkinson had already adapted his machine for a different purpose, one which would help to realise the most important invention of the Industrial Revolution.

In the same year that Wilkinson patented his boring mill, the Scottish steam engine pioneer James Watt entered into a historic partnership with an English businessman to commercialise a new design of engine which he had patented five years earlier in 1769. Watt's condensing steam engine was aimed at improving the efficiency of the Newcomen engine and relied on cylinders with a very accurate bore in order to maintain pressure and prevent the steam from escaping. Having worked for several years as an instrument maker, Watt had little difficulty in producing a working model of his steam engine. However, building a full-scale engine was another matter, as cannon boring mills were unable to provide the necessary accuracy and there were no other machine tools available that could perform this task.

Fortunately, Watt's business partner Matthew Boulton was acquainted with John Wilkinson through their mutual membership of the Lunar Society, a dinner club of prominent industrialists and intellectuals which met in Birmingham during this period, and was aware of Wilkinson's work on precision boring mills. After being introduced to Watt, Wilkinson agreed to adapt his cannon boring mill to make it suitable for machining a cylinder to Watt's exacting requirements. Having proved that it could be done, Wilkinson then devised a new type of boring mill specifically for steam engine cylinders.

Wilkinson was rewarded for his efforts by being appointed sole supplier of cylinders to the firm of Boulton & Watt and James Watt's efficient steam engines went on to play a pivotal role in the Industrial Revolution, providing a ready source of power for all manner of industrial machinery including machine tools themselves.

Precision Machine Tools

Wilkinson's cylinder boring mill of 1776 heralded the dawn of a new era of precision engineering. However, the development of precision machine tools was also driven by clock and watch making, in particular the demand for smaller and more accurate timepieces, and by the manufacture of scientific instruments. At the forefront of these developments was the English mathematical and astronomical instrument maker Jesse Ramsden.

In 1767 Ramsden invented a new type of circular dividing engine, a specialist machine for engraving accurate graduated scales, which incorporated a worm and wheel mechanism to rotate the workpiece through the desired angle. The accuracy of this mechanism was dependent upon the uniformity of the thread pitch of the worm screw and the closeness of the mesh between it and the teeth of the worm wheel. In order to manufacture a worm screw of sufficient accuracy, the talented and resourceful Ramsden also designed and built the first successful precision screw-cutting lathe.

To cope with the hardened steel material of the worm screw, Ramsden's lathe employed rigid metal construction and a diamond tipped cutting tool. It also implemented an idea first seen in Leonardo da Vinci's sketch of a screw-cutting lathe over two hundred years earlier; a set of 'change wheels' which altered the gear ratio between the spindle and the lead screw in order to change the thread pitch of the screw being cut. Ramsden used his new lathe to machine two examples of the worm screw, one of which he then modified by creating sharp notches in the threads. He then used the modified screw as the cutting tool when machining the teeth of the worm wheel, thus ensuring a precise fit.

In 1774 Ramsden constructed a larger circular dividing engine which also incorporated improvements to his original design. For this second machine, Ramsden was awarded the tidy sum of £615 from the Board of Longitude, the British government body established in 1714 to solve the problem of discovering longitude at sea. The award was made in recognition of the new machine's potential for improving the accuracy of maritime navigation instruments such as the sextant. However, a condition of acceptance was that Ramsden must make the design publicly available along with that of his precision screw-cutting lathe.

Ramsden also designed a second screw-cutting lathe in which the lead screw engaged with a circular toothed plate which controlled the movement of the tool. This too was placed in the public domain. His final version, completed in 1778, was the culmination of more than a decade of development, with each model used to machine the lead screw for successive versions, thereby improving the accuracy through each iteration of the design. The Board of

1. Computer Prehistory – Calculating Machines

Longitude award conditions helped ensure that Ramsden's innovative machine tool designs were adopted by other instrument makers of the period, thus advancing the art enormously.

Ramsden's designs may also have influenced the work of Henry Maudslay, the British engineer who was the most influential of all the machine tool pioneers. In his screw-cutting lathe of 1797, Maudslay brought together earlier developments into a single machine tool that is the direct ancestor of all modern lathes. Maudslay's elegant design employed a pair of parallel triangular bars which served as guideways for the slide rest. A lead screw, which was positioned between the guideways, could be coupled to the carriage supporting the slide rest by means of a split nut and clamp arrangement. Maudslay also included an adjustable tailstock and backrest to accommodate long workpieces and to provide adequate support against the thrust of the cutting tool.

Like Ramsden before him, Maudslay continued to refine the design of his screw-cutting lathe in a quest to improve the accuracy of his workmanship. His second version, completed in 1800, featured flat guideways and a set of 28 change wheels to accommodate a wider range of thread pitches. In his third design he combined triangular and flat guideways, and added stepped drive pulleys to allow the speed of rotation to be varied. With this machine tool, he was able to manufacture a precision screw 5 feet long with an extremely fine pitch of 50 threads per inch. This screw was sufficiently accurate that it was used for calibrating astronomical instruments.

Despite the obvious commercial potential of these designs, Maudslay had little interest in producing machine tools for sale, preferring instead to remain in the business of running a successful engineering workshop. This provided an opportunity for another firm to establish a market by introducing the first commercially available machine tools. The first to do so was Holtzapffel & Co., a firm founded in London in 1794 by John Jacob Holtzapffel, who had worked for Jesse Ramsden for two years after emigrating from France. Holtzapffel & Co specialised in lathes for ornamental turning, a popular pastime during this period, and the company sold its first lathe in June 1795.

Machine tools for industrial use were introduced by the Leeds-based firm of Fenton, Murray & Wood around 1800. Originally a competitor of Boulton & Watt in the steam engine business, the company began supplying cylinder boring mills to external customers including Portsmouth naval dockyard and a French engineering works. Other manufacturers soon followed and by 1820 a sizeable machine tool industry had been established.

Measuring Instruments

Progress in precision engineering was also dependent on the development of accurate measuring instruments to ensure that parts were manufactured to the correct size. Instruments based on graduated scales, such as rulers and slide callipers, had been in use for centuries but these were inadequate for measuring to the accuracy that the new precision machine tools were capable of. Fortunately, the advent of precision screws for use in machine tools also facilitated the development of more advanced measuring instruments based on the ability of a screw to effect precise linear dimensions using the rotational movement of the screw to subdivide its pitch. The first of these was the micrometer.

In its simplest form, a micrometer comprises a u-shaped calliper frame fitted with a fixed anvil at one side and a moveable anvil on the end of a fine pitch screw, or spindle, at the other. Measurements are made by adjusting the spindle until the object is lightly held between the anvils. The resulting dimension is then read by adding the value indicated on a circular scale on the thimble of the spindle to the value shown on a linear scale on the barrel.

The principles on which the modern micrometer is based were established by the English astronomer and instrument maker William Gascoigne in 1639. In an effort to measure angular distances through an astronomical telescope, Gascoigne devised a set of callipers that appeared in the field of view of the telescope. A precision screw, incorporating a left hand thread on one half and a right hand thread on the other, allowed the gap between the jaws of the callipers to be finely adjusted. Measurements were made by counting the number of complete revolutions of the screw and reading the fraction of a revolution turned using a circular scale mounted on the end of the screw. By knowing the optical magnification of the telescope, Gascoigne was also able use this device to measure the diameter of the moon and stars.

This combination of a calliper and precision screw was first adopted for engineering measurement by the steam engine pioneer James Watt. In 1772, Watt built a benchtop instrument for his own personal use. It comprised a u-shaped brass frame fitted with fixed and moveable steel anvils, with a large dial and crank handle at one end and a smaller dial on the front. The moveable anvil took the form of a horizontal slide which was adjusted by means of a rack and pinion mechanism. This was operated by turning the crank handle, with the number of complete revolutions of the pinion indicated by a pointer on the smaller dial and fractions by a scale on the large dial with graduations representing increments of one thousandth of an inch. Watt's measuring instrument is the first known example of a bench micrometer.

A bench micrometer was also constructed by the machine tool designer Henry Maudslay in 1805. Maudslay's design was intended for high accuracy measurement and dispensed with the calliper frame in favour of a gunmetal bedplate more reminiscent of a small lathe. This supported two saddles fitted with anvils, one of which was fixed and the other connected through a slot in the bedplate to a very fine pitch screw of 100 threads per inch. A large knob on the end of the screw was engraved with a scale divided into 100 graduations.

1. Computer Prehistory – Calculating Machines

A graduated scale on the bedplate indicated the number of complete revolutions for a given dimension. Capable of measuring to a resolution of one ten-thousandth of an inch, Maudslay nicknamed his micrometer the 'Lord Chancellor' for its ability to serve as an arbiter in disputes between his employees over the accuracy of their workmanship.

The development of the micrometer into the familiar handheld instrument of today was made by the French engineer Jean Laurent Palmer. Palmer designed his screw calliper as an instrument for measuring the thickness of sheet metal and obtained a French patent on his design in September 1848. With a 20 mm range and 0.05 mm resolution, Palmer's design incorporated most of the features of a modern outside micrometer.

Palmer exhibited his screw calliper at the 1867 Paris Exposition where it was seen by Joseph R Brown and Lucian Sharpe of the US machine tool firm Brown & Sharpe on a visit to France in August of that year. The two men had been approached a few months earlier by an inspector from a brass plate manufacturer who had presented a number of alternative designs for a gauge that could reliably measure the thickness of sheet metal in an effort to avoid potential disputes with customers. They realised that Palmer's design offered a better solution to the problem and the following year Brown & Sharpe introduced the world's first mass produced micrometer, an improved version of Palmer's design with a finer pitch screw and a clamp to lock the spindle. Now every machine shop had the means to measure their work accurately, with the result that quality improved significantly.

Interchangeability and Mass Production

The powerful combination of precision machine tools and measuring instruments led to the realisation of the principle of interchangeability, where components are manufactured with such precision that they can be swapped with other examples of the same component without having to alter them to make them fit. Interchangeability simplifies manufacturing by removing the need for custom fitting of parts, thereby speeding up the assembly process and reducing the requirement for skilled labour.

The idea of interchangeability was first proposed by a French army officer, Lieutenant General Jean-Baptiste Vaquette de Gribeauval. In 1765, Gribeauval initiated a process of rationalising and standardising French field artillery which became known as 'le système Gribeauval', the ultimate goal of which was to achieve interchangeability throughout the full range of armaments. After experiencing early success with interchangeable gun carriages, Gribeauval reasoned that if this was extended to small arms, which were more complex, they could be manufactured faster and more economically. Interchangeable parts would also make field repairs much easier to carry out and would extend the service life of small arms considerably.

While in the course of implementing his reforms, Gribeauval learned that a French gunsmith, Honoré Blanc, was already experimenting with interchangeable parts for muskets. Blanc had recently designed a new version of the popular Charleville musket for the French infantry, the Model '77, which had been well received. By now Gribeauval had risen to the post of Inspector General of Artillery for the entire French Army. He used this authority to appoint Blanc inspector of three Royal arsenals and asked him to redouble his efforts on interchangeability.

By 1785 Blanc had succeeded in producing a batch of interchangeable musket locks and was planning to extend his methods to include the other parts of a musket. As precision machine tools were not yet available commercially, he employed a combination of die casting and templates or jigs to guide the hand finishing process. The following year Gribeauval provided the funding for Blanc to set up a factory and increase production but Gribeauval's death in May 1789 and the start of the French Revolution later that same year put an end to Blanc's status as a government contractor and the enterprise foundered.

Fortunately Blanc's work on interchangeability had caught the attention of the American statesman Thomas Jefferson, who served as US Ambassador to France from 1785 to 1789. Jefferson had visited Blanc's workshop in August 1785 and was given a hands-on demonstration in which he was invited to assemble a musket lock from a selection of 50 of each part chosen at random. When the parts fitted perfectly, Jefferson realised that Blanc's interchangeability system offered a solution to the chronic shortage of skilled craftsmen in the fledgling US armaments industry, which required the country to import most of its weapons from Europe, and wrote enthusiastically about it in a letter to the US Foreign Secretary. Three years later, as the political situation in France began to deteriorate, Jefferson approached Blanc with an offer to relocate him and his work to the US but Blanc declined. Instead Jefferson purchased six of his muskets and sent them to the US Secretary of War as an example of what had been achieved along with a recommendation that Blanc's methods be adopted by US armouries. However, Jefferson's recommendation would not be acted upon for several years.

In June 1798, the American inventor Eli Whitney won a major contract from the US government to supply 10,000 muskets to the War Department within a two year delivery period. Whitney was facing financial ruin following a disastrous attempt to protect and exploit his invention of the cotton gin and was the first businessman to take advantage of the US government's willingness to provide generous cash advances on major armaments contracts in preparation for a possible war with France. Despite having no factory, no workers and no experience in gun manufacturing, Whitney secured the $134,000 contract, and its $10,000 advance, on the back of a compelling vision of water-powered machine tools to significantly reduce the amount of skilled labour required.

1. Computer Prehistory – Calculating Machines

Whitney began by setting up an armaments factory in New Haven, Connecticut and visiting established US armouries in order to learn their methods. However, building the necessary machine tools and training a new workforce to operate them was a huge undertaking and he soon fell behind schedule. In an effort to help him, Thomas Jefferson, who was a personal friend, told Whitney about Honoré Blanc's work on interchangeability and another friend, US Treasury Secretary Oliver Wolcott, sent him a copy of a paper written by Blanc in 1790. This information was the lifeline that Whitney needed and he wasted no time in taking Blanc's methods and incorporating them into his own concept of manufacturing which he named the Uniformity System.

Whitney conducted a demonstration of his Uniformity System in January 1801, fitting ten preassembled musket locks to the same musket in front of an audience of senior US government officials which included the now Vice President Thomas Jefferson and President John Adams. The success of this demonstration bought Whitney some time until he was able to deliver the first batch of 500 muskets the following September. However, tests conducted on surviving examples of Whitney's muskets have shown that they do not contain interchangeable parts and it now seems likely that the demonstration was faked.

Whitney continued to struggle with delivery timescales and it took him a further eight years to complete the contract. Nevertheless, he deserves credit for introducing the principle of interchangeability to America. Encouraged by the US government, which now insisted that all armaments contracts specify interchangeable parts, Whitney's Uniformity System was adopted throughout the US armaments industry. It was eventually perfected by the US firearms manufacturer John H Hall at the Harpers Ferry armoury in 1827 and became known as the American System of Manufacturing. Hall developed various special purpose machine tools which he utilised to finish parts produced by die casting and drop forging. This was augmented with a system of carefully designed fixtures to position parts accurately during machining and gauges to inspect the finished parts. The combination of precision machining and accurate measurement allowed Hall to obtain the consistency of manufacture that had eluded Whitney.

By adopting the American System of Manufacturing, individual firms were able to achieve interchangeability for their own products. However, interchangeability between different manufacturers was far more challenging. Achieving this would require not only the introduction of recognised standards for common parts but also standards for the measurement of length to ensure that measurements made with different instruments were consistent. The central figure in these developments was the British mechanical engineer Sir Joseph Whitworth.

Whitworth had learned his trade working for the great Henry Maudslay and in 1833 he established his own machine tool manufacturing business in Manchester, rising to become the foremost machine tool builder in the country. In later years he also moved into armaments manufacture. Throughout his long career Whitworth strove to improve standards of accuracy in engineering manufacture. He was among the first to recognise the importance of calibration, where measuring instruments are compared against one of greater accuracy in order to assess performance or improve accuracy by using the information obtained to correct subsequent measurements made with that instrument.

At the Great Exhibition of 1851 Whitworth exhibited a length measuring machine which could detect differences of one millionth of an inch. He used this machine to compare the length of his own standard yard bar with that of the official government standard. He also developed a technique for producing accurate flat surfaces which could be used as reference planes. This work led to the establishment of a system of traceability to national standards of length.

The advent of screw-cutting lathes and other thread forming tools during the Industrial Revolution allowed threaded fasteners such as screws and bolts to be manufactured by machine rather than by hand, thus ensuring uniformity. However, thread sizes differed from manufacturer to manufacturer which hindered interchangeability. Whitworth proposed a standard system for screw threads in 1841. It featured a fixed thread angle of 55 degrees and a standard pitch for a given diameter. The specifications were based on a sample of screws collected from a large number of manufacturers which were measured to obtain their average values. Whitworth's rationalised system of screw threads was in general use throughout Britain by 1860 and other nations soon followed with their own systems of screw thread standards.

Interchangeability and standardisation provided the foundations for the spread of mass production from armaments manufacture to domestic products such as clocks and watches, sewing machines and bicycles. It also facilitated the successful commercialisation of calculating machines.

The Arithmometer

The new machine tools and mass production techniques introduced during the Industrial Revolution offered the prospect of reducing the manufacturing cost of calculating machines to a level where they would no longer be merely toys for the amusement of wealthy patrons. However, production would only be economically viable if there was also a market need for such devices. Fortunately, the Industrial Revolution provided this too due to the immense increase in commercial activity and in international banking that took place during this period. The stage was now set for the foundation of a new industry, the first champion of which was the French businessman and inventor Charles Xavier Thomas de Colmar.

1. Computer Prehistory – Calculating Machines

The son of a doctor, Charles Xavier Thomas was born in 1785 in the town of Colmar in the Alsace region of France. In 1809 he joined the French military as an administrator, swiftly rising to Inspector of Supply for the entire French Army. His duties as an administrator involved making lengthy calculations and it was while serving in the Army that Thomas first had the idea of building a calculating machine. After leaving military service in 1819, Thomas obtained a position as General Manager of the Phoenix insurance company in Paris. The complex actuarial calculations that he encountered in his new job provided further evidence of the need for mechanising the calculation process and the following year he designed his first calculating machine, a four function decimal calculator which he named the Arithmometer.

The Arithmometer closely resembled Leibniz's Stepped Reckoner, not only in its outward appearance but also internally due to the use of stepped wheels and a sliding carriage. However, given the dearth of published material on the Stepped Reckoner at this time, it seems more likely that Thomas had seen the machines of Philipp Matthäus Hahn rather than being influenced directly by the work of Leibniz. Also, unlike the Stepped Reckoner, Thomas's tens carry mechanism was capable of propagating a carry across multiple digit positions simultaneously.

The input capacity of the Arithmometer was 3 digits which were set by a trio of sliders. An additional slider was provided for setting the single digit multiplicand or divisor. This controlled a mechanism that automatically performed the repeated addition or subtraction operations necessary for multiplication and division. The calculating mechanism was operated by pulling on a silk ribbon and results were indicated by means of a 6 digit display that featured two sets of results windows, one revealing a dial with increasing values for addition and the other showing one with decreasing values for subtraction. A sliding cover was also provided to select which set was visible during a calculation.

Thomas had a prototype of the Arithmometer built for him by a Paris clock and instrument maker named Devrine. He also took steps to protect his invention by applying to the French Minister of the Interior for a patent in October 1820 which was granted the following month. He then submitted it for review to the influential Société d'Encouragement pour L'Industrie Nationale in Paris in 1821, where it received a favourable report in the Société's bulletin.

After a second prototype was completed in 1822, Thomas appears to have set aside work on the Arithmometer in order to concentrate more fully on his business activities. By 1829 he had established his own insurance firm, Compagnie du Soleil, which was based in Paris. Thomas had spotted a gap in the French market for a company that specialised in fire insurance and he was able to build this business up over the next two decades through a series of acquisitions and mergers, becoming a rich man in the process. This wealth would help to bankroll his later calculator development work.

Thomas returned to calculator development in 1844, exhibiting another Arithmometer prototype at the French Industrial Exposition held in Paris that year. He then teamed up with a watchmaker from Neuilly-sur-Seine named Piolaine and the two men developed an improved model which replaced the flimsy silk ribbon with a metal crank handle and helical gear mechanism. Capacity was increased to 5 digits for input and 10 digits for display, and a reversing lever was added to select addition or subtraction. The redesign was completed by July 1848 and Piolaine was given responsibility for setting up a workshop and putting the new model into production. Piolaine also began to refine the design to make the machine more suitable for mass production but he died unexpectedly a few months later.

Fortunately Piolaine's assistants were able to complete the construction of one of the production prototypes that he had been working on before his death. Thomas exhibited this machine at the Exposition of the Second Republic in Paris in 1849, where it was awarded a silver medal, but without any practical engineering experience he was unable to put it into production. A solution was found in November 1850 when Thomas formed a partnership with another skilled craftsman, A M Hoart, for the purposes of manufacturing Arithmometers. Thomas's original 1820 patent had lapsed after 5 years so he obtained a new French patent in December 1850 and also applied for a British patent early in 1851 in anticipation of exhibiting the Arithmometer at the Great Exhibition in London. He then embarked on a publicity campaign, seizing every opportunity to publicly exhibit his machine and also presenting elaborately decorated examples to European royalty. This generated considerable personal recognition for Thomas in the form of numerous awards and honours, including the prestigious title of Officier of the Légion d'Honneur. Sales of the Arithmometer also began to pick up and by 1854 around 250 machines had been manufactured.

A further revision of the Arithmometer design was undertaken in 1858 which introduced a zero setting mechanism for the results display. The intricate mechanism for automatic multiplication and division was also replaced with a much cheaper to manufacture single digit display that merely kept count of the number of repeated addition or subtraction operations performed. Despite these improvements, by 1865 sales were beginning to flag and the total number of machines produced had only reached 500. Sales then received a boost after Thomas exhibited an Arithmometer at the Paris Exposition of 1867, resulting in the production of a further 300 machines by 1870.

Following Thomas's death in March 1870, Arithmometer production continued under the direction of his son, Louis Thomas de Bojano. By 1878 over 1,500 had been produced. More than half of these went to overseas customers, such as the Prudential Assurance Company in the UK which purchased 24 machines.

Promotional literature from 1855 boasted that Thomas had spent the enormous sum of 300,000 francs on developing the Arithmometer. However, with a typical selling price of 500 francs and total sales of only 1,500 units over a lengthy period of 28 years, it is unlikely that he ever made a profit from it. Nevertheless, Thomas had demonstrated the commercial potential of calculating machinery and other manufacturers, such as Burkhardt in Germany and Elliott Brothers in Britain, soon followed, producing and selling similar machines in greater numbers.

The 1870s brought a plethora of alternative calculator designs and the establishment of major calculating machine industries in Germany and the USA. Besides the banking and insurance sectors, early customers included scientists, engineers and civil servants. The industry also benefitted from the gradual mechanisation of office work which began with the introduction of typewriters and cash registers around 1874. By the turn of the century the larger firms were each producing thousands of calculating machines per year and the subsequent reduction in manufacturing costs through economies of scale further served to open up new markets.

The Engines of Charles Babbage

In the same year that Thomas patented his first Arithmometer prototype, a young British mathematician by the name of Charles Babbage was embarking on a journey which would stretch the potential of calculating machinery far beyond that envisaged by Pascal and Leibniz. Babbage had recognised that the technology employed in calculating machines held the promise of being capable of much more than simple arithmetic. However, realising such capability would be a formidable intellectual and engineering challenge that would consume the greater part of his adult life.

Charles Babbage was born in London in December 1791. His father's career path, first as a goldsmith and then as a banker, ensured that the family were far from poor but they became very wealthy after his father accepted a partnership in the newly established bank of Praeds, Digby, Box, Babbage & Company in 1801 which he then sold two years later. Charles was a sickly child and this was reflected in his somewhat haphazard education, a mixture of private tutoring and public schooling conducted between Enfield in north London, Cambridge and the family home in rural Devon where he was often sent to recuperate. His schoolmaster in Enfield was a keen amateur astronomer and is thought to have been responsible for sparking Babbage's lifelong passion for mathematics. Though he was given to occasional flights of fancy, Babbage proved to be an able pupil and made efforts to familiarise himself with many of the classic mathematical texts of the 18th century.

In October 1810, Babbage enrolled at Cambridge University to study mathematics and chemistry. However, he was unenthused by the rigid teaching methods and limited knowledge of his tutors at Cambridge and his attention soon wandered. Babbage possessed a gregarious personality and this led to him becoming involved in various clubs and associations, few of them serious until, in May 1812, he co-founded the Analytical Society with six fellow students and one of their tutors.

The aim of the Analytical Society was to cultivate an analytical approach to calculus which was popular amongst European mathematicians but had been largely ignored in Britain, where mathematical research had stagnated in the giant shadow of Sir Isaac Newton. The Society's first task was to promote the use of the Leibniz notation for infinitesimal calculus as opposed to the more commonly used Newton version. One of his fellow co-founders was John F W Herschel, son of the famous astronomer Sir William Herschel. The Society failed to gain wider acceptance and was dissolved in 1813 but Babbage and Herschel's friendship would become a lifelong one.

Upon graduating with an unremarkable Bachelor's degree (due in part to his extracurricular activities), Babbage left Cambridge in the summer of 1814. A few weeks later he married Georgiana Whitmore, who he had met two years earlier, and set up home with his new wife in the Devon village of Chudleigh. Babbage sought employment in the local mining industry but was unsuccessful and after a few months the couple relocated to London where Babbage proceeded to apply for various academic positions, again without success. However, Babbage was fortunate to possess a private income in the form of a generous allowance of £300 per year from his father plus a smaller one from his father-in-law and this allowed him to indulge his love of mathematics without the worry of having to earn a living from it.

With the help of the Herschels, Babbage soon began to make a name for himself in London intellectual circles. Steady progress in his mathematical studies resulted in a number of well received papers on the theory of functions and he was invited to give a series of lectures on astronomy at the Royal Institution. This culminated in his election as a Fellow of the Royal Society in March 1816, aged only 24.

Babbage's interest in astronomy and astronomical instruments also grew during this period and in February 1820 he and his friend John Herschel were once again involved in the genesis of a learned society. Along with eleven others, they founded the Astronomical Society of London to encourage and promote the science of astronomy, a subject poorly represented by existing organisations such as the Royal Society, which some felt had become more of a gentlemen's club than a serious scientific institution. Renamed the Royal Astronomical Society in 1831 on receipt of a royal charter from King William IV, this would be a much more influential and long-lived body than Babbage and Herschel's earlier effort. It would also provide the impetus for Babbage's first foray into the world of calculating machinery.

1. Computer Prehistory – Calculating Machines

One of the first matters that the Astronomical Society concerned itself with was the production of astronomical tables. In the absence of readily available calculating machinery, numerical tables were an invaluable tool for all sorts of calculations ranging from scientific research to banking and navigation at sea. In astronomy, they were used in the calculation of planetary positions, lunar phases, eclipses and calendrical information. However, the production of numerical tables was an arduous task and errors were common. As the Industrial Revolution progressed and society's reliance on numerical tables increased, the reliability of such tables became an issue of national importance.

As the two most talented mathematicians amongst its founder members, Babbage and Herschel were asked by the Astronomical Society to oversee the task of preparing the new tables. However, despite careful planning and the use of two clerks to calculate each entry twice, Babbage and Herschel found that the completed tables contained numerous errors. While discussing errors in one of the tables, Babbage is reported to have said to Herschel "I wish to God these calculations had been executed by steam", to which Herschel replied, "It is quite possible". Babbage's own account of this epiphany from his memoirs describes an earlier occasion while still a student at Cambridge in which he began daydreaming about calculation by machinery while contemplating a table of logarithms. Some historians have questioned the reliability of Babbage's account which was written many years later but what is agreed is that by 1821 Babbage had begun work on the design of his first calculating machine.

The Difference Engine

Babbage was convinced that machinery was the key to producing error-free numerical tables. He envisaged a machine that would not only calculate correctly but would also print the results automatically, thereby also eliminating transcription and typesetting errors. He took inspiration from the heroic efforts of the French civil engineer Baron Gaspard de Prony to produce logarithmic and trigonometric tables for the French government following the introduction of the metric system in 1791. De Prony had employed a large team of people to perform the calculations manually. These were organised into three groups according to ability and the work divided up to suit. This hierarchical approach was made possible through the use of the method of finite differences, where a continuous mathematical function is represented by a polynomial series, the successive values of which can be calculated by adding the first, second, third and possibly higher order differences between each value until the differences become constant. As this method uses only arithmetical addition and subtraction, it removes the need for multiplication and division, thus reducing the likelihood of errors. Production of the tables remained, however, a major undertaking which took nine years to complete.

Babbage may also have been influenced by the work of the German military engineer Johann Helfrich Müller. Müller had successfully constructed a four function calculator in 1784 which, although capable of calculating to 14 digits, closely resembled the machine built by Philipp Matthäus Hahn a decade earlier. His next idea was much more ambitious. In a letter to the eminent physicist Georg Christoph Lichtenberg written in September 1784, Müller proposed building a more advanced machine for automatically calculating and printing tables of values of an arithmetical function. Müller further developed this idea in a 50-page pamphlet published two years later in which he refers to the method of finite differences, however, details were sketchy and the machine was never built.

Babbage named his new invention the Difference Engine. In order to test his ideas, he began by constructing a working model which was designed to calculate values of the formula $x^2 + x + 41$ to six decimal places using two orders of difference. The machine would have to be able to store columns of numbers, perform addition on them and execute a tens carry where necessary. He initially considered using sliding metal rods to represent the numbers but soon realised that the rotary motion of gearwheels provided a more natural method of resetting to zero following a carry operation and settled on a geared mechanism instead.

Babbage occasionally dabbled in mechanics and had set up a small engineering workshop in a room of his London home which he used to build the model. However, he lacked the specialist skills necessary to manufacture the more complex components, such as the gearwheels, so he employed various tradesmen on a contract basis for these parts. The model was completed in May 1822. The following month he demonstrated it to the Astronomical Society.

In July 1822, having established proof of principle, Babbage wrote an open letter to the President of the Royal Society, Sir Humphry Davy, proposing the construction of a full-scale machine that could be used for calculating astronomical, nautical and other numerical tables and which would automatically print the results. Babbage's letter reached the attention of the British government in April 1823 and was referred back to the Royal Society for advice. The Society's response was resoundingly positive and was bolstered by the Astronomical Society of London which awarded Babbage a gold medal for his proposal. These glowing endorsements helped to secure Babbage a meeting with the Chancellor of the Exchequer in June to discuss the possibility of government funding for the project. The Chancellor was clearly impressed by what he heard and promised Babbage the initial sum of £1,000, with an invitation to apply for further government funding at a later stage to cover the remainder of the estimated £5,000 cost of construction. When the grant was received in August 1823, it was for a more generous sum of £1,500.

In preparation for the construction of the Difference Engine, Babbage embarked on a short tour around Britain, visiting workshops and factories in order to familiarise himself with the latest manufacturing techniques and ascertain the accuracy that could be achieved with each. This was to be the first of a number of such surveys conducted at various intervals over the next few years and the information gathered would prove very useful. However, lacking the practical engineering experience necessary to produce the engineering drawings for the Difference Engine himself and to cope with every element of its construction, Babbage had also hoped to find an engineer that he could work with to help realise his ideas and the tour was a failure in this respect. He then turned for advice to the civil engineer Marc Isambard Brunel, also a Fellow of the Royal Society, who recommended the toolmaker and draughtsman Joseph Clement.

Joseph Clement had an impeccable pedigree, having worked as chief draughtsman for two of the giants of mechanical engineering, the inventor Joseph Bramah and the machine tool pioneer Henry Maudslay. He left Maudslay's employment in 1817 to set up a small engineering workshop of his own at Southwark in southwest London, specialising in work requiring high accuracy. Clement was also a talented machine tool designer in his own right and he would later be awarded a gold medal from the Society for the Encouragement of Arts, Manufactures, and Commerce for a new type of facing lathe.

Babbage and Clement commenced work on the Difference Engine in the autumn of 1823. Babbage's specification called for a machine which could tabulate values of a polynomial to 16 decimal places using up to 4 orders of difference. The design that evolved featured sets of what Babbage called 'figure wheels', a combination of a gearwheel and a cylinder with numbers inscribed on its circumference, to represent each decimal digit. These were stacked in vertical columns, with one column for each order of difference plus an additional column at one end of the machine to store the resulting tabular values. Sandwiched between each figure wheel was an adding mechanism that employed a plunger and cam arrangement to couple an intermediary gearwheel to the figure wheel for the fraction of a revolution that corresponded to the number to be added. Intermediary gearwheels were designed so that they meshed with the corresponding figure wheel in the adjacent column, thus transferring the number to be added to the appropriate digit position in that column. Also located between figure wheels in each column was an elaborate carry mechanism which operated in conjunction with levers mounted on a row of vertical shafts positioned behind the figure wheel columns.

To reduce the mechanical strain on the machine's components when performing a large number of addition and carry operations simultaneously, operations were executed in four separate stages. This was achieved by splitting addition into two stages, one using the odd difference columns and the other using the even, each followed by a separate carry operation. An additional benefit of this scheme was that it reduced all calculations to four steps irrespective of the number of differences involved.

Implementation of the four stage calculation cycle necessitated the development of a sophisticated control mechanism to regulate the timing of the plungers for the intermediary gearwheels and control the overall sequence of operation. This was achieved using a device known as a 'barrel' which replicated the action of a set of cams whose profiles could be altered. It comprised a vertically mounted cylinder drilled with sets of holes in vertical rows spaced at regular angular intervals around its circumference, each of which could be fitted with a removable stud. The cylinder was rotated incrementally then pushed forward so that each stud position was presented in turn to a set of rods which controlled the action of the various parts of the machine. The sequence of operation could be altered by simply changing the positions of the studs. It seems likely that Babbage was inspired by a similar mechanism found in music boxes, where a revolving cylinder fitted with pins is used to pluck the teeth of a comb of tuned metal strips in the correct sequence.

The control mechanism of the Difference Engine also incorporated the apparatus for automatically printing the results. Babbage had conducted a study of printing machinery and initially planned to use moveable type which would be selected one piece at a time by the Engine and automatically assembled into blocks for printing. This was later changed to a system that punched numbers one digit at a time into a soft metal plate which would then be used as a mould for casting stereotype printing plates. A mechanism was designed to sense the final values at each digit position on the results column and select the corresponding digit on a punch mechanism. Another mechanism was provided to shift the plate along or down as required between each punch action. In this way, a complete page of tables could be built up automatically and it would only be necessary to add the headings and other annotations manually before printing. Provision was also made for altering the layout of tables by changing the settings on the pair of wheels that controlled the spacing of numbers.

In 1827, Babbage suffered a series of personal tragedies. The first of these was the death of his father in February, a not entirely unexpected event, given his father's advanced age and declining health, and one that also left Babbage a rich man. However, it was rapidly followed by the sudden death of his second son Charles in July and his beloved wife and newborn baby the following month. These events affected Babbage deeply and he was advised to take a long holiday for the sake of his health. After leaving Clement with instructions to continue work on the Difference Engine in his absence and requesting that John Herschel keep him informed of progress, Babbage embarked on an extended tour around Europe in October 1827, not returning until November 1828. This lengthy absence inevitably affected the project, which had now reached a critical stage, and progress slowed as a result.

With the initial government grant long spent, Babbage had been bankrolling the project with his own funds for some time. On his return from the continent,

1. Computer Prehistory – Calculating Machines

he made it a priority to remind the government of its commitment to provide funding for the construction of the Difference Engine. Unfortunately, the absence of any written agreement made this difficult and he and his supporters had to lobby vigorously to force the government to act. After another referral to the Royal Society, Babbage succeeded in obtaining a government payment of £1,500 in April 1829 to reimburse his expenditure along with assurances that further funding would be forthcoming. However, work on the project came to an abrupt halt a few weeks later following an argument with Clement over the amount of money that Babbage owed him. Work only resumed after representatives of each party had been appointed to monitor the accounts but this had wasted a further 9 months.

Babbage had originally estimated that the Difference Engine would take 3 years to complete but the lack of funding, his prolonged absence abroad and the 9 month dispute with Clement had all put paid to this ambitious schedule. Finalisation of such a complex design also required considerable experimentation and the full set of engineering drawings was not completed until 1830. The final version incorporated two additional orders of difference and called for an estimated 25,000 individual parts. Designed on a relatively large scale with each figure wheel around 125 mm in diameter, the Difference Engine would have weighed several tonnes and stood over 2 metres high. Despite Babbage's earlier wish for calculations to be executed by steam, the machine was designed to be operated by hand.

With the drawings completed, the project now entered the construction phase. This too would be a time-consuming process given the enormous number of high precision components required. After two years, sufficient components were ready to assemble a sizeable section of the calculating mechanism. Babbage, mindful of the protracted timescale, wanted something that would provide the government with evidence of progress so he instructed Clement to build a demonstration piece comprising three columns from the calculating mechanism, each fitted with six figure wheels, along with sufficient ancillary components to allow the mechanism to be operated. The demonstration piece contained around 2,000 parts and was completed in December 1832. Although the control mechanism and printing apparatus were absent, Babbage was able to use this device to successfully calculate portions of a table of logarithms. It also formed the centrepiece of the frequent social gatherings that Babbage had become well known for hosting.

Final assembly of the Difference Engine would require a large dedicated space which Clement's workshop did not possess. Babbage had pressed for and succeeded in having the Engine declared government property two years earlier as part of the negotiations to secure further public funding. Concerned about security against fire and theft, he now persuaded the government to pay for the construction of a fireproof building in which to assemble and house the completed machine. Various sites were considered and rejected until Babbage, who was growing weary of the daily 5 mile coach journey between his home in Marylebone and Clement's workshop in Southwark, insisted that the new building should be located on the site of some unused stables at his home. A small house adjacent which Babbage also owned would be made available for Clement to live in when he was on site. Clement was unhappy with these arrangements and demanded £660 per year as compensation for splitting his business between two locations, an exorbitant sum but one that seems to have reflected the deteriorating relationship between the two men. When this was refused in March 1833, Clement immediately halted construction and fired the ten workers that he had employed on the project. One of these was the engineer Joseph Whitworth, who would later use the knowledge he had gained working on the Difference Engine project to revolutionise standards of accuracy in engineering manufacture.

Clement's actions sounded the death knell for the beleaguered Difference Engine project. Although the majority of the components for the calculating mechanism had been manufactured, these still required to be assembled, a process that would have required considerable hand fitting of parts, and work had not yet started on the control mechanism and printing apparatus. Babbage drew up a list of possible options for the future of the project for consideration by the government. His preferred option was to complete the machine using an alternative engineering partner but Clement refused to release the drawings or the finished parts until he had received payment due for work already done. Several months went by before the Treasury agreed to pay Clement directly for the money he was owed and it would be more than a year before all the items, including the demonstration piece, were safely relocated to the new building next to Babbage's home. Clement was allowed to keep the special-purpose machine tools which he had developed for the construction of the Engine. Despite the loss of his most important customer in highly controversial circumstances, Clement's reputation seems to have remained intact and his business continued to thrive.

The total amount of government expenditure on the Difference Engine project came to £17,479 including the additional £2,190 spent on the fireproof building. This was a huge sum in an age where an entire steam locomotive cost less than £800. Moreover, Babbage estimated that a similar amount would be necessary to complete the machine. He continued to press for a decision from the government on the future of the project but a general election and political reforms delayed this for some years. When the decision finally came in November 1842, it was that no further government support would be forthcoming. Twenty one years after it had begun, the Difference Engine project was finally dead. Fortunately, Babbage had not been idle while awaiting the decision and was already several years into his next calculating machine project.

1. Computer Prehistory – Calculating Machines

The Analytical Engine

The Difference Engine was by far the most technologically advanced calculating machine yet devised. However, as a special-purpose calculator for tabulating the values of a polynomial, it would have had limited applicability and it is perhaps more impressive as a feat of precision engineering than a calculating machine. Babbage's second major design project would be much more ambitious than the Difference Engine in this respect. It would be a general-purpose calculating machine which could be configured to automatically evaluate almost any mathematical expression.

Having recovered the drawings of the Difference Engine from Joseph Clement in July 1834, Babbage began to consider how to improve its operation. He had learned a great deal in designing the Engine's control mechanism and printing apparatus and he now set out to apply this knowledge to the development of new mechanisms that would extend its capability. Babbage had been aware for some time that the Difference Engine would be far more flexible if it was able to store intermediate results and feed them into subsequent calculations. He realised that this should be possible if the figure wheel columns were arranged in a circle so that the values appearing on the results column could be fed back into the highest order difference column. However, the tabulation of certain mathematical functions also required the ability to perform multiplication as well as addition and this could only be achieved by separating out the calculating functionality from the storage of numbers. The figure wheel columns would then become a collection of independently addressable numerical storage units which Babbage called the 'store'. A separate part called the 'mill' would contain the circular calculating mechanism which could be connected through linkages with any column in the store to produce the desired result. These two architectural elements would form the basis of what came to be known as the Analytical Engine.

Fired up by the potential of this new architectural configuration, Babbage set to work on the functional design. He began by designing a mechanism which could perform multiplication using essentially the same method of repeated addition employed by Leibniz in his Stepped Reckoner of 1674. Single digit multiplicands would be dealt with by repeatedly adding the operand to itself, with the number of repeat operations dictated by the value of the multiplicand. For multiplicands with more than one digit, this sequence would be repeated for each additional digit and the column digits shifted up by one place each time to correspond with the position of the digit in the multiplicand.

Babbage also designed an automatic division mechanism for the Analytical Engine which would operate by a process of repeated subtraction and stepping in a similar manner to the method of multiplication. An elaborate mechanism would compare the divisor with the remainder after each subtraction operation to determine whether a further subtraction operation should be performed or the decimal place shifted.

The sequence of operation of the Analytical Engine involved transferring numbers from the store to the mill, performing arithmetical operations on them and transferring the results back to the store. The process of fetching a number from the store entailed rotating each of the figure wheels in that column to the zero position while a geared mechanism, operating in a similar fashion to the adding mechanism in the Difference Engine, transferred this movement to the mill via a rack and pinion arrangement. However, because numbers would be erased during this process through the action of resetting to zero, those that required to be retained for later use would have to be copied back to the store. This could be achieved by simply leaving the rack and pinion gearing engaged when restoring it to its starting position following a transfer.

The introduction of multiplication and division would place additional demands on the tens carry mechanism. This had been designed to operate sequentially in the Difference Engine, making multiple carries propagated across several digit positions a lengthy process. As Babbage intended to use much larger numbers of 40 digits or more in the Analytical Engine, a method had to be found for speeding up the process of propagating a multiple carry across such a large number of digit positions. Babbage solved this by inventing an anticipating carry mechanism, designing and testing more than twenty different versions before arriving at the optimum configuration. Each figure wheel was fitted with an arm which incorporated a circular end with a vertical hole drilled through it. When the wheel rotated to the 9 position, a rod located beneath was able to pass through the hole in the arm, allowing a sector gear attached to be lifted into mesh with the figure wheel above. This mechanism was carefully designed so that the movement of the rod would propagate through each digit position, thereby setting the sector gear to mesh at each position where the arm was in the correct alignment for a carry. When the lowermost figure wheel was rotated from the 9 to the 0 position, the sector gear would transmit this movement to a common shaft which rotated every other sector gear in that column. In this way, each figure wheel in a column that required to be changed as a result of a carry would be rotated simultaneously.

The expanded functionality of the Analytical Engine demanded a highly sophisticated control mechanism that could not only handle the numerous individual steps involved in a multiplication or division calculation but would also cope with the complex interconnections between the store and the mill and would allow changes to the overall sequence of operation to be made for different mathematical expressions. To provide this, Babbage extended the control barrel concept that he had developed for the Difference Engine. Individual barrels with fixed studs were employed to control the operation of each calculating mechanism while a central barrel with removable studs was provided to direct the overall sequence of operation.

1. Computer Prehistory – Calculating Machines

Babbage devoted considerable time and effort to the functional design of the Analytical Engine over the next year or so and by November 1835 he had made sufficient progress to begin the detailed engineering design. He hired an experienced draughtsman, Charles G Jarvis, to oversee the production of the engineering drawings and two assistants to help him. Jarvis was perfectly suited to the task, having served as senior draughtsman on the Difference Engine project before losing his job when Clement halted construction.

Engineering drawings provide all the information necessary to manufacture the parts for a machine and permit it to be assembled correctly but they do not describe its operation. Babbage had first encountered this limitation when designing the Difference Engine some years earlier and had created a form of mechanical notation to describe the movement of parts and how they interact. This consisted of a system of symbols to indicate the source and type of motion and whether the mechanical connection between components was permanent, such as when they are attached to a common shaft, as a result of friction between them or imparted through an intermittent arrangement such as a cam or ratchet. These symbols were incorporated into a table which provided information on the sequence and timing of operations. In this way, it would have been possible to determine the position of any component at any point during the cycle of operation of the machine. Babbage had published a description of his mechanical notation in the Philosophical Transactions of the Royal Society in 1826. He now found it an invaluable tool in designing the intricate mechanisms of the Analytical Engine.

As the design of the Analytical Engine progressed, Babbage turned his attention to how the machine would be used. Through attempting to produce sets of control sequences for the machine, Babbage realised that he needed greater flexibility in the control mechanism. He found an ideal solution in the punched card technology that had been developed in France to control weaving looms. The technology had been introduced in 1801 by a French silk weaver, Joseph Marie Jacquard, based on earlier work by the inventor Jacques de Vaucanson. Jacquard's loom was controlled by pasteboard cards with holes punched into them, each row of which corresponded to one row of thread in the pattern. The presence or absence of a hole dictated the position of a hook which controlled how the fabric was woven by raising or lowering a longitudinal thread. Highly intricate patterns could be produced by stringing together sets of cards in a particular order. Furthermore, the number of cards in a set was not limited, unlike Babbage's control barrels which were restricted to a specific number of steps before the barrel came full circle and the sequence repeated.

Babbage began to incorporate punched card technology into the design of the Analytical Engine in June 1836. He designed a card reading mechanism which sensed the presence or absence of holes by presenting each card in turn to a set of levers located at each hole position. Levers would be unobstructed by the card if a hole was encountered or pushed backwards if there was no hole at that position. The levers were attached to a mechanism that converted their final positions into the appropriate action. Barrels were retained to control the low-level sequence of operations while the punched card mechanism controlled the positioning of the central barrel which now had fixed studs.

Punched cards were also used to input numbers into the Analytical Engine, thus reducing the possibility of operator error when setting numbers manually, and to specify which locations in the store were to be used. These were known as 'variable' cards to differentiate them from the 'operation' cards that directed the calculation functions of the machine. As they differed in size and number of hole positions from the operation cards, a separate card reading mechanism was provided to accommodate them.

Babbage's card readers also possessed the ability to skip forward or backward by a specified number of cards when a particular condition had been met during a calculation, such as a change of sign or a predetermined counter value having been reached. Known in modern computing terminology as a conditional branch, this powerful feature would permit control to be switched to different sections of the card set in order to ignore operations which were no longer required by a calculation or to create loops of repeated operation sequences.

For the output of results, the Analytical Engine would incorporate an improved version of the stereotype printing plate apparatus devised for the Difference Engine but it also included provision for printing results directly onto paper or recording them onto punched cards by means of a card punching mechanism. This latter feature also extended the machine's ability to store intermediate results.

The design of the Analytical Engine was essentially complete by 1838 but Babbage continued to refine it for a further nine years, finally ceasing work in 1847. Much of this later work was aimed at simplifying the design and reducing the number of components but he also took the opportunity to redesign the multiplication and division mechanisms to speed up calculation. This was achieved through the implementation of the method of partial products, where a table of the nine single-digit multiples of the multiplicand or divisor is first calculated then added together or subtracted as dictated by the value of the multiplier or dividend. Adopting this method would allow operations to be overlapped, thereby reducing the time required to perform a multiplication operation to around 1 minute for two 40-digit numbers. Dedicated columns of figure wheels in the mill provided the required storage for the table of multiples.

Babbage paid for the design work out of his own pocket but he never sought reimbursement for this expenditure or attempted to raise the additional funding necessary to construct the Analytical Engine. There are hints that he was hoping the government would approach him but this would have been highly unlikely, given Babbage's poor track record on the Difference

1. Computer Prehistory – Calculating Machines

Engine project and a less obvious need for a machine with the Analytical Engine's capabilities. Government apathy towards the project also prompted Babbage to decline a generous offer of technical and financial assistance from Joseph Whitworth, who was now a successful engineering manufacturer. Consequently, the only parts of the Analytical Engine which were built were a few mechanical assemblies and test models relating to the mill and the anticipating carry mechanism, although Babbage later experimented with the newly introduced die casting process as a means of manufacturing gearwheels more economically.

The completed Engine would have stood over 4 metres high. Its length would have been dictated by the size of the store, which projected out in a line from the side of the mill. Babbage wrote of the store having a capacity of 1,000 numbers but this would have been unrealistic and 100 numbers would have been sufficient for most calculations, giving an overall length of around 4 metres. Requiring more power to operate than the Difference Engine, the machine would indeed have been driven by a small steam engine.

A Second Difference Engine

In 1846, having almost completed his revisions to the design of the Analytical Engine, Babbage turned his attention back to his earlier machine. Using the knowledge he had gained in the design of the Analytical Engine, Babbage undertook a complete redesign of the Difference Engine, simplifying the calculating mechanism and reducing the number of components. This became known as the Difference Engine No. 2.

The general architectural design of the Difference Engine No. 2 was unaltered from his earlier machine but Babbage took the opportunity to upgrade the specification by increasing the numerical precision from 17 to 31 decimal digits and raising the number of orders of difference from 6 to 7. Despite the higher specification, Babbage was able to dramatically reduce the component count from 25,000 to around 8,000. This was achieved by adopting the simpler addition and carry mechanisms developed for the Analytical Engine. The Difference Engine No. 2 also shared the same stereotyping and printing mechanism as the Analytical Engine.

A full set of engineering drawings for the Difference Engine No. 2 was completed by 1849 under the supervision of Charles Jarvis. At the suggestion of the President of the Royal Society, Babbage offered the design to the British government. He wrote to the Prime Minister Lord Derby in June 1852 offering to donate the drawings and notations for the machine as a gift to the nation on condition that it would be built. Letters of support were also submitted by various interested parties, including the inventor and machine tool manufacturer James Nasmyth, who highlighted the considerable advances in machine tool technology made during the first Difference Engine project and the resulting benefits to industry. The matter was referred to the Chancellor of the Exchequer, Benjamin Disraeli, who seems to have allowed his distaste of all things scientific and technological to cloud his judgement. Ignoring expert advice, Disraeli decided that the project would be too expensive and rejected Babbage's offer.

The following year, the Swedish inventor Georg Scheutz and his son, Edvard, succeeded where Babbage had twice failed by completing the construction of their own difference machine. Scheutz had been inspired by an article on Babbage's first Difference Engine which appeared in the July 1834 issue of the Edinburgh Review magazine. Assisted by his engineering student son, Scheutz began work on the machine in 1837 and a working model was completed by 1840 but full-scale construction was only possible following a grant from the Swedish government in 1851. The specification of the Scheutz machine was similar to Babbage's original, with 15 decimal digits and 4 orders of difference. However, it was built on a far smaller scale and differed considerably in the detail of its design and in its method of operation.

Following completion in October 1853, the Scheutz difference machine was brought to London where it received considerable attention. It was examined by a committee appointed by the Royal Society, which acknowledged the machine's potential for recalculating and printing existing numerical tables in order to weed out errors. The machine was then exhibited at the Paris Exposition of 1855, where it was awarded a gold medal in a category that included Thomas de Colmar's Arithmometer. To the great surprise of many who knew him, Babbage welcomed this new development. He befriended the Scheutzs and became an enthusiastic supporter of their work but he also seized every opportunity to remind anyone who would listen that it was he who had first conceived such a device.

After the Paris Exposition, the Scheutz machine was loaned to the Paris Observatory while a buyer was sought. With Babbage's help, it was sold in December 1856 to the Dudley Observatory in Albany, New York, for £1,000. The following year a copy was commissioned by the British General Register Office to assist in the calculation of life expectancy tables. This was built by the firm of Bryan Donkin & Company, a British printing machine manufacturer which had been instrumental in bringing the original Scheutz machine to London. The new machine was delivered in July 1859 but was found to be unreliable in use despite incorporating several improvements over the original Scheutz design.

Babbage's Legacy

Though always a somewhat eccentric character with an obsessive nature, Babbage became more irascible with age and his later years were marked by a bizarre campaign against street musicians, particularly organ grinders.

1. Computer Prehistory – Calculating Machines

These efforts resulted in the passing of an Act of Parliament in July 1864 which outlawed such activities in central London but the new law was not welcomed by the general public. Instead, Babbage was ridiculed in the press and became a target for harassment by every street musician in London. He increasingly withdrew from the London social scene in which he had been such an active participant for much of his life and his final years were spent in relative isolation. Babbage died at the age of 79 in October 1871, only five months after his lifelong friend John Herschel.

Although he had failed to realise either of his two great inventions, Babbage achieved much throughout a long and highly productive life. He was active in several other fields, including cryptography, where he broke the Vigenère cipher, a powerful encryption method which was thought to be unbreakable. He also dabbled in politics, standing unsuccessfully as Liberal candidate for the London borough of Finsbury in the General Election of November 1832 and again in June 1834 when the seat became vacant. An influential book, 'On the Economy of Machinery and Manufactures', which was first published in 1832, was the result of Babbage's lengthy studies of manufacturing techniques. There is also no question that his ten-year collaboration with Joseph Clement advanced the art of precision engineering enormously, a fact borne out by the esteem in which Babbage was held by the eminent engineers Joseph Whitworth and James Nasmyth. His abilities as a mathematician were also recognised by his appointment as Lucasian Professor of Mathematics at Cambridge University from 1828 to 1839, a position once held by no less a figure than Sir Isaac Newton.

After Babbage's death his youngest surviving son, Henry P Babbage, completed several fragments of his father's Engines, including a number of small demonstration pieces assembled from unused parts of the Difference Engine No. 1. Henry also campaigned for the Analytical Engine to be built. However, in 1878 a committee appointed by the British Association for the Advancement of Science concluded that the complexity of the design and insufficient detail in the drawings made it impossible to be certain that the machine would operate as planned if constructed. Instead, they suggested that a simpler machine might be built using selected parts of the mechanism. This prompted Henry to construct a section of the mill of the Analytical Engine which he cleverly reconfigured to operate as a four function calculator. Following completion in 1906, Henry used this machine to successfully calculate multiples of Pi.

In 1985, the Science Museum in London initiated a project to construct a full-scale replica of Babbage's later design for his first machine, the Difference Engine No. 2. Certain assumptions had to be made, however, as the original drawings lacked information on the choice of materials, machining methods, manufacturing precision and finishes. Nevertheless, Babbage's design was followed faithfully and only two corrections were necessary to render the mechanism operational. With the help of a local engineering design firm and various specialist subcontractors, the calculating mechanism was completed by November 1991, in time for an exhibition commemorating the bicentennial year of Babbage's birth. Although modern machine tools were employed to produce the components for the Engine, the quality of manufacture was strictly limited so that it remained within the standards of engineering precision achievable when Babbage designed it.

Further funding to enable the construction of the stereotyping and printing mechanism for the machine was provided by Nathan Myhrvold, a senior executive with the US software giant Microsoft. This was completed and added to the calculating mechanism in March 2002. The whole machine contains over 8,000 parts and weighs 5 tonnes. Myhrvold also provided additional funding for a second replica to be built for his private collection. This was completed in 2009 and is currently on loan to the Computer History Museum in Mountain View, California. The fact that these replicas exist and operate according to specification is vindication of Babbage's reputation and proof that the engineering technology available to him in the 1840s was indeed capable of producing working versions.

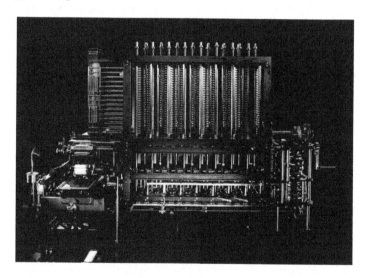

Replica of Babbage's Difference Engine No. 2 (Photo © Copyright xRez Studio)

Babbage's greatest legacy is the Analytical Engine. He left approximately 300 engineering drawings and more than 600 notations and 6,000 pages of notebooks describing the Engine and its components. These are now in the collections of the Science Museum. However, Babbage never provided a detailed description of the operation of the Analytical Engine nor discussed his ideas for its use in any detail and this has hampered our understanding of it. Fortunately, we have the valuable contribution of Augusta Ada Gordon, later Ada King, Countess of Lovelace.

1. Computer Prehistory – Calculating Machines

The daughter of the infamous Romantic poet Lord Byron, Ada Lovelace showed an early aptitude for mathematics. She first met Babbage as a teenager in 1833 and was captivated by his description of the Difference Engine. Babbage became her mentor, offering advice on her mathematical studies and introducing her into his circle of intellectual friends. In August 1840, Babbage attended a scientific meeting held in the Italian city of Turin at which he gave a series of presentation on the Analytical Engine. Among the attendees was the Italian military engineer and mathematician Luigi Federico Menabrea. Menabrea was so impressed by Babbage's work that he wrote an article describing the Engine which was published in a Swiss journal in October 1842. In an effort to generate some positive publicity for Babbage in the wake of the November 1842 decision to withdraw government support for the Difference Engine project, Ada translated Menabrea's article from French into English and submitted it for publication in a British journal. She also added copious notes which were prepared in consultation with Babbage.

Ada Lovelace's notes on the Analytical Engine choose to ignore the engineering details of the mechanism and focus instead on how the machine would be used to solve mathematical equations. They show that her understanding of the limitations and the capabilities of the Engine went deeper than Babbage's. In them, she discusses the possibility of referring to storage locations numerically and of using the machine to manipulate symbols, two important concepts that predate their rediscovery by the pioneers of digital computers by almost a century, and there is little doubt that she would have played a pivotal role in putting the machine to work had the Analytical Engine been built.

With the potential to perform complex mathematical calculations automatically, the Analytical Engine was the first true general-purpose computing device and its development has rightly been called one of the great intellectual achievements of humankind. However, despite a striking similarity between the functional design of the Analytical Engine and the architecture of today's electronic digital computers, there is no direct evolutionary link between the two. Babbage's failure to complete any of his Engines and the dearth of written material on the operation of the Analytical Engine meant that his astonishing achievements went largely unrecognised for many years and only fully came to light after the pioneers of digital computers had reinvented much of his work. We can only wonder as to how much further ahead progress in computer technology would have been, and the different path that it might have taken, had Babbage actually succeeded in constructing this incredible machine.

Further Reading

Aspray, W. (ed.), Computing Before Computers, Iowa State University Press, Ames, Iowa, 1990.

Bedini, S. A. and Maddison, F. R., Mechanical Universe: The Astrarium of Giovanni de' Dondi, Transactions of the American Philosophical Society, New Series, 56 (5), 1966, 1-69.

Belanger, J. and Stein, D., Shadowy vision: spanners in the mechanization of mathematics, Historia Mathematica 32 (1), 2005, 76-93.

Bromley, A. G., The Evolution of Babbage's Calculating Engines, IEEE Annals of the History of Computing 9 (2), 1987, 113-136.

Field, J. V. and Wright, M. T., The early history of mathematical gearing, Endeavour New Series 9 (4), 1985, 198-203.

Fuegi, J. and Francis, J., Lovelace & Babbage and the Creation of the 1843 'Notes', IEEE Annals of the History of Computing 25 (4), 2003, 16-26.

Hounshell, D. A., From the American System to Mass Production, 1800-1932, Johns Hopkins University Press, Baltimore, Maryland, 1984.

Hyman, A., Charles Babbage: Pioneer of the Computer, Princeton University Press, Princeton, New Jersey, 1985.

Johnston, S., Making the Arithmometer Count, Bulletin of the Scientific Instrument Society 52, 1997, 12-21.

Kistermann, F. W., Abridged Multiplication – The Architecture of Wilhelm Schickard's Calculating Machine of 1623, Vistas in Astronomy 28 (1-2), 1985, 347-353.

King, H. C., Geared to the Stars: The Evolution of Planetariums, Orreries, and Astronomical Clocks, Adam Hilger Ltd, Bristol, 1978.

Martin, E., The Calculating Machines: Their History and Development, MIT Press, Cambridge, Massachusetts and Tomash Publishers, Los Angeles, California, 1992.

Merzbach, U. C., Georg Scheutz and the First Printing Calculator, Smithsonian Institution Press, Washington, 1977.

Moore, W. R., Foundations of Mechanical Accuracy, Moore Special Tool Company, Bridgeport, Connecticut, 1970.

Rolt, L. T. C., Tools for the Job: A Short History of Machine Tools, B T Batsford, London, 1965.

Swade, D. D., The Construction of Charles Babbage's Difference Engine No. 2, IEEE Annals of the History of Computing 27 (3), 2005, 70-88.

Woodbury, R. S., The Legend of Eli Whitney and Interchangeable Parts, Technology and Culture 1 (3), 1960, 235-253.

2. The Analogue Alternative

"I often say that when you can measure what you are speaking about and express it in numbers, you know something about it; but when you cannot measure it, when you cannot express it in numbers, your knowledge is of a meagre and unsatisfactory kind", Lord Kelvin, 1867.

The Attraction of Analogue

Early calculating machines were not exclusively digital. The properties of certain mechanisms, in which the relationship between input and output is continuously variable rather than operating in discrete quantities, encouraged an alternative approach known as analogue computing. Computing devices which employed such mechanisms could be used in the solution of complex mathematical problems that were beyond the capability of early digital calculating machines. They were also inherently faster than their digital counterparts.

These advantages prompted considerable development activity, much of which took place in parallel with digital calculating machine development. Although the two approaches differ fundamentally, they shared many of the same basic components and some cross fertilisation of ideas took place. The success of analogue computing technology in a number of demanding application areas also provided the inspiration for some of the most important of the early digital computer projects. The starting point for this development activity was the invention of the mechanical integrator.

Building Blocks – Mechanical Integrators

In 1814, a Bavarian land surveyor by the name of Johann M Hermann invented the first analogue computing device to employ the principle of mechanical integration. Hermann's work as a surveyor involved mapping the boundaries of plots of land and calculating the area contained within them. However, these plots were seldom regularly shaped, which required the surveyor to perform tedious mathematical integration calculations in order to determine their area. To simplify this task, Hermann designed an instrument which provided a mechanical analogue of the integration function using the movement of a wheel rolling against a rotating cone.

Hermann's device comprised a wheel-and-cone arrangement mounted vertically on a carriage fitted with a pointer for tracing the perimeter of the area to be measured. A rack and pinion mechanism converted the longitudinal movement of the pointer into rotational movement of the cone. Transverse movement of the pointer was transferred via the carriage to a wedge-shaped slider which pushed the integrating wheel vertically up and down the side of the cone. As the cone rotated, this rotation was transferred to the integrating wheel, with the speed of rotation of the wheel varying with its height on the cone. Assuming there was no slippage between the wheel and the cone, the total number of revolutions of the integrating wheel corresponded to the area within the closed curve traced by the pointer. This concept of calculation through measurement is fundamental to analogue computing and would reappear in many of the mathematical instruments of the 19th century.

Hermann's device was the first in a series of mathematical instruments based on the principle of mechanical integration, known as Planimeters, which were developed over the following 70 years. It is thought that Hermann may have had assistance with the design of his Planimeter from Steuerrat Lämmle, an academic based at the University of Munich. In any case, the instrument had been developed for Hermann's own personal use and he appears to have made no attempt to publicise it. The Planimeter itself was lost in 1848 and no contemporary accounts of it are known to exist. The only evidence we have of its existence is an article published in an obscure German technical journal several years later in 1855. Fortunately, the concept of mechanical integration had already been reinvented by the Italian mathematician Tito Gonnella in 1824.

Gonnella was a professor of mathematics and mechanics at the Accademia di Belle Arti in Florence. Although unaware of Hermann's work, the design of his Planimeter was strikingly similar, incorporating the same wheel-and-cone arrangement. Gonnella had a prototype instrument manufactured locally but it did not work very well. As an academic, Gonnella was expected to disseminate his ideas and he published an article describing the device in 1825 which was entitled 'Description of a machine to square plane surfaces'. Later that same year, he received a request from the Court of the Grand Duke of Tuscany to supply a Planimeter for the Duke's collection of mathematical instruments. Gonnella was keenly aware of the reputation of Swiss craftsmen for precision engineering so he sent copies of his design to several Swiss manufacturers in an effort to have the instrument manufactured in Switzerland. It seems that these efforts were unsuccessful, although he is known to have exhibited a Planimeter in London many years later at the Great Exhibition of 1851, where it was awarded a Council Medal. He also worked on the development of digital calculating machinery, inventing a key-driven adding machine in 1859.

Following Gonnella's failed attempt to have his Planimeter manufactured in Switzerland, it may be no coincidence that the development path of the Planimeter became dominated by Swiss inventors. In 1827, a Planimeter

2. The Analogue Alternative

was invented by the Swiss surveyor and engineer Johannes Oppikofer. It also followed the same basic wheel-and-cone design of Hermann's original device but featured a horizontally mounted cone which dispensed with the requirement for a wedge-shaped slider. Two indicator dials were also provided to display the total number of revolutions plus any fraction of a revolution rotated by the integrating wheel.

Oppikofer arranged to have a prototype of his Planimeter constructed by the mechanic Johannes Pfäffli but Pfäffli died unexpectedly before it could be completed. The work was then taken up by the instrument maker Heinrich Rudolf Ernst who completed the prototype around 1828 and presented it the following year to the Naturforschende Gesellschaft, a learned society in Bern. Ernst later relocated to Paris where he exhibited an improved device at the Académie des Sciences in 1834. A number of these instruments were then manufactured and sold commercially.

A major improvement to the basic design of the Planimeter was made in 1849 by the Swiss engineer Kaspar Wetli. One limitation of the wheel-and-cone integrator mechanism was an inability to deal with negative values, which restricted its use in certain applications, so Wetli set out to create a new design that would solve this problem. His initial experiments involved adding a second cone which was positioned tip to tip with the first cone but with its axis tilted so that the sides of both cones were parallel and formed a straight line. The range of movement of the integrating wheel along the side of the cone was extended to allow it to travel beyond the tip of the first cone and onto the surface of the second cone which rotated in the opposite direction, thus reducing the total number of revolutions of the integrating wheel and permitting negative values to be represented. Wetli then added a flat disk positioned across the top of the cones in order to smooth the path of the integrating wheel when moving from one cone to another. The disk rotated through contact with the cones beneath but Wetli soon realised that the cones could be eliminated altogether if the disk was driven directly by the movement of the pointer. This was achieved by means of a wire drive arrangement which employed a metal wire wrapped around a spool to rotate the disk.

Wetli patented his wheel-and-disk Planimeter in 1849 and went into partnership with the Austrian instrument maker Georg C Starke to manufacture it on a commercial basis. As the demand for Planimeters grew, boosted by their use for the calculations required to determine steam engine performance, the wheel-and-disk integrator was gradually adopted by other manufacturers and became the standard mechanism for instruments of this type.

The commercialisation of Planimeters received another boost following the introduction of the Polar Planimeter in 1854 by the Swiss mathematician and physicist Jakob Amsler. Amsler's clever design dispensed with the rotating disk by placing the integrating wheel directly into contact with the surface on which the instrument was being used. The wheel and indicator dial mechanism were incorporated within the arm of the pointer which was connected through a rotating joint to another arm that pivoted around a fixed point. As the pointer traced the perimeter of the area to be measured, the integrating wheel would undergo a combination of rotation and sliding movements which together produced the same result as an orthogonal Planimeter.

Being of simpler design than an orthogonal model, Amsler's Polar Planimeter would be much cheaper to manufacture. Enthused by the commercial potential of such an instrument, Amsler took the bold step of setting up a workshop in the Swiss town of Schaffhausen to manufacture Polar Planimeters for commercial sale. The venture was an immediate success and within three years he had given up a promising academic career in mathematical physics to concentrate on the business full time. By 1884, Amsler's factory had produced and sold over 12,000 Polar Planimeters. This stimulated competition from other manufacturers, with the result that the Planimeter became the most common type of mathematical instrument available in the 19th century. Mechanical integrator mechanisms of the type pioneered by Hermann, Gonnella and Wetli would also provide the building blocks for more sophisticated analogue computing devices.

The Integraph

The mechanical integrator was first exploited for more complex mathematical tasks in a device known as the Integraph. An Integraph is a mathematical instrument for deriving the integral curve corresponding to a given curve. By manipulation of the integral curve and successive approximations, it is possible to use an Integraph to solve graphically certain commonly encountered classes of differential equations.

The Integraph consists of a pair of parallel bars supported on two rollers linked by a common axle. The bars serve as the guideways for two carriages, one of which carries a pointer arm and the other an integrating wheel and pen, both of which are in contact with the surface on which the instrument is being used. The pointer arm and integrating wheel mechanism are connected by a rod which pivots around the midpoint of the guideways to form a parallelogram arrangement. As the user traces the curve with the pointer, the connecting rod transfers these movements through the two carriages and integrating wheel to the pen which draws the corresponding integral curve.

The Integraph was invented in 1878 by the Lithuanian mathematician and electrical engineer Bruno Abdank-Abakanowicz. Abdank-Abakanowicz patented his invention in 1880, however, his design was mechanically complex and several engineering problems had to be overcome before a practical instrument could be constructed. One of the most significant problems was how

2. The Analogue Alternative

to maintain the correct angle of the integrating wheel so that it always steered in the same direction as the pointer. This was solved in 1885 by David Napoli, the Director of the workshop of the French Eastern Railway, who Abdank-Abakanowicz had met at an international exhibition in Vienna in 1883. Napoli added a system of transfer gears which transferred any angular change of the connecting rod to the frame holding the integrating wheel. Soon after this Abdank-Abakanowicz began a collaboration with the Swiss instrument maker Gottlieb Coradi who made further improvements to the Integraph design for a production version which was introduced in 1888.

The Integraph was difficult to use, expensive to buy and required a large drawing board or similar surface on which to operate. Nevertheless, the computation of integrals had become a common requirement in many engineering and scientific disciplines and other instrument manufacturers also began to produce Integraphs. Some of these were based on an alternative design which was conceived independently by the British physicist Sir Charles Vernon Boys in 1881 while a student at the Royal School of Mines.

Although its use as a computing device was limited to graphical solutions, the Integraph demonstrated the potential of precision instruments for complex mathematical tasks, paving the way for a number of special-purpose analogue computing devices that became collectively known as continuous calculating machines.

The Work of Lord Kelvin

The introduction of an improved design of mechanical integrator mechanism by the British engineer James Thomson around 1864 led to the development of a new class of analogue computing devices. Unlike the Integraph, these instruments were designed to produce numerical solutions to complex mathematical problems.

Thomson was professor of civil engineering and surveying at Queen's College (now Queen's University) Belfast. He devised his integrator mechanism while working on the design of a new type of Planimeter that would be better suited to certain meteorological applications than the commonly available models. Thomson's design retained the rotating disk from a wheel-and-disk Planimeter but replaced the integrating wheel with a ball and cylinder arrangement. The ball was placed on the upper surface of the disk and held between the prongs of a fork-shaped bracket and guideway which allowed the ball to rotate freely but restricted its travel across the face of the disk to a single axis. The position of the ball along this axis was controlled by the transverse movement of the pointer in the same way as an integrating wheel on a conventional Planimeter. The cylinder was located above the disk with its axis of rotation parallel to the disk surface and positioned in such a way that it was in contact with the ball but not the disk. As the disk rotated, the rolling action of the ball transferred this movement to the cylinder.

Thomson's disk-ball-and-cylinder integrator was based on earlier work by the brilliant Scottish physicist James Clerk Maxwell. Maxwell was still an undergraduate student at Cambridge University when he visited the Great Exhibition of 1851. One of the exhibits which caught his attention was a wheel-and-cone Planimeter. On examining the instrument, he observed that the sliding action of the integrating wheel as it moved up and down the side of the cone might cause slippage to occur, thus introducing an error in the reading. He devised a new integrator mechanism which was based on the pure rolling motion of two balls and incorporated this into two separate Planimeter designs in order to illustrate the principle. Maxwell described these instruments in a paper to the Royal Scottish Society of Arts in January 1855 whereupon the Society agreed to award him a grant of £10 towards the construction of a working prototype. However, it seems that he was unable to secure the additional funding necessary, as construction never went ahead, and it was only when Thomson built his disk-ball-and-cylinder integrator more than a decade later that Maxwell's concept was finally realised.

It was James Thomson's younger brother, the eminent physicist and mathematician William Thomson, who first recognised the potential of the disk-ball-and-cylinder integrator for continuous calculating machines. The two brothers came from a family of academics. Their father, who was also called James, had taught mathematics at the Royal Belfast Academical Institution in their native Northern Ireland before bringing his two sons to Scotland in October 1833 following his appointment as professor of mathematics at the University of Glasgow. Both sons began their undergraduate studies in engineering and natural philosophy at the University the following year despite being aged only 12 and 10 at the time. James graduated in 1839 but William chose to complete his studies at Cambridge University before returning to Glasgow in 1846, aged 22, to take up a position as Professor of Natural Philosophy, a post he held until his retirement more than 50 years later in 1899.

The Tide Predictor

William Thomson was a keen sailor and much of his research work had a nautical theme. As an island nation, Britain was heavily reliant on the sea for trade and communications. However, the movement of shipping between ports was governed by the tides. This made tidal prediction an issue of national importance. Although the physics of tidal patterns was already well understood, predicting tides was notoriously difficult. Thomson began a serious study on the subject of tidal prediction in 1867. By applying the principles of harmonic analysis, he discovered that tidal patterns could be modelled by combining a series of trigonometric functions with independent periods and amplitudes representing the gravitational influences of the moon and the sun. Thomson

2. The Analogue Alternative

was also aware that if these functions could be replicated using mechanisms, then it should be possible to use machinery to assist with the computation of tide tables for use in ports and harbours.

Mechanisms which generate sinusoidal motion, such as cranks and cams, were readily available but Thomson was unable to find a suitable mechanism that could combine their movements. He discussed this problem with the inventor and railway engineer Beauchamp Tower in 1872, who suggested adopting a method employed by the physicist Sir Charles Wheatstone in his alphabetic telegraph instrument of 1840. Wheatstone had used a chain passing around a number of pulleys which rotated on shafts that were able to move from side to side in a direction perpendicular to their axes of rotation. The chain was fixed at one end and weighted at the other to keep it taut. As each shaft moved up or down, it would take up or release a corresponding length of the chain. These changes in the length of the free end of the chain represented the sum of the movements of the different shafts. Thomson recognised immediately that this was the solution to his problem and set about finalising the design of his machine.

Kelvin's First Tide Predicting Machine (Photo © Science Museum / Science & Society Picture Library)

Thomson's tide predicting machine comprised a row of 8 pulleys suspended from a frame and linked together by a common wire which was fixed at one end and threaded over and under each pulley in turn. Each pulley was also connected by another wire to a geared crank mechanism located beneath it. The machine was operated by turning a handle at one end of a gear shaft which rotated each of the crank mechanisms at the appropriate rate governed by the gear ratio, moving the pulleys up and down and causing a corresponding tightening and slackening of the common wire. The amplitude of each pulley movement could be adjusted by altering the length of the crank arm. The sum of the harmonic components represented by the motions of the pulleys was obtained by connecting the free end of the common wire to a pen and recording its movement on a strip of paper fitted to a rotating drum which was also driven by the gear shaft. The times of high and low tides, as indicated by the peaks and troughs on the resulting chart, could then be tabulated.

Construction of the prototype tide predicting machine was funded by the British Association for the Advancement of Science and was completed by 1873. An improved version with 10 pulleys was designed by Thomson in 1875 with the assistance of Edward Roberts of the Nautical Almanac Office. This was built by Alexander Légé & Company of London the following year and exhibited at the Paris Exposition Universelle in 1878. Roberts and Légé then proceeded to develop a much larger 20-component machine for the British Government's India Office in 1879 which was successfully used to compute tide tables for the Indian ports from 1880 to 1906.

The Harmonic Analyser

The harmonic components necessary in order to set up Thomson's tide predicting machine were extracted from graphs of actual tide heights obtained using a tide gauge. However, extraction involved a lengthy manual process known as Fourier analysis which was based on the evaluation of integrals. Thomson began looking at the possibility of mechanising this process in January 1876. He discussed his ideas with his brother James, who recalled the disk-ball-and-cylinder integrator that he had devised more than a decade earlier. Thomson could see that his brother's integrator device would simplify the design of such a machine considerably. He also realised that by back coupling the fork of one integrator mechanism to the cylinder of a second integrator so that any transverse movement of the ball resulted in a corresponding rotation of the cylinder, it would be possible to solve second order linear differential equations. Furthermore, if an 'endless chain' of integrators was created by linking several pairs and coupling the first to the last, this concept could be extended to solve differential equations of any order.

Within days of their inspirational discussion, the two brothers drafted no less than four scientific papers describing the disk-ball-and-cylinder integrator and its potential uses for submission to the Royal Society. William proceeded to

2. The Analogue Alternative

incorporate the integrator mechanism into the design of a bench-top prototype machine for the Fourier analysis of tide height graphs which he called the Harmonic Analyser. He then had the machine constructed by the Glasgow-based optician and mathematical instrument maker James White.

The Harmonic Analyser was operated by placing the tide height graph onto a horizontally mounted rotating drum and tracing the curve of the graph with a pointer which could be moved transversely along the length of the drum using a rack and pinion mechanism. A rod transmitted this pointer movement to the fork-shaped brackets that controlled the lateral positions of the 5 integrator balls which were mounted in a row along the length of the machine. The drum was driven by a hand crank and connected by an arrangement of gear shafts and bevel gears to the integrator disks, four of which rotated clockwise and anticlockwise in an oscillatory motion at different phases to produce the required harmonic components. The fifth disk rotated continuously in the same direction to provide the constant term required for the equation. The resulting values were read off from indicator dials fitted to each of the cylinders to record the number of revolutions turned.

The completed prototype was demonstrated to the Royal Society in May 1878. However, as it was equipped with only two pairs of integrators, the machine could only be used to extract the first and second harmonics of a waveform. It also contained a design flaw due to its compact dimensions which restricted the range of movement of the pointer, making the machine unsuitable for the majority of tidal graphs. Fortunately, this was less of an issue in other mathematical applications, such as the analysis of thermograms and photographic barograms, graphical records of daily changes in atmospheric temperature and pressure. Thomson had the prototype modified to make it suitable for these applications. He then donated it to the Meteorological Office, the British government department responsible for weather forecasting. This led to the development of a larger instrument with 7 integrators which was specifically designed to analyse thermograms and barograms. Named the Meteorological Analyser, this larger machine was built by R W Munro, a London-based firm which specialised in the manufacture of meteorological instruments, and was completed in December 1879. It was later acquired by the Science Museum in London and is now part of the Museum's permanent display of mathematical instruments. In order to fulfil the requirements of the original application of tide prediction, an 11 integrator machine was constructed with the help of a grant from the Royal Society. This was known as the Tidal Harmonic Analyser and was also completed in 1879.

The Story of the Computer

Kelvin's Meteorological Analyser of 1879 (Photo © Science Museum / Science & Society Picture Library)

Their large physical size and reliance on high precision components made Thomson's harmonic analysers very costly to manufacture, which tended to limit their use to one-off prototypes built for a specific application. However, an ingenious redesign by the German mathematician Olaus Henrici created a tabletop version that could be manufactured more cheaply and sold commercially. Henrici was professor of mechanics and mathematics at the Central Technical College in London. His inspiration for an alternative design of harmonic analyser came from the British mathematician and philosopher William Kingdon Clifford, who had devised a graphical representation of Fourier analysis which reduced the curve to be analysed to a series of simpler curves whose area is equal to the harmonic components of the original curve. Henrici designed a prototype machine based on this principle in 1889. He then worked with the Swiss instrument maker Gottlieb Coradi, who had become the leading manufacturer of precision mathematical and drawing instruments, to develop a production version which was introduced in 1894.

Clifford's method of Fourier analysis allowed Henrici to dispense with the rotating drum mechanism by placing the machine directly on top of the graph to be traced in the same manner as that of a Planimeter or Integraph. The central component was an adaptation of James Thomson's disk-ball-and-cylinder integrator mechanism and comprised a glass sphere surrounded by a frame that rotated about a vertical axis which supported two rubber wheels spaced 90 degrees apart in the horizontal plane and held in contact with the sphere. Up to 10 of these rolling sphere integrator mechanisms could be fitted to a single machine.

2. The Analogue Alternative

Henrici's harmonic analyser was operated by turning knobs at each end of the machine which moved a cursor in two axes in order to trace the curve. Longitudinal movement of the cursor was transferred through an arrangement of wires and pulleys into a rotational movement of the integrator mechanism frames. Different sized pulleys allowed each of the frames to rotate at different rates according to the order of the harmonic represented. Transverse movement of the cursor caused the whole machine to move across the graph and this movement was converted into a rotation of the integrator spheres through drums driven by the shaft connecting two of the three wheels which supported the machine. The resulting Fourier coefficient values were read from indicator dials fitted to the pairs of wheels on each of the integrators.

William Thomson was never able to realise his idea for extending the capability of the disk-ball-and-cylinder integrator to allow the solution of differential equations. Practical difficulties with the torque output from an integrator wheel being insufficient to drive further integrating mechanisms would have made the machine unusable. However, this would not be the end of Thomson's pioneering work on continuous calculating machines. In 1878 he proposed a pulley-based machine for solving simultaneous linear equations in another paper submitted to the Royal Society. The machine comprised a set of 6 tilting platforms suspended by an arrangement of wires and pulleys which operated together to constrain the relative tilts possible in such a way that they provided a solution to a set of 6 simultaneous equations. Thomson never constructed this machine either, although it is not clear why as there appear to be no major flaws in the design. Remarkably, both concepts would come to fruition almost 50 years later through work at the Massachusetts Institute of Technology in the USA.

William Thomson's many inventions would make him one of the most famous scientists of the Victorian age. He received a knighthood from Queen Victoria in 1866 for his contribution to the laying of the first successful transatlantic telegraph cable. This was followed by his elevation to the peerage in 1892. The title he took, Baron Kelvin of Largs, provided him with the name by which he is more familiar today.

Lord Kelvin published more than 600 scientific papers in an academic career lasting 53 years but it was his entrepreneurial spirit and willingness to engage with industry that set him apart from his fellow scientists. This was evident in his fruitful association with the instrument maker James White & Company which led to the formation of a series of marine instrument companies, one of which (Kelvin Hughes Limited) still exists today. He died, aged 83, in December 1907 and is buried in Westminster Abbey in London. The location of his grave, in the nave of the Abbey alongside the tomb of Sir Isaac Newton, is testament to the esteem in which he was held.

Gunnery Instruments and Fire Control Computers

The effective range of naval gunnery increased dramatically in the late 19th century to the point where it was no longer possible for gunners to aim directly at an enemy ship and expect the shell to hit it. With a time of flight of 30 seconds or more and fast moving steam-driven ships, the target would no longer be in the same place as when the shell was fired. Unless the ship on which the gun was mounted was stationary, its own course and speed would further complicate matters. External factors, such as the speed and direction of the wind and the rotation of the earth, also influenced the trajectory of the shell.

For longer range weapons to be effective, better observation instruments would be necessary in order to accurately estimate the range and compass bearing of a distant target. However, this was only part of the problem. Complex trigonometric calculations were also required to determine the necessary direction and elevation of the gun but such calculations would take too long to perform manually in the heat of battle. Fortunately, the recent advances in analogue computing technology made by Lord Kelvin and others offered a potential solution.

Range Clocks

The earliest gunnery instruments to take advantage of these advances in analogue computing technology were range clocks. A range clock is a clock-like device that provides a continuous estimate of the range to the target which is constantly changing due to the relative motion of firing and target ships.

The development of the range clock was first proposed in December 1903 in a letter to the Admiralty written by Captain Percy Scott, who was in charge of HMS Excellent, the Royal Navy's School of Gunnery at Portsmouth. Scott was an early advocate for the introduction of new technology for naval gunnery and had devised an improved method of aiming known as continuous aim. The key to this method was having a continuous indication of the range to the target. After failing to obtain a response to his proposal, Scott used personal links he had with the British defence contractor Vickers Sons & Maxim to encourage the company to develop and manufacture the device. It was designed by the superintendent of the firm's ordnance works, Arthur Trevor Dawson, a former Lieutenant in the Royal Navy who would later become Managing Director of Vickers, and engineer James Horne.

The Vickers Range Clock gave a continuous indication of target range when set with the initial range and the rate of change. It comprised a wheel-and-disk integrator mechanism housed within a circular case fitted with an indicator dial and pointer on the upper face. The integrator mechanism incorporated twin rotating disks which were mounted in parallel, with the integrating wheel positioned in the gap between them. A clockwork motor rotated the integrator

2. The Analogue Alternative

disks in opposite directions at a constant speed. The position of the integrating wheel across the faces of the disks was controlled by a lead screw connected to a gearbox and crank handle with a graduated scale for setting the rate of change. Rotational movement of the integrating wheel was transferred by gearing to the indicator dial pointer which rotated in either direction to display target range. The initial range was set by rotating the graduated scale of the indicator dial so that the correct value corresponded with the rest position of the pointer.

After seeking patent protection for the Range Clock in the UK and the USA, Vickers constructed a number of prototypes for evaluation during the Royal Navy's annual long-range gunnery practice exercise held in 1905. The following year Vickers received an order for 246 units for use on Royal Navy ships. This encouraged rival manufacturers such as Barr & Stroud to introduce similar instruments.

In use, the initial range was obtained using an optical rangefinder and the rate of change obtained using a Dumaresq, a special-purpose mathematical instrument invented in 1902 by Lieutenant John S Dumaresq, a gunnery officer with the Royal Navy. The Dumaresq employed a circular dial and sliding scale to model relative course and speed and resolve it into two components, one along the line of sight, giving the rate of change of range, and the other across the line of sight, giving the 'Dumaresq deflection'. Dumaresq patented his invention in August 1904 and it was manufactured by the London-based firm of instrument makers Elliott Brothers.

The Vickers Range Clock had been designed so that the initial settings would remain unaltered during operation but it was soon discovered that the rate of change varied in practice. However, constantly resetting the rate of change while the clock was running caused excessive wear on the integrating wheel. This prompted the introduction in 1907 of a revised version which incorporated a clutch mechanism for disengaging the integrating wheel from the rotating disks when repositioning it.

Arthur Pollen and the Argo Clock

The range clock was a relatively simple though highly effective solution for maintaining a continuous estimate of target range but range estimation was only one of the problems faced by naval gunners in the dreadnought era. The course and speed of the target ship were equally important parameters. The first person to develop an integrated solution for naval gunnery fire control was the British businessman Arthur Joseph Hungerford Pollen.

Arthur Pollen was an unlikely computing pioneer. Born in Southwater, Sussex in 1866, he originally trained as a lawyer and also dabbled in politics and journalism. His move into industry came in 1898 when he was appointed through his wife's family connections as Managing Director of the Linotype & Machinery Company, the UK's leading manufacturer of typesetting machines.

In February 1900, he was invited by his cousin, an officer in the Royal Navy, to witness a naval gunnery exercise off the coast of Malta. Even with his inexpert eye, Pollen could see that the long range accuracy of the guns was extremely poor in comparison with similar guns used on land. Pollen was appalled at this and despite having no nautical or engineering experience, he pledged to use all the resources at his disposal to help improve the accuracy of naval gunnery.

On studying the problem, Pollen discovered that a major contributory factor was the lack of accuracy in measuring the range to the target and so this was where he concentrated his initial efforts. In February 1901, he wrote to the Admiralty proposing that they adopt a new triangulation-based rangefinder system he had devised in association with an independent inventor named Mark Barr which he called the Pollen System of Telemetry. The system comprised two telescopic sights on mounts separated by a distance of 150 feet which were used by two observers to obtain target bearings simultaneously. Electrical signalling transmitted this information to a pantographic instrument which automatically performed the simple trigonometric calculation required to obtain the range. This proposal was rejected but Pollen continued his studies into the problem, making use of Linotype engineering staff to analyse a hypothetical naval engagement. The analysis provided a valuable insight into the many other factors involved and showed the importance of the rate of change of range when estimating the range to the target.

Working with Linotype design engineer Harold Isherwood, Pollen next turned his attention to the development of a system for plotting the movement of the target relative to that of the firing ship. Such a system would not only produce a more accurate estimate of the range to the target, and its likely position when the shell landed, but it would also allow the firing ship to match its course and speed with that of the target vessel, thus allowing it to keep an enemy ship within the effective range of its guns. In the spring of 1904, Pollen sought the advice of Lord Kelvin, who happened to be a member of the Linotype board. Kelvin encouraged Pollen to explore the use of analogue computing mechanisms for calculation but warned him that the development of a complete solution to the problem of accurate long range gunnery might take 10 years or more.

Pollen continued to lobby the Admiralty to adopt his inventions and in May 1905 his determination paid off when he was awarded the sum of £4,500 to construct prototypes of the two-observer rangefinder, a true-course plotting unit and a clockwork powered instrument which calculated change of range. The plotting unit comprised a moving carriage and a swinging arm which was fitted with pens at both ends. The movement of the target ship was recorded on a sheet of paper by driving the carriage at a speed corresponding to that of the firing ship and manually setting the angle of the swinging arm to the target bearing and its length to the target range. The resulting two-line plot was then measured to obtain the true course and speed of the target and the range to it.

2. The Analogue Alternative

Although similar in function to a range clock, Pollen's change of range instrument did not employ an integrator mechanism. Instead, an alternative arrangement of mechanical linkages was employed to produce a continuous indication of range based on the relative motion of the two ships. However, the engineering design of this instrument proved difficult to execute and it was excluded from a series of sea trials of the equipment which began in November 1905. These trials revealed practical difficulties with the speed at which the plotting unit could be manually updated, and with the detrimental effect of the motion of the ship on the accuracy of target bearings obtained by the two-observer rangefinder when sailing in heavy seas. Further trials planned for January 1906 were cancelled while Pollen and Isherwood embarked on a series of design changes intended to solve these issues. This led to the development of a significantly enhanced change of range instrument which they called the Argo Clock.

The central component of the Argo Clock was a Thomson disk-ball-and-cylinder integrator. An electric motor was employed to rotate the integrator disk at a constant speed. The position of the ball across the face of the disk was controlled by changes in the target bearing, with the starting position dictated by the initial change of range rate which was set using a dial. Rotational movement of the integrating cylinder was used to drive the pointer of a dial displaying the target range. Pollen's first prototype calculated target bearing manually using a similar method to the Dumaresq but a later version added a second integrator mechanism to provide a continuous indication of target bearing. This was coupled to the range integrator via a mechanical linkage that provided the initial bearing settings and corrected for the time of flight of the shell. The later version also featured an improved design of integrator mechanism which added a second cylinder in contact with the ball in order to minimise slippage and increase the output torque available.

The Argo Clock was the central component of an integrated naval gunnery fire control system which Pollen called the Aim Correction (AC) Battle System. This included a single-observer rangefinder supplied by Barr & Stroud, which was augmented with a gyroscopically stabilised mounting designed by Isherwood, plus a redesigned true-course plotting unit that automatically generated the input data required for the Argo Clock. The system also incorporated a gyroscopic correction mechanism to compensate for the motion of the ship.

Pollen submitted a proposal to the Admiralty for the development of his new system in August 1907. The following February he was given the go ahead to conduct trials of the system. However, before work could begin the Admiralty decided to withdraw its offer as a result of a misunderstanding over the terms of the agreement. Instead, Pollen was paid the sum of £11,500 to compensate for his losses and in recognition of his contribution to the development of fire control theory. Pollen refused to rest and after much lobbying he eventually secured a £6,400 contract in April 1909 to supply a set of trial instruments

The Story of the Computer

including a prototype Argo Clock.

The trial instruments were completed in February 1910 and were evaluated in a series of sea trials held in May and June of that year. Despite encountering the usual issues with the reliability and robustness of prototype equipment in a hostile environment, Pollen's instruments showed considerable promise. A committee appointed to supervise the trials recommended adoption of both the Argo Clock and the Barr & Stroud rangefinder but called for further development of the true-course plotting unit. However, the committee's recommendations were dismissed by the Admiralty which decided to reject the plotting unit completely and requested improvements to the design of the Argo Clock to be submitted for further evaluation.

Pollen had resigned as Managing Director of Linotype in December 1905 to devote more time to his gunnery developments and had taken on Harold Isherwood as his employee. This arrangement remained in place until January 1909 when Pollen announced the formation of the Argo Company which he had established some time earlier in order to provide a firmer financial footing for their work. Following months of negotiation, the company secured its first order from the Admiralty in April 1910 for 45 gyroscopically stabilised rangefinder mountings. The order came with a generous advance of £15,000. However, most of this was spent on acquiring a manufacturing capability by purchasing a substantial shareholding in another firm. In order to raise the additional funds necessary to complete the improvements to the Argo Clock, Pollen was left with no choice but to sell shares in his company. He did this in early 1911, raising over £19,000.

The design for the improved version of the Argo Clock was submitted to the Admiralty in May 1911. Designated the Mark II, a working prototype was ready by the end of January 1912. Following a series of successful pre-trial experiments, the Admiralty placed an order for 5 production units in September. These incorporated further improvements and became known as the Mark IV. The improved design featured a total of 4 disk-ball-and-cylinder integrator mechanisms which the physicist Sir Charles Vernon Boys described as "the most perfect of all that has been done in this direction". A redesigned true-course plotting unit was completed in April the following year and the final or Mark V version of the Argo Clock was completed a few months later in October 1913. After more than a decade of development, Pollen had finally perfected his system of naval gunnery fire control. Lord Kelvin's estimate of development time had been proved correct.

The Dreyer Fire Control Table

Despite the obvious benefits of his Argo technology, the Admiralty's support for Pollen's completed AC Battle System was not forthcoming. A rival fire control system had recently emerged which was less automated but much less

2. The Analogue Alternative

expensive to manufacture. Unfortunately for Pollen, the inventor, Lieutenant Frederic C Dreyer of the Royal Navy, was also an influential figure in naval gunnery circles, having served as Assistant to the Director of Naval Ordnance.

Dreyer had first encountered Arthur Pollen's ideas for improving the accuracy of naval gunnery in 1905 while serving on a committee chaired by the inventor of the range clock, Captain Percy Scott. Pollen's technology was not yet proven nevertheless it appears to have inspired Dreyer to embark on the development of his own system for naval gunnery fire control. The resulting system was named the Dreyer Fire Control Table.

Dreyer's system was based on an assumption that the rate of change of range remains fairly constant and that the relationship between target range and bearing is sufficiently weak that they can be plotted separately. The Dreyer Fire Control Table comprised a large iron table fitted with a Dumaresq, two plotting units (one for range and the other for bearing) and a range clock. These were coupled together using rotating shafts, chain drives and various other linkages and operated by a team of 8 crewmen in a coordinated effort. The system was manufactured by Elliott Brothers, which also manufactured the Dumaresq, and the first prototype was completed in 1911.

Dreyer had been assisted in the early stages of development by his brother, Captain John T Dreyer of the Royal Artillery, who had designed a number of gunnery devices for artillery use. One of John Dreyer's devices, a range corrector, was incorporated into the Fire Control Table but the bulk of the engineering design was carried out by senior Elliott Brothers engineer Keith Elphinstone. For the prototype system, a standard Vickers Range Clock was used. However, this was soon replaced with a new clock designed by Elphinstone which retained the wheel-and-disk integrator mechanism from the Vickers model but incorporated improvements to facilitate continual readjustment while running. Like the Argo Clock, a second integrator mechanism was also added to provide a continuous indication of target bearing.

Five of the production Fire Control Tables were ordered by the Admiralty in February 1912 and delivered by the end of that year. The following year, the Admiralty took the decision to adopt the Dreyer system and ceased all further communication with the Argo Company. Dreyer's influence within the Admiralty seems to have been a factor in this decision but his system's reliance on manual operation also appealed to gunnery officers who were distrustful of the highly automated Pollen system.

Monopoly and secrecy clauses in the Admiralty contracts awarded to the Argo Company had prevented Pollen from seeking customers in other countries. These conditions were finally lifted in December 1912. Argo Company sales brochures were produced and distributed to the UK-based representatives of various foreign navies in October 1913. However, promising negotiations were disrupted by the outbreak of the First World War. A small order which included two Argo Clocks came from the Russian Navy in September 1914. This kept the company solvent but with no further orders on the horizon, Pollen put the business on hold in early 1915 and became a full-time war journalist, specialising in articles on naval strategy and tactics. After the war, he accepted a position as director of the Foreign Department of the Birmingham Small Arms Company and also served as a Council member for the Federation of British Industries. In 1926 Pollen's business career came full circle when he returned to his original job as Managing Director of Linotype after the post became vacant.

The First World War would also provide some vindication of Pollen's system due to the poor performance of the Dreyer Fire Control Table in battle, as demonstrated by a punishing encounter with the German fleet at the Battle of Jutland in May 1916. Despite having superior numbers of ships, most of which were equipped with the new fire control technology, the Royal Navy suffered heavy losses. The Dreyer system was unable to cope with rapidly changing range and bearing rates, the only exception being one ship which was fortunate to have a Dreyer table fitted with an Argo Clock in place of the Elphinstone range clock. Naval historians disagree over whether the Pollen system would have fared better despite its more advanced technology. Other factors played a part, not least of which was that the German ships were equipped with rangefinders manufactured by the renowned optical firm Zeiss which performed better than their British counterparts in the conditions of poor visibility encountered during the battle.

Further vindication for Pollen came in 1925 when he was awarded the considerable sum of £30,000 by the Royal Commission on Awards to Inventors in recognition of elements of the Argo Clock which had been incorporated into the Dreyer fire control system without his permission. Dreyer's Royal Navy status had given him access to Pollen's design for the Mark II version of the Argo Clock which had been submitted to the Admiralty in May 1911. Elphinstone had also visited the Argo Company a few months later in October 1911 where he was shown the Mark II prototype, which was then in the final stages of construction, along with the engineering drawings for it. A government tribunal concluded that these two opportunities had led to plagiarism on the part of Dreyer and Elphinstone. This was a harsh judgement, given the significant differences in the internal workings of the two clocks due to the use of different integrator mechanism designs. A more blatant example of plagiarism of Pollen's technology can be seen in the development of the first successful US fire control system.

The Ford Rangekeeper

The Argo Company's 1913 marketing campaign had alerted the US Navy to the existence of Pollen's advanced fire control technology. Information on the AC Battle System was relayed to the US Navy Bureau of Ordnance through the

2. The Analogue Alternative

Assistant US Naval Attaché in London, who also suggested that one should be purchased for evaluation purposes. However, the Bureau of Ordnance was already working with a US firm, the Sperry Gyroscope Company, on the development of fire control technology and decided instead to contract Sperry to develop an equivalent system.

Already a trusted supplier of navigational equipment to the US Navy, Sperry had been established in April 1910 to manufacture gyroscopic stabilisers and gyro-compasses devised by the company's founder, the prolific inventor Elmer A Sperry. Following successful trials of the Sperry gyro-compass in a US battleship in 1911, the Navy Bureau of Ordnance encouraged the company to turn its engineering expertise to the problem of fire control. Initial efforts centred on the development of an electrical signalling system to transmit gyro-compass readings between different parts of a ship and to synchronise the operation of the guns. This worked well but the company lacked the technology to provide target range and course data.

In 1913 Sperry created a British subsidiary and set up a small factory in London to manufacture gyro-compasses for the European market. Sperry executives based in London were able to gather detailed information on the Pollen technology and send it back to the parent company in the US. The executives recommended that Sperry obtain a license to manufacture the AC Battle System in the US but this was soon dismissed when the Bureau of Ordnance contract to develop an equivalent system was received and it also became clear that the company could acquire the necessary design information through official US Navy channels without having to license it.

The Sperry Fire Control System was ready for testing by early 1915. The main element of the system was a true-course plotting unit or 'Battle Tracer' designed by the company's Chief Engineer Hannibal C Ford. However, the prototype had no range clock which severely compromised performance. While Sperry worked to address this shortcoming, the Bureau of Ordnance was forced to contract with a foreign supplier, purchasing a quantity of Vickers range clocks for installation alongside the Sperry system.

Hannibal Ford was a talented and resourceful engineer and he had left Sperry in 1914 to branch out on his own. He founded the Ford Instrument Company in Long Island, New York the following year, having raised an impressive $50,000 in capital, with ambitious plans to compete with Sperry by producing gyro-compasses to a British design. However, these plans were abandoned following receipt of a Bureau of Ordnance request for proposals in June 1915 for a new range clock which would allow continual readjustment of the rate of change. Keen to encourage competition and avoid the risk of being tied to a single supplier, the Bureau had sent the request to both Sperry and Ford. Prototype range clocks from both companies were demonstrated to the US Navy in May 1916 but only the Ford clock was sufficiently complete to undergo sea trials which were held in July. By the time that the Sperry clock was ready a few months later, the decision had already been taken to adopt the Ford device.

Like the Argo Clock, the Ford Rangekeeper was much more than a range clock, also providing a continuous indication of target bearing. Another feature common to both instruments was the use of integrator mechanisms of the Thomson disk-ball-and-cylinder type, although Ford chose to add a second ball and a spring-loaded cylinder rather than use Isherwood's twin cylinder arrangement to minimise slippage and increase output torque. Ford's clock also differed in overall design from the Argo Clock by indicating three different range values and separating firing and target ship motion.

The Ford Rangekeeper was highly sophisticated mechanically but was not as tightly integrated with the other elements of the fire control system as the Argo Clock, requiring more manual input, and was considered to be more difficult to use. However, as with the Dreyer table, this lack of automation seems to have appealed to gunnery officers. In May 1917, the US Navy placed an initial order for 25 production units at a cost of $8,000 each and a total of 97 Mark I Rangekeepers were eventually produced.

Although there is no evidence of direct plagiarism, Ford certainly believed that he may have infringed Pollen and Isherwood's patents with the design of the Rangekeeper and in April 1918 he sought assurances from the US Navy that his company would be protected in the event of any legal action against it. These assurances were given and when Isherwood later attempted to sue the Bureau of Ordnance for patent infringement, the case was delayed for several years before being dropped altogether as a condition of US lend-lease cooperation with the Royal Navy during World War II.

Other navies soon followed the Royal Navy and US Navy's lead in adopting gunnery fire control technology, prompting other defence contractors to develop fire control products. In the US, the Navy Bureau of Ordnance formed a close working relationship with three domestic suppliers; the Ford Instrument Company, Arma Engineering and General Electric. Ford supplied the fire control systems, Arma the gyroscopic correction equipment and GE the electrical transmission equipment which connected the various systems together. Like Ford, Arma had been established by former employees of Sperry Gyroscope and the two young companies eventually succeeded in replacing Sperry as the preferred suppliers of fire control equipment to the US Navy, both benefiting from technical problems with the Sperry equipment and the larger company's apparent inability to respond quickly to requests for improvements.

In the UK, the poorly performing Dreyer Fire Control Table was replaced in 1925 by a new generation of table manufactured by Elliott Brothers to a design produced by an Admiralty committee of experts. One of these experts was the Argo Company's Harold Isherwood who worked to ensure that the new design incorporated key elements of the Argo Clock. Variants of the Admiralty

Fire Control Table (AFCT) were used in Royal Navy capital ships throughout World War II and on into the 1950s. Although Arthur Pollen was ultimately unsuccessful in commercialising his inventions, the leading naval gunnery fire control systems in the UK and the US both had their origins in Pollen's work and his vision of integrated fire control would come to fruition in successive generations of equipment.

Early Analogue Computer Development at MIT

The pioneering efforts of Lord Kelvin and others during the second half of the 19th century also provided the inspiration for a programme of research in analogue computation at the Massachusetts Institute of Technology which was led by the American electrical engineer Vannevar Bush. This culminated in the development of the most influential of all the analogue computing machines, the Differential Analyzer.

Vannevar Bush was born in Everett, Massachusetts in March 1890, the son of a minister in the Universalist church. After leaving high school in 1909 he won a scholarship to study engineering at Tufts College (now Tufts University) in nearby Medford. Bush was a natural engineer and was awarded his first patent in December 1912 while still an undergraduate student for the invention of a profile tracer, a surveying instrument for charting the profile of roads. This incorporated a type of wheel-and-disk integrator mechanism which operated in combination with a pendulum to vary the speed of rotation of the chart recorder in accordance with the slope of the surface being traversed. He graduated in 1913 with both Bachelor's and Master's degrees, the latter obtained as a result of his invention.

Bush had hoped to stay on at Tufts and try for a doctorate but he lacked the funds to do so. Instead, he took a job in the electrical industry as an equipment tester for General Electric but quit after a year and returned to Medford for a mathematics teaching post at Jackson College for Women. Bush finally returned to academic study two years later, gaining admission to the prestigious Massachusetts Institute of Technology in nearby Cambridge as a doctoral student in September 1915. He demonstrated his academic prowess in grand style by taking less than a year to complete his doctorate in electrical engineering which he obtained in April 1916. Following 3 years as an assistant professor at Tufts College, Bush secured a position as an associate professor in the electrical engineering department at MIT in 1919 and was appointed professor of electrical power transmission 4 years later in 1923.

The Continuous Integraph

At MIT, Bush's research on the behaviour of electrical power transmission networks involved lengthy calculations and he encouraged his research students to explore the use of calculating machinery to help reduce the manual effort involved. In 1925 Bush suggested to one of these students, Herbert R Stewart, that he should devise a machine to perform continuous integration, a mathematical function used in the analysis of transient oscillations in electrical networks. Stewart began by familiarising himself with earlier work on mechanical integrator mechanisms, including James Thomson's disk-ball-and-cylinder integrator. Then a fellow student, Frank D Gage, suggested that it might be possible to devise an electrical solution to the problem. This led Stewart to the realisation that a standard watt-hour meter could be used as an integrator mechanism. A watt-hour meter is an electromechanical device that measures electrical energy using a spinning disk which rotates in proportion to the voltage and current passing through it. An integrating counter coupled to the disk displays the value of the total energy consumed over time. It is commonly used to measure domestic electricity consumption in homes.

Under Bush's guidance, Stewart and Gage worked together on the construction of the new machine which they called the Continuous Integraph. It cleverly combined a watt-hour meter with a moving carriage to continuously plot the integral of the product of two functions. Graphs of the two functions to be integrated were placed alongside each other on the flat top of the carriage which was driven at a constant speed along an axis parallel to the x-axis of the graphs by means of a lead screw and electric motor. The machine required two operators to trace each graph simultaneously using two pointers mounted on transverse slider arms. Each pointer arm incorporated a slide-wire potentiometer, an electrical device that produces a change in electrical resistance which is proportional to the position of a sliding contact along its length. The potentiometers controlled the voltage and current received by the watt-hour meter which had been modified to allow it to rotate in either direction.

To avoid errors due to excessive mechanical load on the delicately spinning disk, a follow-up servo mechanism was developed that sensed the rotational movement of the disk through mercury-tipped contacts and conveyed this to an electric motor which mimicked the movement of the disk. This was connected through a second lead screw to the plotter pen in such a way that rotational movement of the disk caused the pen to move a corresponding distance along the y-axis of the machine. The pen plotted a graph representing the resulting integral of the two functions on a sheet of paper placed on the moving carriage.

A multiplying attachment was also developed to allow the product of two functions to be plotted or the results of an integration to be multiplied by another function. This took the form of a link motion mechanism which was attached to the moving carriage of the machine. A complicated arrangement of lever arms, lead screws and a chain drive transferred the movements of the plotter pen and one of the tracing pointers to the attachment and combined them to produce the desired plot.

2. The Analogue Alternative

The Product Integraph

The Continuous Integraph was completed in 1926 and was eagerly used by Bush's research students for a range of computational tasks. One student by the name of King E Gould successfully adapted the machine to solve first order differential equations by back coupling the output from the watt-hour meter to one of the tracing pointers in a similar way to that proposed by Lord Kelvin 50 years earlier for his brother's disk-ball-and-cylinder integrator. Another student, Harold L Hazen, who was engaged in a study of electronic circuits involving calculations which required two successive integrations to be performed, was encouraged by Bush to extend the design of the Continuous Integraph. He did so by adding a second integrator mechanism plus several other features intended to increase the machine's versatility. This revised version became known as the Product Integraph.

For the Product Integraph, the moving carriage was split into four separate platens that could be independently driven. The electromechanical watt-hour meter was retained for the initial integration stage and a mechanical wheel-and-disk integrator added for the second stage of integration. The disk of the mechanical integrator was mounted in the vertical plane on a sliding carriage and made to rotate at uniform speed using an electric motor. The position of the integrating wheel across the face of the disk was controlled by the vertical position of the sliding carriage which was driven by an electric motor and lead screw mechanism connected to the follow-up servo mechanism of the watt-hour meter. Rotational movement of the integrating wheel was picked up by another follow-up servo mechanism and used to drive the plotter pen.

Completed in 1927, the Product Integraph was capable of solving second order differential equations to an accuracy of between 2 and 3%. Bush's achievements in analogue computation were recognised by the Franklin Institute, which awarded him its Louis E Levy medal in May 1928, with honourable mentions given to Stewart, Gage and Hazen. However, this award was somewhat premature as Bush's greatest achievement in analogue computation was still to come.

The Differential Analyzer

The hybrid design of the Product Integraph, with its awkward combination of electromechanical and mechanical integrating technologies, had revealed the shortcomings of the watt-hour meter as an integrator mechanism. Bush was attracted by the elegance and simplicity of the mechanical wheel-and-disk integrator, a mechanism he had first employed over 15 years earlier in his profile tracer. In the autumn of 1928 he secured funding from MIT to build a new machine for solving differential equations which was intended to address these shortcomings by being based entirely on the wheel-and-disk integrator. This would be called the Differential Analyzer.

The design of the Differential Analyzer was based around a flexible arrangement of rotating shafts which permitted the principal components to be interconnected in any combination. A set of 18 longitudinal or bus shafts were provided which extended the full length of the machine. These were bisected with a series of cross shafts that connected with the principal components located along two sides of the machine. Repositionable gearboxes allowed adjacent shafts to be coupled together as required using appropriate gear ratios or pairs of shafts to be combined to produce the sum of two rotations. A variable speed electric motor supplied the motive power required to operate the machine.

Six wheel-and-disk integrator units were provided which permitted the solution of differential equations up to the sixth order or a system of 3 second order equations. These units were of similar design to the mechanical integrator used in the Product Integraph except that they were mounted with their disks in the horizontal plane and the integrating wheel on top. For each unit, the position of the sliding carriage supporting the disk was controlled by a lead screw connected to one of the cross shafts, while two other cross shafts were coupled to the rotational drive mechanism of the disk and the integrating wheel.

A key feature of the Differential Analyzer was the use of Bethlehem torque amplifiers to connect the integrating wheels to the system of shafts. The Bethlehem torque amplifier had been invented in 1925 by Henry W Nieman of the Bethlehem Steel Corporation for use in automotive power steering applications. The mechanism's input and output shafts were each fitted with drums on bearings which were free to rotate independently of the shaft. These were made to rotate in opposite directions by means of an electric motor and belt drive but were connected to each other by a friction band. Any rotation of the input shaft caused the friction band to tighten on one drum and loosen on the other, thus causing the output shaft to be rotated in phase with the input shaft but with an increased torque supplied by the energy from the continuously rotating drum.

The Differential Analyzer was equipped with four input tables for tracing the graphs of the functions to be integrated. Unlike the moving platens on the Product Integraph, these were fixed, with the x-axis movement provided by a slide that supported the transverse arm containing the pointer which was coupled to a lead screw driven from one of the cross shafts. The y-axis position of the pointer was manually adjusted by the operator using a hand crank and lead screw mechanism which was coupled to another cross shaft that conveyed these movements back to the machine. An output table with two independent plotter pens was also provided for plotting the solution. The machine also featured a multiplying attachment similar to that developed for the Continuous Integraph.

2. The Analogue Alternative

The design of the Differential Analyzer was conceived by Bush and Hazen with additional contributions from other research students including Gordon S Brown, Samuel H Caldwell and Harold E Edgerton, all of whom would go on to make names for themselves in their chosen fields. Great care was taken with the manufacture of the principal components to ensure the highest possible accuracy. Other mechanisms devised by Henry Nieman were incorporated into the design to eliminate any backlash, or play, within the gearing and the integrating wheels were fitted with jewelled bearings to minimise friction. These measures resulted in an impressive accuracy figure of 0.12% despite the enormous mechanical complexity of the machine, although this would have been reduced in practice due to operator error when manually tracing graphs.

The Differential Analyzer was operational by early 1931, with the final cost of construction totalling around $25,000, but the design did not remain static and the machine was continually extended and adapted to serve the needs of a succession of research projects. Built on a large scale, the substantial frame supporting the system of shafts was over 7 metres long and 1 metre wide. The machine was operated by first setting it up so that every term in the equation to be solved was represented by the revolutions of a shaft. These shafts were then interconnected so that their revolutions summed to zero. Another shaft was assigned to represent the independent variable. As this was rotated, it drove the other shafts in accordance with the equation. Planning and reconfiguring the machine for a new calculation could take several days but the time required to run the machine once set up was typically only a few minutes.

Vannevar Bush with his Differential Analyzer

The high construction cost and specialist nature of the Differential Analyzer made it an unappealing proposition for commercial exploitation. However, it was a powerful research tool and the design was eagerly copied by several other academic groups. The first of these was the University of Pennsylvania's Moore School of Electrical Engineering which built a 10-integrator version with Hazen's help in 1934. British academics from Manchester and Cambridge universities soon followed suit but, without the benefit of direct assistance from MIT, both groups wisely decided to build working models first before embarking on full-scale machines. Construction of the full-scale machines, both of which were equipped with 8 integrators, was subcontracted to the Metropolitan-Vickers Electrical Company, a large electrical engineering firm based in the city of Manchester. The Manchester University machine was completed in March 1935 and the Cambridge machine in October 1939. The largest copy was built at Oslo University in Norway in 1938. Featuring 12 integrators, this machine incorporated several improvements to the MIT design. It also prompted the development of a similar machine at Moscow State University in Russia in 1940.

The MIT Differential Analyzer was the physical embodiment of Lord Kelvin's concept of a machine which employed an endless chain of integrators to solve differential equations of any order. The advent of the torque amplifier had allowed Bush to surmount the problem of the torque output from an integrating wheel being insufficient to drive further integrator mechanisms. However, Bush insisted that he was unaware of Kelvin's seminal 1876 paper until after the construction of the Differential Analyzer had begun and it is Bush's name rather than Kelvin's which has become synonymous with the development of analogue computing machinery for mathematical and scientific applications.

Vannevar Bush was appointed MIT's first Dean of Engineering in 1932. He would go on to become a pillar of the US scientific establishment and was to play a leading role in determining US national policy in science and technology during World War II. Analogue computation would fare less well, however, eventually falling out of favour for mathematical and scientific applications due to issues over the accuracy of computing devices which perform calculation through measurement and the difficulty in designing analogue machines that were sufficiently general-purpose. Nevertheless, analogue computing machinery was very well suited to real-time computational problems such as gunnery fire control and it would remain the first choice for this type of highly specialised application for some years to come. Its initial use for naval gunnery was extended to anti-aircraft guns in the mid 1920s, through the work of Vickers in the UK and Sperry Gyroscope in the US, and the mechanical integrator remained a central component of such systems until they were eventually rendered obsolete by the introduction of radar for target identification in the 1950s.

The enduring legacy of the Differential Analyzer was in raising awareness of the use of computing machinery for mathematical and scientific applications, and in demonstrating what could be achieved with large-scale computing machinery projects. It is no coincidence that three of the groups who built copies of the MIT Differential Analyzer would go on to play leading roles in the development of digital computers.

Further Reading

Brooks, J., Dreadnought Gunnery at the Battle of Jutland: The Question of Fire Control, Routledge, Abingdon, 2005.

Bush, V., Gage, F. D. and Stewart, H. R., A Continuous Integraph, Journal of the Franklin Institute 203 (1), 1927, 63-84.

Bush, V., The Differential Analyzer: A New Machine for Solving Differential Equations, Journal of the Franklin Institute 212 (4), 1931, 447-488.

Clymer, A. B., The Mechanical Analog Computers of Hannibal Ford and William Newell, IEEE Annals of the History of Computing 15 (2), 1993, 19-34.

Friedman, N., Naval Firepower: Battleship Guns and Gunnery in the Dreadnought Era, Seaforth Publishing, Barnsley, 2008.

Henrici, O., On a new Harmonic Analyser, Proceedings of the Physical Society of London 13, 1894, 77-89.

Holst, P. A., Svein Rosseland and the Oslo Analyzer, IEEE Annals of the History of Computing 18 (4), 1996, 16-26.

Mindell, D. A., Between Human and Machine: Feedback Control and Computing before Cybernetics, The Johns Hopkins University Press, Baltimore, 2002.

Owens, L., Vannevar Bush and the Differential Analyzer: The Text and Context of an Early Computer, Technology and Culture 27 (1), 1986, 63-95.

Scott, R. H. and Curtis, R. H., On the Working of the Harmonic Analyser at the Meteorological Office, Proceedings of the Royal Society of London 40 (242-245), 1886, 382-392.

Sumida, J. T., In Defence of Naval Supremacy: Finance Technology and British Naval Policy 1889-1914, Unwin Hyman, London, 1989.

Thomson, W. and Tait, P. G., Treatise on Natural Philosophy, Cambridge University Press, Cambridge, 1879.

Zachary, G. P., Endless Frontier: Vannevar Bush Engineer of the American Century, The Free Press, New York, 1997.

3. The Electrification of Calculating Machinery

> *"At present there exist problems beyond our ability to solve, not because of theoretical difficulties, but because of insufficient means of mechanical computation"*, Howard H Aiken, 1937.

The Importance of Automation

After almost 250 years of intermittent development, the 19th century finally saw the widespread adoption of mechanical calculators for office use. These machines were a boon for simple arithmetic calculations but they offered little improvement over manual methods for the more complex types of calculation required in many scientific and engineering disciplines. Such calculations might involve complex mathematical formulae with several intermediate stages to work through before arriving at the final result. For calculating machines to be suitable for these types of calculations, they would have to be capable of automatic operation.

Automation would require not only a convenient source of power but also a massive increase in the sophistication of the calculating mechanism to allow for automatic control and sequencing. With his design for the Analytical Engine, Charles Babbage had shown that mechanical machines were capable in principle of this level of sophistication but their practicality was questionable due to the limitations of power and speed imposed by the use of mechanical technology. The construction costs would also have been prohibitive. Fortunately, progress would come through the application of electrical power to calculating machinery.

Electrically Powered Calculators

The development of the electric motor in the 1870s and the introduction of a public electricity supply by most industrialised nations in the decade that followed opened up the possibility of electrically powered calculating machinery. The earliest electrically powered calculator was the Autarith, which was designed by the Czechoslovakian inventor Alexander Rechnitzer and manufactured by Autarit GMBH in Vienna, Austria. Based on the Leibniz stepped wheel principle, the Autarith was the first mechanical calculator to feature a fully automatic multiplication and division mechanism and it was this key improvement that made the machine suitable to be motor-driven.

Rechnitzer's initial design, which was patented in the US in 1906, employed a clockwork motor but he made a number of improvements to the design to allow the machine to be driven using an electric motor.

Despite being first to market, the Autarith was a commercial failure and production ceased in 1913. Nevertheless, Rechnitzer had shown the way forward in the automation of calculating machines and his innovative design was soon adapted for use by other calculator manufacturers. By the 1920s, motor-driven desktop calculators were widely available from leading manufacturers such as Marchant, Monroe and Burroughs. However, these machines still required the user to enter numbers and select the appropriate arithmetic operators manually. They also lacked the ability to store intermediate results. Without the means to perform whole sequences of operations automatically, such machines were likely to come to a technological dead end.

The Electromechanical Arithmometer

In 1920, the Spanish civil engineer, mathematician and inventor Leonardo Torres y Quevedo exhibited an electromechanical arithmometer at an exhibition held in Paris to commemorate the centenary of the introduction of Thomas de Colmar's Arithmometer. Torres y Quevedo had worked for a number of years on the development of electromechanical machinery and had successfully demonstrated the potential of this technology in control applications by means of a series of highly sophisticated chess playing automata. In 1914, he published a major essay on the application of electromechanical technology to automatic calculating machinery and his arithmometer was created in order to demonstrate some of the ideas proposed.

Torres y Quevedo's arithmometer featured an ingenious stepped commutator which sensed the position of the digit wheels in the calculating mechanism and compared them with the value of the operand during each cycle of the calculation. Calculations were entered on a typewriter device in the same way that they would be written down, with the answer also printed automatically on the typewriter once the calculation had been completed. The machine possessed some sequencing capability, having the ability to add columns of five-digit numbers automatically, but could only cope with one calculation at a time.

Although somewhat limited in flexibility, Torres y Quevedo's machine represented a significant advance in automatic computation. However, perhaps because it came towards the end of his career, he seems to have had no interest in pursuing it further and his work attracted little attention beyond his own country. Instead, progress in this field would come through an entirely different route.

3. The Electrification of Calculating Machinery

Building Blocks – Switching Circuits and Symbolic Logic

The application of electrical power to calculating machinery was merely the first step in the move towards fully automatic computation. What was needed was a system which could facilitate the representation of complex mathematical formulae by mechanisms. If such a system could also improve sequence control, rapid progress would be made.

An invention devised almost 100 years earlier by the eminent American physicist Joseph Henry was to provide the foundation for the move from manual to automatic computation. Henry had begun experimenting with electromagnetism in 1827 while working at the Albany Academy in New York, where he was Professor of Mathematics & Natural Philosophy. In 1835, he demonstrated that electricity could be used to control equipment remotely by means of electromagnets, a type of magnet in which the magnetic field is produced by the flow of electric current through a coil of wire. As part of his apparatus, Henry used a horseshoe-shaped electromagnet to actuate an electrical switch, thus creating the first electromagnetic relay. A committed educator, Henry declined to patent any of his inventions and several, including the relay, were taken up by Samuel Morse for use in his highly successful electromagnetic telegraph system of 1836.

Switching Circuits

Electromagnetic relays were developed primarily as a means of controlling a high power electrical circuit using a lower power circuit. However, by connecting them together in various ways they could also be used to create highly complex electrical switching circuits. The invention of the telephone by Alexander Graham Bell in 1876 created a demand for electrical switching equipment to connect telephone calls and the telephone industry would drive developments in switching circuits for the next 60 years.

Early telephone networks relied on human operators located in each telephone exchange to connect calls manually using a switchboard. The operator would connect the caller with the desired number using an array of jack sockets and patch cables to complete the circuit for the duration of the call. The first practical automatic telephone exchange arose from a desire to eliminate the operator, but for reasons of privacy rather than efficiency. It was developed in 1889 by Almon B Strowger, an American businessman who ran a firm of undertakers in Kansas City, Missouri. Remarkably, Strowger had no formal education in science or engineering and his motivation for inventing an automatic telephone exchange appears to have been to prevent competitors stealing business from him as a result of unscrupulous telephone operators routing calls to rival firms.

Strowger's system allowed telephone subscribers to make calls to other subscribers directly by tapping out the number they wanted on a four-button keypad. The central component was an electromagnetic stepping switch which used the sequence of electrical pulses from the keypad to select a particular circuit by means of an electromagnetically controlled drum. The drum was coupled to an indexing mechanism that employed an arrangement of ratchets and levers actuated by electromagnets to rotate the drum in a series of steps and also move it incrementally along its rotational axis. These actions enabled a wiper arm to connect automatically with any one of up to 1,000 electrical contacts fitted to the inside of the drum. The telephone keypad was replaced by the more familiar rotary dial in 1896.

Strowger obtained a US patent for his system in March 1891. Later that same year he formed a company, the Strowger Automatic Telephone Exchange, to commercialise his invention and the company's first exchange was installed in La Porte, Indiana in November 1892. Over the next few years Strowger's stepping switch was adopted throughout the telephone industry, making him a very wealthy man. However, as the industry grew and telephone networks became more complex, the switching equipment also increased in complexity. Eventually, it would become necessary to employ an obscure branch of mathematics known as symbolic logic in order to analyse and optimise telephone switching circuits.

Binary Arithmetic

The simplest types of electrical switch can adopt one of two positions, off or on, and are therefore easily represented numerically using the binary number system, with a zero representing the closed position of the switch and a one representing the open position. Although binary numbers have been used throughout history, most notably in 2nd century India for the classification of musical meters, it was the 17th century mathematician Gottfried Wilhelm Leibniz who first realised they could also represent any decimal number and hence could be applied to arithmetical calculations.

Leibniz developed his system of binary arithmetic in 1679 and first published it in 1701 in a paper to mark his election to the Paris Academy entitled 'Essay d'une nouvelle science des nombres'. He also considered the use of the binary number system in calculating machines, envisaging a machine in which binary numbers are represented by marbles and holes, but he never took this idea any further. Instead, in his quest for a universal metaphysical language, Leibniz saw the binary number system as a means of characterising human reasoning or 'logic', reducing all thought to a number of simple ideas that could be represented by one of two values. What Leibniz had proposed was a rudimentary form of symbolic logic but his ideas were largely ignored by the philosophical and scientific communities of the day and another 150 years would pass before a complete logical system was developed.

3. The Electrification of Calculating Machinery

Boolean Algebra

In 1847, the British mathematician George Boole published an 82-page pamphlet entitled 'The Mathematical Analysis of Logic', which introduced the concepts of symbolic logic. Boole, a gifted scholar who was almost entirely self-taught, had considered logic from a strictly mathematical, rather than philosophical, perspective. He showed that logical statements, such as "all men are mortal" could be reduced to an algebraic form which allowed them to be added, subtracted and multiplied in almost exactly the same manner as numbers.

Although apparently unaware of Leibniz's earlier work, Boole's method followed that of Leibniz by using a two-valued system, where a 'proposition' can be either true or false. He combined this with a series of logical operators, the most basic ones being logical AND, where a value is true only if both of its operands are true and which Boole symbolised with the multiplication sign (x), and logical OR, where a value is true if either or both of its operands are true and which he symbolised with the addition sign (+).

Boole followed up his pamphlet in 1854 with a book entitled 'An Investigation of the Laws of Thought on which are Founded the Mathematical Theories of Logic and Probabilities'. This became the standard text for what would later be termed Boolean algebra.

Logic Machines

The connection between Boolean algebra and electrical switching circuits was first recognised by the American scientist and philosopher Charles Sanders Peirce. Peirce was an early champion of Boole's ideas, espousing them in a series of lectures on the logic of science given at Harvard University in the spring of 1865, a few months after Boole's death. He then spent much of the following 20 years of his career building upon the work of Boole and others, making his own major contributions to the field that included the truth table, a mathematical table used to list all possible input and output values for a logical expression. During this period Peirce became interested in the work of another champion of Boolean algebra, the British economist and logician W Stanley Jevons.

In his position as Professor of Logic and Philosophy at Owens College, Manchester, Jevons had devised a series of teaching aids to illustrate logical principles. This led him to explore the use of mechanical devices to solve logical expressions and in 1869 Jevons constructed a machine that he called the Logical Piano. Resembling a small upright piano, the machine comprised an arrangement of levers and pulleys operated by a keyboard which was used to enter propositions. Results were displayed in the form of a truth table on the front face of the machine which was created by means of sliding bars behind windows. On pressing a key, those truth table values that were logically inconsistent with the premise entered would be automatically removed from the display, thereby leaving only the correct table entries when all the terms had been entered.

The Logical Piano could be used to simplify Boolean expressions of up to 4 terms but did not provide a complete solution. Nevertheless, it inspired Allan Marquand, an American art historian who had studied philosophy under Peirce at Johns Hopkins University, to develop his own logic machines and it was this work which would lead Peirce to make the connection between symbolic logic and electrical switching circuits.

Marquand's logic machine designs borrowed liberally from those of Jevons but aimed to improve upon them. His first machine, constructed in 1881, attempted to increase the number of Boolean terms from 4 to 10 but was unsatisfactory. Fortunately his second machine, which was completed the following year, was more successful. It rationalised the design of Jevons' Logical Piano by replacing the sliding bars with a system of 16 pointers to display the truth table values and by reducing the number of keys from 21 to 10.

Marquand moved to the College of New Jersey (now Princeton University) in 1881 to take up a post lecturing in logic and assisting in Latin. He continued to correspond with his former tutor and it was in a letter written to Marquand in December 1886 that Peirce first proposed the use of electrical switching circuits to emulate Boolean conditions. Peirce also suggested that electromechanical technology might be used to build a machine for solving complex mathematical problems. Guided by Peirce, Marquand designed an electrical circuit for a new logic machine which would have similar functionality to his second machine but would be operated electromechanically. It comprised 16 electromagnets configured in a 4 by 4 array and controlled by 8 key-operated switches. The electromagnets controlled the positions of the truth table pointers and were wired in such a way that more than one winding had to be energised simultaneously for the pointer to move away from its rest position.

Alas, there is no evidence that Marquand actually constructed this circuit and the diagram remained hidden amongst Marquand's papers until it was rediscovered in 1951. Seemingly oblivious to the huge potential of this work, neither he nor Peirce took these ideas any further.

Tesla's Logic Gate

The electrical implementation of symbolic logic took another important step forward a few years later through the work of the brilliant Serbian scientist Nikola Tesla. Tesla was a major contributor to the emergent field of electrical engineering, a fact which is often overlooked due to his increasingly eccentric behaviour and bizarre scientific predictions in later life. One of the many areas of scientific study in which Tesla excelled was electromagnetism. While working on the development of an early system of radio control for vehicles, he

3. The Electrification of Calculating Machinery

designed an electrical circuit that performed a logical AND operation.

Anticipating problems with electromagnetic interference, Tesla's system involved transmitting radio signals on two separate frequencies. A circuit in the receiver ensured that the vehicle responded only if both frequencies were present. The action of this circuit mimicked a logical AND operation, such that only if both inputs were present would an output be generated. Tesla described the circuit in a US patent application for a radio-controlled boat, which was filed in July 1898 and granted later that same year, and refined his ideas in two further patents granted in 1903. He also demonstrated a working model of the boat, which he called a 'Teleautomaton', at an electrical exhibition held in Madison Square Garden in New York in September 1898. Newspaper reports of Tesla's demonstration describe it creating a sensation amongst those who witnessed it.

Tesla's radio control patents would become the focus of international attention following the efforts of the UK-based Italian inventor Guglielmo Marconi to patent his own radio technology in the USA in 1900. After a long legal battle, Tesla's US patents were eventually granted priority by the US Supreme Court in June 1943, six months after his death, although Marconi's patents continued to be upheld in the UK. Tesla's logic gate was only one component of his radio control system but its importance was recognised when priority was subsequently upheld by the US Patent Office in several patent infringement cases when, many years later, computer manufacturers attempted to patent logic gates for digital computer applications.

A Blueprint for the Digital Age

Peirce and Marquand had recognised the potential of electrical switching circuits in solving mathematical problems and Tesla's work on control systems had shown how electrical circuits could be configured to operate as logic gates. However, the concept of switching circuits as a tool for automatic calculation had not yet been fully developed. The final step towards this goal was made by the celebrated American electrical engineer and mathematician Claude E Shannon.

Shannon was introduced to Boolean algebra as a mathematics undergraduate at the University of Michigan in the early 1930s. In 1936, he secured a post as a research assistant at the Massachusetts Institute of Technology, working part-time in the Department of Electrical Engineering while studying for his Master's degree. He became interested in electrical switching circuits while working on the relay-based control circuitry for the Rockefeller Differential Analyzer, a giant electromechanical version of Vannevar Bush's earlier machine which was under construction at MIT during this period, courtesy of an $85,000 grant from the Rockefeller Foundation. Shannon required a suitable topic for his electrical engineering Master's thesis and the complexity of the Analyzer's switching circuits seemed to offer the necessary intellectual challenge. A short period spent at Bell Telephone Laboratories in the summer of 1937 served to increase his awareness of the practical problems associated with complex switching circuits and reinforced the need for such work.

Shannon's thesis, 'A Symbolic Analysis of Relay and Switching Circuits', was submitted on 10 August 1937. In what would be the first of many important contributions to computer science, Shannon set out to prove that Boolean algebra and binary arithmetic could be used to analyse and simplify such circuits. To illustrate his methods and at the same time demonstrate the versatility of relay and switching circuits, he also included several example circuits. One of these was a circuit which employed electromagnetic relays and switches to add two numbers automatically. Shannon realised that the use of the binary number system could simplify the circuitry enormously and so the calculation was performed in binary. He also provided an example of a design for a machine that could print tables of prime numbers using a method known as the Sieve of Eratosthenes.

With these examples, the 21-year-old student had shown in principle that a combination of binary arithmetic and symbolic logic, embodied in electromagnetic relays and electrical switching circuits, could be used as the basis for the construction of digital calculating machines capable of performing complex sequences of mathematical operations. In his remarkable Master's thesis, and the paper derived from it that appeared in the Transactions of the American Institute of Electrical Engineers the following year, Claude Shannon had unwittingly brought together the work of Peirce, Marquand and Tesla to create a blueprint for the digital age.

The Punched Card Tabulating Machine

The first successful application of automation to a computational process was made several years before the arrival of electrically powered calculators but it did involve electrical machinery. It was made by the American engineer Herman Hollerith in answer to a requirement to process the vast amount of statistical information gathered every 10 years as part of the US population census.

Herman Hollerith was born in Buffalo, New York in February 1860. His parents had emigrated to the USA some years previously as a result of political upheavals in their native Germany. His father, a classics teacher, died in an accident when Hollerith was only 7 years old, forcing his mother to take on the role of breadwinner for her family of five children, which she did in an impressive manner by turning her hobby of hat making into a successful millinery business. This entrepreneurial spirit would also be shown by her son some years later.

3. The Electrification of Calculating Machinery

After experiencing problems at school due to his poor spelling ability, Hollerith was tutored at home. Although his spelling ability would remain a lifelong weakness, the home schooling seems to have had the desired effect. He entered the College of the City New York at the age of 15 and graduated with an Engineer of Mines degree from the Columbia School of Mines (now part of Columbia University) four years later in 1879, obtaining a distinction in his final examinations.

Soon after graduating, Hollerith followed his former professor to Washington DC to work for the US Census Office in preparation for the national population census of 1880. Hollerith's job involved compiling statistical data on the use of steam and water power by the iron and steel industries, and he experienced at first hand the huge manual effort required to tabulate the information collected. This entailed examining each questionnaire returned and recording a mark on a tally sheet for each data item, or combination of data items, to be tabulated. Totals were then determined by adding the counts from individual tally sheets. Processing the data manually was not only error prone but highly time-consuming. With a population of over 50 million to contend with, the 1880 US census would take 7 years to process.

During this period Hollerith met the eminent physician John Shaw Billings, who had been appointed head of the Department of Vital Statistics for the 1880 census. While discussing the difficulties of tabulating statistical data, Billings suggested to Hollerith that it might be possible to mechanise the process using Jacquard loom technology. Hollerith's engineering curiosity was aroused and he began to study the problem in more detail.

Early Efforts

In September 1882 Hollerith moved to Boston to take up a position as an instructor of mechanical engineering at the nearby Massachusetts Institute of Technology. The move was encouraged by the economist and statistician Francis A Walker, who had held the post of Superintendent of the census before becoming President of MIT in 1881. At MIT, Hollerith began conducting experiments in his spare time in an effort to find a practical solution to Billings' suggestion. However, he found his teaching duties unfulfilling and decided to leave MIT at the end of the academic year.

Hollerith's next step was to return to Washington where he obtained a job as an assistant examiner in the US Patent Office, a move which may have been prompted by a desire to learn how the patent system operated so that he would be in a better position to protect his inventions. Having acquired sufficient knowledge of patent law and practice, he resigned from the Patent Office a few months later in March 1884 and set himself up as an independent patent agent, earning a living through offering his services to other inventors while continuing to pursue his own work on machinery for tabulating statistical data.

Hollerith's initial efforts involved technology borrowed from the electric telegraph. He developed a system in which statistical information was represented by holes punched in a continuous strip of paper tape. Each row of 26 hole positions was used to represent a person, with the presence or absence of holes at these positions denoting the person's sex, race, age, etc. Age was represented by a combination of 2 holes out of 20 hole positions, the first hole denoting tens and the second units. The holes were punched manually using a hand punch and die plate but were counted automatically by means of a specially designed counting device. As the tape was drawn through the counting apparatus, each row of holes was positioned in turn under a set of wire brushes which made contact with the metal surface beneath the paper tape if a hole was present, thus completing an electrical circuit which advanced a corresponding electromechanical counter by one digit. Having processed the entire tape, the totals for each data item recorded could simply be read off the counter dials. Hollerith set out these ideas in his first patent application which was filed in September 1884 under the somewhat grandiose title of 'The Art of Compiling Statistics'.

The Change to Punched Cards

Hollerith soon realised that punched paper tape was a less than ideal recording medium for statistical data, as the data could not be easily subdivided and the order in which it was held was fixed. Inspired by the sight of a railway conductor punching holes in tickets to describe the ticketholder's physical appearance, Hollerith decided to change to a system of punched cards where each person would be represented by a separate card. In doing so, Hollerith was following in the footsteps of Charles Babbage, apparently unaware that Babbage had proposed the use of punched cards for data input in the Analytical Engine almost 50 years earlier.

The change to punched cards necessitated modifications to Hollerith's counting apparatus. He designed a hand-operated press arrangement which featured a rubber bedplate with a set of holes that mirrored all possible hole positions on the cards, beneath each of which was a small cup filled with mercury. Located above the bedplate was a moveable plate fitted with spring-loaded pins at each hole position. Cards were placed in turn onto the bedplate and the moveable plate lowered onto the card. If a hole was present, the pin would pass through the card, enter the mercury in the cup beneath and complete an electrical circuit. If no hole was present, the pin would simply retract on encountering the card.

The electromechanical counters and dials from the earlier apparatus were retained but Hollerith introduced intermediate circuitry that facilitated the counting of groups of data items. This was achieved using electromagnetic relays which were wired up so that a counter would only be advanced when a specific combination of holes representing the group was present. A wiring

3. The Electrification of Calculating Machinery

panel was provided to allow these circuits to be easily reconfigured for different combinations of holes.

Hollerith also added another feature to simplify the task of subdividing sets of cards. This comprised a wooden 'sorting box' with 22 separate compartments, each of which was fitted with a spring-loaded lid held closed by a catch connected to an electromagnet. The electrical signals that advanced the counters according to the presence of punched holes were also fed to the sorting box. On reading a card, the electromagnet holding the lid of the corresponding compartment closed was actuated and the lid sprang open. The operator would then place the read card in the open compartment and close the lid. An additional wiring panel was provided to allow compartments to be set up for different sorting operations.

Hollerith Electric Tabulating System (Photo © Adam Schuster)

Hollerith named his invention the Hollerith Electric Tabulating System. By 1886, he had completed a prototype and with Billings' help, a full-scale test was arranged, tabulating mortality statistics for the City of Baltimore Department of Health. Hollerith designed a rectangular card approximately 215 mm by 82 mm in size which contained 192 hole positions arranged in 3 rows of 32 along each long edge of the card to facilitate punching using a railway conductor's hand-held punch. The test was successful in proving that the system worked but it also revealed the impracticality of punching each card manually using a hand-held punch. Hollerith solved this problem by designing a pantograph device that minimised the physical effort required to punch a card. It was operated by positioning an index pin mounted on a lever arm over the appropriate hole to be punched on a drilled guide plate and pushing the pin into the hole, whereupon the lever arm transferred this movement to a punch mechanism positioned appropriately over the card blank. The greater reach of the lever arm punch mechanism also allowed the whole card area to be utilised, thus increasing the number of possible hole positions and allowing the overall size of the card to be reduced to that of a US banknote.

A second successful test in 1888, tabulating mortality statistics for the State Health Office of New Jersey, resulted in the adoption of Hollerith's system by both the New York City Health Department and the Office of the Surgeon General of the US Army. However, rather than sell his equipment to these organisations, Hollerith offered to rent it for $1,000 per year, a relatively common business practice amongst US companies at this time. Manufacture of the equipment was subcontracted to the Western Electric Company and the Pratt & Whitney Machine Tool Company, leaving Hollerith free to concentrate on new product development and on winning business from his principal target customer, the US Census Office.

In September 1889, a commission was appointed to evaluate equipment for improving the processing of census data in preparation for the 1890 US census. The Hollerith Electric Tabulating System was chosen for evaluation alongside two others, one based on the use of colour-coded cards and the other on slips of paper with coloured ink. These rival systems were intended to enhance manual processing and neither involved machinery of any kind. The evaluation involved the tabulation of 1880 census data for four districts of the City of St Louis. Although Hollerith's system also involved a manual stage to transcribe the data onto punched cards, once this was done it was much faster due to the mechanised counting and sorting operations, resulting in an overall performance more the twice as fast as the other two contenders. It was also far more flexible in permitting different types of analysis to be performed on the same data. The evaluation had produced a clear winner and Hollerith's equipment was duly selected by the US Census Office for use in the forthcoming census.

Hollerith received an initial order for 6 systems, which were installed by March 1890, plus a contract for a further 100, 40 of which he agreed to supply at a reduced annual rental of $500 each on the understanding that they would be operated for only 8 hours per day. He also secured orders from census agencies in Austria, Italy, Norway and Canada as a result of exhibiting his system at several European exhibitions. However, business declined sharply four years later following the completion of the 1890 census, as the equipment was no longer required and the rental agreement with the US Census Office was not renewed. Hollerith was placed in the agonising situation of having to fire all but 4 of his employees. He began looking for a potential new market that would provide a steadier income than censuses and found it in the railway industry.

The Integrating Tabulator

The railway freight business in the Victorian era relied on the manual processing of millions of waybills, documents that recorded information such as the consignor and consignee, the point of origin, destination, weight, route and the amount charged for carriage. Hollerith knew that his tabulating system had the potential to dramatically reduce the manual effort required to process waybills. However, the use of numeric data for items such as weight and cost would require a major redesign of the counting apparatus so that it could add, or accumulate, integer values from successive cards.

Hollerith achieved this by devising the electrical equivalent of the Leibniz stepped wheel mechanism, which was by now a common feature in mechanical calculators. This was used to generate the appropriate number of electrical pulses received by the electromechanical counters for a given integer value. Hollerith replaced the 9 gear teeth of increasing length on the cylindrical drum with corresponding strips of conducting material. The meshing pinion gear was replaced with 9 wire brushes positioned along the length of the drum such that they each made contact with a different number of strips as the drum rotated. An electric motor rotated the drum one complete revolution during a reading operation. The pulses generated by this mechanism during rotation were then routed via electromagnetic relays to a set of electromechanical counters which corresponded to the units, tens, hundreds and thousands positions. These relays were set by the detection of specific holes on a redesigned punched card, thus determining which of the counters would receive pulses from a particular brush during the reading operation.

Hollerith incorporated his ingenious new 'accumulator' mechanism into a re-engineered counting apparatus which he called an Integrating Tabulator. He then approached the New York Central & Hudson River Railroad, which was the second largest railway company in the USA, with an audacious proposal to supply his new tabulators free of charge on a trial basis so that the company could evaluate their effectiveness. His proposal was duly accepted. Following some teething troubles which necessitated the inclusion of three extra accumulators in each tabulator to allow the addition of multiple numeric quantities, he finally received a contract from the Railroad in September 1896 to supply tabulating equipment for a combined rental of $5,000 per month. Hollerith's gamble had paid off. Securing a major contract from a customer whose needs would be less cyclic than those of the census agencies gave him sufficient confidence to incorporate his business as the Tabulating Machine Company in December 1896.

Competition

Over the next few years Hollerith introduced a series of refinements aimed at increasing the performance and ease of use of his machinery. An automatic card feed mechanism was added in 1900 followed by the introduction of an automatic card sorter in October 1901. To simplify the task of reconfiguration, wiring panels were replaced by 'plugboards', arrays of jack sockets and patch cables similar to the switchboards used in telephone exchanges. These new features first saw service in processing data for the 1900 US census. In the absence of any direct competitors, Hollerith's company had again secured the contract to supply tabulating equipment for the US census. However, political concerns over the spiralling cost of renting Hollerith's equipment, the rate of which was now based on the number of cards processed, resulted in a decision in 1905 that the US government should make efforts to develop its own tabulating machinery.

With Hollerith's primary patents due to expire, the Director of the newly created US Census Bureau, Simon N D North, spotted an opportunity to replicate Hollerith's core technology but develop it along different lines in order to circumvent the more recent patents. Using $40,000 of government funding, he set up a workshop and recruited engineering staff to work on the project, some of whom had previously worked for Hollerith. Hollerith's initial response was to mount a legal challenge aimed at proving that this was an infringement of his patents but he soon decided not to pursue it, turning his attention instead to finding new commercial opportunities for tabulating machinery.

By 1908, the Tabulating Machine Company had around 30 major customers, including railroads, life insurance firms, public utilities and industrial manufacturers. However, within three years Hollerith was suffering health problems brought on by the stress of personally managing almost every aspect of an increasingly successful business. Following an approach by the financier Charles R Flint, Hollerith agreed to step down as general manager and allow his company to be merged with two other firms, International Time Recording Company and Computing Scale Company, both of which were operating in complementary sectors of the office equipment market. The merger took place in June 1911, creating the Computing-Tabulating-Recording Company (C-T-R). Hollerith was retained as an engineering consultant to the new firm, eventually retiring 10 years later, aged 61, in 1921.

The efforts of the US Census Bureau to develop tabulating equipment spawned a direct competitor to C-T-R in the form of the Powers Accounting Machine Company which was founded in 1911 by James L Powers. Other competitors had also emerged from within the calculating machine industry to serve the growing data processing needs of commerce. Despite no longer having the tabulating machinery market to itself, C-T-R remained the market leader

3. The Electrification of Calculating Machinery

under the deft leadership of Thomas J Watson, who had joined the company in May 1914. In February 1924, Watson changed the name of the company to one that would become synonymous with the computer industry, International Business Machines Corporation, commonly known as IBM.

Herman Hollerith died in November 1929. By the time of his death, IBM had almost 6,000 employees and was generating substantial revenues of $18M per year. Hollerith had not only created a new industry, one which would shape the development of the computer like no other, but he had also helped to create the company that would dominate this industry for the next 60 years.

The Early Machines of Konrad Zuse

Around the same time that Claude Shannon was discovering Boolean algebra while studying mathematics at the University of Michigan, another undergraduate studying civil engineering 4,000 miles away in Germany was contemplating the use of machinery to reduce the effort involved in performing tedious stress calculations. His name was Konrad Zuse and his ideas would result in the development of the world's first binary digital sequence-controlled calculator.

Konrad Zuse was born in Berlin in 1910. As a child, he demonstrated a precocious talent for painting but a deepening fascination with mechanisms and structures led him to suppress his artistic side and pursue a career in engineering. On entering the Technische Hochschule in Berlin-Charlottenburg in 1927, he initially chose to study mechanical engineering, however, his artistic nature soon resurfaced and Zuse began to harbour thoughts of becoming an architect. After a short time studying architecture, he finally found an acceptable compromise and switched to a civil engineering course.

It was while studying civil engineering in 1934 that Zuse began to develop ideas for machines which could solve complex engineering calculations. Unaware that Babbage had trodden the same path a century before, Zuse reasoned that such machines would have to be capable of performing long sequences of arithmetical operations automatically and to deal with intermediate results during lengthy calculations. Considering the problem from a visual perspective, his first thoughts were to employ a graphical approach which he intended to apply to the calculation of bending moments. This would involve specially designed printed forms coupled to a special-purpose machine that could read input values represented by holes punched in the form, with the position of the holes dictating both the type of arithmetical operation and the order in which it should be performed.

Further thought established that the positions of the holes were unnecessary if the sequence of operations was somehow held in the correct order. Also, rather than using punched holes, the input values and intermediate results could be stored in 'registers'. Not unlike the adding counters found in punched card machines, these would be individual storage areas within the machine where numbers could be held until required but which were designed instead to be reset to a different value whenever necessary rather than to accumulate sums. This approach would also lend itself to a more general-purpose machine that could cope with a range of engineering formulae.

The use of the binary number system appealed to Zuse for reasons of mechanical simplicity but he soon realised that it also possessed certain mathematical and logical advantages. Zuse was familiar with the work of Leibniz but unaware of subsequent developments which had taken place in symbolic logic, so he conceived his own system of 'conditional propositions' that corresponded roughly to Boolean algebra. Slowly, a design specification began to take shape for a binary digital sequence-controlled calculator.

The Versuchsmodell-1

Upon graduation in July 1935, Zuse secured a job as a structural engineer for the Henschel Aircraft Company in Berlin but he had now found his true vocation and quit after less than a year in order to concentrate on his calculator development work. He began the construction of his first machine, which he named the Versuchsmodell-1 or V-1, in the living room of his parents' home in the summer of 1936. Funding for materials came from family and friends, several of whom were also recruited to help with the manufacture of the large number of custom-made components necessary for the machine. Although Zuse recognised the suitability of electromagnetic relays for a binary calculator, he was wary of using them, being far more comfortable with visually comprehensible mechanical technology, so the V-1 was entirely mechanical in construction with the exception of an electric motor that cranked the machine along at a leisurely 1 cycle per second.

Zuse ensured that the specification of the V-1 would accommodate his fellow engineers' preference for floating-point numbers rather than fixed-point numbers or integers. In order to represent these numbers to a precision that would be sufficient for most engineering calculations, he chose to represent them in units of 22 binary digits or 'bits'. This 22-bit 'word length' comprised 1 bit for the algebraic sign, 7 bits for the exponent and 14 bits for the mantissa. To further increase the range of numbers that could be represented, Zuse used a normalised semi-logarithmic form of binary representation where the first digit of the mantissa, which was not actually stored, is always one.

Storage was implemented by means of slots in sliding metal plates with steel pins that could be positioned at either end of the slot, one end representing a zero and the other representing a one. A linkage transmitted the positions of each of the pins to a sensing mechanism. Using this simple but effective method of construction, Zuse and his team of willing helpers built a storage

3. The Electrification of Calculating Machinery

unit or 'memory' with a capacity of 16 words.

The calculation or arithmetic unit comprised two binary adders, one each for the exponent and mantissa, and two 22-bit registers. Zuse had observed that in binary arithmetic, multiplication is essentially the same as repeated binary addition and division similarly equivalent to repeated subtraction, so all four arithmetical operations were reduced to additions or subtractions. These were implemented using a method called 'carry look-ahead' which significantly reduced the number of steps required for the calculation, thereby increasing the calculation speed.

A third unit controlled the operation of the machine. Zuse formulated a set of eight machine instructions which included the four basic arithmetical operations plus instructions to read a word from and write a word to memory, one to read a value from the keyboard and one to display a result. Memory instructions incorporated a 6-bit value that provided the location or 'address' of the word in memory to be used, thus giving a maximum 'address space' of 64 words. Sophisticated control logic was required to interpret these instructions and perform the required operations. This was implemented using a series of mechanical logic gates which operated in a similar manner to the storage unit. The basic building block was an ingenious mechanical switching element that employed sliding metal plates linked by pins to create the mechanical equivalent of an electromagnetic relay. It could assume one of two positions depending on the position of a perpendicularly mounted control plate. Hundreds of these switching elements were combined to create the control logic of the V-1.

Zuse V-1 in Living Room (Photo © Horst Zuse)

Sequences of instructions were held on punched tape. However, to save having to buy rolls of paper tape, Zuse's friend and fellow student Helmut T Schreyer had the idea of using discarded 35 mm motion picture film stock, which was readily available through Schreyer's part-time job as a film projectionist. A punched tape reader was constructed to feed the instructions into the machine, which were punched by hand onto the cellulose acetate film using an 8-bit binary code. The input values for calculations were entered manually as decimal numbers by means of a numeric keyboard and converted into binary by the machine. Results were converted automatically back into decimal and displayed using an arrangement of indicator lamps.

The V-1 was operational by June 1938. It was the size of a small car and contained over 20,000 components, the majority of which were handmade. Despite the inevitable lack of manufacturing precision, the mechanical memory worked perfectly. However, the more complex control unit was much less reliable and Zuse, who by this time had familiarised himself with symbolic logic, became resigned to the fact that his mechanical switching technology was too inflexible for implementing complex logical designs. Particularly problematic was the necessity to transmit 'signals' from one part of the mechanism to another, a task which would be eased considerably by the adoption of electrical technology. Fortunately, he had already begun work on his next machine.

The Move to Electromechanical Technology

It had become clear to Zuse during development of the V-1 that electromagnetic relays would have certain advantages but, as an engineer, he needed hard evidence that they would be sufficiently reliable as a sticking relay could compromise accuracy. He set out to build an interim machine to test the reliability of electromechanical switching technology. Unfortunately, such components could not be handmade and the funding he had received from family and friends would not be sufficient to buy them in the quantities required. Zuse needed an industrial sponsor. He approached Kurt Pannke, a manufacturer of analogue computers for gunnery applications, who agreed to provide funding of 7,000 Reichsmarks (equivalent to around $1,750 at the time) to allow Zuse to continue his work.

Because the V-2 would be an interim machine, Zuse was able to pare down the specification. Hence, the floating-point arithmetic of the V-1 was dropped in favour of a fixed-point representation and the word length reduced to a more manageable 16 bits. As the mechanical memory of the V-1 had functioned reliably, it was pressed into service again and reconfigured to provide 16 words of storage for the new machine. With Claude Shannon's work not yet in the public domain, Zuse had applied Boolean algebra to switching circuits to create his own rudimentary system of switching algebra and he used this to translate the logical design of the V-1 into one that employed electromagnetic relays.

Zuse's friend Helmut Schreyer, who was studying for a diploma in electrical engineering, suggested that the high specification electromagnetic relays

3. The Electrification of Calculating Machinery

and stepping switches used in telephone switching circuits would make ideal components for the new machine. However, these were relatively expensive so they scoured second-hand shops in search of suitable used examples, eventually collecting around 200 telephone relays with which to construct the arithmetic and control units. To ensure that the relays triggered in synchronisation, a rotating drum mechanism was developed which used an arrangement similar to a commutator found in an electric motor. A trigger pulse was generated when conducting segments, which were evenly spaced around the circumference of the drum, came into contact with spring-loaded carbon brushes. The trigger pulse frequency, equivalent to the clock speed in an electronic computer, was set by the rotation speed of the drum and could be varied up to a maximum of 3 Hz. This drum mechanism also served to reduce reliability problems caused by sparking at the relay contacts by momentarily cutting off the electrical current flowing through the relay during switching.

Following the German invasion of Poland in September 1939, Zuse was drafted into military service and work on the V-2 came to an abrupt halt. Zuse was sent to the infantry while his friends campaigned in vain to have him released. In December 1939, Zuse wrote to the German Army Ordnance Office suggesting that his calculating machine technology could be applied to cryptography. On the strength of this letter, he was allowed to visit Berlin for a meeting with Herbert Wagner, who was the head of Special Division F of the Henschel Aircraft Company, the firm that Zuse had briefly worked for after leaving university. Wagner was intrigued by Zuse's calculator development work but had a more pressing need for an experienced structural engineer. At Wagner's request, Zuse was exempted from active duty and assigned instead to the Henschel Aircraft Company to work on guided weapons development. Zuse was free again to continue work on the V-2 but now only able to do so in his spare time. Progress slowed considerably as a result and the V-2 was not completed until mid 1940. Nevertheless, the finished machine proved sufficiently reliable and Zuse could now proceed with plans to develop a fully specified electromechanical model.

Work on the new machine, designated the V-3, began before the V-2 was fully completed. In an effort to raise further funds, Zuse contacted the Deutsche Versuchsanstalt für Luftfahrt (DVL), the German Aeronautics Research Institute in Berlin. Following a successful demonstration in September 1940 of the recently completed V-2, the Institute agreed to partially fund the development of the new model. In April 1940, Zuse had registered a company called Zuse Ingenieurbüro und Apparatebau. With the DVL support, he was now able to employ a handful of staff on a part-time basis to help with the construction work.

With a 22-bit word length and floating-point arithmetic, the V-3 was very similar in specification to the V-1. The only major changes were the addition of a square root function and extra logic to perform arithmetic exception handling. An electromechanical memory was constructed using 1,408 relays to provide 64 words of storage. A further 400 relays were required for the construction of the control unit and 600 for the arithmetic unit. The punched motion picture film input and indicator lamp output display of the earlier machines were retained.

Every instruction cycle in the V-3 was divided into five steps by microsequencers in the control unit; decoding the instruction, fetching the operand, executing the instruction, writing the result and preparing the argument for the next instruction. This allowed Zuse to overlap the steps at each end of the cycle to improve performance and was the earliest example of what would come to be known as instruction pipelining. With a clock speed of just over 5 Hz, the V-3 was able to multiply two floating-point numbers in approximately 4 seconds.

Zuse also paid particular attention to the handling of arithmetic exceptions in the design of the V-3. Calculations that resulted in an overflow or underflow condition were detected automatically by a monitoring circuit and the operator alerted by illuminating an exception lamp and halting the operation of the machine. The number zero, which posed a particular problem due to the semilogarithmic numeric representation used by Zuse, was also dealt with by this circuit.

The V-3 was operational by May 1941. Shortly afterwards, Zuse was drafted again and posted to the newly opened up Eastern Front, where the German Army would sustain heavy losses. Fortunately, his exempted status was renewed on route and he was permitted to return to the relative safety of his work as a structural engineer for the Henschel Aircraft Company in Berlin. Gradually, he was able to reduce his working hours to part-time, which freed up more time to devote to calculator development.

A Production Prototype

With his first three designs, Zuse had established both the architectural principles and the practicalities of digital sequence-controlled calculators. He was now ready to create the prototype for a machine which could be built in production quantities and sold commercially. However, in order to make it suitable for a wide range of scientific and engineering applications, it would have to be equipped with a much larger memory than his previous designs. A longer word length would also be advantageous. Zuse envisaged a storage capacity of up to 1,000 words but this would require a huge number of relays, making the machine hideously expensive to manufacture and also very large physically. His solution was to return to the relatively compact and inexpensive mechanical storage technology of the earlier machines.

The new model, designated the V-4, featured electromechanical arithmetic and control units coupled to a mechanical memory manufactured from parts

3. The Electrification of Calculating Machinery

cannibalised from the V-1 and V-2. The 22-bit word length of the earlier machines was increased to 32 bits by increasing the mantissa from 14 to 24 bits. To allow greater flexibility, the V-4 also had provision for up to 6 punched tape readers, including punched tape input of data which would permit fully automatic operation. The set of 9 machine instructions used in the V-3 was extended to include a comparison operation, a necessary feature for the solution of algebraic expressions, and instructions to control the additional tape readers.

To provide improved performance over the V-3, the clock speed was increased to 30 Hz. Zuse also incorporated a clever 'look-ahead' feature that read instructions from the punched tape two steps ahead of the calculation and, wherever possible, changed the order in which instructions were executed to further increase calculation speed. Memory operations could also be initiated up to two steps in advance to compensate for the slow access speed of the mechanical memory and numbers that were needed again immediately could be placed in a temporary store within the control unit to shorten access time.

The V-4 also featured an additional unit, the 'Planfertigungsteil', which simplified the creation of the instruction tapes. This comprised a special keyboard which provided the input to a tape punch and an indicator panel to display the instructions stored on punched tapes, thereby allowing the operator to check the instructions and make corrections if required. Using this unit, it was possible to learn how to create sequences of instructions for the V-4 in as little as 3 hours.

Construction of the V-4 began in 1942 and was expected to take between 12 and 18 months to complete. However, progress was soon being regularly interrupted by the allied air raids which targeted Berlin with increasing frequency from November 1943 onwards. Furthermore, with the focus on using his machines for engineering calculations, the wider impact of Zuse's work was not fully appreciated by the German military and it was not regarded as vital to the war effort. Consequently, materials became more and more difficult to obtain. Nevertheless, Zuse and his small team persevered and the machine was operational by early 1945, albeit with a reduced storage capacity of 64 words and only 2 punched tape readers.

In the final weeks of the war in Europe, it became much too dangerous for civilians to remain in Berlin so Zuse and his team were evacuated along with the V-4. The authorisation for the move was granted by the Ministry of Aviation as a result of some convenient confusion between the machine and the similarly named V-1 flying bomb and V-2 rocket. In April 1945, the V-4 was reassembled in the Aerodynamischer Versuchsanstalt (AVA), an Aerodynamics Research Institute in Göttingen, where it was briefly demonstrated to senior staff before being dismantled once again and shipped to Hinterstein, a small village in Bavaria where Zuse and his family had chosen to relocate to escape the last of the hostilities. With the war now coming to an end, it was no longer wise for Zuse's work to be associated with the feared vengeance weapons of the Third Reich, so the machine was renamed the Z4 by substituting the letter V with a Z (for Zuse) and the names of Zuse's earlier machines were retrospectively changed also.

Zuse Z4 on display at the Deutsches Museum, Munich (Photo © Clemens Pfeiffer)

For the next two years, the chaotic situation in post-war Germany made it impossible to engage in any practical development work and Zuse's team began to disperse. Zuse himself eked out a living producing oil paintings and woodcuts for tourists, his calculator work confined to the theoretical aspects. However, by 1948, conditions had improved sufficiently that he was able to rent space in a former flour store in the Bavarian village of Hopferau and resume the task of reassembling the Z4.

In 1949, Eduard Stiefel of the Eidgenössische Technische Hochschule (ETH) in Zurich, the Swiss Federal Institute of Technology, visited Hopferau to investigate unconfirmed reports of an advanced computing device. Stiefel was given a demonstration of the partially rebuilt Z4 and saw that, despite its mechanical memory and incomplete state of development, the ease of use would make the machine an ideal acquisition for a new Institute of Applied Mathematics which he was in the process of setting up at ETH. He struck a deal with Zuse to acquire the Z4 on a 5-year lease for the sum of 50,000 Swiss Francs (equivalent to around $12,000) with an option to purchase the machine outright for an additional 20,000 Swiss Francs. Zuse used this money to

3. The Electrification of Calculating Machinery

set up a new company, Zuse KG, relocated to the town of Bad Hersfeld near Göttingen and spent the following year rebuilding the Z4 to ETH requirements, incorporating a number of modifications specified by ETH in the light of recent advances in computing made in the USA. In July 1950, the fully refurbished and modified Z4 was delivered to Zurich, where it was used successfully for several years.

Zuse's Legacy

Even before he began building his first machine, Zuse was routinely recording his ideas in a series of German patent applications, the first of which was filed in April 1936. Sadly, most of these were rejected either due to insufficient disclosure or prior art in the form of Babbage's work. His most important patent application, which contained 51 claims covering the Z3, was filed in 1941 but only made public in 1952 and was finally rejected in 1967. Also, there was little physical evidence of Zuse's work as the Z1, Z2 and Z3 had all been destroyed in allied bombing raids in 1944 and 1945. Only the Z4 survived the war intact. Consequently, Zuse's incredible achievements were unknown outside Germany until after the end of World War II and it would be many years before his rightful place in history was acknowledged.

War is often a catalyst for technological progress but this was clearly not so in Zuse's case, the German military showing an uncharacteristic lack of imagination when it came to applications of calculating machinery. Furthermore, his isolation from parallel developments in computing technology taking place in other parts of the world during the same period meant that not only did he fail to benefit from them but he was also prevented from influencing them.

Britain, the USSR and the USA all came to recognise the tremendous advances made by German scientists and engineers in several fields under the Nazi regime and the most important were tracked down after the war with a view to securing their services. A US initiative, codenamed Project Paperclip, was particularly successful in obtaining German rocket scientists for the US space programme. The British also had a similar aim under Operation Surgeon and Zuse was brought to London in 1948 to be interviewed by a representative of the British Tabulating Machine Company. Unfortunately, the language barrier and the specialist nature of Zuse's work meant that the interviewer failed to fully comprehend the magnitude of Zuse's achievements and what would have been a golden opportunity for the company went unrecognised.

Zuse would go on to experience modest commercial success in post-war Europe with a series of electromechanical and electronic calculators but his time as a leading innovator had passed. The war years had seen a shift in focus towards Britain and the USA and developments in these countries would now dominate the story of the computer.

The Story of the Computer

The Bell Labs Relay Calculators

Bell Telephone Laboratories was created in 1925 as the research and development arm of the giant American Telephone & Telegraph Company. By the 1930s, it had become an innovation powerhouse for telephone and communications technology. Bell Labs would also play an important role in the story of the computer. Its entry into the realm of computer development came as a result of the scientific curiosity of one of its employees, George R Stibitz.

George Stibitz was born in York, Pennsylvania in 1904. Like his parents, a mathematics teacher and a Professor of Theology, Stibitz was academically inclined and showed an early interest in science. After obtaining a PhD in Mathematical Physics from Cornell University in 1930, Stibitz took a job as a research mathematician at Bell Labs in New York. In 1937, he was given an assignment to study the magneto-mechanics of telephone relays, so in order to understand their operation more fully, he liberated a pair of telephone relays from a pile of discarded components and took them home. Using these relays, two torch bulbs, two dry cell batteries and a few strips of metal from a tobacco tin, Stibitz constructed a rudimentary electromechanical binary adder circuit on his kitchen table during a weekend in November 1937. The circuit operated by adding two binary digits, which the user keyed in manually using two switches fashioned from the metal strips. The result was displayed by means of the torch bulbs. In recognition of its kitchen origins, the completed device was later dubbed the Model K by Stibitz's wife.

As a mathematician engaged in telephone switching research, Stibitz was fully aware that binary notation could be used to mathematically represent telephone relays and so he was able to make the intellectual leap to binary calculation using relays independently of Claude Shannon, who by this time had completed his Master's thesis but not yet published his work. Realising that he had made an important discovery, Stibitz proceeded to design more sophisticated relay circuits capable of performing all four basic arithmetical operations on a larger number of binary digits and automatically handling decimal to binary conversion. He then presented these ideas to his boss, Thornton C Fry, the head of the Mathematics Department.

Fry was also in charge of a large team of people who were permanently engaged on the task of performing lengthy calculations to determine the characteristics of filter circuits and transmission lines in telephone networks. These calculations required the use of complex numbers, a type of numerical representation which combines a real component with an imaginary part that is equal to the square root of minus 1. The calculations were performed using standard mechanical desktop calculators which were unable to store intermediate results, making the task very labour intensive and prone to transcription errors. Fry proposed the development of a relay-based machine to perform complex number

3. The Electrification of Calculating Machinery

calculations automatically. After some deliberation, Bell Labs management agreed to fund the project. Stibitz was partnered up with an experienced Bell Labs design engineer, Samuel B Williams, and the two men set about the task of designing and building the new machine, which would be named the Complex Number Calculator.

The Complex Number Calculator

Stibitz's design for the Complex Number Calculator featured two calculating units to allow multiplication and division to be performed simultaneously on each component of the complex number. Addition and subtraction were accomplished using an accumulator which operated in a similar way to Hollerith's punched card tabulator mechanism, permitting successive results to be added or subtracted to any digit column rather than incrementing the counter one digit at a time. Numbers were input and output in a fixed-point decimal format with a precision of eight decimal digits in the range of plus 1 to minus 1. Internally, numbers were held to a precision of ten decimal digits in order to minimise rounding errors. Stibitz had originally planned to use a pure binary representation internally but opted instead for binary-coded decimal (BCD), a hybrid encoding scheme where each decimal digit is encoded using 4 binary digits, in order to speed up the large number of binary to decimal conversion operations that would be required for complex number calculations. This is shown in the following table:-

0	1	2	3	4	5	6	7	8	9
0000	0001	0010	0011	0100	0101	0110	0111	1000	1001

Stibitz found that the number of relays required for certain operations could be reduced using a variation of this scheme known as excess-three BCD, where 3 is added to each number before encoding, and this was the form of encoding employed in the finished design.

The Complex Number Calculator was controlled and operated using a Teletype, an electromechanical typewriter used for transmitting and receiving typed messages over telegraph lines. The keyboard of a standard Teletype was modified to include the control commands for the calculator. To facilitate wider access to the machine, three Teletypes were installed in different parts of the Bell Labs building and connected to the calculator through a simple interlock arrangement which prevented more than one from operating at a time.

Construction began in November 1938 and was completed by October 1939. The machine was built largely using standard telephone system components and contained approximately 460 relays. Input values and intermediate results were stored in a rudimentary memory made up of 10 crossbar switches, a new type of electromechanical switch which had been recently introduced to replace Strowger stepping switches in telephone exchanges. Despite having a slow memory access speed due to the use of the crossbar switches, the calculator was able to multiply two complex numbers, a calculation requiring 13 separate steps, in around 45 seconds.

The Complex Number Calculator was not an experimental machine but was developed to fulfil a specific need. Consequently, it contained none of the logical sophistication of Konrad Zuse's designs. Unlike Zuse's machines, the sequence of instructions that the calculator could handle was limited to the intermediate steps required to perform a complex number calculation and could not be altered. However, no trigger pulse or clock was required to synchronise operation. Successive operations were simply triggered by the completion of the one before, a technique subsequently termed 'asynchronous timing'.

Bell Labs Complex Number Calculator (Photo © Lucent Technologies, Inc.)

Following thorough testing, the machine was placed into routine service in January 1940 and remained in use for over 9 years. In September 1940, it was publicly demonstrated at a meeting of the American Mathematical Society held at Dartmouth College in Hanover, New Hampshire. Rather than transport the bulky machine to the meeting, Stibitz and Fry took advantage of the ability of

3. The Electrification of Calculating Machinery

a Teletype to communicate over telegraph lines by having one of the machine's modified Teletypes temporarily installed at the College and using it to access the calculator over 200 miles away in New York. Following the presentation of a paper by Stibitz describing the Complex Number Calculator, attendees were invited to try the machine out for themselves. Nothing like this had ever been attempted before and the demonstration made an indelible impression on all those who witnessed it.

Buoyed with success, Stibitz proposed the development of a more advanced relay calculator with improved sequence control and error detection capabilities. However, the Complex Number Calculator had cost $20,000 to build, which was a relatively large sum of money at that time. Without a specific application to help justify the cost of construction, Bell Labs management were unconvinced of the need for another computing machine and the project was shelved.

The Relay Interpolator

Following the USA's entry into World War II in December 1941, Bell Labs became increasingly involved with work in support of the war effort. Like many US academic and industrial research organisations, it began to receive funding from the US National Defense Research Committee (NDRC), which had been created in June 1940 to sponsor industrial research into military problems. To facilitate stronger links with the NDRC, Stibitz was seconded as a Technical Aide and assigned to the NDRC's fire control section where it was thought his growing expertise in electromechanical computation might be put to good use.

In 1940, Bell Labs scientists developed a new type of electronic analogue computer for directing anti-aircraft guns. Attracted by the promise of improved performance over conventional electromechanical gun directors, the US Army persuaded the NDRC to fund the development of a production version, which was codenamed the T-10. The production prototype was completed in December 1941. In order to test the performance of the T-10 against the electromechanical competition, Stibitz was given the job of designing an automatic testing machine. Known as the Tape Dynamic Tester, this machine simulated the flight path of an aircraft in a form suitable for input to gun directors and was controlled by punched tape. However, preparing the punched tapes for the machine was a far from trivial task, requiring the calculation of a large number of intermediate values of a mathematical function through a process of linear interpolation, so Stibitz also designed a relay calculator for the job which became known as the Relay Interpolator.

The NDRC contracted the detailed engineering design and construction of the Relay Interpolator to Bell Labs. The design was a relatively simple one built around a 5-digit adder with 2 internal registers for arithmetic and 5 additional registers for the storage of intermediate results. To improve performance, these registers were implemented using telephone relays rather than crossbar switches. Subtraction was performed by a technique known as the method of complements in which all the binary digits of the number are inverted and then added to the original number. The machine could also multiply a number by a small integer through a process of repeated addition.

The Relay Interpolator was the first machine to benefit from Stibitz's ideas on sequence control and error detection. Operation was controlled by a loop of punched paper tape created on a Teletype, therefore, the sequence of instructions could be easily altered. Despite having a very limited instruction set, this facility allowed alternative methods of interpolation to be tried. A separate punched tape reader was used to input the flight path coordinates for interpolation.

To guard against intermittent errors caused by relays failing to trigger, numbers were encoded using a biquinary-coded decimal scheme, where each decimal digit is encoded using 7 binary digits arranged into two groups and in such a way that only a single relay in each group is set to 1 at a time, as shown below:-

0	1	2	3	4	5	6	7	8	9
10-10000	10-01000	10-00100	10-00010	10-00001	01-10000	01-01000	01-00100	01-00010	01-00001

An error detection circuit continually checked the status of the relays and halted the machine if anything other than one relay in each group was set, indicating that an error had occurred.

The Relay Interpolator was completed in July 1943. It contained a total of 493 telephone relays. The completed machine was placed into operation at Bell Labs in September 1943 then transferred to the US Naval Research Laboratory in Washington DC at the end of the war where it remained in use until 1961.

The Ballistic Computer

Having completed the design of the Relay Interpolator, Stibitz began working on the design for a more advanced model which could also calculate the ballistic trajectory and final detonation position of an artillery shell from the aiming and fuse setting information supplied by the gun director. Comparing this data with an aircraft's simulated position at the time of detonation provided a means of determining the accuracy of the gun director without having to test it physically using trial runs of actual aircraft and live ammunition or photographic analysis. This new machine was known as the Ballistic Computer.

Stibitz's design for the Ballistic Computer retained the biquinary-coded decimal architecture of the Relay Interpolator but featured a slightly longer word length of 6 decimal digits and double the number of registers. A new multiplier unit, designed by E L Vibbard at Bell Labs, made use of the 2 additional arithmetic registers to perform multiplication and division by the accumulation of partial

3. The Electrification of Calculating Machinery

products in a similar way to longhand multiplication. This was considerably faster than the method of repeated addition used in the Relay Interpolator, with the result that a multiplication operation took around 1 second.

The machine also featured two additional punched tape readers, one for the aiming and fuse setting data from the gun director and another containing tables of ballistic functions. Although not quite a general-purpose machine, the additional registers and punched tape devices made the Ballistic Computer considerably more versatile than Stibitz's earlier designs. Consequently, it was used for a number of computational tasks in addition to those which it been designed for.

The Ballistic Computer was built for the NDRC by Bell Labs at a cost of $65,000 and installed in the US Army Anti-Aircraft Artillery School at Camp Davis, North Carolina in June 1944. The completed machine contained approximately 1,300 telephone relays, any one of which could adversely affect the operation of the machine if it failed to trigger, but the use of biquinary-coded decimal coding in combination with the error detection circuitry ensured reliable operation. In 1948, the Ballistic Computer was moved to another US Army facility at Fort Bliss, Texas, where it continued in service until 1958. A second machine of the same basic design, known as the Error Detector Mark 22, was also constructed for the US Naval Research Laboratory in Washington DC.

A General-Purpose Calculator

The success of the Relay Interpolator and Ballistic Computer encouraged the NDRC to support a much more ambitious project to develop a large-scale relay machine that could be used by the US military for a variety of computational tasks related to the production of artillery firing tables. The resulting $500,000 contract was awarded to Bell Labs in 1944. Stibitz acted as a consultant on the design of the new model but was not closely involved in its development.

The name chosen for the new machine, the Model V, came about through a renaming of the four earlier Bell Labs relay calculators, which were retrospectively named the Models I to IV. The Model V featured floating-point decimal arithmetic with a word length that comprised a 7 decimal digit mantissa, a 1 decimal digit plus 1 binary digit exponent and the algebraic sign, giving a range of plus 19 to minus 19. However, its most impressive feature was an advanced modular design which provided for up to 6 separate arithmetic units, each equipped with its own storage registers and input/output devices and capable of operating as an independent computer.

The Model V's arithmetic units contained a 10-digit adder plus 3 arithmetic registers and 15 storage registers. Also included were 4 additional registers used for control purposes, a special-purpose register for manipulating individual digits within a number and 8 smaller registers capable of storing only a plus or minus sign. Subtraction was performed by the method of complements.

Multiplication was performed by repeated addition, with division and a square root function both performed by repeated subtraction. External to the arithmetic units was a shared unit which included a set of up to 5 'lookup tables' containing fixed values of commonly used trigonometric functions.

Two operator consoles could be connected to each arithmetic unit. Additional circuitry ensured that only one operated at a time and was also capable of automatically switching to the next available console at the end of a session. Each console contained up to 12 punched tape readers; 1 for mathematical constants and overall sequence control, 5 for instructions and 6 for tabular data. A special section within each arithmetic unit known as the Discriminator permitted control to be switched between different sections of an instruction tape on the result of a comparison operation, thus providing the ability to dynamically alter a sequence of instructions during execution. This is the first documented implementation of a very powerful feature that would later be termed a 'conditional branch'.

Two examples of the Model V were built, both fitted with 2 out of a possible 6 arithmetic units. Each was the size of a large room, weighed around 10 tons and incorporated over 9,000 telephone relays and up to 55 items of Teletype equipment. The first was completed in July 1946 and delivered to the National Advisory Committee on Aeronautics (NACA) at Langley Field, Virginia. The second went to the US Army's Ballistic Research Laboratory (BRL) at Aberdeen Proving Ground, Maryland. Although they were not completed until after the war had ended and the need for artillery firing tables had subsided, both machines were fully utilised on a wide variety of applications and remained in service until 1958.

The Bell Labs Model V was impressively engineered with excellent reliability and many powerful features. However, with a performance of 0.3 seconds for an addition and a fraction under 1 second for a multiplication, it was no faster than the earlier Bell Labs machines. One more relay calculator would later be built, the Model VI, a trimmed down version of the Model V for internal use but with the Model V, Bell Labs had taken electromagnetic relay technology as far as it could go.

After the war, Stibitz chose not to return to Bell Labs. Instead, he set himself up as an independent consultant in applied mathematics, later specialising in biomedical applications of computing technology. Nevertheless, his relay calculators provided the first proof that automatic computing machinery, when carefully designed and solidly engineered, could be used routinely in demanding real-world applications.

The Harvard Mark I

The same year that George Stibitz began tinkering with a pair of electromagnetic

3. The Electrification of Calculating Machinery

relays on his kitchen table, Howard H Aiken, a postgraduate student at Harvard University with an interest in the work of Charles Babbage, was also pondering the application of electromechanical technology to automatic calculation. Howard Aiken was born in Hoboken, New Jersey in 1900. When he was 12 his family moved to Indianapolis, Indiana. Shortly afterwards, his father left home, abandoning his family and forcing Aiken to leave school at the earliest opportunity to seek employment. Fortunately, one of his teachers recognised a keen aptitude for mathematics and arranged for Aiken to continue his studies while working in the evenings as a switchboard operator for a local public utility company.

After completing 4 years at high school, Aiken continued to work to support his mother while putting himself through university, studying electrical engineering at the University of Wisconsin where he received a bachelor's degree in 1923. Aiken then spent a number of years working as an electrical engineer for various public utilities but the intellectual rigours of academic life beckoned once more and in 1932 he enrolled as a postgraduate student in physics at the University of Chicago. However, Chicago failed to live up to Aiken's expectations and he quit after less than a year for another postgraduate studentship in the Department of Physics at Harvard University in Cambridge, Massachusetts.

Aiken's doctoral research in electron physics at Harvard led to a system of nonlinear differential equations which could only be solved by numerical techniques involving large numbers of repetitive calculations. Like his German counterpart Konrad Zuse, Aiken began to consider the use of calculating machines to mechanise the task. He quickly established that none of the commercial calculators available at that time were suited to scientific computation and that a different kind of machine would be required. A review of earlier work in this field led him to discover the writings of Charles Babbage. Drawing inspiration from Babbage's Analytical Engine, Aiken produced a document outlining the specification for a large-scale automatic calculator capable of tackling the kind of scientific calculations that he himself needed to perform. However, despite his years of practical engineering experience, Aiken now saw himself as a career academic and had absolutely no desire to build the machine. Instead he would seek out a suitable manufacturer with which to collaborate. By April 1937, he was ready to commence the search.

An Ideal Collaborator

Aiken's first port of call was the New Jersey based Monroe Calculating Machine Company where he was interviewed by the company's Chief Engineer, George C Chase. Chase greeted Aiken's ideas with great enthusiasm, immediately recognising the long-term commercial potential of such advanced technology, which could perhaps even extend to the traditional accounting machines that were the mainstay of Monroe's business. Unfortunately, the company's top management were less enthusiastic and after several months of deliberation, they decided not to proceed with the project. Chase then suggested to Aiken that he contact IBM and gave him the name of Harlow Shapley, a professor of astronomy at Harvard with good links into the company. Through Shapley's connections and those of another Harvard professor, Theodore H Brown, Aiken was able to approach IBM at the appropriate level.

In November 1937, Aiken met with IBM's most senior engineer, James W Bryce. Bryce had assumed legendary status within the company, having been honoured as one of the ten greatest living inventors at the centennial celebration of the US Patent Office the previous year. Bryce was intrigued by Aiken's ideas but suggested that he first pay a visit to Columbia University in New York to see how the astronomer Wallace J Eckert had been using standard IBM punched card tabulating equipment for scientific computation. With the aid of a specially designed control switch based on plugboards and rotating cams, Eckert had successfully interconnected a multiplying punch, numeric printing tabulator and duplicating punch, and was using them to solve the differential equations necessary for the production of astronomical and lunar tables.

Following his visit to Columbia, Aiken reported back to Bryce that his computational requirements went beyond the capability of Eckert's facilities, impressive though they were. However, Eckert's work had given Aiken a much clearer idea of how to build his machine. He redrafted his original document into a 23-page memorandum describing how an automatic calculating machine could be built using electromechanical modules lifted from existing IBM punched card equipment. Calculations would be performed using a fixed-point decimal representation, with 10 digits to the left of the decimal point and 12 to the right, and the machine would take the form of a giant switchboard incorporating panels for each arithmetic operation.

In order to progress the project, Bryce needed an estimation of the cost. He assigned three experienced IBM engineers, Clair D Lake, Francis E Hamilton and Benjamin M Durfee, to work with Aiken on fleshing out the design of the machine. This was completed by September 1938. Bryce's support for Aiken's work held great sway within the company and the project duly received the formal approval of IBM Chairman Thomas Watson in February 1939. Watson was keen for IBM to establish a strategic relationship with one of America's prestigious Ivy League universities and so agreed to cover the estimated $100,000 cost of development on the basis that the collaboration would generate considerable publicity and goodwill for the company.

Development

Development of the IBM Automatic Sequence Controlled Calculator (ASCC), as the new machine was officially known, began in May 1939 at IBM's North

3. The Electrification of Calculating Machinery

Street Laboratory in Endicott, New York. Lake was confirmed as Chief Engineer for the project and chose Hamilton and Durfee to assist him. Aiken, who had received his doctorate in February and was now an Instructor of Physics at Harvard, also remained closely involved, providing valuable input on the functional specification and circuit design.

The architectural design of the ASCC was based on the use of multiple accumulators which were built using the adding counter mechanisms from IBM punched card tabulators. The design incorporated a total of 72 accumulators, each comprising 24 electromechanical counter wheels that corresponded to the 23 decimal digit word length plus an additional digit for the algebraic sign. Because these mechanisms possessed the ability to add numbers automatically, there was no need for a separate adder or arithmetic unit. Arithmetic operations were performed in fixed-point decimal format, with addition and subtraction accomplished directly using the accumulators. Multiplication and division were performed with the aid of a separate unit which automatically generated a multiplication table containing all 9 integer multiples of the multiplicand and then selected the appropriate entries for adding together while shifting the decimal place, as required, after each addition.

The feature-packed design also included separate logarithm, exponential and sine units which operated in conjunction with the multiply-divide unit and 4 additional accumulators to calculate tables of logarithms, antilogarithms and trigonometric functions. In addition were 3 mechanical interpolators that performed non-linear interpolation on equidistant values of a function, which was supplied on punched tape together with the interpolation coefficients necessary to obtain the desired accuracy.

At the heart of the ASCC was a sequence control unit that directed the operation of the machine. Instructions were fed into this unit from a punched tape reader which accepted unusually wide paper tape fabricated from uncut IBM punched card stock. The 3¼-inch width tape accommodated 24 hole positions in each row. These were divided into 3 groups of 8, one each for the address of the accumulator to be read from, the address of the accumulator to be written to and the machine instruction itself. A series of relay circuits converted the instructions into electrical signals that controlled the action of the electromechanical counter wheels in each of the accumulators.

Numerical data could be entered either on punched tape, by means of three dedicated tape readers which employed the same 24-hole format as the instructions, or on punched cards via two punched card units. This was the earliest example of automatic input of both instructions and data in a general-purpose machine, predating Zuse's Z4 by over a year. The tape and card readers were supplemented by 60 sets of 24 rotary dial switches which were used to manually enter mathematical constants. Results were printed using a pair of IBM 'Electromatic' electric typewriters or output onto punched cards for further processing.

The ASCC also featured an unusual 'reference to previous results' function in which accumulator number 72 could be used to determine when a specific number of iterations of a calculation had been reached by detecting when no end-around carry took place and then stopping the machine. Although not as powerful as a conditional branch, this feature permitted mathematical functions to be evaluated by repeated iterations.

The physical layout of the ASCC closely followed Aiken's switchboard concept. It comprised 7 main units which were arranged in a horizontal line and linked together by a common driveshaft that synchronised operation and supplied mechanical power to the counter wheels. A powerful 5 horsepower electric motor rotated the driveshaft at a constant 200 rpm, giving a basic cycle time or clock speed of 3.3 Hz.

Aiken spent the summers of 1939 and 1940 at Endicott but in April 1941 he was called up for active service in the US Naval Reserve, given the rank of Lieutenant Commander and assigned to teaching duties at the Naval Mine Warfare School in Yorktown, Virginia, which curtailed his continued involvement in the project. In September 1942, Hamilton was reassigned to war-related projects but continued to work on the ASCC in his spare time. Development had been expected to take 2 years but these wartime interruptions delayed progress. Nevertheless, construction was essentially complete by the end of 1942 and the machine performed its first calculation on 1 January 1943. Following a lengthy period of testing and refinement, the ASCC was finally demonstrated to Harvard faculty members in December 1943. Shortly afterwards, it was dismantled and shipped to Harvard where it was installed in the University's Research Laboratory of Physics.

IBM ASCC at Harvard (Photo © IBM Corporation)

3. The Electrification of Calculating Machinery

At 51 feet long and with more than 760,000 components, including 3,304 electromagnetic relays, the ASCC was the largest electromechanical calculator ever built. It also looked stunning, sheathed in a futuristic enclosure specially created by the celebrated industrial designer Norman Bel Geddes. Performance was not quite so impressive, requiring 5.7 seconds for a full precision multiplication and 15.3 seconds for division. Nevertheless, with automatic calculating machinery in great demand for the war effort and still something of a rarity, the ASCC was commandeered by the US Navy following commissioning at Harvard and used exclusively by the Navy Bureau of Ships during the remainder of the war. Aiken was transferred back to Harvard in April 1944 to take charge of the operation and the Navy also provided additional personnel to operate the machine.

Parting Company

A formal dedication and acceptance ceremony took place at Harvard on 7 August 1944. Unfortunately, the press release which the University had prepared for the event named Aiken as the sole inventor of the machine and failed to fully acknowledge IBM's role in its development. Watson, who had made the journey to Harvard for the dedication, was understandably furious. The concept was clearly Aiken's but the realisation owed a huge debt to the IBM engineers. IBM had also footed the substantial bill for development, which, at $200,000, was double the estimated cost, and Watson was planning to announce the donation of a further $100,000 to Harvard to cover the operating costs.

Watson's initial reaction was to boycott the event but he was persuaded by Aiken and Harvard President James B Conant to attend and the ceremony went ahead as planned. In his speech, Aiken acknowledged Lake, Hamilton and Durfee as co-inventors and all four men were subsequently named as inventors on a US patent application filed in February 1945. However, the damage had been done. Aiken's stubborn refusal to apologise for the slight meant that Watson forever remained convinced that it had been a deliberate act on Aiken's part rather than an unfortunate oversight. The 7-year relationship between Harvard and IBM was at an end.

On completion of his wartime Naval service, Aiken was appointed Professor of Applied Mathematics at Harvard. He went on to oversee the development of three more large-scale calculators, all without IBM involvement. These were named the Harvard Mark II, III and IV, which also resulted in the ASCC becoming better known as the Harvard Mark I. Aiken also organised two important symposia on large-scale digital calculating machinery in January 1947 and September 1949, which were jointly sponsored by the US Navy Bureau of Ordnance and Harvard University. He also established one of the first postgraduate research programmes in computer science.

The ASCC remained in operation until 1959. With its plethora of accumulators and special-purpose calculating units, the ASCC was a powerful and versatile machine, and the quality of IBM's engineering ensured that it operated faultlessly despite the mind boggling mechanical complexity of the design. Aiken's efforts as an evangelist for scientific computing also resulted in the machine being seen by, and influencing, many of the other computer pioneers of the day, and the separation of signal pathways for instructions and data in the ASCC resulted in the coining of the term 'Harvard Architecture' to describe machines with separate data and instruction buses. However, Aiken's greatest achievement was in lending academic respectability to the new discipline of computer science. The machine itself, with its sluggish electromechanical innards, was a technological dinosaur, a situation made all the more apparent by the exciting developments in electronic computation already taking place elsewhere.

Further Reading

Austrian, G. D., Herman Hollerith: Forgotten Giant of Information Processing, Columbia University Press, New York, 1982.

Buck, G. H. and Hunka, S. M., W Stanley Jevons, Allan Marquand, and the Origins of Digital Computing, IEEE Annals of the History of Computing 21 (4), 1999, 21-27.

Ceruzzi, P. E., Reckoners: The Prehistory of the Digital Computer from Relays to the Stored Program Concept 1935-1945, Greenwood Press, Connecticut, 1983.

Chase, G. C., History of Mechanical Computing Machinery, Proceedings of the 1952 ACM National Meeting, 1952, 1-28.

Cohen, I. B., Howard Aiken: Portrait of a Computer Pioneer, MIT Press, Cambridge, Massachusetts, 1999.

Heide, L., Shaping a Technology: American Punched Card Systems 1880-1914, IEEE Annals of the History of Computing 19 (4), 1997, 28-41.

Irvine, M. M., Early Digital Computers at Bell Telephone Laboratories, IEEE Annals of the History of Computing 23 (3), 2001, 22-42.

Randall, B., From Analytical Engine to Electronic Digital Computer: The Contributions of Ludgate, Torres and Bush, IEEE Annals of the History of Computing 4 (4), 1982, 327-341.

Rojas, R., Konrad Zuse's Legacy: The Architecture of the Z1 and Z3, IEEE Annals of the History of Computing 19 (2), 1997, 5-16.

Speiser, A. P., Konrad Zuse's Z4: Architecture, Programming and Modifications at the ETH Zurich, in The First Computers – History and

3. The Electrification of Calculating Machinery

Architectures, MIT Press, Cambridge, Massachusetts, 2000.

Zuse, K., The Computer – My Life, Springer-Verlag, Berlin Heidelberg, 1993.

4. The Dawn of Electronic Computation

"The ENIAC could have been invented 10 or 15 years earlier and the real question is, why wasn't it done sooner?", J Presper Eckert.

The Appeal of Electrons

The automatic calculating machines of Zuse, Stibitz and Aiken had demonstrated the huge potential of digital sequence-controlled calculators but they had also revealed the practical limitations of the electromechanical switching technology on which they were based. Reliability was clearly an issue, with sticking relays causing intermittent errors which could go undetected unless the machine design incorporated elaborate error detection features. However, the main weakness of electromechanical technology was in terms of switching speed which severely limited calculation performance.

A faster and more reliable type of electromagnetic relay, known as a reed relay, had been developed at Bell Telephone Laboratories in the late 1930s for use in telephone exchange equipment. Reed relays increased the maximum switching frequency from a few cycles per second to several hundred cycles per second but even this was not fast enough for some. Digital electronics offered the prospect of calculation at electronic speeds and this would be the main driving force for the move from electromechanical to electronic computation.

Building Blocks – Electronic Devices

The development of electronic devices began with a chance discovery made by the prolific American inventor Thomas Alva Edison in February 1880. Having created the first practical electric light bulb the previous year, Edison set out to improve the durability of his somewhat fragile device to make it suitable for commercial use. One of the problems with Edison's prototype was a gradual blackening of the inside surface of the glass bulb during operation which eventually rendered it useless. While investigating methods to reduce this blackening effect, he discovered that an electric current could be made to flow in a vacuum.

Edison correctly deduced that the blackening was caused by carbon atoms being discharged from the carbonised filament, so he introduced a metal electrode into the evacuated bulb in an effort to attract the carbon atoms away from the glass. Experimenting with the polarity of the electrode, he observed that a current flowed between the filament and electrode when the electrode was connected to the positive terminal of the bulb, despite the electrode being electrically isolated from the filament, but no current flowed when the electrode was connected to the negative terminal. He also noted that this current increased in proportion to an increase in the brightness of the bulb. The phenomenon became known as the Edison Effect.

In November 1883, Edison filed a US patent application for a device which employed the effect to indicate changes in current in electrical distribution circuits but he never fully understood the phenomenon. A proper scientific explanation was not provided until 1897 when the British physicist Joseph J Thomson discovered the electron. Described as thermionic emission, the Edison Effect was found to be caused by electrons emitted from the hot filament being attracted to the positively charged electrode. Although Edison had first discovered it, his astute business sense seems to have deserted him on this occasion and it would be left for others to profit from this highly important discovery.

Thermionic Valves

In 1904, another British physicist, John Ambrose Fleming, developed the first practical electronic device based on the Edison Effect. Fleming was based at University College London, where he held the post of Professor of Electrical Engineering. He was also retained as a scientific advisor to Guglielmo Marconi's Wireless Telegraph Company and it was whilst investigating methods for improving the detection of radio signals in long distance wireless communication that he developed the device.

The weak link in early wireless communication systems was the coherer, a primitive electromechanical device which functioned as the detector. Fleming was familiar with the phenomenon of thermionic emission, having studied it at length during the 1880s while working as a consultant to the Edison Electric Light Company. These studies had shown that the ability of thermionic devices to permit current to flow in only one direction could, under certain conditions, be harnessed to convert alternating current to direct current, a process known as rectification. Because radio signals are wave-based, rectification would also greatly simplify the measurement of signal strength when testing different detectors, so Fleming set about the task of improving Edison's thermionic bulb device, redesigning the electrode in the shape of a hollow aluminium cylinder suspended on platinum wires and positioned within the glass bulb so that it surrounded the filament. By connecting this device to an antenna using an induction coil, he found that it was indeed capable of rectifying the weak high-frequency oscillations found in radio signals.

Fleming named his device an 'Oscillation Valve' because it acted like a water valve in only allowing the current to flow in one direction. He patented it

4. The Dawn of Electronic Computation

in Britain in November 1904 and in the USA the following year. The device subsequently acquired the name 'diode' due to its two principal elements, the filament and the electrode. Fleming had not only found a practical use for thermionic emission but his device, and the March 1905 Royal Society paper describing it, are considered by many to be the birth of electronics.

The First Prototype Fleming Valves

The next important step in the development of electronic devices was made by the controversial American inventor Lee de Forest. De Forest had developed an interest in wireless telegraphy while studying for a doctorate in Physics at Yale University, which he received in 1899. After completing his studies, he embarked on a career in wireless communications but drifted from one to another of the established companies in the field before deciding to set up on his own with a succession of short-lived business ventures which culminated in the incorporation of the American De Forest Wireless Telegraph Company in 1903. In 1905, he began experimenting with a thermionic device that employed a glass bulb filled with halogen gas at low pressure and fitted with a second filament in place of the electrode. By heating the gas within the bulb, he found that the device became sensitive to radio signals, which caused perturbations in the flow of current in an electrical circuit containing the filaments.

De Forest's assistant Clifford Babcock named the new device the 'Audion', an amalgamation of the words audible ions, and a US patent application was duly filed in January 1906. This also included an alternative arrangement where one of the filaments was replaced by a disk-shaped electrode. Remarkably, the application was granted in June of the same year despite a striking similarity to Fleming's Oscillation Valve. Influenced by experimental work on the electrical conductivity of gas flames, de Forest had mistakenly assumed that the Audion worked because a heated gaseous medium passed a direct current. Although later disproved, this argument and an emphasis on the application of the device as a detector were deemed sufficient to differentiate his invention from that of Fleming.

The Audion performed poorly as a detector and was a commercial failure as a result, so de Forest set out to improve the device. He did this by introducing a third element in the form of a grid of fine nickel wire positioned between the filament and electrode. He discovered that if he applied the signal from the wireless antenna to the grid instead of the filament, he could obtain a much more sensitive detection of the signal. The grid had the effect of boosting the signal by superimposing upon it the current flowing from the filament to the electrode. De Forest filed a US patent application for his three-element Audion in January 1907 which was granted in February 1908.

The three-element Audion was sufficiently successful that in 1913 de Forest was able to license the device to the American Telephone & Telegraph Company (AT&T) for commercial use, excluding wireless applications, for a fee of $50,000. AT&T obtained the wireless rights the following year for an additional $90,000 and the remaining rights in 1917 for a further sum of $250,000. In July 1920, the Radio Corporation of America (RCA) also acquired rights to the Audion through cross-licensing agreements with AT&T and soonafter began to offer three-element Audion devices for commercial sale.

The commercial success of the three-element Audion was not based on the device's ability as a radio signal detector. In the summer of 1912, Edwin H Armstrong, an electrical engineering student at Columbia University, was studying the operation of the Audion when he hit upon the idea of feeding the amplified signal back through the grid to further strengthen the signal. Armstrong built a circuit to accomplish this and found that it could amplify the signal by several orders of magnitude. He also observed that when the 'feedback' was increased sufficiently, the circuit began to generate a sustained high-frequency oscillating signal, thereby showing that the Audion could not only function as a powerful amplifier but also as a high-frequency oscillator.

Armstrong submitted a US patent application for his regenerative feedback circuit in October 1913 which was granted the following year. However, de Forest subsequently claimed that he had accidentally discovered feedback in August 1912 when he noticed a howling sound through headphones. He proceeded to file his own application for a patent on the invention in March 1914, prompting a bitter legal battle which de Forest eventually won 20 years later when the US Supreme Court ruled in his favour in May 1934.

Further research conducted at the General Electric Research Laboratory showed that the three-element Audion functioned more effectively as an amplifier if the halogen gas was replaced with a relatively high vacuum. This modification was adopted and led to the devices becoming popularly known as vacuum tubes in the USA. Three-element thermionic devices acquired

4. The Dawn of Electronic Computation

the generic name 'triode' and became a crucial component in the evolution of wireless telegraphy into commercial radio and public broadcasting in the 1920s.

Despite the controversy surrounding the originality of de Forest's Audion, it was Fleming's earlier patent for his Oscillation Valve which was invalidated by the US Supreme Court in a seemingly partisan decision made in June 1943. The Court maintained that the use of the device for rectifying low-frequency currents was known art when filed and that a subsequent disclaimer to this effect was invalid due to the length of time taken to issue it.

Flip-Flops and Digital Counters

The earliest applications of thermionic valves were in analogue electronic circuits for wireless telegraphy and radio but thermionic valves could also be used to build digital circuits. The first example of a digital electronic circuit was invented by the British physicist William H Eccles and his associate Frank W Jordan in 1918. Working at the City & Guilds College in Finsbury, London, Eccles and Jordan designed an electronic circuit to generate rectangular waveforms (a square-wave signal) of relatively low frequency. Their circuit contained a pair of triodes which were cross-connected in such a way that the grid of one was connected to the positive electrode of the other. In this configuration, the circuit could assume only two stable states and would 'flip' from one state to the other on the application of a trigger pulse.

Eccles and Jordan realised that they had created the electronic equivalent of an electromagnetic latching relay. They obtained a British patent for their invention, which they referred to as a trigger relay, in 1918 and described it in an article published in the December 1919 issue of the journal The Radio Review. Their circuit became more widely known as the Eccles-Jordan Circuit or 'flip-flop'. Having similar functionality to the electromagnetic relay but with the prospect of greatly increased speed of operation, the flip-flop would provide the basic building block for electronic memory devices and logic circuits. However, the first practical application of digital electronics was not in computing but in atomic physics.

In 1930, Charles E Wynn-Williams, a British theoretical physicist working at Cambridge University's prestigious Cavendish Laboratory, constructed an instrument to perform the automatic high-speed counting of subatomic particles in nuclear disintegration experiments. Conventional relay-based counters were unable to keep pace with the millisecond rate at which the particles appeared, so Wynn-Williams turned to digital electronics in an effort to find a solution. He used an electronic amplifier to amplify the tiny electrical currents produced by the particles as they entered an ionisation chamber. These were fed to a binary digital counter which operated by scaling down the rate of counting until it was slow enough for an electromechanical counter to keep up. This was achieved using 3 pairs of bi-stable electronic devices arranged in a cascade so that a successive device would only be triggered following the triggering of both devices in the previous pair, which gave a speed reduction of 23 or 8 times. The instrument became known as a scale-of-two counter.

Wynn-Williams could have used flip-flops as his bi-stable devices but chose instead Thyratron valves. These were gas-filled triodes specifically developed for high-voltage switching applications which operated in a similar manner to a flip-flop. Based on the mercury-vapour discharge tube invented by the American physicist George W Pierce at Harvard University in 1914, the first commercial examples appeared around 1928.

Although Wynn-Williams had devised his scale-of-two counter for use as a measuring instrument, the ability to count digits at electronic speeds had an obvious application in electronic computation. Researchers at the Massachusetts Institute of Technology soon developed a ring counter, where the last flip-flop in the cascade is connected to the first to form a closed circuit, thus providing the electronic equivalent of a mechanical counter wheel.

Electronic Pioneers

Many of the building blocks for electronic computation were already in place by the time Zuse, Stibitz and Aiken had embarked on their groundbreaking development projects. However, all three pioneers of digital sequence-controlled calculators remained firmly wedded to electromechanical technology and it would be a different set of protagonists who would become the first to apply electronics to automatic computation. This was achieved through three landmark academic projects, two in the USA and one in Germany, which were conducted independently during the same period. The first of these was led by the analogue computing pioneer Vannevar Bush at the Massachusetts Institute of Technology.

The Rapid Arithmetical Machine

Vannevar Bush had first speculated on the possibility of digital electronic applications in a scientific paper published in the Bulletin of the American Mathematical Society in October 1936. This gave rise to discussions with his colleague Samuel Caldwell on the practicalities of digital electronics in computing, such as the ring counters recently developed at MIT and elsewhere. Over the next two years a conceptual design evolved for an electronic digital calculator which they called the Rapid Arithmetical Machine.

The Rapid Arithmetical Machine would be a sequence-controlled four-function decimal calculator with the instruction sequence supplied by means of punched paper tape. Two additional punched tape units would be employed for input of numerical data and constants. Arithmetic operations would be performed

4. The Dawn of Electronic Computation

electronically but a small number of electromagnetic relays would be required in the control and input/output circuitry.

In 1938, Bush began working on the engineering design with William H Radford, a research assistant in MIT's Department of Electrical Engineering. The following year, the National Cash Register Company (NCR) agreed to provide financial support for the project and Wilcox P Overbeck was hired as a research assistant to work on the hardware implementation. Radford's main contribution was in the design of electronic circuits and much of Overbeck's work related to the development of special-purpose thermionic valves and improving the reliability of components.

Bush wrote a memorandum reviewing progress in March 1940 in which he estimated that the machine would be capable of multiplying two 6-digit numbers in 0.2 seconds. However, as America's inevitable entry into World War II drew nearer, external factors began to undermine the project. Bush, already thinly stretched, was becoming increasingly drawn into war-related activities and in June 1940 he was appointed by US President Franklin D Roosevelt to chair the newly established National Defense Research Committee (NDRC), which had been created at Bush's own suggestion. Caldwell took over the supervision of the project but other key project personnel were also lost to more urgent wartime priorities. In 1941, Radford was promoted to the rank of assistant professor and became involved in setting up the MIT Radar School. The final blow came in early 1942 when Overbeck was transferred onto atomic weapons research. With no research staff remaining to continue the work, the project has to be shelved.

In the early stages of the Rapid Arithmetical Machine project, the MIT team led the world in the application of digital electronics to computation but the project was inadequately resourced, despite the NCR funding, and the rate of progress slow. During the same period Bush and Caldwell were also building the Rockefeller Differential Analyzer, which became operational in December 1941, and this project is likely to have been given a higher priority due to the intended use of the machine by the US Navy in the production of artillery firing tables and radar antenna design. With no specific military application for the Rapid Arithmetical Machine, progress stalled following the USA's entry into World War II and MIT's lead would be brief. By the end of the war, the technology proposed for the machine had become outdated and Caldwell's efforts to resurrect the project came to nothing.

Although the Rapid Arithmetical Machine was never built, circuit designs were completed and various components assembled and tested. Bush subsequently played down his role in the history of the electronic digital computer, perhaps due to embarrassment over the project's ultimate lack of success. However, the project did have some influence on research into high-speed electronic counters which was being conducted in parallel by Joseph R Desch and Robert

E Mumma at NCR's Electrical Research Laboratory in Dayton, Ohio, work that led to the development of cryptanalysis devices for the US Navy. Desch and Mumma described their efforts in a US patent application of March 1940 entitled 'Calculating Machine' which describes a valve-based binary multiplier and accumulator. The Rapid Arithmetical Machine was also the inspiration for at least one successful post-war computer development project at MIT.

The Work of Helmut Schreyer

In Germany, Konrad Zuse's friend and occasional collaborator, the electrical engineer Helmut Schreyer, had also begun working on the application of electronic technology to calculating machinery. Schreyer's interest in automatic computation had been kindled through his association with Zuse but it was driven by his research work in the field of high-frequency electronics and his aim was to prove that high-speed calculation could be accomplished through electronic means. The two men would continue to collaborate but Schreyer chose to remain in academia and pursue an academic career.

In 1938, having obtained a diploma in electrical engineering, Schreyer secured a position as a research assistant in the telephonic and telegraphic engineering department of the Technische Hochschule in Berlin-Charlottenburg. Schreyer had been instrumental in persuading Zuse to use electromagnetic relays in the Z2 machine but he was also keenly aware of the limitations of electromechanical switching technology. His first goal was to develop an electronic replacement for the electromagnetic relay. He devised a basic circuit containing a triode and a neon gas discharge lamp. The characteristics of the neon lamp were such that it would only begin to conduct electricity when the voltage reached a certain well-defined threshold but would then continue to conduct with a lower 'holding' voltage flowing through it. With two stable states, Schreyer's circuit was functionally equivalent to the flip-flop although considerably slower, having a maximum switching frequency of around 5 KHz. Nevertheless, the circuit was a considerable improvement on electromagnetic relays. Schreyer called this circuit a 'Röhrenrelais' or tube relay.

Schreyer and Zuse gave a lecture on the possibilities of electronic computation to a small group at the Technische Hochschule in 1938. To help illustrate their ideas, they included a demonstration of the prototype of Schreyer's tube relay device. However, their suggestion that several thousand of these power-hungry devices could be used to build an electronic calculator was regarded as ludicrous by the audience. Nevertheless, Schreyer was encouraged by his Head of Department, Professor Stäblein, to use the device as the basis for his doctoral research.

Schreyer first proposed the use of electronic technology for calculating machinery in a memorandum written to the German military authorities in October 1939 in an attempt to release Zuse from active service. The document

4. The Dawn of Electronic Computation

outlined an ambitious plan to build an electronic computing machine which would calculate at a speed of 10,000 operations per second and could be used in a variety of military applications, including the production of artillery firing tables. However, with no track record to speak of, Schreyer's proposal was not acted upon.

Schreyer's PhD thesis, which was entitled 'Das Röhrenrelais und seine Schaltungstechnik' (The Tube Relay and its Circuit Technology), was completed in August 1941. Given the audience reaction of a few years earlier, Schreyer thought it best to avoid the subject of automatic computation entirely but he did include a frequency divider as an example circuit. This was analogous to a shift register, a group of bi-stable devices connected together in such a way that the bits of data are transferred along the line when the circuit is activated.

After obtaining his PhD, Schreyer continued his research into the tube relay device. His next step was to build a series of circuits to perform basic logic operations but he soon encountered problems with inconsistencies in the electrical characteristics of the dual triodes that he was using. Fortunately, he had a freind who worked at the German radio and television company Telefunken who introduced him to the company's head of thermionic valve development. Schreyer took the opportunity to outline his specification for a new design of dual valve which would incorporate both a triode and a tetrode (a four element thermionic device) within the same package. This specification was adopted by the company for a new valve model and Schreyer was given 150 examples of the prototype device for his own use.

By early 1942 Schreyer had established that his circuit technology had the capability for use in electronic computing and he now needed funding to build a machine. Together with Zuse, he approached the German Army High Command with a proposal to construct a large-scale electronic calculator incorporating 1,500 thermionic valves for use in anti-aircraft defence applications. However, their estimated development timescale of 2 to 3 years was unacceptable to the OKH officials, who believed that the war would be over before the machine could be completed. Schreyer and Zuse's next move was to approach the DVL, the German Aeronautics Research Institute which had supported the development of the Z3. The Institute was unable to commit to a full development project but agreed to fund the construction of a small-scale prototype.

Development work was carried out at the Technische Hochschule. Zuse contributed to the project by producing the overall design for the machine, which would be a special-purpose calculator for converting 3-digit decimal numbers into 10-bit binary. Schreyer then took Zuse's design and implemented it in electronic circuitry. At the heart of the machine was his tube relay device. By combining the new Telefunken dual valve with faster switching neon lamps, Schreyer was able to create an improved version of the device that could operate at up to 10 KHz. A binary storage unit was also developed which employed Thyratron valves configured in such a way that they permitted parallel access.

During this period Schreyer was also working part-time for the Heinrich Hertz Institut für Schwingungsforschung (the German Institute for Radio Communication) on war-related research. This took priority over his electronic computing work and slowed progress. Nevertheless, with the help of Zuse's firm, which carried out some of the assembly work, the prototype was completed within a year. It contained around 100 valves and comprised an arithmetic unit, storage unit, instruction unit and keyboard. Initial problems with the reliability of the tube relay devices were overcome by bathing the neon lamps in ultraviolet light. However, alterations to the specification of components in the tube relay circuits were necessary in order to realise the full 10 KHz switching frequency.

In November 1943, the machine was badly damaged during a bombing raid on Berlin. Schreyer spent the remainder of the war attempting to repair it but his other work increased in priority, the allied air raids caused frequent interruptions and further research into electronic computing came to a halt. Towards the end of the war, Schreyer was transferred to the city of Erlangen following the relocation of the Heinrich Hertz Institut and the machine went with Zuse to Göttingen where it remained after Zuse subsequently moved to Bavaria. However, when Zuse and Schreyer attempted to retrieve the machine in 1947 it had disappeared.

Schreyer had applied for a German patent on the tube relay device in November 1940 but this was not granted until August 1954. With the loss of his prototype and almost no published material, Schreyer was left with very little to show for his years of pioneering work in electronic computation. Unlike Zuse, Schreyer had joined the Nazi Party, which made life more difficult for him in post-war Germany. In 1949 he emigrated to Brazil, where he established a new career in Brazil's burgeoning telecommunications industry.

The Atanasoff-Berry Computer

The third pioneering contribution to the development of electronic digital calculators came as a result of a modest research project conducted by an unassuming theoretical physicist at an obscure college in Iowa. John V Atanasoff was born in the New York town of Hamilton in 1903, the son of a mathematics teacher and an electrical engineer who had emigrated from Bulgaria. An early interest in mathematics was nurtured by his parents and resulted in him excelling at school and choosing an academic career. Having obtained a Batchelor's degree in electrical engineering, a Master's in mathematics and a PhD in theoretical physics, he was invited to return in the autumn of 1930 to the institution where he had obtained his Master's degree, Iowa State College in Ames, Iowa, and accepted a position there as an Assistant

4. The Dawn of Electronic Computation

Professor of Mathematics. In September 1935, he was promoted to Associate Professor of Mathematics and Physics.

Like Howard Aiken, Atanasoff became interested in automatic computation as a result of performing the tedious calculations required for his doctoral thesis, and the early years of his academic career were characterised by a search for practical solutions to the extended systems of linear algebraic equations that he regularly encountered in his research work. His initial efforts concerned the use of IBM punched card equipment. Atanasoff and a colleague from the Statistics Department obtained access to one of the IBM tabulators that the College leased for statistical applications and were able to reconfigure it so that it allowed numbers to be entered in pairs, with one number obtained from a card and the second number entered from the plugboard. The machine was then used to analyse complex atomic spectra. However, standard IBM tabulators had insufficient numerical accuracy to solve large systems of linear equations and it would have been impossible to make any permanent modifications to the machine without invalidating the terms of the lease. Atanasoff's thoughts then turned to the possibility of connecting a large number of Monroe desk calculators together using a continuous shaft to drive all the machines in synchronisation. However, he soon realised that adapting conventional equipment was unlikely to deliver a satisfactory solution.

In 1936, Atanasoff and one of his students built an analogue instrument as an aid to the calculation of Laplace Transforms, a type of integral transform used in applied mathematics to solve differential equations. The instrument employed a cube of paraffin wax which would be sculpted into a two-dimensional surface and the surface tested in order to determine if it satisfied certain mathematical conditions. They named the device a 'Laplaciometer'. Although useful, Atanasoff's concerns over the accuracy of analogue machines for computation convinced him that a digital solution would be a far better option. He then began to consider a suitable number system, weighing up various possibilities before choosing binary on the basis of speed of operation and simplicity of implementation in hardware.

As a theoretical physicist, Atanasoff was familiar with Wynn-Williams' scale-of-two counter. This confirmed to him that binary switching circuits could be built using electronic components. He had also been thinking about data storage and had devised a scheme where the direction of polarisation in a magnetic material could be used to represent a 0 or 1. Atanasoff now had the main elements of an electronic digital calculator but he could not envisage how to put it all together.

After dinner one night towards the end of 1937, these thoughts began to overwhelm him. He went for a drive in his car to clear his head and ended up 190 miles away in Illinois. While spending some time in a roadside bar pondering the problem, he arrived at the basic specification. The machine would be an electronic sequence-controlled calculator employing the binary number system and with the capacity to perform Boolean algebra. Instead of magnetic storage, he decided that a more practical solution would be to employ a capacitive memory technology where the presence or absence of charge in a capacitor would represent each binary digit.

Over the following year, Atanasoff worked to complete the detailed design of the machine, much of which was performed in his spare time, and by early 1939 he was ready to begin construction. In March 1939, he applied to the Iowa State College Research Council for funding to cover the cost of building a prototype and was duly awarded a research grant of $650. Atanasoff decided to use most of this grant to hire a research assistant to help with the construction. He asked a colleague in the Electrical Engineering Department to recommend someone suitable and was given the name of Clifford E Berry, a final year electrical engineering degree student who was adept at practical electronics work. Following graduation and a temporary job over the summer months, Berry joined the project in September 1939.

The two men found that they worked very well together and initial progress was swift. Within a few weeks they had not only completed the prototype but had also worked out most of the engineering detail for a full-scale machine. The prototype comprised a binary add-subtract unit coupled to a small rotating memory device with a storage capacity of two 25-bit words. The memory device featured a rotating Bakelite disk fitted with 25 capacitors on each face. These were charged or discharged to represent a 1 or a 0, a process which had to be performed manually as the prototype had no input device. Wiper brushes on both sides of the disk made contact with the outer terminals of each capacitor during rotation, thus allowing the state of the capacitor to be 'read' and its value fed to the add-subtract unit. A single rotation of the disk caused the 25-bit binary number stored on one side of the disk to be added to or subtracted from the number stored on the opposite side, with the arithmetical operation selected manually by a toggle switch on the add-subtract unit. Although simple in construction and incorporating only 13 thermionic valves, successful operation of the prototype proved Atanasoff was on the right track and a demonstration in January 1940 helped to secure a further grant of $700 from the College Research Council to begin construction of a full-scale machine.

The full-scale machine was designed to solve up to 29 simultaneous equations with 29 unknowns through a process of successive elimination, using only subtraction, addition and shifting of digits to reduce the coefficients. To perform this task, the machine contained a total of 30 add-subtract units that operated in parallel to support up to 30 simultaneous operations and featured a 50-bit word length to maintain accuracy throughout the calculation process. Although capable of complex logical operations, the machine's sequence of operation was limited to those steps necessary to carry out the process of successive elimination and was essentially fixed. A numeric keyboard was

4. The Dawn of Electronic Computation

provided to allow the values for calculations to be entered manually, or via a punched card reader that accepted standard IBM punched cards. These were entered as decimals and converted into binary by the machine. Results were converted back into decimal format by the machine and displayed using a set of 15 electromechanical indicator dials.

The capacitor storage technique was again employed for the memory device which took the form of two 8-inch diameter drums rotating on a common shaft, each drum containing banks of miniature paper capacitors that stored 30 numbers of up to 50 bits in length. The drums rotated continually at 60 rpm, giving a basic clock speed of 1 Hz. In order to prevent loss of data due to the charge decaying, a regeneration cycle refreshed each charged capacitor immediately after reading.

The capacitor memory device had insufficient capacity to store intermediate results so Berry designed a secondary storage device. This took the form of a novel punch unit which used an electric spark to burn small holes in specially treated paper at a rate of 60 holes per second. Reading was accomplished by detecting the drop in voltage which occurred across regions of the paper that contained a hole. This unit had the capacity to store the complete contents of a capacitor memory drum on a single sheet of paper.

In keeping with the modest construction budget, the overall design of the machine was very economical. The add-subtract units were implemented using only 7 triodes each and the total number of thermionic valves numbered around 300. The components were housed in a welded steel frame about the same size as a large desk. Although the arithmetic circuitry was fully electronic, the control circuitry involved relays and the 1 Hz clock speed was provided by a mechanical clock driven by an electric motor.

The small research grant obtained from the College was only sufficient to fund the early stages of the development work so in April 1940 Atanasoff approached various companies, including IBM, in an effort to obtain further funding for the project. However, none of the office equipment firms he contacted showed any interest in the project. Only the US electronics firm Raytheon provided some support by supplying thermionic valves free of charge. In August 1940, with funding almost gone, Atanasoff wrote a paper describing the project and sent it off to three research foundations along with a request for a grant of $5,330 to fund the completion of the machine. Fortunately, the Research Corporation, a US foundation for the advancement of science, recognised the potential of the work and the grant was awarded in March 1941.

In early 1942, the machine successfully calculated 2 equations with 10 unknowns, marking an important milestone in the application of digital electronics to high-speed calculation. However, as construction neared completion, the war began to take its toll on the project. Berry had successfully obtained his Master's degree with a dissertation which described his development of the secondary storage device and was now at risk of being drafted into US military service. In July 1942, in an effort to avoid the draft and find a better wartime use of his talents, Berry accepted a war-related job designing mass spectrographs at Consolidated Engineering Corporation, a scientific instrument manufacturer based in Pasadena, California. Shortly afterwards, Atanasoff decided the time was right to make his own contribution to the war effort and transferred to the US Naval Ordnance Laboratory in Washington DC in September 1942 to conduct research on the testing of acoustical mines. With both men now absent, construction came to a grinding halt before the intermediate storage device and punched card reader could be completed and the machine made fully operational.

Close-Up of the Atanasoff-Berry Computer (Photo © Iowa State University)

Neither Atanasoff nor Berry returned to Iowa after the war. Following his appointment as head of a new Computer Division at the Naval Ordnance Laboratory in the summer of 1945, Atanasoff became involved with another computer development project but progress was unsatisfactory and the project was cancelled in late 1946. He eventually moved into industry, co-founding a successful weapons research and engineering firm, but was never again associated with computer development. Berry remained in the field of mass spectrography, rising to become Consolidated Engineering's Assistant Director of Research in 1952, but died in tragic circumstances in October 1963. From then on, Atanasoff insisted that the Iowa machine be referred to as the Atanasoff-Berry Computer (ABC) in honour of Berry and to reflect his major contribution to the machine's development.

4. The Dawn of Electronic Computation

The College had hired a Chicago-based patent lawyer, Richard R Trexler, in July 1941 with the intention of filing a US patent application on the machine but College administrators decided to use a proportion of the Research Corporation grant to cover the patent costs, which Atanasoff resisted, and the application was never completed. On a return visit to Iowa State College in 1948, Atanasoff was dismayed to find that the ABC had been dismantled in order to clear storage space in the basement of the Physics Department where it was situated. The welded steel frame was found to be too wide to go through a doorway which had been added since the machine was first installed so the machine was taken apart and its components discarded or cannibalised. All that remained of the ABC was one of the capacitor memory drums.

With no follow-on work, no patent and no machine, Atanasoff and Berry's achievements went unrecognised until a high-profile patent infringement case in 1967 brought them to public attention. Only at that point did Atanasoff realise that their work may well have had an unintentional but significant influence on the subsequent development of the electronic digital computer through a landmark project initiated at the University of Pennsylvania in June 1943. However, this project would be preceded by some remarkable achievements in electronic computation which took place at a secret government facility on the other side of the Atlantic.

The Colossus of Bletchley Park

In December 1942, a project was initiated by British cryptanalysts which resulted in the development of the first large-scale electronic digital calculator. The project took place at Bletchley Park, a top secret facility operated by the Government Code and Cypher School (GCCS) at a Victorian country mansion and estate situated approximately 40 miles north-west of London. Due to the nature of the work, the project was conducted under the strictest secrecy and the existence of this remarkable machine only came to light more than 30 years after it had been built.

Bletchley Park became the main location for British code-breaking activity in August 1939 and gradually increased in size from a handful of government employees who called themselves Captain Ridley's shooting party in order to disguise their true purpose, to a complement of over 9,000 at its peak. Staff were recruited from a wide range of disciplines and talents, ranging from mathematicians and linguistics experts to crossword puzzle champions. Much of the activity at Bletchley Park concerned the decryption of radio messages sent by the German military which had been encrypted using the Enigma machine, a typewriter-like device that employed a plugboard and rotors to substitute plain text for a seemingly random sequence of letters. The decryption methods developed by the Bletchley cryptanalysts, which were derived from earlier work by the Polish Cipher Bureau in breaking Enigma messages in 1932, were extremely labour-intensive, requiring the checking of many thousands of possible rotor settings in order to determine which one had been used for a particular message. The repetitive nature of this work lent itself to automation and it was only a matter of time before machinery was developed to assist with the process.

One of the people who will be forever associated with Bletchley Park is the mathematician Alan M Turing. Born in London in 1912, Turing acquired a reputation as a brilliant scholar when still a junior academic at Cambridge University in 1936. Turing was among the first of the new recruits to Bletchley Park, arriving on 4 September 1939, the day after Britain declared war on Germany. One of his earliest contributions was the design of an electromechanical logic machine named the Bombe which emulated the Enigma rotors, using a motorised mechanism to step rapidly through every possible combination and apply a logical test to each one to determine if it met the conditions for a possible solution that had been wired into the machine before operation. Turing's Bombe was closely based on an earlier device developed by the Polish Cipher Bureau. However, Turing improved the design by incorporating 36 sets of rotors to allow multiple checks to be made in parallel. The prototype was operational by March 1940 and more than 200 examples of this workhorse machine and its successors were manufactured during the war by the British Tabulating Machine Company.

Heath Robinson

Electromechanical technology, as exemplified by Turing's Bombe, initially served the GCCS cryptanalysts well. However, this situation changed in December 1942 following the introduction of a new cipher machine, the Lorenz SZ-40, which the Germans were using to encrypt teleprinter messages transmitted by radio between command centres of the German Army.

The Lorenz machine was designed to operate as an attachment to a standard teleprinter and automatically encrypted each 5-bit teleprinter character by combining it with an obscuring character using modulo-2 addition. The sequence of obscuring characters used for each message was generated using a pseudo-random process controlled by a set of 12 pinwheels. The pinwheel start positions were manually set by the operator at the start of each message and were critical to the deciphering of messages. The mathematician Max H A Newman, who had been Turing's tutor at Cambridge University, devised a machine to automate a method for finding the pinwheel start positions which had been conceived by his Bletchley Park colleague William T Tutte. As speed of operation was of prime importance, Newman specified that the machine should have electronic circuitry.

The new machine was designed to Newman's specification at the

4. The Dawn of Electronic Computation

Telecommunications Research Establishment (TRE) in Malvern by the inventor of the scale-of-two counter, Charles Wynn-Williams. It operated by comparing messages to a 'keystream', a possible sequence of obscuring characters obtained through manual cryptanalysis techniques, and counting the number of instances where a particular set of Boolean conditions were satisfied. The relative position of the keystream to the message was then incremented by one character and the process repeated until every position had been compared. If the keystream was correct, then the position which yielded the highest count would correspond to the pinwheel start positions.

Wynn-Williams' design incorporated electronic logic circuits and a set of digital counters implemented using Thyratron valves. In order to maintain reliability, Wynn-Williams sought to minimise the total number of thermionic valves by using electromagnetic relays for some of the slower functions. A second counter kept a separate count of the total number of characters processed. The machine was equipped with two of each type of counter, all of which had a capacity of 4 decimal digits to give a maximum value of 9999. Counter values were displayed via a panel of indicator lamps, although a printer was later added to eliminate reading errors. Messages were input to the machine in the form of continuous loops of punched paper tape containing the sequences of 5-bit characters from the Lorenz transmissions. The machine was also equipped with two high-speed photoelectric tape readers, one for the message tape and the other for a tape containing the keystream data.

Wynn-Williams' machine was built by the General Post Office (GPO), the UK government agency responsible for telecommunications, at the Post Office Research Station in Dollis Hill, London, and delivered to Bletchley Park in June 1943. Its bizarre appearance resulted in the machine being named Heath Robinson after the celebrated British cartoonist best known for his drawings of fanciful or ridiculous inventions. However, a problem with maintaining synchronisation between the two punched tapes made the machine unreliable in service, and it achieved only a fraction of its intended operating speed of 2,000 characters per second. Nevertheless, it remained a useful tool for reducing the time taken to decipher Lorenz messages and at least two additional examples were constructed.

Colossus

Thomas H Flowers, a senior GPO electronics engineer based at Dollis Hill, was one of several people consulted in an effort to solve the Heath Robinson synchronisation problem. Flowers had spent almost a decade working on the application of electronic switching technology to telephone exchange equipment and knew that the keystream tape could be eliminated by using electronic processing techniques. Unlike Wynn-Williams, Flowers was also undaunted by the prospect of using large numbers of thermionic valves in an application which demanded high reliability. Experience had taught him that valves were much less likely to fail if they were kept permanently powered up.

Flowers proposed a radical re-engineering of the Heath Robinson machine, eliminating the majority of the electromechanical components with an ambitious design that would require at least 1,500 valves, a huge number which had only been surpassed in experimental telephone exchange equipment that he himself had built. Predictably, his confidence in the operational reliability of such equipment was not shared by GCCS and the proposal was rejected. However, reasoning that necessity would eventually force a change in attitude, Flowers decided to push ahead with development. He obtained the permission of his GPO bosses to proceed and work on the new machine began at Dollis Hill in March 1943.

Flowers was assisted in the design of the machine by two GPO colleagues, Sidney W Broadhurst and William W Chandler. Their design broadly followed the specification laid down by Max Newman for Heath Robinson. It employed a binary adder and logic circuitry for performing Boolean operations on 5-bit character data. However, a bit-stream generator, comprising a bank of 12 Thyratron ring counters of different scales, was developed to generate electronically the data previously held on the keystream tape. This device was controlled by a bank of rotary selector switches. The photoelectric tape reader from Heath Robinson was also redesigned to operate at up to 5,000 characters per second. As the clock pulse for the electronic circuitry was generated from the sprocket holes in the punched tape, the speed of operation of the machine could be easily reduced for testing purposes by simply reducing the tape speed.

A set of four 4-digit decimal counters were implemented using Thyratron valves. Unlike the counters in Heath Robinson, these operated on a biquinary-coded decimal scheme as they could not be made to operate reliably as standard decade counters at the faster speed of operation. The machine was also equipped with a series of logic gates (AND and OR) which could be combined in any sequence using a plugboard. It was also capable of a primitive form of conditional branching where the contents of a counter at the end of each pass of the tape loop could be compared with a preset value set by switches and would only print the total in the counter if it exceeded the preset value. These features gave the new machine a powerful capability that went well beyond that of Heath Robinson.

The machine was completed by early December 1943, the breakneck speed of development made possible by the project having been assigned the highest wartime priority and by the liberal use of off-the-shelf GPO components. It comprised eight equipment racks arranged in two large cabinets which housed the electronics plus a large frame known as the bedstead that contained the high-speed tape reader and a system of pulleys for handling the long loops of message tape. An IBM Electromatic electric typewriter was used for printing results. Codenamed Colossus by the Bletchley Park cryptanalysts on account

4. The Dawn of Electronic Computation

of its physical size, it became operational in February 1944.

Colossus Mark 2

Soon after Colossus went into operation at Bletchley Park, GCCS ordered the construction of an additional 11 machines, with delivery of the first batch of three required within 4 months. Fortunately, Flowers had anticipated just such a request and the development of a production version designated Colossus Mark 2 had been initiated before the prototype was completed. Development was led by Allen W M Coombs, who had joined the Dollis Hill team in September 1943 and who took over the leadership of the project when Flowers moved on to other work following a promotion. Other contributors to the design of the Mark 2 included the statistician I Jack Good and Donald Michie, a classical scholar who would later become an important contributor to the field of artificial intelligence.

With the majority of the critical functions implemented electronically, the limiting factor in the performance of the prototype machine was the maximum speed at which the message tape could be read. Experiments had shown that it was impractical to increase tape speed significantly so the Mark 2 design team adopted a parallel processing approach in order to improve performance. They increased the number of logic units to 5 and added a set of six-stage shift registers in order to buffer the stream of data 5 pairs of consecutive characters at a time. Each character pair was then fed to a separate logic unit and processed simultaneously.

The Mark 2 was also equipped with an additional bedstead for setting up the next message tape without having to halt the operation of the machine. Switches and plugboards were also redesigned to enhance functionality and the opportunity was also taken to make some improvements to the circuitry. The extra logic units plus an additional 4-digit counter brought the total number of thermionic valves in the Mark 2 to around 2,400.

The first Colossus Mark 2 was installed at Bletchley Park in June 1944. A total of nine machines were completed before the war ended and production ceased. The prototype was also upgraded to Mark 2 specification, bringing the total number of fully specified machines to ten. Following the end of the war eight were dismantled and the remaining two machines transferred to the newly established GCCS headquarters at Eastcote on the outskirts of London, which was renamed Government Communications Headquarters (GCHQ) in June 1946. Both were subsequently transferred to Cheltenham following the relocation of GCHQ in 1952. Remarkably, they continued to operate until 1960, after which they were destroyed along with all the design drawings. The UK Government's Official Secrets Act prevented disclosure of the work until the mid 1970s when, under the '30-year rule', certain documents were gradually declassified and placed in the public domain through the UK Public Record

Office. A key figure in this process was the computer historian Brian Randell, who lobbied vigorously for their release.

Colossus Mark 2 and Operators

The development of Colossus was a significant milestone in the history of electronic computation. However, it was a special-purpose machine created to perform Boolean operations on characters represented as 5-bit data and was never intended to perform arithmetical calculations. Colossus was capable in principle of operating as a calculator, given that logic operations are used to perform arithmetic in binary digital calculators, but its dedicated architecture meant that the basic sequence of operation was essentially fixed. Despite a plethora of plugs and switches, sequence control was limited to the redirection of the data stream or counter output values to different parts of the logic circuitry.

Although Colossus was a major step forward, the extreme secrecy under which it was created prevented the project from contributing directly to the evolution of electronic computers. Flowers himself returned to his pre-war work for the GPO on the application of electronics to telephone exchange equipment. Fortunately, several key members of the Colossus development team became involved in civilian computer development projects after the war and their experience in electronic switching and logic circuitry would prove invaluable in facilitating the establishment of a British computer industry.

Despite the huge momentum of the work at Bletchley Park, the first truly sequence-controlled large-scale electronic digital calculator would originate in the USA. Like Colossus, it would also be developed to fulfil a wartime need.

4. The Dawn of Electronic Computation

ENIAC

By the start of World War II, the Moore School of Electrical Engineering at the University of Pennsylvania in Philadelphia had established a reputation as a centre of excellence in computational research. Founded in 1923, the School had built its own differential analyzer in 1934 under the guidance of MIT's Harold Hazen and had developed a close working relationship with the US Army Ordnance Department, which had helped fund the analyzer's construction alongside that of a smaller model for the US Army Ballistic Research Laboratory (BRL) at Aberdeen Proving Ground in Maryland.

In early 1940, BRL set out to update its artillery firing tables in preparation for America's likely entry into the war. These tables provided the settings that gunners require to compensate for the effects of external factors, such as wind speed and air temperature, on the trajectory of an artillery shell. The preparation of artillery firing tables was extremely labour intensive, with a single trajectory taking two full working days to calculate using a mechanical desk calculator. However, this could be reduced to around 30 minutes by using a differential analyzer, excluding the time required to set the machine up. The Aberdeen differential analyzer was quickly put to work but the urgency of the situation prompted BRL to approach the Moore School with a request to use their differential analyzer, the construction of which had been sponsored by the Ordnance Department on the understanding that it could be commandeered for military use in the event of war.

In order to satisfy demand, the Moore School analyzer was put to work on a two-shift system. Mathematics students were also recruited to perform trajectory calculations using mechanical desk calculators and by late 1942, more than 100 students were working in shifts at the Moore School. These were almost exclusively female, due to a shortage of male students caused by the large numbers joining up for military service.

It was this wartime pressure for computing resources which prompted the development of the most important electronic digital calculator of this period. This watershed project would be the vision of two remarkable men, John W Mauchly and J Presper Eckert.

Mauchly and Eckert

John Mauchly was born in Cincinnati, Ohio in 1907. As the son of a physicist, Mauchly was encouraged in his academic pursuits and in 1925 he was awarded a scholarship to study at the Johns Hopkins University School of Engineering in Baltimore. However, he found that engineering offered little to satisfy a growing interest in natural phenomena. After two years he decided to follow in his father's footsteps and transferred to the Physics Department, but took the unusual step of electing to work directly towards a PhD in physics without taking any intermediate degrees. Mauchly's doctoral research was in the field of molecular spectroscopy and he obtained a PhD in 1932. The following year, after a short period working as a research assistant at Johns Hopkins University, he accepted a position as Head of the Physics Department (and its only member of staff) at Ursinus College, a small school in Collegeville, Pennsylvania which specialised in the liberal arts.

At Ursinus, Mauchly became interested in automatic calculating machinery for weather prediction and in 1940 he constructed a harmonic analyser for calculating weather variables as part of his research into the effects of solar activity on precipitation. Eager to learn as much as possible about computing technology, Mauchly attended the influential demonstration of the Bell Labs Complex Number Calculator given by George Stibitz at Dartmouth College in September 1940. In December 1940, while presenting a paper at an American Association for the Advancement of Science meeting, he also met John Atanasoff. Atanasoff told Mauchly of his work to develop an electronic digital calculator at Iowa State College and invited Mauchly to see the work for himself. Mauchly was unable to take up the invitation until June 1941, whereupon he spent several days as a guest of Atanasoff and his family, during which he was shown the partially built ABC.

Mauchly had dabbled with digital electronics but needed to learn more if he was to use it for his computational work. A few weeks after his trip to Iowa, he enrolled on a summer electronics course at the Moore School of Electrical Engineering. The intensive 10-week course had been created at the request of the US government to produce electronic engineers for the war effort by retraining people working in related scientific or technical disciplines. One of the laboratory instructors on the course was Presper Eckert, an Electrical Engineering postgraduate student who undertook occasional teaching duties while studying for his Master's degree.

At 22 years old, Eckert's engineering genius was already evident. He had filed his first patent application the previous year and had also amassed considerable experience in electronic circuit design through his involvement in a string of consultancy projects while still an undergraduate at the Moore School. Mauchly found that he and Eckert shared a common interest in electronic computation. Moreover, they formed a perfect partnership, with Eckert's detailed practical knowledge of electronics providing the ideal sounding board for Mauchly's visionary ideas.

On completion of the course, Mauchly and another attendee, Arthur W Burks, who held a PhD in philosophy and mathematical logic, were offered fixed-term positions as instructors in electrical engineering at the Moore School in order to replace staff who had been called up for military service. Despite the lower academic status of the post and similarly miniscule salary, Mauchly decided to accept the offer on the basis that it might provide opportunities to work on the development of computational systems. There was also the prospect of

4. The Dawn of Electronic Computation

supplementing his income through lucrative consultancy work.

Following his move to Philadelphia, Mauchly began work on a project to improve the design of radar antennae for the US Army Signal Corps. The work involved a large number of repetitive calculations but, despite the Moore School's reputation for computational research, Mauchly was surprised to learn that there was actually very little calculating equipment available within the School. The School's differential analyzer was already fully engaged on artillery firing table calculations for the BRL and would remain so for the foreseeable future. Mauchly came to the conclusion that his needs would be best served by building an electronic computing device. He discussed the feasibility of this idea with Eckert who, informed by his recent experience of digital electronics as part of a project on the timing of radar pulses for the MIT Radiation Laboratory, agreed that a digital solution would be preferable.

The Proposal

In August 1942 Mauchly drafted a memorandum, 'The Use of High Speed Vacuum Tube Devices for Calculating', in which he proposed the construction of an electronic "computor" (sic) that could replace a mechanical differential analyzer in the solution of differential equations. The memorandum described how electronic digital counters could be used to perform basic arithmetical operations on decimal numbers by counting pulses at a rate of 100,000 per second. It emphasised the speed and accuracy of such a device in comparison to a differential analyzer, particularly when applied to the calculation of ballistic trajectories. He submitted the memo to the School's Director of War Research, John G Brainerd, who circulated it to Faculty members for comment but received no response.

By early 1943, the Army's demand for firing tables had increased beyond the ability of both BRL and the Moore School's facilities to cope. Lieutenant Herman H Goldstine, a US Army ballistics expert who held a PhD in mathematics from the University of Chicago, had recently been appointed as BRL liaison officer for the Moore School. Goldstine heard of Mauchly's memorandum from Joseph Chapline, the engineer in charge of the School's differential analyzer. Goldstine recognised a potential solution to the growing backlog of trajectory calculations in Mauchly's proposal and proceeded to champion the project. He brokered preliminary discussions between BRL and the Moore School and asked the School to prepare a more detailed proposal for submission to BRL.

The proposal, entitled 'Report on an Electronic Diff. Analyzer', was written by Mauchly and Eckert with an introduction by Brainerd. It described a fully electronic digital machine which could assume the role of a differential analyzer in the solution of ballistics problems but also possessed the flexibility to be applied to other general-purpose computational tasks. The deliberately ambiguous title referred to the use of the method of differences to solve differential equations, a numerical method similar to that used by Babbage for the Difference Engine. The proposal was presented by the three men at a meeting with BRL and US Army officials on 9 April 1943. Their response was resoundingly positive and within a few weeks the Ordnance Department had issued a $61,700 research and development contract to cover the first six months of a project to build the new machine, which would be called the Electronic Numerical Integrator and Computer (ENIAC).

Project PX

Work on the project, codenamed Project PX, began towards the end of May 1943. Brainerd took on the role of project supervisor and Eckert was appointed as Chief Engineer. However, Mauchly's teaching duties meant that he was only able to contribute to the project in an advisory capacity. The other members of the project team included Mauchly's fellow student on the summer 1941 electronics course, Arthur Burks, plus recent engineering graduates T Kite Sharpless and Robert F Shaw. Joseph Chedaker, who had worked in manufacturing industry before joining the Moore School, was responsible for construction. The team gradually increased in size as other Faculty staff became available and were assigned to the project, eventually numbering approximately 50 people, of which 12 were engineers.

The architectural design of ENIAC was based on the use of multiple decimal accumulators, echoing that of the Harvard Mark I. However, in ENIAC the electromechanical counter wheels were replaced with electronic ring counters. Initial experimentation showed that flip-flops were more capable of operating at higher pulse rates than Thyratron valves so the ring counters were implemented using a set of ten flip-flops plus a carry mechanism which sent a carry pulse to the ring counter representing the next highest digit upon completion of a full cycle. Each accumulator contained ten ring counters, giving a word length of 10 decimal digits, along with an additional circuit for the algebraic sign.

Addition and subtraction were accomplished by simply transferring electronic pulses between two accumulators. A method of repeated addition or subtraction was used for the multiplication and division of integers but separate units were provided to improve performance for real numbers. A high-speed multiplier unit was designed which operated in conjunction with four accumulators to perform multiplication by the accumulation of partial products, based on a technique Mauchly had seen implemented mechanically in a Remington Rand printing multiplier. This unit generated a table of partial products by multiplying the multiplicand by each digit of the multiplier, which were then added together using the accumulators. The divider unit was designed to perform division, also in conjunction with four accumulators, by a method known as non-restoring division, where successive subtractions of the denominator to the numerator's remainder are performed until a change of sign occurs, whereupon the numbers are then summed. This unit was also capable of performing square root operations using a similar technique.

The design of ENIAC incorporated a set of 10 accumulators and these also served as the primary storage medium for intermediate results. They were supplemented by a Function Table unit which was developed in order to store fixed values of commonly used mathematical and ballistic functions. This unit had a capacity of 104 12-digit numbers, which were input manually using banks of rotary selector switches and the values held in a resistor matrix.

Numerical data was entered into ENIAC using a standard IBM punched card reader. To compensate for the relatively slow speed of this device, a unit called the Constant Transmitter was designed to provide temporary storage for up to eight 10-digit numbers using high-reliability electromagnetic relays supplied by Bell Labs. This unit acted as a buffer between the electronics and the electromechanical card reader, allowing numbers held in temporary storage to be fed into the machine at a much faster rate than they could be read from a punched card. A printer and IBM card punch were also provided for data output.

Sequence control was handled by a unit known as the Master Programmer which operated in combination with a plugboard that permitted units to be connected in various ways, the design of which was derived from the mechanical interconnection system of a differential analyzer. The Master Programmer was equipped with a set of 10 'steppers', each equipped with a separate input line, a 2-digit decade counter and 6 output lines. These steppers could be used to control repetitive operations by counting the number of clock pulses that had taken place and halting the operation on reaching a preset value or when the algebraic sign of a specific accumulator had changed. The stepper would then activate the next output line to initiate a new sequence on a different part of the machine and the counter would be automatically reset. Highly complex sequences could be set up by using one stepper to control the operation of several others.

A set of 11 wires carried control pulses around the various units, with selector switches on each unit used to determine the action of that unit on receiving an activation pulse. A similar set of wires also carried the data pulses around the machine. Because the accumulators were self-contained units, this arrangement allowed them to be controlled independently and operated in parallel. A unit known as the Cycling Unit generated clock pulses at 10 microsecond intervals, giving a clock speed of 100 KHz. Overall control of the machine was governed by the Initiating Unit, which contained the controls for turning the power on and off, initiating a computation and clearing the contents of the accumulators, as well as program controls for the punched card reader and printer.

In June 1944, two accumulators and a section of the Cycling Unit were assembled and tested in order to prove the design. The design was then frozen to allow the main construction phase to begin. Construction was expected to be complete by early 1945 but changes to the specification which had been requested by BRL during the design phase had resulted in the addition of two Function Table units and a doubling of the number of accumulators to 20. Problems with external suppliers also delayed progress but by October 1945 all the units had been constructed and final assembly could take place. ENIAC was fully operational by November 1945. The completed machine was arranged as a U-shaped collection of 40 panels in 30 units which together weighed over 30 tonnes and lined the walls of a room 50 feet long by 30 feet wide.

ENIAC Publicity Photograph (Photo © University of Pennsylvania)

Despite numerous delays and persistent criticism from prominent members of the National Defense Research Committee, the US Army's faith in the project remained strong, bolstered by Goldstine's regular progress reports to BRL. Nine supplements extended the initial 6-month contract and took the Ordnance Department's expenditure on the project to a grand total of $486,804, a considerable increase over the original cost estimate of $150,000.

The additional accumulators and Function Table units resulted in the number of thermionic valves used in ENIAC increasing from an estimated 5,000 to an eventual total of 17,468, more than seven times the number used in the Colossus Mark 2. Like their electric light bulb ancestors, standard thermionic valves were designed for a lifespan of around 2,000 hours. However, with such a large number within the same system, one valve would be expected to fail on average every few minutes. Fortunately, Eckert realised from the outset that

4. The Dawn of Electronic Computation

reliability would be a major issue. Following advice from RCA, the company that manufactured the valves, which suggested that reliability would be improved if they were operated under a reduced load, he ensured that ENIAC's electronic circuits were designed for no more than half their rated voltage and a quarter of their rated current.

Another problem with the use of such a large number of thermionic valves was the amount of heat they produced and a large proportion of ENIAC's massive 150 Kilowatts power consumption manifested itself as heat produced by the machine's circuitry. This was dissipated by a forced-air cooling system which used an array of industrial fan blowers to pump air through each unit. The cooling system was designed and supplied by Eggley Engineers, a local engineering firm.

A dedication ceremony was held at the University of Pennsylvania on 15 February 1946 and ENIAC was formally accepted by the Ordnance Department in July of that year. However, the planned move to Aberdeen Proving Ground was postponed due to delays in preparing a new air-conditioned building at BRL in which to house the machine. The need for artillery firing tables had subsided following the end of World War II but the US military had other pressing computational needs and ENIAC was put to work on a succession of complex problems, the first of which was performing design calculations for the hydrogen bomb. The machine was eventually transferred to Aberdeen Proving Ground in January 1947 where it continued to operate until October 1955.

ENIAC's Impact

ENIAC's performance was far superior to any previous sequence-controlled calculator, requiring a mere 200 microseconds for an addition and 2.8 milliseconds for a multiplication. It also possessed the flexibility to cope with a wide range of computational problems. However, reconfiguring the machine for a different set of calculations was a major task which involved the careful rewiring and resetting of hundreds of plugs and switches. Planning and implementing a reconfiguration could take several days.

ENIAC had been designed to support parallel operation but this was very difficult to implement in practice due to the complexity of the task and the need to maintain overall synchronisation. Its flexibility was also limited by the lack of a conditional branch feature that would allow the sequence of operation to be altered dynamically, such as the one implemented in the Bell Labs Model V Relay Calculator which was under development during the same period. However, soon after ENIAC was completed, it was discovered that the machine could be made capable of performing a rudimentary conditional branch by connecting the data lines of one accumulator to the control lines of another in such a way that an operation on the second unit would only be triggered if the output from the first unit was non-zero. By doing so, ENIAC's operations could, in principle, be controlled based on the content of its data.

Following the end of the war, the US military became keen to publicise some of the technological developments it had funded through classified projects and ENIAC's confidential security classification was relaxed. The day before the dedication ceremony, the War Department issued a press release and held a press conference and public demonstration. This generated a large amount of newspaper coverage and spawned several magazine articles in influential publications such as Newsweek and Scientific American.

With its accumulator-based architecture and a clunky sequence control system that betrayed its differential analyzer origins, ENIAC was a somewhat poor example of computer design. However, it was large-scale, electronic and general-purpose, and with Colossus remaining shrouded in secrecy, people naturally assumed that ENIAC was the first successful example of an electronic digital calculator. Hence, it would become the most influential of the early electronic computer development projects but it would also be the most controversial.

A comprehensive 207-page patent application was filed in June 1947, naming Eckert and Mauchly as co-inventors. The application had been drafted at the behest of the Ordnance Department which was keen to see a patent issued in order to ensure the Army's license-free rights to ENIAC. However, the all-encompassing nature of the claims provoked allegations of infringement, firstly from IBM in respect of patents granted on its tabulating machines and then from Bell Labs with its relay calculators. Consequently, the application was not granted until February 1964.

The newly issued ENIAC patent remained contentious. During this period, the patent rights had passed to the Sperry Rand Corporation through a series of acquisitions and mergers. In May 1967, another US computer manufacturer, Honeywell Incorporated, filed a suit against Sperry Rand on the grounds of anti-trust violations and unjustified claims to the invention of the electronic computer. This triggered a protracted legal dispute in which Honeywell's lawyers seized upon Atanasoff's work as evidence of prior art.

Much has been written about the influence Atanasoff's work may or may not have had on Mauchly and it remains a highly controversial subject. Mauchly clearly took inspiration from Atanasoff's example but there is no firm evidence that he actually took any of Atanasoff's ideas. As a research physicist, Mauchly was already aware of developments in digital electronic counters before he met Atanasoff. What is less clear is whether he had considered their use in computation before the meeting.

Although both machines were electronic, ENIAC was markedly different from the ABC. It was decimal rather than binary and computed enumeratively by counting electronic pulses rather than logically, its logic circuitry confined to sequence control. More importantly, it was also general-purpose, with a sequence of operation which could be altered. Despite these significant differences, the presiding Judge, Earl R Larson of the US District Court for Minnesota, decided to rule the ENIAC patent invalid in April 1973. The primary reason given was Atanasoff's prior work and his influence on Mauchly but another factor was a delay in filing (as the hydrogen bomb calculations, which took place more than a year before filing, were not considered to be a test run but 'public use').

Ironically, ENIAC's design deficiencies led directly to the intellectual breakthrough that would herald the next age of automatic computation, the stored-program concept. Unlike Atanasoff, Eckert and Mauchly were to remain at the centre of these developments and their partnership would go on to become of one of the most successful in the history of the computer.

Further Reading

Aspray, W. (ed), Computing Before Computers, Iowa State University Press, Ames, 1990.

Aspray, W., Was Early Entry a Competitive Advantage? – US Universities That Entered Computing in the 1940s, IEEE Annals of the History of Computing 22 (3), 2000, 42-87.

Atanasoff, J. V., Advent of Electronic Digital Computing, IEEE Annals of the History of Computing 6 (3), 1984, 229-282.

Burks, A. W., The invention of the universal electronic computer – how the Electronic Computer Revolution began, Future Generation Computer Systems 18 (7), 2002, 871-892.

Bush, V., Instrumental Analysis, Bulletin of the American Mathematical Society 42 (10), 1936, 649-669.

Coombs, A. W. M., The Making of Colossus, IEEE Annals of the History of Computing 5 (3), 1983, 253-259.

Flowers, T. H., The Design of Colossus, IEEE Annals of the History of Computing 5 (3), 1983, 239-252.

Goldstine, H. H. and Goldstine, A., The Electronic Numerical Integrator and Computer (ENIAC), Mathematical Tables and Other Aids to Computation 2 (15), 1946, 97-110.

Mackintosh, A. R., The First Electronic Computer, Physics Today 40 (3), 1987, 25-32.

McTiernan, C. E., The ENIAC Patent, IEEE Annals of the History of

Computing 20 (2), 1998, 54-80.

Randell, B., The Colossus, in A History of Computing in the Twentieth Century, Academic Press, New York, New York, 1980.

Schreyer, H. T., An Experimental Model of an Electronic Computer, IEEE Annals of the History of Computing 12 (3), 1990, 187-197.

PART TWO
INDUSTRIALISATION

5. The Metamorphosis of the Calculator – Stored-Program Machines

> *"The machines now being made in America and in this country will be 'universal' – if they work at all, that is, they will do every kind of job that can be done by special machines"*, Max H A Newman, March 1948.

A Fundamental Limitation

The pioneering efforts of Zuse, Stibitz and Aiken had led to the creation of fully automatic calculating machines capable of performing long sequences of operations without manual intervention. The adoption of punched tape as an input medium provided a convenient method of storing instruction sequences, allowing them to be easily changed to suit different computational tasks, and the addition of a conditional branch feature significantly improved flexibility by permitting the order in which the instructions were executed to be altered under machine control. The performance limitations imposed by the electromechanical technology employed in these early machines were swept away by the introduction of digital electronics, which brought greatly increased speed of operation. However, it also revealed a fundamental limitation with the design of sequence-controlled calculators.

During the development of Colossus, the GPO engineers had found that the operating speed of the photoelectric tape reader, which also dictated the clock speed of the machine, could not be taken above approximately 9,700 characters per second or the paper tape would begin to disintegrate. This prompted a decision to set the operating speed of the machine to a leisurely 5,000 characters per second despite the ability of the electronic circuitry to perform at faster speeds. The Bletchley Park cryptanalysts were able to compensate for this performance limitation by simply requesting the construction of additional machines but the staff at the Moore School of Electrical Engineering had no such option when building the giant ENIAC. Having realised from the outset that the use of punched media for sequence control would severely constrain the machine's performance, the ENIAC design team chose to return to a more primitive form of control using switches and plugboards. This permitted the full performance of the machine to be attained but it also compromised flexibility and ease of use.

For electronic calculators to become truly powerful, they would need the ability to execute and dynamically alter sequences of instructions, programs, at electronic speeds. The key to achieving this ability was the stored-program concept.

The Stored-Program Concept

The essence of the stored-program concept is the ability to store instructions internally so that they can be executed at the full internal speed of the machine. In order to do this, it is necessary to represent the instructions in the same basic format as data. By doing so, it then becomes possible to perform computations on the instructions themselves in order to modify them or to create entirely new sequences. Known as program self-modification, this important aspect of the stored-program concept was first suggested by Charles Babbage in a brief passage in one of his notebooks dated 9 July 1836. In this passage, Babbage comments on the possibility of using the Analytical Engine to compute a modified set of instructions and output them onto punched cards ready for use in a further computation.

Had Babbage actually succeeded in constructing the Analytical Engine, it seems unlikely that he would have been able to take this concept further, given his apparent lack of awareness of programming. Ada Lovelace might have fared better, given her deeper understanding of the subject. In her notes on the Analytical Engine, she discusses the use of numbers to denote operations and hints at how these might be held in the same part of the machine as numerical data. The mechanism of the Analytical Engine would have prevented this but Ada's ideas may have triggered further developments if she had been given the opportunity to put them into practice.

The stored-program concept was next touched upon by the brilliant mathematician Alan Turing. Several years prior to his wartime work on the design of cryptanalysis machinery at Bletchley Park, Turing had considered the use of machines for computation from a strictly theoretical perspective during his time as a research fellow at King's College, Cambridge. In May 1936, he submitted an academic paper to the London Mathematical Society entitled 'On computable numbers, with an application to the Entscheidungsproblem'. The Entscheidungsproblem, or decision problem, was a problem posed by the German mathematician David Hilbert in 1928 which related to the existence of a general method that could be applied to mathematical problems in order to determine if they are capable of being solved. In his paper, Turing set out to challenge Hilbert's assertion that all mathematical problems are computable by showing that there are problems which are incapable of a definite solution. To illustrate this, he introduced the concept of a universal computing machine which could be applied to any mathematical function that could be represented algorithmically. Turing's abstract machine did not differentiate between instructions and numerical data, both of which were represented as symbols

5. The Metamorphosis of the Calculator – Stored-Program Machines

and stored internally. The machine could also compute a sequence of symbols that, when fed into another Turing machine, would make it replicate the operation of the first.

What Turing had described was a stored-program computer with the ability to perform program self-modification. However, Turing's paper contained no details of how such a machine might be constructed and its relevance was lost on those pioneers of sequence-controlled calculators who were active in the period immediately following publication in November 1936. Nevertheless, the paper served to establish Turing's formidable reputation within the field of mathematical logic and also provided a solid theoretical foundation for later developments.

Around the same time that Turing was writing his paper, electromechanical computing pioneer Konrad Zuse was also pondering the concept of storing instructions within the machine's memory but from an emphatically practical perspective. Zuse described this idea in his first German patent application which was filed in April 1936. The application contained several statements relating to this concept, including "The computation plan can also be stored so that the commands can be transmitted to the control device at the computation phase". Zuse's use of an 8-bit binary code for machine instructions lent itself well to the storage of instructions within memory alongside data. However, the advantages of the concept were much less obvious in electromechanical calculators, where there is little difference in speed between input and calculation, and it was also a much less attractive proposition when memory access speed is already an issue. Consequently, Zuse never took this idea any further and his patent application was also rejected.

The Moore School

The stored-program concept finally came to fruition as a result of the groundbreaking ENIAC project at the University of Pennsylvania's Moore School of Electrical Engineering. Faced with the prospect of using switches and plugboards for sequence control in order to maintain performance, the ENIAC project team were only too aware that their design would have serious limitations. These were eased slightly by the development of the Master Programmer unit which, through its myriad steppers and counters, increased flexibility by permitting complex looping of instructions but reconfiguring the machine for a new set of calculations remained a lengthy and tedious operation. The team were also concerned about the different forms of storage within ENIAC and their varying speeds. ENIAC's limitations were tolerable because the main purpose of the machine was the calculation of ballistic trajectories for artillery firing tables, a highly repetitive task. However, in order for Presper Eckert and John Mauchly's vision of a general-purpose computer to be fully realised, a more flexible method of sequence control would have to be found.

An analysis of a range of computational problems suitable for ENIAC revealed that some problems would benefit from additional accumulators whereas others would benefit more from extra storage capacity. The restrictions imposed by ENIAC's accumulator-based decimal arithmetic and the machine's inability to perform logic operations were also apparent. Eckert and Mauchly were becoming aware of the need for a problem-independent configuration which could cope with a wide variety of computational tasks. A common storage area capable of storing operating instructions and the contents of the Function Table units alongside numerical data would also be of great benefit.

Some of these ideas were set out in an internal memorandum written by Eckert in January 1944 under the title 'Disclosure of Magnetic Calculating Machine'. Eckert's 3-page document described a small-scale electronic calculator with binary arithmetic and various forms of temporary or permanent storage involving magnetic memory devices with rotating drums or disks. What made this unassuming little document important historically was that it also suggested that instruction sequences and function values could be stored internally by means of the same devices.

In July 1944, with the design of ENIAC now frozen, it was agreed that the team would turn their attention to the development of a new machine as a vehicle for these ideas, construction work on ENIAC permitting. Designed on a similarly grand scale to ENIAC, the new machine would feature a binary architecture with a 32-bit word length, a 1 MHz clock speed and provision for 1,000 words of electronic storage. The following month Ballistic Research Laboratory liaison officer Herman Goldstine proposed to BRL management that additional funds be granted to the Moore School for research and development work aimed at constructing the new machine. Goldstine's proposal was accepted and in October 1944, the US Army Ordnance Department issued a $105,600 supplement to the ENIAC contract to allow this work to begin. The new machine would be called the Electronic Discrete Variable Automatic Calculator (EDVAC).

A central figure in the development of EDVAC was the Hungarian mathematician and chemical engineer, John von Neumann. A brilliant scholar with a charming personality and an uncanny ability to perform complex calculations in his head, von Neumann had moved to the USA in 1930 to take up a teaching position in mathematical physics at Princeton University in New Jersey. In 1933 he was appointed alongside Albert Einstein as a one of five inaugural Professors of Mathematics at the recently established Institute for Advanced Study (IAS), an independent research institution also located in Princeton. As von Neumann's scientific reputation grew, so too did his influence within US government and academic circles, and during World War II he was regularly called upon to advise on a range of scientific matters. This culminated in an invitation in 1943 to participate in the effort to develop the first atomic bomb, the Manhattan Project. Von Neumann's crucial contribution was in solving

5. The Metamorphosis of the Calculator – Stored-Program Machines

the complex design calculations for the implosion which triggered detonation.

Von Neumann learned of the existence of ENIAC in August 1944 through a chance conversation struck up with Goldstine while waiting for a train at Aberdeen railway platform after attending a scientific advisory committee meeting at BRL. Recognising the great mathematician from lectures that he had attended, Goldstine approached von Neumann and introduced himself. Von Neumann had acquired a keen interest in computing technology through his work on the Manhattan Project and the conversation soon turned to Goldstine's involvement in the development of ENIAC. On hearing that the new machine would be capable of automatic calculation at previously unheard of speeds, von Neumann made it his business to find out more. He obtained security clearance to visit the Moore School to see the machine for himself, which he did for the first time in early September 1944. Following this visit, he offered his services free-of-charge as a consultant to the Moore School team.

Von Neumann began making regular consultancy visits to the Moore School. With ENIAC already under construction, he became actively involved in the technical discussions surrounding the design of the new machine. His initial effect on the team seems to have been a positive one and significant progress was made over the following weeks. A series of four full-day project meetings were held in March and April of 1945 during which much of the design of EDVAC was finalised. The following month von Neumann sent Goldstine a 101-page handwritten document entitled 'First Draft of a Report on the EDVAC' as a summary of the technical discussions.

Von Neumann's report eloquently described the principal elements, logical structure and instruction set of the new machine. With his extraordinary intellect and enviable talent for assimilating and presenting information, von Neumann was able to express the stored-program concept with a clarity of thought that had been absent from earlier musings on the subject, using the analogy of neurons in a human brain borrowed from recently published work in the field of neural network modelling to help explain the operation of the machine's binary circuits. Von Neumann may also have been influenced by the work of Alan Turing, although the report does not reference Turing's 1936 paper directly, as he was certainly familiar with it and is known to have had a number of secret meetings with Turing during the war. What was entirely original, however, was the introduction of a practical method of program self-modification through the use of an address-substitution mechanism which allowed the instruction sequence to change its own memory address references during execution. Implemented by means of a dedicated machine instruction, this would allow a program to jump to a different location in memory, the new location given by the result of a calculation, an important feature when dealing with large arrays of numbers.

Although intended by von Neumann as a working draft, Goldstine thought so highly of the document that he decided to have it typed up in its incomplete form, with sections on programming and the input/output system missing, no cross references and only von Neumann's name on it. He then issued 32 copies to project team members and other Moore School staff on 25 June. Unsurprisingly, this provoked considerable resentment within the project team, with several members of the team unhappy that the report failed to acknowledge their contributions. The strongest reaction came from Eckert and Mauchly, who both felt that von Neumann had deliberately taken their ideas and passed them off as his own. To make matters worse, Goldstine also began to distribute copies of the report to interested parties from beyond the Moore School, despite the project's confidential security classification. Through such recipients, copies of the report became widely available throughout both the US and the UK research communities. Goldstine's motivation for doing so may have been to ensure that the ideas would become public domain and could not be patented, as was the intention with ENIAC. In any case, his actions fuelled the fire of yet another controversy concerning Eckert and Mauchly.

In fairness to von Neumann, he never actually claimed to have invented the stored-program concept but he appears to have declined a later opportunity to set the record straight and with no other names appearing on the 'First Draft of a Report on the EDVAC', his is the name that has become most strongly associated with it. An example of this is the term 'von Neumann Architecture', which was later coined to describe machines that store data and instructions within the same memory and process them sequentially (in contrast to Harvard Architecture machines in which data and instructions are held separately and can be accessed in parallel). Following von Neumann's untimely death from cancer at the age of 53 in February 1957, Goldstine continued to make the case that his mathematical hero was the originator of the stored-program concept, most notably in the book 'The Computer from Pascal to von Neumann' which was published in 1980.

The stored-program concept was the final piece in the architectural jigsaw that would turn calculators into fully-fledged computers capable of carrying out all manner of complex tasks. However, realisation of the concept would also require the development of new high-speed memory technology, a far from trivial task.

Building Blocks – High-Speed Memory Devices

The electromagnetic relays and electromechanical counter wheels used for storage in the early machines were slow and unreliable. As performance soared following the adoption of electronic circuitry, the need for memory devices that could operate reliably at electronic speeds became increasingly important. The advent of stored-program machines created a demand for memory

5. The Metamorphosis of the Calculator – Stored-Program Machines

devices that were not only high-speed but also high capacity. Hence, memory devices that could meet this specification became the subject of considerable development effort in the race to create the first stored-program machines and their characteristics shaped the design of these machines.

Colossus and ENIAC had shown that thermionic valves, when configured as Thyratron ring counters or flip-flops, could be used to provide high-speed electronic storage but literally thousands of valves were required for only a few words of memory. The prospect of adding thousands of extra valves to machines whose operational reliability was already in question due to the large number of valves used in their electronic circuits was not appealing. Thermionic valves were also heat generating, power hungry and costly, and these factors also counted against the use of large numbers of additional valves as main memory in stored-program machines. A more reliable but less expensive solution was needed.

The initial breakthrough in high-speed memory technology would come out of the intense effort devoted to the development of radar on both sides of the Atlantic during World War II.

The Mercury Delay Line

The first high-speed memory device to gain widespread acceptance was the mercury delay line. A mercury delay line consists of a hollow metal cylinder filled with liquid mercury and fitted with quartz crystal acoustic transducers at each end. The transducer at one end converts electronic pulses into sound waves which propagate through the mercury. On reaching the opposite end of the cylinder, the sound waves are detected by the second transducer and the pulses amplified, reshaped and fed back into the delay line, thus creating a sustainable pattern of electronic pulses that can be held in the device for as long as necessary. These pulses can be made to correspond to a bit pattern representing a binary number, the value of which can be altered by changing the pattern of electronic pulses fed into the delay line. Storage capacity is dependent upon the length of the cylinder and the frequency at which the pulses are generated, with higher capacity delay lines of around 1 metre in length capable of storing up to 1,000 pulses simultaneously.

The first acoustic delay line was developed by the eminent physicist William B Shockley at Bell Telephone Laboratories in the late 1930s as a device for implementing a predictable delay in information transmission applications. It was refined by Presper Eckert and fellow ENIAC project team member Kite Sharpless in 1943 as part of a contract research project on the timing of radar signals conducted at the Moore School for the MIT Radiation Laboratory. By replacing Shockley's water and ethylene glycol mixture with mercury and adding a non-reflective backing to the quartz crystal transducers, Eckert and Sharpless were able to significantly improve stability and increase the

The Story of the Computer

maximum frequency at which the device could operate.

Mercury Delay Line Memory (Photo © Corbis)

Despite having proposed the use of rotating magnetic memory devices for high-speed storage in his January 1944 memorandum, Eckert became concerned that mechanical technology might limit performance due to access speed. Recalling the work he had done the previous year on delay lines for radar, he set out with another Moore School colleague, C Bradford Sheppard, to further develop the mercury delay line for use as a computer memory device. However, the delay lines developed for telecommunications and radar applications were simple time delay devices and lacked the ability to store information for longer than the time it took for the sound waves to decay. In order for mercury delay lines to be suitable for use in computers, they would require the ability to sustain a pattern of pulses indefinitely. This was accomplished by the introduction of a regeneration circuit which reconditioned the decayed pulses and fed them back into the device.

Having added a regeneration circuit, Eckert and Sheppard also had to overcome several engineering problems in order to turn the mercury delay line into a practical computer storage device. Because the speed of sound through mercury varies with temperature, the device had to be tightly temperature controlled using heating elements operating under thermostatic control. This necessitated a lengthy warm up period for the device to become stable before use. Any residual errors in timing caused by inaccurate temperature control were corrected electronically using a standard timing pulse. Cylinder length, which determined storage capacity, was also limited due to the sound waves

5. The Metamorphosis of the Calculator – Stored-Program Machines

decaying as they travelled through the mercury. However, the most serious limitation related to data access, as the serial nature of the device meant that individual pulses could not be read on demand. The computer would have to wait for the entire sequence of pulses ahead of the desired pulse to reach the end of the cylinder before access to an individual pulse could be obtained. This also determined the access time of the device, which varied depending upon the position of the pulse in the sequence and could be as much as 1 millisecond on higher storage capacity devices. Despite these shortcomings, the mercury delay line was sufficiently reliable and cost-effective that it became the most popular of the early high-speed memory devices.

Magnetostrictive Delay Lines

A variation on the design of the delay line, which was developed by the British computer pioneer Andrew D Booth in 1951, replaced the mercury filled cylinder with a length of solid nickel rod. Pulses were propagated electromagnetically rather than acoustically by means of the magnetostrictive effect in which ferromagnetic materials change shape by tiny amounts when subjected to a magnetic field. This removed the need for bulky mercury cylinders which improved robustness and significantly reduced the cost of manufacture. Later developments replaced the nickel rod with a coil of wire, further reducing the size of the device. The low cost and compactness of magnetostrictive delay line devices would ensure their continued popularity for more than two decades and they were still being used for temporary storage in devices such as display terminals well into the 1970s.

Electrostatic Storage Tubes

The limitations of the mercury delay line were largely overcome by the development of memory devices based on the principle of electrostatic storage. The main component in electrostatic storage is the cathode ray tube (CRT), a type of thermionic valve invented by the German physicist Karl Ferdinand Braun in 1897 and further developed in the 1920s for use in electronic television cameras and display screens. CRT displays operate by firing a beam of electrons at a phosphor coated glass screen, creating a spot of light where the beam strikes the screen. By controlling the direction and intensity of the beam, complex patterns such as television pictures can be produced. The action of electrons striking the screen also creates an electrostatically charged area around the spot which persists for a fraction of a second before decaying. It is this property of CRTs that makes them suitable for use as a storage device. By detecting and continually refreshing these charged spots, it is possible to store binary data, with the presence or absence of a spot denoting the binary value. A single CRT can store a relatively large amount of data simply by filling the screen with a 2-dimensional array of spots. Furthermore, as individual spots can be created and detected independently of their neighbours, electrostatic storage technology is not constrained by the need to access data serially.

The earliest work on electrostatic storage was carried out at the MIT Radiation Laboratory in the early 1940s as part of a joint effort by US and British scientists to develop microwave radar for the war effort. Standard CRTs were adapted to provide short-term non-regenerative storage for the cancellation of ground echoes in radar systems. Experiments in electrostatic storage based on the Iconoscope, a type of CRT developed by television pioneer Vladimir K Zworykin for use as a television camera tube which could differentiate 500 by 500 individual picture elements, were also conducted at the Moore School by Eckert and Sheppard in 1945. However, the first practical realisation of electrostatic storage was achieved by the British electrical engineers Frederic C Williams and Tom Kilburn while working at the UK Telecommunications Research Establishment (TRE) in 1946.

The Williams Tube

Frederic Calland Williams was born in the village of Romiley near Stockport in 1911. In 1929, he won a scholarship to study electrical engineering at nearby Manchester University, where he obtained bachelor's and master's degrees. Following a brief period as a college apprentice with the Metropolitan-Vickers Electrical Company, he obtained a second scholarship, this time to Oxford University where he was awarded a DPhil in 1936. He returned to Manchester University in September 1936 to take up a post as an assistant lecturer in the Department of Engineering.

In early 1939 Williams was recruited by the British government to work on radar research at the Royal Air Force's Bawdsey Research Station in Suffolk. As the work on radar grew in importance and the number of researchers increased, the radar research group was relocated, first to the village of Worth Matravers in Dorset in May 1940 and then to Malvern in Worcestershire in August 1942 whereupon it was renamed the Telecommunications Research Establishment. Williams remained at TRE for the duration of the war, rising to become the leader of a group that served as trouble-shooters for electronic design problems which other engineers were unable to solve.

Williams garnered an international reputation for his wartime work at TRE and this led to an invitation from the MIT Radiation Laboratory to co-edit two volumes in an important series of textbooks on electrical engineering. As part of this work, Williams visited MIT in November 1945 and then again in June 1946 where he learned of the Radiation Laboratory's work on the use of CRTs for radar echo cancellation. During his second trip, he also visited the Moore School where he was shown ENIAC and was impressed by the scale and reliability of the machine. Although his primary interest remained radar, Williams was now also becoming aware of the needs of computer developers for high-speed memory devices.

5. The Metamorphosis of the Calculator – Stored-Program Machines

On his return to TRE in July 1946, Williams began experimenting with a pair of CRTs and found that it was possible to regenerate data by continually transferring it from one CRT to the other. He then succeeded in eliminating one of the CRTs by attaching a wire mesh pick-up plate to the outer face of the CRT screen which sensed the presence of the charged spots through an increase in capacitance that occurred in the plate. This signal was fed into a regeneration circuit which ensured that the spots were refreshed for as long as necessary in a similar manner to the regeneration of pulses in mercury delay lines.

Williams was assisted in this work by Tom Kilburn, a young Cambridge mathematics graduate who had joined TRE in September 1942 after completing a 6-week crash course in electronics. By November 1946, Williams and Kilburn had succeeded in storing a single bit of binary data. The following month Williams recorded his ideas in the first of a series of British and US patent applications covering electrostatic storage. News of their achievement soon reached Colossus architect Max Newman who, following his release from war duties at Bletchley Park in September 1945, had moved to Manchester University to take up a prestigious appointment as Fielden Professor of Pure Mathematics. Newman had been awarded a substantial grant of £35,000 from the Royal Society in the summer of 1946 for the creation of a calculating machine laboratory at Manchester, the bulk of which would be used to fund the construction of a stored-program computer for mathematical research. He needed someone with the requisite engineering expertise to take on this challenging project and was given Williams' name by the experimental physicist Patrick M S Blackett, who had worked with Williams at Manchester University before the war and had followed his career with great interest. Blackett encouraged Williams to apply for the recently vacated position of Edward Stocks Massey Professor of Electro-Technics at the University. With both Blackett and Newman on the appointment committee, Williams secured the position with little difficulty.

On taking up his new post in January 1947, Williams was able to persuade his former employers at TRE to continue their support for the work on electrostatic storage. They agreed to second Kilburn to the University for an initial period of one year and to provide the necessary materials and equipment to allow the two men to continue their efforts. At Manchester, Kilburn was now free to work on electrostatic storage full time and he began to play a more prominent part in its development. He also took advantage of his move to academia by registering for a PhD under Williams' supervision. Kilburn was assisted by another TRE secondee, Arthur Marsh, who left after a few months and was replaced in June 1947 by Geoff C Tootill.

The next step in the development process was to increase the number of bits of data that could be stored on a single electrostatic storage tube. However, this would require a reliable method of discharging, or erasing, a charged spot without affecting adjacent spots. Kilburn spent some time exploring various alternatives for the shape and pattern of the spots in an effort to find a suitable method, including one that involved focusing and defocusing the spot, before settling on a dot-dash scheme, where the spot that was required to be erased would be extended in shape from a dot to a dash, thus dissipating the charged area.

By November 1947, Williams and Kilburn had succeeded in storing 2,048 bits of data for several hours on a standard 6-inch diameter CRT. The following month, Kilburn submitted a progress report to TRE management in an effort to secure a second year of secondment to Manchester. His report, entitled 'A Storage System for use with Binary-Digital Computing Machines', was also circulated internally within the Department of Electro-Technics and a number of copies were given to interested external parties, with the result that the information was disseminated widely throughout Britain and the USA.

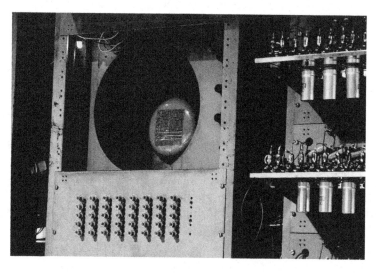

Williams Tube (Photo © Ben Green)

The principal advantage of electrostatic storage over delay line memory was the ability to access individual bits of data, an attribute known as random access. This improved access time considerably and also opened up the possibility of parallel transfer, where entire words of memory could be accessed simultaneously by spreading the data over a number of storage tubes and using a separate tube for each bit of the word. The technology was not without problems, however, and various practical issues had to be overcome, not least of which was a strong susceptibility to electromagnetic interference which resulted in the bit pattern changing randomly when triggered by external factors such as passing vehicles. This was solved by encasing the tube in a metal housing which screened it from the external interference. The use of

5. The Metamorphosis of the Calculator – Stored-Program Machines

standard CRTs intended for television use also revealed wide variations in the quality of manufacture, particularly in the uniformity of the phosphor coating, which compromised reliability and required the tube to be replaced regularly.

Despite continued concerns over reliability, the Williams Tube became the memory device of choice for many of the early high-performance computer development projects and the technology was successfully licensed to a number of external organisations by Manchester University, most notably to IBM in July 1949 where it was further developed for use in the company's first stored-program computer. Electrostatic storage technology was also taken up by several of the electronic component manufacturers in the US and commercial storage tubes were introduced by both RCA and Raytheon.

The Dissemination of the Stored-Program Concept

Following the successful completion of ENIAC towards the end of 1945, the Dean of the Moore School of Electrical Engineering, Harold Pender, began making plans to capitalise on the School's growing reputation as the leading research institution in the exciting new field of electronic digital computers. In January 1946, he appointed Irven Travis as Supervisor of Research. Travis, who had been an Assistant Professor at the School before the war, set about his new role by establishing a more businesslike approach which was intended to improve the poor financial management of the School and put an end to the bitter arguments over intellectual property that had arisen following the issuing of von Neumann's draft EDVAC report.

One of the reasons behind Eckert and Mauchly's strength of feeling regarding intellectual property rights was their growing awareness of the commercial opportunities for electronic digital computers. Mauchly in particular, his interest in automatic calculating machinery having stemmed originally from research into weather prediction, was becoming increasingly convinced of the commercial potential of computers in statistical applications. During consultancy trips to Washington DC in 1944 and 1945, he had initiated exploratory discussions with the US Weather Bureau and the US Census Bureau, both of which had been very receptive to the idea of developing a computer for processing statistical data.

In March 1946, Travis introduced a new patent policy that required Moore School research staff to assign inventors' rights to the university, thereby relinquishing any opportunity to derive financial benefit from their work. This was an unusually prescriptive policy but one that Travis believed was necessary in order to align staff terms and conditions of employment with the School's contractual obligations to grant patent rights to the US government on government sponsored research. However, it was clearly at odds with Eckert and Mauchly's growing business interests so when the Moore School staff were subsequently requested to sign a patent release form, Eckert, Mauchly and several of their colleagues refused to sign it. Eckert and Mauchly met with Travis several times in an attempt to resolve the issue but Travis refused to back down and within weeks both men had tendered their resignations.

The Moore School Lectures

The distribution of von Neumann's draft EDVAC report followed by the public announcement of ENIAC in February 1946 generated a huge amount of interest in the work of the Moore School. In keeping with the School's role as a place of learning, Pender and Travis decided that the best way to respond to this would be to organise a series of lectures. The lectures would be aimed mainly at the growing computer research community and would also involve presenters from other pioneering institutions in order to give a more complete picture of developments in the field. To cover the fees and travel expenses of the presenters, Pender and Travis turned to their benefactors in the US military, successfully securing the sponsorship of the US Army Ordnance Department and the Office of Naval Research.

Entitled 'Theory and Techniques for the Design of Electronic Digital Computers', 48 lectures were held over an 8-week period from 8 July to 31 August 1946. These took the form of a summer school in which formal lectures were held on weekday mornings, with the afternoons set aside for informal seminars. Topics ranged from machine architecture, circuit design and electronic components through to programming techniques and applications.

Eckert, Mauchly, Goldstine and other members of the EDVAC project team delivered the majority of the lectures. However, Eckert and Mauchly were no longer employed by the School, having left in March to pursue their commercial interests, so they had to be paid a fee in order to secure their participation. Von Neumann, who by this time had returned to his research activities at the Institute for Advanced Study, also played his part, delivering a lecture on new problems and approaches. Presenters from other institutions included two of the giants of electromechanical computing, Howard Aiken from Harvard University and George Stibitz, who was by this time operating as an independent consultant, having left Bell Telephone Laboratories in 1945. Another external participant was Calvin N Mooers from the US Naval Ordnance Laboratory who lectured on the Laboratory's ultimately unsuccessful project to develop its own computer which was led by another computing pioneer, John Atanasoff.

The UK was represented by the influential mathematical physicist Douglas R Hartree, who lectured on the use of computer machinery to solve problems in applied mathematics. Hartree was well qualified to do so. In 1934, he and his research student, Arthur Porter, had built a Differential Analyzer at Manchester University. Although based on Vannevar Bush's design, it was constructed using the children's model construction set Meccano in order to

5. The Metamorphosis of the Calculator – Stored-Program Machines

minimise costs. Hartree had also been one of the first civilian users of ENIAC, using the machine in April 1946 to perform fluid dynamics calculations, and had subsequently been invited by the US government to advise on non-military applications of ENIAC.

Attendance at the lecture series was strictly by invitation only. With the US military footing the bill, the 28 attendees comprised mainly researchers from US government agencies and defence contractors plus a handful of research staff from MIT and two academics from the UK, David Rees from Max Newman's group at Manchester University and Maurice V Wilkes from the University of Cambridge. The attendees also included the originator of switching circuit theory, Claude Shannon, who had moved to Bell Labs from MIT in 1941.

The Moore School Lectures were highly effective in disseminating information on the latest developments in automatic computation, not least of which was the stored-program concept. They came at exactly the right time, pointing the way for computer development projects that had recently begun at MIT and Manchester University, and also serving as the catalyst for important new developments which would take place at the US National Bureau of Standards and at Cambridge University. Although the emphasis was firmly on the development of the technology, course attendees were also made keenly aware of the commercial potential of computers through the opinions of Mauchly, Eckert and their associates.

The First Stored-Program Machines

With the conceptual groundwork laid and the engineering building blocks firmly in place, the way was now clear for the development of stored-program computers. The next few years would witness an explosion of activity in the race to construct the first stored-program machine. However, EDVAC would not be the first machine to realise this fundamental concept. Goldstine's evangelical efforts to disseminate von Neumann's report and the Moore School Lectures had helped to plant the seeds of a number of stored-program computer development projects both in the US and the UK, and several of these would bear fruit before EDVAC.

The Fate of EDVAC

The departure of Eckert and Mauchly in March 1946 was a major blow to the EDVAC project and one from which it never fully recovered. Following their departure, another founder member of the ENIAC project team, T Kite Sharpless, was appointed as project leader but he only remained until September the following year before leaving to take up a full-time position with the Technitrol Engineering Company, a commercial venture that he had co-founded with three Moore School colleagues a few months earlier. By this time most of the other original members of the team had also drifted away and stability did not return to the team until the appointment of Richard L Snyder as project manager in 1948. However, the numerous changes in personnel had resulted in a severe loss of momentum and EDVAC was not fully operational until February 1952, more than seven years after the project had begun.

The IAS Electronic Computer Project

By the time Eckert and Mauchly had left the Moore School, John von Neumann had also lost interest in the EDVAC project, having decided in the summer of 1945 that he would build his own stored-program computer for scientific and mathematical research. However, the Institute for Advanced Study, as its name might suggest, had a strict policy forbidding experimental work and no facilities for construction, so he began by looking for another institution with which to collaborate. Two potential partners, MIT and the University of Chicago, also happened to be seeking to revitalise their mathematical research capabilities which had languished during the war. Both now offered von Neumann senior academic positions that included complete freedom to engage in new projects of his own choosing, which put von Neumann in a strong bargaining position with his employers. He told the IAS trustees that he would leave unless they allowed him to build a computer. Anxious not to lose such a valuable member of staff to a rival institution, the trustees relented and the project was approved in October 1945, along with a generous contribution of $100,000 towards the $300,000 estimated cost of construction.

Von Neumann's next task was to secure the additional funding necessary for the construction of the new machine. His first port of call was the Rockefeller Foundation, the charitable body that had sponsored the development of the Rockefeller Differential Analyzer at MIT, but the Foundation had made the rebuilding of post-war Europe its top priority and was no longer interested in funding speculative technology development projects. Von Neumann also initiated discussions with Vladimir Zworykin at RCA Laboratories in Princeton regarding the development of a high-speed memory for the machine based on Zworykin's Iconoscope camera tube. For a time it looked like RCA might also be willing to contribute funding to the project, however, this did not materialise and the company's input, while substantial, would be technical rather than financial.

Von Neumann then turned to his influential friends in the US military. In a series of discussions with representatives of various US military organisations, he presented a compelling case for the potential of electronic computers in applications such as atomic weapons research and meteorology. He was aided in this activity by Herman Goldstine, who had arrived at the IAS in March 1946 to take up von Neumann's offer of a position as Assistant Director of the project. Their combined efforts resulted in the surprise agreement of the

5. The Metamorphosis of the Calculator – Stored-Program Machines

US Army Ordnance Department to support the project, despite an ongoing commitment to the work at the Moore School and in the knowledge that the new machine would not be handed over after completion. Instead, it would serve as a prototype for a number of similar computers to be built by a selection of US government agencies. The Ordnance Department was eventually joined in its support for the project by the Office of Naval Research and the Office of Air Research through a tri-service agreement that replaced the original contract.

Having secured the necessary funding for the project, the next step was to assemble a suitable project team. Von Neumann and Goldstine knew from firsthand experience that some of the most experienced computer engineers in the world were to be found at the Moore School. They also knew that the EDVAC team was in the process of breaking up so they moved quickly to take advantage of the situation. Seemingly oblivious to their own key roles in the break up and the ill feeling that their actions had caused, they decided to approach Eckert himself and offer him the position of Chief Engineer on the IAS project. Eckert inevitably declined but he took some time in giving his answer. In the meantime, they secured the services of Arthur Burks, one of the few members of the EDVAC team who had sided with von Neumann and Goldstine in the dispute over intellectual property, and the three men set about the task of drafting a design specification for the new machine.

Their goal was to create a machine which could handle computational problems several orders of magnitude more complex than was possible using existing computing machinery. To accomplish this, they set out plans for a state-of-the-art electronic stored-program machine with a binary architecture, a generous 40-bit word length and an ambitiously capacious 4,096 word memory. Memory would be implemented using a new type of electrostatic storage tube under development at RCA Laboratories as part of RCA's contribution to the project. The machine would be equipped with 40 such tubes, one for each bit of a word, to facilitate parallel access. To supplement this high-speed memory, they also proposed the development of an auxiliary storage device which would use magnetic wire or tape as the storage medium.

The design specification was completed by June 1946 and published in the form of a weighty IAS report entitled 'Preliminary Discussion of the Logical Design of an Electronic Computing Instrument'. That same month Julian H Bigelow, an electrical engineer and mathematician, was appointed as Chief Engineer for the project on the recommendation of the eminent mathematician Norbert Wiener, who had worked with Bigelow on anti-aircraft fire control systems at MIT during World War II. The project team was completed by the appointment of James H Pomerene and Willis H Ware, two highly experienced electrical engineers who had worked on radar technology at the Hazeltine Electronics Corporation during the war, plus Ralph J Slutz, an MIT graduate who had recently obtained a PhD in theoretical physics at Princeton University, and a further two ex-EDVAC team members, Robert F Shaw and Jack Davis.

With the project team now in place and the design specification drawn up, the project entered the construction phase. The first year was spent developing the principal components and associated test equipment. In the absence of any laboratory facilities, construction began in the boiler room of the IAS main building until a new building was provided in which to house the project in the spring of 1947 and by the end of the second year the team had successfully completed the design and construction of the arithmetic unit. Further progress would now be dependent on the availability of a working high-speed memory, the responsibility for which fell to RCA Laboratories.

Work had started on the development of an electrostatic storage tube for the IAS machine at RCA Laboratories in late 1945 and was led by the British-born electrical engineer Jan A Rajchman. Rajchman had worked on electronic analogue computers for fire control during the war and had also contributed to the development of various components for digital computers, including ENIAC's Function Table units. Casting his experienced eye over conventional CRTs, Rajchman could see that a weakness in their design was a reliance on the consistency of beam deflection for accurate spot placement. He conceived a new design of tube which eliminated the need for consistency in beam control by bombarding the whole screen with electrons and using a grid of electrodes placed in the path of the beam to block the flow of electrons to every 'window' in the grid except the one in front of the required spot position. Selection of individual spots then became a purely digital process, which also simplified the control circuitry. This new design was named the selective electrostatic storage tube or 'Selectron'.

The design of the Selectron offered improved reliability and enhanced performance over the Williams Tube. However, the introduction of large numbers of additional electrodes into the delicate glass envelope created unprecedented manufacturing difficulties and development took much longer than expected. Rajchman had originally planned to store 4,096 bits within a single 3-inch diameter tube but this proved too ambitious, despite the use of a novel method of addressing individual spots which he devised to reduce the number of control wires entering the tube, and the storage capacity had to be reduced to 256 bits in an effort to maintain the pace of development. By the spring of 1948, it was becoming painfully obvious that a working electrostatic storage device was still a long way off. The IAS project team now had no choice but to explore possible alternatives for their high-speed memory.

It was at this point that the IAS team learned of Williams and Kilburn's achievements in electrostatic storage using conventional CRTs. Goldstine had recently obtained a copy of Kilburn's TRE progress report from Douglas Hartree and had circulated it to members of the team. They immediately recognised a potential solution so Goldstine arranged for Bigelow to visit Manchester to see the work for himself while Pomerene would remain in Princeton and begin experiments aimed at replicating it. Within a few weeks Pomerene

5. The Metamorphosis of the Calculator – Stored-Program Machines

had succeeded in storing 16 bits, proving to the team that the Williams Tube was indeed a practical alternative to the Selectron. His efforts gave the green light to an intensive period of activity to construct a full-scale Williams Tube memory. Despite encountering the same problems with electromagnetic interference and CRT manufacturing quality that had plagued Williams and Kilburn, two 1024-bit devices based on standard 5-inch diameter oscilloscope tubes were operating successfully by July 1949 and the entire 40 tube memory was completed by January 1950. The final memory capacity was 1,024 words, only a quarter of that originally specified due to the use of 1024-bit storage tubes instead of 4096-bit tubes.

The change of storage device from Selectron to Williams Tube presented an additional engineering challenge due to the absence of a system clock in the design of the IAS machine. Bigelow had chosen to use a method of asynchronous timing for the machine, where successive actions are triggered by the completion of the preceding one. Unlike the Selectron, Williams Tubes required a regeneration cycle to continually refresh the memory, the timing for which would normally be provided by the machine's clock pulse. This was solved by the provision of a separate 100 KHz clock for the memory and the development of logical interlock circuitry that protected the memory from interruption by a request from the arithmetic unit until the regeneration cycle had been completed.

With the hardware development in the capable hands of Bigelow and his experienced team of engineers, von Neumann and Goldstine were able to turn their attention to how the machine would actually be used. During 1947 and 1948 they developed a complete programming methodology for the machine which they published in an influential IAS report entitled 'Planning and Coding Problems for an Electronic Computing Instrument'. Published in three parts, the report described how a stored-program computer could be programmed to find the solution to various common mathematical problems. It also introduced the use of flowcharts as a means of visualising program control, a concept borrowed from industrial engineering. In the absence of any existing literature on computer programming, the report was widely disseminated and became the first standard text on the subject.

Technical problems were also encountered with the development of the magnetic wire auxiliary storage device, which could not be made to operate reliably due to the high rate of wear caused by the action of the thin metal wire sliding at high speed over the magnetic pickup. The slow pace of progress was also attributed to Bigelow's perfectionism, which became of greater concern as the project inched towards completion and users began queuing up to use the machine. This situation was eased when Bigelow was granted leave of absence in August 1951 to take up a Guggenheim Fellowship which he had been awarded the previous year. This provided the perfect opportunity to replace him as Chief Engineer with Pomerene, who had proved himself through his efficient handling of the development and construction of the Williams Tube memory.

Under Pomerene's leadership, the IAS electronic computer was completed in January 1952, one month earlier than EDVAC. A combination of careful design and the use of a parallel transfer architecture made for an economical circuit design which used only 2,600 thermionic valves. Consequently, reliability was relatively good, as shown by one of the first tasks for the new machine, a large thermonuclear problem for the Los Alamos Scientific Laboratory in New Mexico, which necessitated running the machine for 24 hours per day over a period of 60 days, during which time only a handful of interruptions were experienced. Performance was also considerably enhanced by the parallel transfer architecture, with an addition operation requiring only 56 microseconds, including memory access time, compared with more than 800 microseconds for EDVAC.

John von Neumann with the IAS Computer (Photo © Alan Richards)

RCA's problems with the Selectron tube and the other technical problems encountered during development ensured that the IAS electronic computer would not be the first stored-program machine to be completed but it would become one of the most influential. A condition of the military funding for the project was that the machine's design details be made available to other US government-funded computer development projects. Consequently, engineering drawings for the IAS machine were sent out to five other projects and this resulted in it serving as the prototype for the Ballistic Research

5. The Metamorphosis of the Calculator – Stored-Program Machines

Laboratory's ORDVAC, Los Alamos Scientific Laboratory's MANIAC, both of which were completed in March 1952, the University of Illinois' ILLIAC (completed in September 1952), Argonne National Laboratory's AVIDAC (completed in January 1953), Oak Ridge National Laboratory's ORACLE (completed in June 1953) and the RAND Corporation's JOHNNIAC (completed in the summer of 1953). JOHNNIAC was also the first machine to incorporate the Selectron tube, which was eventually manufactured successfully, albeit in limited quantities.

With the IAS report containing the design specification now in the public domain, the IAS electronic computer also became the model for a further nine computer development projects worldwide, including projects in Australia, Israel, Japan, Russia and Sweden. As the only computer that John von Neumann personally oversaw the development of, the IAS machine most closely embodies the von Neumann architecture, which makes it particularly fitting that it was also the most popular of the early electronic computer designs.

The Automatic Computing Engine

That other great mathematical figure in the evolution of the stored-program concept, Alan Turing, was also associated with the construction of one of the first stored-program machines. Turing's involvement came about as the result of a post-war initiative to create a British national computing resource.

Towards the end of World War II, a number of senior figures in UK scientific and government circles began calling for a national effort to coordinate activities in scientific computing and capitalise on wartime technological advances in the field. One of the strongest supporters of this idea was the Director of the National Physical Laboratory (NPL), Sir Charles Galton Darwin, grandson of the famous naturalist. Darwin was determined that NPL, which is the UK's national measurement standards laboratory located in Teddington near London, should play a leading role. In April 1945, he established a Mathematics Division at NPL with a remit to undertake research into new computing methods and machines, and to provide a computing service to government departments, industry and universities. The centrepiece of this new facility would be a state-of-the-art electronic stored-program digital computer.

The Superintendent of the new NPL Mathematics Division was John R Womersley. Womersley needed someone with specialist knowledge who could lead the design and development of the new machine. Although a statistician by training, Womersley had clearly done his homework. He was familiar with Turing's 1936 paper and may also have known about his wartime work on the design of cryptanalysis machinery at Bletchley Park. Having decided that Turing was the right man for the job, Womersley sought the help of Turing's former tutor and fellow Bletchley Park cryptanalyst Max Newman in arranging an introduction. A meeting was duly arranged, during which Womersley showed Turing a copy of von Neumann's controversial EDVAC report. This was the perfect bait and Turing was hooked.

Turing began working at NPL in October 1945. By the end of that year, he had produced a 48-page report entitled 'Proposed Electronic Calculator' which contained an outline of the architecture of the proposed machine, an analysis of the requirements for a range of typical computational problems, a description of the logical design and a detailed discussion of the electronic components required to build it. Turing had taken the EDVAC design as his starting point, his report explicitly referencing von Neumann's draft EDVAC report, but went considerably further in defining a stored-program computer. Until the publication of the design specification for the IAS machine six months later, Turing's report was the most complete specification for a stored-program computer in existence.

Named the Automatic Computing Engine (ACE) by Womersley in a nod to Babbage, the report described a large-scale electronic stored-program machine with a binary architecture, 32-bit word length, 1 MHz clock speed and a gargantuan 6,400 word memory that would require 200 individual mercury delay lines to implement. Despite the high specification, Turing estimated the cost of construction at an optimistic £11,200 and, clearly influenced by his wartime experiences at Bletchley Park, suggested a similarly optimistic timescale for the work.

Turing's design philosophy for the ACE was to keep the hardware as simple as possible in order to maximise performance and for the clever work to be done through programming with the aid of a library of reusable sequences of instructions that would nowadays be called subroutines. Consequently, the instruction set only covered the most basic of operations and there was no separate circuitry for division or for the handling of floating-point numbers. To minimise the effect of the serial delay lines on performance, the ACE design also employed a form of distributed processing in which there was no central accumulator. Instead, calculations would be performed by transferring data between a set of 14 temporary stores implemented using smaller delay lines of one or two words capacity each, making use of locally available arithmetical and logical functionality to reduce the need for transferring data to and from memory.

Womersley submitted the report for approval to the NPL Executive Committee in February 1946, along with a more realistic cost estimate of £60-70,000, and the go ahead was given the following month. Turing continued to refine his design specification while a small project team was put in place. The team would be led by mathematician James H Wilkinson, who arrived in May 1946 from the Ministry of Supply where he had worked on numerical methods for ballistic calculations. Wilkinson was soon joined by Michael Woodger, a young mathematician who would go on to play an influential role in the development

5. The Metamorphosis of the Calculator – Stored-Program Machines

of programming languages. Their principal task was to develop the extensive library of subroutines that the ACE design required.

As NPL's in-house electronic engineering capabilities were very limited, a decision was made to subcontract construction of the machine to the Post Office Research Station at Dollis Hill. The engineering team would be led by Colossus creator Tommy Flowers. A contract was placed in June 1946 and work started on the delay line memory but resources at Dollis Hill were severely stretched due to a backlog of urgent work on the national telephone network which had built up during the war. By the autumn of 1946, it was clear that the GPO could not commit the necessary resources to the project and NPL began looking for an alternative subcontractor. One possibility was Frederic Williams' group at the Telecommunications Research Establishment. TRE management initially responded positively to the idea but Williams was already busy making arrangements for his move to Manchester University. The NPL team then approached Williams' new employers at Manchester and a collaboration was mooted but the discussions came to nothing.

In January 1947, the ACE project team was joined by Harry D Huskey, who arrived from the USA on a one-year visiting appointment. Huskey had begun his career as a mathematics instructor at the University of Pennsylvania where he had become involved in the latter stages of the ENIAC project, gaining valuable experience working on the input/output equipment. Huskey's appointment was made on the recommendation of Douglas Hartree, who had met him during his visit to the Moore School in the summer of 1946. Soon after his arrival at NPL, Huskey suggested that it would be more sensible to build a small prototype machine first and persuaded Womersley that the team should construct the machine themselves in order to gain experience. This would be called the ACE Test Assembly. Turing took the view that the development of a prototype was an unnecessary distraction and distanced himself from the project, though he did continue to contribute advice on the electronic implementation.

Work on the ACE Test Assembly started in April 1947. The goal was to construct the smallest machine that could physically implement Turing's logical design, based on version V of his ACE specification but with simplified control logic. It would be equipped with 8 delay lines of 32 words capacity, giving a memory capacity of 256 words, plus 6 one-word temporary stores. To help with the increased workload, the ACE project team was expanded in the summer of 1947 by the arrival of two new recruits, Donald W Davies, who had recently graduated in mathematics from Imperial College and also held a degree in physics, and Gerald G Alway, a recent graduate in mathematics from Cambridge. However, despite the additional staff resources, the project would soon fall apart, triggered by a chain of events that Turing himself had inadvertently set in motion some months earlier.

In January 1947 Turing had visited several of the computer development projects in the US and had been impressed by the rate of progress that could be achieved when construction was handled in-house by engineers working in close cooperation with the designers. On his return, he lobbied for the establishment of an Electronics Section at NPL. His wish was granted in August 1947. A group of around a dozen staff was set up under the control of the NPL Radio Division. It included Edward A Newman and David O Clayden, two electronics engineers with radar and television development experience who were recruited from the firm Electric & Musical Industries Limited (EMI).

As significant progress had been made with the Test Assembly, Womersley suggested that the NPL Electronics Section take over its construction but the group's leader, Horace A Thomas, was determined to do things his way. He successfully persuaded Darwin that it was inappropriate for a team comprised solely of mathematicians to build a computer and that work on the Test Assembly should cease forthwith. Thomas then began planning for the construction of a full-scale machine based on an earlier version of Turing's specification but with only 3 delay lines and 2 temporary stores. However, his newly formed group lacked the necessary experience in digital electronics and progress ground to a halt.

Frustrated by the lack of progress, Turing left NPL in October 1947 for a one-year sabbatical at King's College, Cambridge. He returned the following spring but resigned permanently in September 1948 to take up a position as Deputy Director of Max Newman's new Royal Society Computing Machine Laboratory at Manchester University. Huskey returned to the US in December 1947, his one-year appointment having come to an end, and joined the National Bureau of Standards' newly formed Institute for Numerical Analysis in Los Angeles. With the departure of two key team members and with no further role to play in the construction of the machine, morale within the ACE project team collapsed.

Fortunately for the ACE project, Electronics Section leader Horace Thomas also left NPL a few months later in March 1948. His replacement, Francis M Colebrook, made efforts to establish closer relations with the Mathematics Division and suggested that they return to the idea of building a small prototype machine to demonstrate that they could work together. The Test Assembly was resurrected and the design substantially improved and updated under the new name of Pilot Model ACE. The 32-bit word length and 1 MHz clock speed were retained, as was the mercury delay line memory but with its capacity increased by the addition of a further two 32-word delay lines plus two additional 2-word temporary stores.

In February 1949, Womersley requested funding from the UK government's Department of Scientific and Industrial Research (DSIR) to engage an industrial contractor in order to assist with the completion of the Pilot Model ACE and

5. The Metamorphosis of the Calculator – Stored-Program Machines

help towards the construction of the full-scale machine. The contractor proposed was the English Electric Company, a large electrical engineering firm whose interests ranged from aircraft and railway locomotives to consumer electronics. Womersley's request was granted the following month and a contract was placed with English Electric in May 1949. The company seconded a number of engineers and technicians to NPL from its Nelson Research Laboratories in Stafford in order to help complete the machine.

Pilot Model ACE (Photo © Science Museum / Science & Society Picture Library)

With the assistance of English Electric, the Pilot Model ACE was operational within a year, running its first program on 10 May 1950, albeit with only one delay line fitted. Construction was completed over the next few months and the machine was publicly demonstrated during three open days held at NPL in November 1950. Despite containing only 800 thermionic valves, the newly completed Pilot Model ACE was the fastest computer yet constructed, with an addition operation requiring as little as 64 microseconds depending on the position of the next instruction in memory. This was largely due to Turing's ingenious logical design and the use of a programming technique known as 'optimum coding' that placed instructions in delay line storage in such a way that they could be fetched immediately when needed. The Pilot Model ACE might also have been the world's first operational stored-program computer were it not for the hesitant manner in which it was developed.

The full-scale version of the ACE was eventually built. Completed in late 1957, it featured a 48-bit word length, a 1.5 MHz clock speed and contained approximately 7,000 valves. However, Turing's original plan of a large memory was never realised and the machine was equipped instead with a delay line memory of much smaller capacity supplemented by four magnetic drum storage devices. In the 12 years which had elapsed since Turing first envisaged the ACE design, computing technology had advanced at a phenomenal pace and delay line storage now appeared very antiquated. Consequently, the machine made little impact despite its revolutionary architecture. What made this sadder still was that Turing never lived to see it, having taken his own life in June 1954. Two years earlier, he had been convicted of gross indecency as a result of a homosexual act which at that time was illegal in the UK. His punishment, for which he chose chemical castration rather than a prison sentence, and subsequent loss of security clearance for his cryptanalysis work are thought to have led to this tragic decision.

The Manchester Mark 1

By November 1947, the combined efforts of Frederic Williams and Tom Kilburn at Manchester University had resulted in the construction of a working electrostatic storage tube memory. They now needed to test the Williams Tube to confirm that it could operate satisfactorily at electronic speeds. In the absence of any suitable test equipment, their only option was to build a small experimental computer. Not only would this be the only way to test the memory fully, by permitting data to be read from and written to the device at electronic speeds, but its development would also contribute towards fulfilling Max Newman's prime requirement of a stored-program computer to provide the centrepiece for his Royal Society Computing Machine Laboratory.

Neither Williams nor Kilburn had any experience of computer design but they were fortunate in having access to the wealth of expertise available within Newman's group at Manchester, much of it through former cryptanalysts who had worked with Newman on Colossus at Bletchley Park. Newman himself provided no direct input to the design of the machine but was helpful in explaining the basics of computer architecture to the two men. However, Jack Good, who had been a key member of the team that designed the Colossus Mark 2, contributed to the logical design and also formulated a basic set of 7 instructions for the machine. David Rees, another Bletchley Park cryptanalyst who had followed Newman to Manchester and who had been one of the two British attendees at the Moore School lectures the previous summer, was also part of this group.

With the help and encouragement of his Manchester colleagues, Kilburn produced an outline design for the new machine which he included in his TRE progress report of December 1947. Kilburn and his assistant Geoff Tootill then proceeded to build it. The work continued to receive the generous support of TRE, which supplied the majority of the electronic components used. The official name given to the machine was the Small-Scale Experimental Machine (SSEM) but it soon acquired the nickname 'Baby'.

The main element of the Manchester Baby was the electrostatic memory, a

5. The Metamorphosis of the Calculator – Stored-Program Machines

single Williams Tube of 1024 bits capacity. Around this Kilburn and Tootill created a simple stored-program machine with a binary serial architecture, a 32-bit word length and a clock speed of 117 KHz. No adder circuit was provided so all arithmetical operations had to be performed by means of subtraction. Two additional Williams Tubes were used to implement a 32-bit accumulator and two 32-bit control registers, one to hold the instruction to be executed and the other for its address in memory. A fourth electrostatic tube without a pick-up plate or metal housing was also provided to allow the contents of any of the storage tubes to be displayed, each of which could be selected via a set of pushbuttons. Input of instructions and data was by means of an array of 32 pushbuttons that required information to be written to the memory a word at a time.

Progress with construction was rapid and the Manchester Baby ran its first program on 21 June 1948, a date that ensured its place in history as the world's first operational stored-program electronic computer. However, with only 32 words of memory available for both instructions and data, the program, which was written by Kilburn to find the highest factor of a number, was tiny and comprised only 17 instructions. Over the following weeks, two other programs were run on the Baby, one of which was written by Alan Turing, who arrived at Manchester in September. The successful execution of these early programs proved unquestionably that the Williams Tube memory could operate at electronic speeds, thus giving the green light to further development work.

At over 5 metres long and with 550 thermionic valves, the Baby's large physical size belied its nickname. However, the machine had been conceived and built as a test rig for the Williams Tube and as such was not well suited to further development. By October 1948, the decision had been made to proceed with the construction of a full-scale prototype that, while based on the same technology as the Baby, would involve a substantial redesign to make it better suited to the computational requirements of Newman and Turing.

To provide extra staff resources, the size of the project team was increased by the addition of Dai B G Edwards and Gordon E Thomas, two research assistants who had graduated in physics and electronic engineering from Manchester in the summer, plus Alec A Robinson, a PhD student who had provided occasional assistance with the Baby. The subject of Robinson's doctoral research was the design of a high-speed parallel multiplier, a highly desirable feature in mathematical and scientific computing, therefore this would be incorporated into the design of the new machine. The nagging issue of Kilburn's extended secondment from TRE was finally resolved in December 1948 when, having recently obtained a PhD for his work on electrostatic storage, he was made a permanent member of staff and given a lecturer post in the Department of Electrical Engineering.

The official name for the new machine was the Manchester Automatic Digital

Machine (MADM) but it became better known as the Manchester Mark 1. The binary serial architecture and 117 KHz clock speed of the Baby were retained but the word length was increased to 40 bits, full arithmetic circuitry was included and extra registers were provided to make the machine more powerful. Memory comprised two high density Williams Tubes of 2,560 bits capacity each, giving a total of 128 words of electrostatic storage. As this was rather small for a full-scale machine, it was supplemented by a magnetic drum memory device of 1024 words capacity that rotated at 200 rpm. Four additional Williams Tubes provided the storage for an 80-bit accumulator, two control registers, two index registers and two additional registers required by the multiplier unit. Input and output were performed using a punched tape unit and teleprinter.

A novel feature of the Mark 1 was the inclusion of index registers, which the team referred to as B-lines. An index register is a special-purpose register for storing an offset memory address, the value of which can easily be incremented within a program in order to address a contiguous block of data. This feature significantly enhanced the machine's ability to perform operations on tables and arrays of numbers, a common requirement in scientific computing applications.

The Baby was gradually dismantled and its components cannibalised for the new machine, the first version of which, known as the Intermediary Version, was operational by April 1949. At this stage it still lacked a punched tape unit. Also absent were the input/output instructions necessary to control the transfer of data to and from the magnetic drum memory device. Eager to begin using the machine for his mathematical research, Turing led efforts to acquire suitable punched tape equipment and then worked with Edwards and Thomas to couple it to the machine. Turing and Newman also influenced the specification for the final set of 26 machine instructions through their research on Mersenne prime numbers.

The Manchester Mark 1 was fully operational by October 1949. The notoriously poor reliability of the Williams Tubes had been improved significantly by the use of custom CRTs supplied by the General Electric Company. Although the Williams Tube is a random access memory device, they were not configured for parallel transfer operation, therefore, the full performance capability of the electrostatic storage could not be realised. Consequently, the basic performance of the Mark 1 was unexceptional, also constrained by the machine's modest clock speed, with an addition operation requiring 1.8 milliseconds. However, this was tempered by the time taken to perform a multiplication operation which, at 2.16 milliseconds, was uncommonly quick due to Robinson's parallel multiplier unit.

In less than 3 years, Williams, Kilburn and their small team of engineers and mathematicians had completed the development of a new type of high-speed

5. The Metamorphosis of the Calculator – Stored-Program Machines

memory device, built the world's first stored-program computer and developed a full-scale machine for mathematical research. During this period, the Manchester group filed 34 patent applications, an indication of the incredible level of innovation achieved. It is very unlikely that Williams and Kilburn would have been able to make such rapid progress had it not been for the knowledge and experience of their Manchester colleagues who had worked on Colossus at Bletchley Park. Therefore, although Colossus did not directly influence later machines due to the extreme secrecy surrounding its development, the early Manchester University computers owe it a considerable debt and they may be considered the true legacy of those wartime developments at Bletchley Park.

The Electronic Delay Storage Automatic Calculator

Despite the remarkable progress made at Manchester University, the Manchester Mark 1 was not the first full-scale stored-program computer to become fully operational. The Manchester team were beaten to this milestone by a group from a rival UK academic institution, the University of Cambridge, which was led by the pioneering computer scientist Maurice V Wilkes.

Maurice Wilkes was born in the town of Dudley, Worcestershire in 1913. After obtaining a bachelor's degree in mathematical physics from the University of Cambridge in 1934, Wilkes was given an opportunity to stay on at Cambridge as a research student and joined a group engaged in radio research at the University's prestigious Cavendish Laboratory. In October 1937, he secured an academic post as a Demonstrator in the University's newly established Mathematical Laboratory and this led to his involvement in efforts to construct a large-scale differential analyzer at Cambridge. Following the start of World War II, his doctoral research on the propagation of radio waves in the ionosphere led to him being recruited to work on radar at the Telecommunications Research Establishment. On completion of his war service, Wilkes returned to Cambridge and he was appointed as acting Director of the Mathematical Laboratory in September 1945.

Wilkes had first become interested in computing machinery during his time as a University Demonstrator and was fascinated by news of the recent developments in electronic computers. Inspired by the effort to establish a national scientific computing resource at NPL, Wilkes began formulating his own plans for a similar but smaller scale facility at Cambridge. In May 1946, he was shown a copy of von Neumann's draft EDVAC report by the astronomer and scientific computation pioneer Leslie J Comrie. Wilkes recognised immediately that this was the kind of machine that he needed for his own plans. By an opportune coincidence, he was also offered a place at the Moore School Lectures which were due to begin a few weeks later in July.

Wilkes succeeded in securing funding for the trip but his application for passage across the Atlantic, which was still tightly controlled by the British government in the aftermath of the war, was delayed and this resulted in him missing the first six weeks of the course. Fortunately, he was able to compensate for this by extending his stay in the US by three weeks and arranging visits to Harvard University, where he met Howard Aiken and was shown the ASCC, and to MIT, where he met Samuel Caldwell and was shown the Rockefeller Differential Analyzer. While in Philadelphia, he also met with Herman Goldstine and was given the opportunity to spend some time in the company of John Mauchly, having been allocated accommodation next door to where Mauchly was living.

Fired up by what he had seen and heard during his visit to the US, Wilkes began working on the design of a stored-program computer during the voyage home. Unlike Manchester, the Cambridge Mathematical Laboratory had little interest in electronic engineering. Instead, the main remit of the Laboratory was to provide the tools for mathematical research and teaching. Therefore, in order to minimise the time and effort necessary for the development, he decided to base his design on that of the Moore School's EDVAC and to stick with proven technology as far as possible rather than attempt to develop anything new. The machine would be called the Electronic Delay Storage Automatic Calculator (EDSAC) in recognition of its origins.

Shortly after returning to Cambridge, Wilkes was confirmed as Director of the Mathematical Laboratory. The Laboratory had acquired Departmental status along with a generous research budget so funding for the new computer would not be an issue. Work started in October 1946. Wilkes possessed a wealth of experience in electronic circuit design from his time at TRE but the construction of a mercury delay line memory was new territory. Looking around the University for someone who could guide him, he found Thomas Gold, a research student in the Cavendish Laboratory who had worked on the design of mercury delay lines for radar echo cancellation at the Admiralty Signal Establishment during the war. Gold was able to prepare a set of engineering drawings for a mercury filled cylinder that would be suitable for storing approximately 500 bits of data. Wilkes then arranged for the cylinder to be manufactured by the University's workshops and, supported by workshop assistant Philip J Farmer, he successfully constructed the necessary pulse generation and control circuitry to make it operational.

Now that Wilkes had a working high-speed memory device, he was ready to commence the detailed design and construction of the machine, a task that would require a much larger team of people. Encouraged by Douglas Hartree, who had recently moved to Cambridge to take up an appointment as Plummer Professor of Mathematical Physics, Wilkes wrote to NPL's John Womersley in December 1946 outlining his design for EDSAC and suggesting that the two groups might collaborate. This approach coincided with NPL's difficulties in finding a replacement subcontractor for the ACE development, prompting Womersley to consider Cambridge as a potential partner with a plan that would involve using the EDSAC design as a pilot model. Womersley

5. The Metamorphosis of the Calculator – Stored-Program Machines

then consulted Alan Turing on the practicalities of this idea but Turing was scathingly critical of Wilkes's wholesale adoption of the Moore School design, which he considered inferior to his own minimalist approach. By the spring of 1947, work on the ACE Test Assembly had already begun and Womersley, having also consulted Hartree, decided that the EDSAC project had progressed too far to be changed and that it would be better for everyone concerned if both groups were left to continue on their separate paths.

One positive outcome of the discussions with NPL was an offer from the DSIR of government funding to support an electronics engineer post at Cambridge. Wilkes used this to hire William A Renwick, a former wartime colleague of Thomas Gold. He also recruited Gordon J Stevens, an instrument maker from a local scientific instrument manufacturer, and Sidney A Barton, an electronics technician who had served in the Royal Air Force. These three men, along with Farmer and Wilkes himself, formed the core of the new project team.

In November 1947, the team received some unsolicited but very welcome industry sponsorship from the British catering giant J Lyons & Company. In return for advice on the company's plans for computerisation of its data processing operation, Lyons provided a donation of £3,000 and the services of an electronics technician for 1 year to support the Cambridge group in the development of EDSAC. The technician chosen for the secondment, Ernest H Lenaerts, quickly became a valued member of the team and the relationship with Lyons continued for many years, culminating in the company's adoption of the EDSAC design for their own use. As the workload increased, further additions were made to the project team and by 1948 the team numbered 19 staff. Wilkes also attempted to recruit Harry Huskey following the end of his 1-year stint at NPL in December 1947 but Huskey, though interested, was already committed to returning to the US and was unable to accept the offer.

The design specification for EDSAC broadly followed that of EDVAC, with a binary serial architecture and mercury delay line memory. However, in order to simplify construction Wilkes decided on a shorter word length of 17 bits and a reduced clock speed of 526 KHz. To compensate for the shorter word length, EDSAC possessed the ability to handle long words of 35 bits and the accumulator was capable of holding twice this number of bits so that accuracy could be maintained during arithmetical operations. Memory comprised 32 delay lines in two banks of 16, each line with a capacity of 32 words, giving a total storage capacity of 1,024 words. An additional bank of smaller capacity delay lines was used to implement the accumulator, a counter and two registers for storing the multiplicand and multiplier during a multiplication operation. Input and output requirements were served by a punched tape unit and a modified Creed teleprinter.

As construction progressed, the EDSAC team's thoughts turned to how the machine might be used. In September 1948, newly graduated mathematician

The Story of the Computer

David J Wheeler joined the team as a postgraduate student to begin writing programs for the machine. Wheeler developed a system called 'initial orders' in which the sequence of instructions required to load the program from punched tape and commence operation, nowadays known as the boot sequence, was hard wired into the machine using a set of uniselector stepping switches. The original capacity of the uniselectors was 31 instructions but this was increased to 41 in August 1949, allowing Wheeler to extend his system to provide a set of higher level instructions called coordinating orders that simplified the use of subroutines. These instructions were the earliest example of what is now called assembly language, where machine instructions and memory addresses are represented by mnemonics and symbolic notation in order to aid the task of programming.

Maurice Wilkes and William Renwick standing in front of EDSAC (Photo © Computer Laboratory, University of Cambridge)

Construction was relatively swift as a result of an early decision to subcontract much of the wiring work to a local electrical engineering firm and EDSAC ran its first program on 6 May 1949. However, only one of the banks of delay lines had been finished therefore memory capacity was limited to 512 words until the other bank was added towards the end of 1949. EDSAC was publicly demonstrated to an audience of 100 attendees at a conference on High-Speed Automatic Calculating Machines held at Cambridge University on 22-25 June 1949. The completed machine contained 3,050 thermionic valves housed in 12 large racks of electronics. Capable of executing an addition operation in 1.4 milliseconds and a multiplication in 5.4 milliseconds, the basic performance of

5. The Metamorphosis of the Calculator – Stored-Program Machines

EDSAC was similar to that of the Manchester Mark 1 despite the use of slower memory devices.

Wilkes's policy of modest design goals and readily available technology had paid off handsomely. Not only did the Cambridge team produce a working computer from a standing start in less than 3 years, they had also caught up with and passed far more experienced teams from the Moore School, the IAS, NPL and Manchester University in the race to construct the world's first full-scale stored-program machine. EDSAC provided a regular computing service, the first of its type in the world, from early 1950 until the machine was replaced in July 1958. The availability of a working machine also allowed research on programming methods to begin in earnest. This resulted in the first textbook on computer programming, 'The Preparation of Programs for an Electronic Digital Computer', which was written by Wilkes, Wheeler and research student Stanley Gill, and published by Addison-Wesley in July 1951. The Mathematical Laboratory's teaching activities were also given a boost and the first ever summer school on computer programming was held at Cambridge in 1950.

The next few years would see electronic computer technology move out of the research laboratory and into industry with the establishment of a commercial market for computers. Remarkably, EDSAC, along with the IAS Electronic Computer, NPL Pilot Model ACE and Manchester Mark 1, would remain at the forefront of these developments, as they would become the prototypes for the first commercial stored-program machines.

Further Reading

Auerbach, I. L., Eckert, J. P., Shaw, R. F. and Sheppard, C. B., Mercury Delay Line Memory Using a Pulse Rate of Several Megacycles, Proceedings of the IRE 37 (8), 1949, 855-861.

Bigelow, J., Computer Development at the Institute for Advanced Study, in A History of Computing in the Twentieth Century, Academic Press, New York, New York, 1980.

Ceruzzi, P., Crossing the Divide: Architectural Issues and the Emergence of the Stored Program Computer 1935-1955, IEEE Annals of the History of Computing 19 (1), 1997, 5-12.

Copeland, B. J. (ed.), Alan Turing's Automatic Computing Engine: The Master Codebreaker's Struggle to Build the Modern Computer, Oxford University Press, Oxford, 2005.

Croarken, M., The Beginnings of the Manchester Computer Phenomenon: People and Influences, IEEE Annals of the History of Computing 15 (3), 1993, 9-16.

Huskey, H. D., From ACE to the G-15, IEEE Annals of the History of Computing 6 (4), 1984, 350-371.

Lavington, S. H., Early British Computers: The Story of Vintage Computers and the People Who Built Them, Manchester University Press, Manchester, 1980.

Randell, B., On Alan Turing and the Origins of Digital Computers, Machine Intelligence 7, 1972, 3-20.

Turing, A. M., On computable numbers, with an application to the Entscheidungsproblem, Proceedings of the London Mathematical Society 2 (42), 1936, 230–265.

Von Neumann, J., First Draft of a Report on the EDVAC, IEEE Annals of the History of Computing 15 (4), 1993, 27-75.

Wilkes, M. V., Memoirs of a Computer Pioneer, MIT Press, Cambridge, Massachusetts, 1985.

Wilkes, M. V. and Renwick, W., The EDSAC – an Electronic Calculating Machine, Journal of Scientific Instruments 26 (12), 1949, 385-391.

Williams, F. C., Kilburn, T. and Tootill, G. C., Universal High-Speed Digital Computers: A Small-Scale Experimental Machine, Proceedings of the IEE 98 II (61), 1951, 13-28.

Williams, M. R., The Origins, Uses and Fate of the EDVAC, IEEE Annals of the History of Computing 15 (1), 1993, 22-38.

6. Data Processing and the Birth of the Computer Industry

"I think there is a world market for maybe five computers", attributed to Thomas J Watson Sr., 1943.

Commercial Potential

When the Chairman of IBM made his infamous remark, he was not alone in thinking that a handful of machines would be sufficient to satisfy the computing needs of the whole planet. Watson's old enemy Howard Aiken, concerned at the limited number of trained mathematicians available to program computers, made a similar remark in 1949. However, these opinions were formed in the context of scientific and mathematical applications of computers, as they were the main focus of the early developments. What both men had failed to recognise was the enormous potential for computers in statistical and business data processing applications.

From the first application of Herman Hollerith's punched card tabulating equipment in the 1890 US Census, there had been steady growth in the use of tabulating and accounting machines for statistical and business data processing purposes which continued throughout the first half of the 20th century and increased dramatically in the post-war economic boom. The office equipment industry had followed the developments in electronic computation with great interest from the outset and by the early 1950s it was becoming clear that computers had a promising commercial future. The scene was now set for the commercialisation of the computer by office equipment manufacturers. However, there was some initial reluctance on the part of the established firms to enter unfamiliar territory and it was only after a handful of pioneering start-ups and specialist contractors had created a commercial market for electronic computers that they fully took the plunge.

Building Blocks – Magnetic Memory Devices

Just as the mercury delay line and electrostatic storage tube had facilitated the development of the earliest stored-program computers, memory devices were the principal technology driver for the first wave of commercial stored-program machines. Two memory technologies in particular contributed to the initial growth in commercial computers, magnetic drum storage and magnetic core storage. Both were based on the magnetic properties of materials but the former involved an electromechanical access technique whereas the latter was fully electronic.

Magnetic Drum Storage

The origins of recording signals by means of a magnetic medium can be traced back to an article by the American mechanical engineer Oberlin Smith which appeared in the journal 'The Electrical World' in September 1888. Smith suggested that mechanical sound recording devices such as Thomas Edison's phonograph could be improved by using a magnetic recording process. The article went on to describe a device which employed a length of string woven with metal filings as the recording medium. However, Smith failed to complete the development of his device and the first magnetic recording device was patented 10 years later in 1898 by the Danish telephone engineer Valdemar Poulsen.

Poulsen's invention, known as the 'Telegraphone', consisted of steel wire coiled around a brass cylinder which rotated under an electromagnet connected to a microphone. Sound was recorded by the current from the microphone energising the electromagnet which then magnetised the wire passing under it. The amplitude of the signal was preserved in the pattern of magnetisation and recordings could be played back by simply replacing the microphone with a telephone earpiece. In the years following Poulsen's landmark invention, all manner of magnetic media recording equipment was developed for the capture of speech and music. Given the ready availability of this technology, it was inevitable that its application to computer memory devices would be considered early in the development of electronic computers.

The use of rotating magnetic drums and disks for the storage of digital information was first proposed by Perry O Crawford, a postgraduate student at the Massachusetts Institute of Technology, as part of an electronic digital anti-aircraft fire control system in his 1942 Master's thesis entitled 'Automatic Control by Arithmetic Operations'. Crawford's concept was then refined by Presper Eckert at the Moore School of Electrical Engineering and described in the controversial internal memo entitled 'Disclosure of Magnetic Calculating Machine' in January 1944. However, Eckert chose not to implement the concept, as he was concerned that a mechanical device would compromise performance, and it was to be three more years before the first practical realisations of a magnetic drum memory would be achieved, more or less simultaneously, by two groups working independently.

A magnetic drum storage device consists of a continuously rotating cylinder with a magnetic material on its external circumferential surface. Small electromagnets, known as read/write heads, are placed very close to the surface of the drum and used to record data in the form of magnetically coded pulses

6. Data Processing and the Birth of the Computer Industry

on a series of parallel circumferential bands or tracks. These same heads are also used to read the recorded pulses, which remain stored within the magnetic material even when the power is switched off but can also be erased and rewritten as often as necessary. Usually read/write heads are fixed in position, with one head per track. To reduce cost, a moving head arrangement can also be used, with a single head mounted on a beam parallel to the rotational axis of the drum and a positioning mechanism to drive the head to the appropriate track on the drum.

The first working example of a magnetic drum storage device was constructed in early 1947 by a team of engineers at the pioneering US start-up firm Engineering Research Associates (ERA). It was built as part of Project Goldberg, a project to develop a special-purpose cryptanalysis machine for the US Navy. The prototype device used strips of magnetic sound recording tape glued to the outer surface of a 34-inch diameter aluminium drum which rotated at a sedate 50 rpm. Later versions replaced the unwieldy magnetic tape by spraying an iron oxide emulsion directly onto the surface of the drum. Considerable effort also went into increasing performance and reliability. ERA's early progress in this field was reported by John M Coombs at the US National Electronics Conference in Chicago in November 1947 and the company went on to become the industry leader in the manufacture of high-speed magnetic drum devices for general-purpose computers.

ERA Magnetic Drum Memory Products

During the same period, a magnetic drum memory was developed independently by the British computer pioneer Andrew D Booth at the University of London's Birkbeck College. Booth's work as a mathematical physicist had led him to explore the use of electromechanical calculating technology for x-ray crystal structure analysis. His efforts to construct a relay calculator attracted the attention of the influential mathematical physicist Douglas Hartree at Manchester University in 1945 who told Booth of the developments in electronic computation then taking place in the US. Booth was intrigued by the prospect of electronic computing technology and in late 1946 he began to consider storage techniques based on magnetic properties of materials. He was given an opportunity to refine these ideas on an extended visit to the US in the spring and summer of 1947 under a Rockefeller Foundation study fellowship, during which time he was based at the Institute for Advanced Study in Princeton. On his return to London, Booth was single minded in his determination to construct his own stored-program computer but he also knew that the meagre resources available at Birkbeck College would dictate more modest design goals and lower cost solutions than he had seen at the IAS.

Booth's initial experiments, conducted in late 1947, involved the use of ferrous oxide coated paper disks from the recently introduced Mail-A-Voice audio recording machine but he was unable to make this device sufficiently stable at the high rotational speeds necessary for computer memory purposes. Had these experiments proved successful, Booth's device would have been the first implementation of a floppy disk drive. He then switched to an alternative design based on a small 2-inch diameter nickel-plated brass drum and by January 1948 had successfully constructed a working prototype magnetic drum memory for use in his first machine, the Automatic Relay Computer (ARC).

The parallel achievements of ERA and Andrew Booth in proving the concept of magnetic drum storage paved the way for other groups in the development of their own magnetic drum devices. Booth's technology was taken up by Williams and Kilburn at Manchester University for use in the Manchester Mark 1. At the Harvard Computation Laboratory, Howard Aiken's group began experimenting with magnetic drum storage techniques in early 1948 and various magnetic drum devices were constructed for the Harvard Mark III machine in 1949. The design of the Harvard Mark III was notable for its reliance on magnetic drum storage devices to implement several important functions and in the use of separate drums for data and instructions. Also in 1948, following an inspirational visit to ERA, engineers at the Northrop Aircraft Company in Hawthorne, California developed a magnetic drum memory for the MADDIDA, an early digital differential analyser.

Elsewhere in California, at the University of California Berkeley, a high-capacity magnetic drum memory was developed over the period 1948-51 as part of the California Digital Computer (CALDIC) project. Although much of the funding for this project came from the US Office of Naval Research, the main objective was to provide a teaching tool for students. Consequently, the design drawings for the magnetic drum memory were made publicly available and other groups were encouraged to take up the Berkeley technology.

6. Data Processing and the Birth of the Computer Industry

By 1951, the magnetic drum had become firmly established as the storage technology of choice for designers of affordable medium-scale computers. The prohibitive cost of high-speed memory in early stored-program machines had led to a hierarchical approach with two or sometimes three levels of storage implemented. The expensive fast-access technology would be used to provide a small amount of primary storage, also known as main memory, which was sufficient to store only the data and instructions in active use. Slower, less costly technology would then be used to provide the bulk of the storage capacity. This was known as backing or auxiliary storage. Magnetic drum technology was sufficiently versatile that it could be used as either primary or backing storage. It could even be used to implement arithmetic registers by fitting extra read/write heads to certain tracks in order to create short recirculating loops of memory with a faster access time.

Magnetic drum storage opened up the market for low cost medium-scale computers. It was reliable, high capacity, non-volatile and relatively inexpensive to implement. However, its electromechanical nature made it significantly slower than the electronic memory devices of the period, such as the mercury delay line and electrostatic storage tube, and it was inevitable that as machine performance increased, magnetic drums would eventually become unable to keep pace. Nevertheless, the ingenuity of the engineers in finding new ways to improve access speed ensured that it would take almost a decade before this performance limitation was finally reached and Presper Eckert's initial concerns vindicated.

Magnetic Core Storage

Conceived prior to the development of the first magnetic drum and electrostatic storage tube devices, magnetic core storage took longer to reach the market due to its greater engineering complexity. However, when it did arrive, it swiftly succeeded both devices for use as main memory in large-scale computers.

Magnetic core memory comprises an array of miniature doughnut-shaped rings or 'cores' manufactured from a ferromagnetic material, usually ceramic. Information is stored by controlling the polarity of the magnetic field contained within the core and this is achieved by passing a current through wires which are threaded through each core. If the current is above a certain threshold, the polarity of the magnetic field is reversed or 'flipped'. The polarity of a core is used to represent the value of a binary digit or 'bit' of data, where a particular polarity represents a 1 and the opposite polarity a 0. The polarity can be read by applying a flip current to the core and sensing the presence or absence of an induced voltage in an additional wire which is also threaded through each core. However, this process also destroys the magnetic state of each core and a refresh cycle is required after each read operation to reset the cores in the array to their previous values.

The Story of the Computer

The use of a wiring configuration based on a 2-dimensional matrix permits individual cores in an array to be flipped or read. The threshold current required to flip a core may be set so that it is only exceeded when current is applied to both the 'x' and 'y' wires simultaneously, the hysteresis characteristics of the ferromagnetic material preventing a core from flipping when only one wire is activated. This is known as the 'coincident-current' method of selection.

The use of magnetic cores for the storage of binary data was first proposed by the prolific Presper Eckert in 1945. Eckert and his Moore School colleague Jeffrey Chuan Chu also conducted the earliest experimental work on magnetic core storage but they were unable to obtain suitable materials and the concept was never fully implemented. However, their work was documented in an EDVAC progress report in June 1946 and Eckert also discussed it openly with various external parties. Consequently, by 1947 the idea of magnetic core storage had begun to occupy the minds of several leading computer developers, including magnetic drum pioneer Andrew Booth.

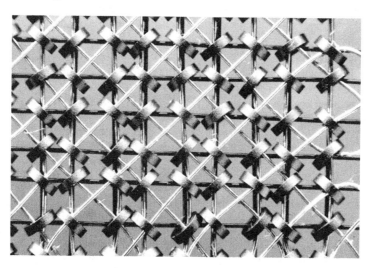

Close-Up of Magnetic Core Memory (Photo © Konstantin Lanzet)

An unlikely protagonist in the history of magnetic core storage was Los Angeles public works inspector and amateur inventor Frederick W Viehe. Viehe filed a US patent application for an 'Electronic Relay Circuit' in May 1947 which included a description of a storage device that used a single magnetic core. Although he never applied his invention to computers, the patent was deemed sufficiently important that the rights to it were purchased by IBM in November 1957 for an undisclosed sum during a subsequent patent dispute between Viehe and another magnetic core pioneer, the Chinese-born physicist An Wang.

An Wang became the first person to successfully implement a magnetic core memory in 1948 while working for Howard Aiken at the Harvard Computation

6. Data Processing and the Birth of the Computer Industry

Laboratory. Wang and his colleague Way Dong Woo built a small shift register, a type of sequential logic circuit, using a string of magnetic cores which operated in a similar way to a mercury delay line. Although Wang's implementation of a magnetic core memory lacked the ability to directly address individual cores, his innovation of the refresh cycle to eliminate the problem of data being destroyed during reading allowed him to file a US patent application in October 1949 which was entitled a 'Pulse Transfer Controlling Device'. This was granted in May 1955 and a license to it was acquired by IBM in March the following year for an initial fee of $400,000 plus a 70% share of royalties from sublicensing. In an effort to pre-empt a potentially damaging dispute over ownership, IBM also bought out Woo's claim of co-inventor's rights to Wang's patent.

Another important figure in the development of magnetic core storage was the electrical engineer, Jay W Forrester. Forrester's is the name most strongly associated with magnetic core memory due to his vigorous promotion of the technology and the high profile he enjoyed as Director of the Massachusetts Institute of Technology's Digital Computer Laboratory during this period. Forrester and electrical engineering postgraduate student William N Papian constructed a small magnetic core memory as part of the highly influential Whirlwind project in the summer of 1950. Forrester's key innovation, conceived in 1949, was the coincident-current method of selection, which was a considerable improvement on Wang's serial addressing method as it permitted individual cores to be addressed, thereby providing a true random access memory.

Forrester's prototype magnetic core memory was operational by October 1950. The work was described in Papian's 1950 Master's thesis, entitled 'A Coincident-Current Magnetic Memory Unit', and in an academic paper published in the Journal of Applied Physics in January 1951. Forrester also filed a US patent application in May 1951 for a 'Multicoordinate Digital Information Storage Device', which IBM secured a license to in 1956 for an eye-watering one-time fee of $13 million. However, this bold commercial endorsement of Forrester's work failed to prevent a looming conflict with another leading figure in magnetic core research, the electrostatic storage tube pioneer Jan Rajchman.

Coincident-current magnetic core storage had also been conceived independently by Rajchman at RCA Laboratories in 1949. However, progress in constructing a practical magnetic core memory at RCA was painfully slow, as Rajchman's small team was also working on the development of the troublesome Selectron storage tube at the same time, and it took until February 1952 before they were able to build a successful magnetic core memory unit. Nevertheless, Rajchman was able to file a US patent application for a 'Magnetic Device' in September 1950 which included a description of the coincident-current concept. The lengthy patent dispute which followed the filing of Forrester's May 1951 application was eventually settled in favour of Rajchman for the majority of the claims made in Forrester's patent. Rajchman's case conceded that Forrester was the first to conceive coincident-current selection but successfully argued that Papian's construction of a small magnetic core memory in mid 1950 did not constitute a reduction to practice. IBM acquired license rights to the Rajchman patent in July 1957 through a cross-licensing agreement with RCA.

The speed, reliability, non-volatility and low power requirements of magnetic core storage made it an attractive proposition for designers of stored-program computers. Consequently, the leading computer manufacturers of the day soon began to take up the new technology and by 1954 magnetic core storage had become sufficiently mature to be a viable option for production machines. However, the manufacture of magnetic core memories remained a difficult and time-consuming operation, requiring careful hand-assembly of the delicate wires and tiny cores. The high fabrication cost initially precluded its use as main memory in all but the highest performance special-purpose computers and its first appearance in mainstream machines was restricted to small amounts of fast storage for input/output buffers. Although never fully automated, the manufacturing process for magnetic core storage was streamlined considerably and as fabrication costs began to fall, it replaced mercury delay lines and electrostatic storage tube devices as the standard technology for main memory in stored-program computers, a position that it retained for over 15 years. The dramatic performance and reliability gains that magnetic core storage brought also served to open up new markets for digital computers in other application areas.

The First to Market

The achievements of Maurice Wilkes at the University of Cambridge, Williams and Kilburn at Manchester University and the ACE project team at NPL had resulted in the UK taking an early lead in electronic computer development. The UK also had a sizeable electrical and electronic engineering industry, with firms such as English Electric and GEC prominent among the largest and most successful companies in the UK manufacturing sector, and it was this industry which would take the initiative in bringing electronic computers to market.

The Ferranti Mark 1

In October 1948, the British government contracted the electrical engineering firm Ferranti Limited to build a production version of the Manchester Mark 1, which, although still under development, was already showing promise. Founded by Sebastian Ziani de Ferranti and C P Sparks in 1883, Ferranti was a successful family-run business with factories in several parts of the UK. The company had been involved in the development of radar during World

6. Data Processing and the Birth of the Computer Industry

War II and had already dabbled in electronic computer technology through the technical assistance it had provided to Williams and Kilburn's group in developing magnetic drum and electrostatic storage devices. Ferranti's formal entry into the computer industry was prompted by the UK Ministry of Supply's chief scientific advisor, Sir Ben Lockspeiser, who, having seen a demonstration of the Small-Scale Experimental Machine, persuaded the company to collaborate closely with the group on the development of a production machine and also provided government funding to encourage the link. The first product of this collaboration was the Ferranti Mark 1, the world's first commercially available general-purpose electronic computer.

A computer development group was set up at Ferranti's Moston factory in 1949 under the direction of instrument department manager James D Carter and two key members of the Manchester team, Geoff Tootill and Alec Robinson, transferred to Ferranti to work on the project. By November 1949, the specification of the production machine had been finalised and construction could begin. The logical design was carried out by Tom Kilburn at Manchester University. In order that programs developed for the prototype could be run on the production machine, the architecture of the Ferranti Mark 1 was closely modelled on the Manchester University prototype.

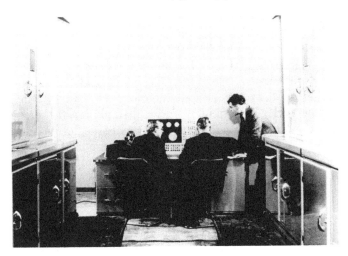

Alan Turing inspects the Ferranti Mark 1 Console (Photo © University of Manchester)

Kilburn's design retained the binary serial architecture and 40-bit word length of the prototype but the number of index registers, or B-lines, was increased from two to eight, 24 additional machine instructions were provided and main memory was doubled from 128 to 256 40-bit words. Main memory was implemented using eight Williams electrostatic storage tubes of 1,280 bits capacity each and was augmented by a high capacity magnetic drum device capable of storing 16,384 words. Input was via a paper tape reader and output was provided by means of a paper tape punch and printer. The enhanced specification of the production model resulted in significantly increased performance over the prototype, despite a slight reduction in clock speed from 117 KHz to 100 KHz.

The Ferranti Mark 1 was the first general-purpose electronic computer to reach the market. Introduced at a purchase price of £95,000, the first production machine was delivered to Manchester University in February 1951. However, reliability problems with the first two production machines necessitated modifications to the design of the magnetic drum and power supplies and these were incorporated into a revised model which was known as the Mark 1*. A total of 9 machines were sold between 1951 and 1957, three of which went to customers outside the UK, one each in Canada, Holland and Italy.

Teashop Technology – The Lyons Electronic Office

Although subsequently marketed as suitable for business applications, the Ferranti Mark 1 had been developed primarily as a scientific computer and its earliest customers had been universities or research organisations. The first project to develop a computer specifically for business use grew out of a UK catering company's need to more accurately monitor sales of cakes throughout its chain of teashops.

In May 1949, J Lyons & Company initiated a project to build an electronic computer to serve the company's growing data processing needs. This bold decision had been prompted by a fact-finding visit to the US by two senior managers of the company, Raymond Thompson and Oliver Standingford, two years earlier in May 1947. Despite the mundane nature of the company's products, Lyons had a refreshingly forward-thinking attitude to business and had pioneered innovative management techniques in its quest for office efficiency. Thompson and Standingford's original remit had been to investigate wartime developments in office systems and equipment but this was extended to include computers following the tantalising snippets of information that were beginning to appear in the press about US developments in electronic computation.

During their trip, Thompson and Standingford met Herman Goldstine at the Institute for Advanced Study who informed them of efforts to develop a stored-program machine taking place much closer to home at the University of Cambridge. Goldstine arranged for them to be contacted by Douglas Hartree and Maurice Wilkes on their return to the UK. Following positive discussions, Lyons agreed to donate £3,000 in cash plus the service of a talented electronics technician, Ernest Lenaerts, for 1 year to support the Cambridge group in the development of EDSAC in return for advice on the company's plans for computerisation of its data processing operation. Therefore, when Lyons finally took the decision to build its own computer some time later, it was

6. Data Processing and the Birth of the Computer Industry

obvious to all concerned that this should be based on EDSAC.

Following the decision to proceed, a project team was quickly assembled and work began on the design of the Lyons Electronic Office or LEO. The team was led by Cambridge-educated physicist John M M Pinkerton, whose wartime experience in radar research and personal links with Wilkes made him the ideal candidate for the job. One of the main objectives was to construct a machine in the shortest possible time so departures from the EDSAC logical design were kept to a minimum. However, considerable effort went into providing the additional devices necessary for the input and output of large amounts of business data, and in developing a programming methodology suitable for business applications.

LEO closely followed the binary serial architecture of EDSAC, featuring the same 526 KHz clock speed and 17-bit word length but with increased memory capacity and provision for multiple input/output devices. Memory was provided by 64 individual mercury delay lines of 32 words capacity each, giving a total capacity of 2,048 words, double that of EDSAC. To improve data flow between main memory and the relatively slow input/output devices, LEO was also equipped with small amounts of fast temporary storage known as buffers. Three input and two output buffers were provided and these were also implemented using mercury delay lines.

Although Lyons possessed substantial in-house engineering resources, the specialist nature of the work made it necessary to subcontract the construction of some of the critical components, such as the circuit boards and delay line cylinders, to other firms. The UK telecommunications company, Standard Telephones & Cables, was contracted to develop a novel magnetic tape device which used an endless loop of quarter-inch tape for high-speed input but development delays and concerns over the reliability of this equipment necessitated a switch to conventional punched card and paper tape devices.

Built at a cost of £150,000, LEO ran its first live data processing job in November 1951, calculating the value of cakes, pies and pastries for despatch to Lyons' retail and wholesale outlets. However, the delays encountered with the development of the input and output devices prevented the machine from running large-scale jobs until February 1954. In the interim, in what was probably the first ever example of a computing bureau service, the company took the opportunity to make LEO available to a number of government and industrial organisations that had expressed an interest in using the machine for a variety of scientific applications.

LEO had been conceived for internal use only but Pinkerton's submission of a design for a new model, the LEO II, in June 1954 prompted a decision by the company to enter the commercial market and a subsidiary company, LEO Computers Limited, was created in November 1954 to manufacture computers "for sale or hire".

Establishing an Industry

The UK's early lead in electronic computer development after World War II was short-lived. The larger potential domestic market in the US and the ready availability of the considerable technological and financial resources necessary to mass-produce computers resulted in an inevitable shift across the Atlantic.

The US electronics industry would come to play a key role in the commercialisation of the computer but its earliest contributions were in the development of electronic components or special-purpose machines rather than mainstream models. Instead, another US-dominated industry would take the initiative. With an existing customer base, established sales channels and mature service networks, office equipment firms were ideally placed to succeed in the computer business. By the mid 1950s, the commercial development of the computer was dominated by three of the big four US office equipment manufacturers; Remington Rand, IBM and Burroughs.

The largest of the big four, the National Cash Register Company (NCR), was also the slowest to address the commercial market for computers and therefore significantly less influential. Despite having been active in electronic computation from the earliest days through its wartime work on electronic fire control systems and cryptanalysis devices for the US Navy, NCR chose to concentrate its substantial product development resources on introducing electronic versions of electromechanical machines, the first of which was the Post-Tronic bookkeeping machine in 1956. Consequently, the first of the big four to make an impression on the computer industry was NCR's closest rival, Remington Rand Incorporated.

Remington Rand – The First Industry Leader

Remington Rand was created in 1927 by the merger of the Remington Typewriter Company with the Rand Kardex Bureau. In the same year, the newly formed company acquired both the Dalton Adding Machine Company and the Powers Accounting Machine Company, IBM's arch rival in the accounting and tabulating machine market. Remington Rand became a household name for its typewriters and electric shavers but it was also one of the world's largest office equipment manufacturers, second only to NCR in market share.

Remington Rand's interest in computers can be traced back to 1943 when engineer Loring P Crosman was hired to establish a modest research effort in electronic computation. By 1950, Crosman's team had built a working prototype of a small sequence-controlled electronic punched card calculator and had also commenced the development of the production version, which was unveiled in July 1951 as the Remington Rand Model 409. The 409 featured

6. Data Processing and the Birth of the Computer Industry

a 10-digit word length and could be programmed in up to 60 steps by means of a plugboard. It was capable of reading a punched card and performing a calculation at a rate of 150 cards per minute, which was considerably faster than similar models from rival firms. Deliveries began in 1952 and more than 1,000 of the Model 409 and its many derivatives were eventually built.

From this inauspicious start, Remington Rand would go on to become the dominant force in the formative years of the US computer industry. However, this was not as a result of its own development efforts but instead through the acquisition of two pioneering start-up companies, Eckert-Mauchly Computer Corporation and Engineering Research Associates.

Eckert-Mauchly Computer Corporation

Having left the Moore School in the spring of 1946, Presper Eckert and John Mauchly were now free to continue their work in any way that they saw fit. Eckert had already turned down an offer from John von Neumann of a position at the Institute for Advanced Study in Princeton. Both men now received and declined a generous offer from IBM's Thomas Watson to establish their own computer research laboratory. Increasingly convinced of the commercial potential of computers, Eckert and Mauchly chose to remain independent and go into business for themselves.

Mauchly immediately rekindled earlier discussions with the United States Census Bureau that he had initiated while at the Moore School regarding the development of an electronic computer for statistical applications. The Census Bureau reaffirmed its interest in the project and allocated a budget of $300,000 but explained that it was not permitted to place contracts for research and development work. A proposed solution was conceived in which the contract would be placed through another US government agency with an interest in electronic computation, the National Bureau of Standards (NBS), which would also provide technical assistance with the specification and contract negotiations in return for a 15% overhead fee.

Electromechanical computing pioneer George Stibitz was consulted to provide an independent opinion on the technical feasibility of the project. He concluded that major uncertainties existed with the project but that the proposal had sufficient merit to justify a feasibility study. However, at Stibitz's recommendation and in line with US government procurement policy, a small number of established contractors were also invited to bid for the work. Of these, only Raytheon submitted a bid and this was priced higher than Eckert and Mauchly's. Therefore, despite Stibitz's reservations, the NBS took the bold decision to award a $75,000 contract for the research and study phase of the project to Eckert and Mauchly.

The NBS decision gave Eckert and Mauchly the green light to set up a company. They formed a partnership, the Electronic Control Company, in Philadelphia in June 1946 and began to assemble a team of engineers, including three former colleagues from the Moore School; Herman Lukoff, Robert F Shaw and C Bradford Sheppard. However, with no business track record to speak of, Eckert and Mauchly were unable to attract investors so the company was initially financed using funds from friends and family. The fact that they were able to raise several hundred thousand dollars in this manner was testament to the faith that those who knew them had in the two men.

The research and study phase of the project took longer than expected to complete but the outcome was positive and the detailed design work on the new computer, which was now named the Universal Automatic Computer or UNIVAC, could begin. Eckert and Mauchly had estimated the development costs for UNIVAC at $413,000 but were prepared to absorb the expected six-figure loss as they confidently expected to sell additional machines at a profit. However, they had failed to foresee the huge strain on cash flow that payment on completion terms would cause and had foolishly agreed to a fixed-fee contract in the mistaken belief that a cost-plus-fixed-fee contract, which would have covered any increase over the estimated development costs and allowed the company to make a small profit, meant relinquishing patent rights. Furthermore, bureaucratic delays resulted in the contract for the design phase not being issued until June 1948, almost a year after the research and study phase had ended. Consequently, despite having pocketed the $75,000 payment for the feasibility study, Electronic Control soon began to experience serious financial problems. NBS officials, concerned at the financial position of their new supplier, sought to alleviate the situation by inserting staged payments into the forthcoming design contract. They also encouraged the US Air Force and the US Army Map Service to each place an order for a UNIVAC. Electronic Control now had firm orders for 3 machines but with no sign yet of the promised $169,600 design contract, Eckert and Mauchly were forced to look to the private sector for work.

Electronic Control was thrown a lifeline in April 1947 when the Northrop Aircraft Company hired Mauchly as a consultant to appraise the feasibility of electronic digital computing technology for in-flight navigational control of a new long-range guided missile, the SNARK. Over the next few months, not only was Mauchly able to convince Northrop of the feasibility of the concept, he also persuaded the company to place a contract with Electronic Control for the design and construction of a prototype computer.

The contract for the work was placed in October 1947. It called for the development of a machine referred to as the Binary Automatic Computer or BINAC, a compact, lightweight computer that would be suitable for use in airborne experiments. Having learned their lesson with the NBS contract, Eckert and Mauchly successfully negotiated a sizeable upfront payment of 80% of the $100,000 value of the contract, with the remainder to be paid upon completion. However, they once again settled for a straight fixed-fee

6. Data Processing and the Birth of the Computer Industry

arrangement rather than pressing for a cost-plus-fixed-fee contract which, although more common in government than the private sector, would have better protected Electronic Control against cost overruns.

The Northrop contract presented Eckert and Mauchly with a valuable opportunity to use the BINAC development as a test-bed for many of the technologies intended for UNIVAC. Despite the special-purpose nature of BINAC and a specification which dictated a binary architecture rather than the decimal architecture planned for UNIVAC, both machines were to be stored-program and would require similar main memory and auxiliary storage subsystems. Eckert and Mauchly's design for BINAC employed a binary serial architecture with a 31-bit word length, two arithmetic registers, a compact set of 25 machine instructions and a remarkable 4 MHz clock speed. Memory would be provided by a mercury delay line unit of 512 words capacity comprising 18 individual delay lines housed within a single mercury tank. BINAC would also feature the first commercial use of magnetic tape for high-speed input and output.

To minimise errors during operation, BINAC was implemented with two identical arithmetic units, each with its own memory unit. The machine performed calculations in parallel and compared results. Only if the results agreed would the next instruction be executed.

Delivery of BINAC was originally scheduled for May 1948 but this proved overly optimistic and development delays pushed completion back by more than a year. During development, the clock speed had to be reduced to 2.5 MHz. Nevertheless, following a satisfactory acceptance test, BINAC was dismantled and shipped to Northrop's California headquarters in September 1949. However, upon reassembly, problems were encountered in getting both arithmetic units to operate simultaneously which were exacerbated by frequent component failures and shoddy construction suggestive of a hurried completion. Budget overruns had pushed the development cost up to $278,000, almost three times the price of the contract, and Electronic Control simply could not afford to spend any additional time on the project.

Northrop's own engineers, who had amassed considerable experience in electronic computation through the in-house development of the MADDIDA digital differential analyser, had resented the decision to use an external contractor from the outset and were reluctant to take on what they saw as someone else's mess. Consequently, Northrop began to lose faith in BINAC and the company eventually decided on an analogue solution for the guided missile application.

Opinions differ over whether BINAC was ever fully operational after delivery but what is certain is that the machine was never used in anger. Nevertheless, the successful acceptance test on 22 August 1949 secured BINAC's place in history as the first operational stored-program computer in the USA. The project was not entirely disastrous for Electronic Control either, as it served to prove Eckert and Mauchly's stored-program technology, thereby providing assurance to the company's prospective UNIVAC customers. The funding also helped to keep the company afloat financially during a critical period.

A few months into the BINAC project, Electronic Control had again begun to run out of money. Despite having recently secured a $20,000 contract from the Prudential Insurance Company for consultancy services and an option to purchase a UNIVAC, it was estimated that an additional $500,000 in working capital would be required to complete the UNIVAC development and remain in business. Therefore, in an effort to attract investors, the partnership incorporated as the Eckert-Mauchly Computer Corporation (EMCC) in December 1947. Within only a few weeks, an offer was received from another prospective UNIVAC customer, market research firm the A C Nielsen Company. However, Nielsen's offer was in return for a controlling interest in EMCC and was rejected without hesitation as Eckert and Mauchly remained stubbornly convinced of the company's ultimate commercial success.

In August 1948, Eckert and Mauchly finally received the investment they desperately needed. It came from the American Totalizator Company, the leading manufacturer of pari-mutuel machines, the relay-based calculators used for registering bets and posting odds at racecourses. A deal was struck with American Totalizator in which EMCC would receive a $50,000 initial payment, a $62,000 loan to be repaid by January 1950 and $438,000 in staged payments to be completed by June 1950, in return for 40% stock and four seats on an expanded Board of Directors. However, 14 months into this agreement, the chief architect of American Totalizator's relationship with EMCC, vice-president Henry Straus, was killed in a plane crash and his untimely death effectively ended the company's interest in EMCC.

By November 1949, EMCC was again at crisis point. Mauchly attempted to secure other loans but was unsuccessful. Severely undercapitalised and saddled with a number of poorly negotiated contracts on which they now expected to lose money, Eckert and Mauchly finally bowed to the inevitable and sold EMCC to the first of the big four US office equipment manufacturers to make a serious offer, Remington Rand. The deal was concluded by February 1950. Remington Rand paid $438,000 for American Totalizator's 40% share of the business plus an additional $100,000 divided amongst the remaining stockholders. In return, EMCC became the company's UNIVAC Division. After only four years in business, Eckert and Mauchly had lost their independence.

Remington Rand's acquisition of EMCC provided the necessary financial stability to complete the UNIVAC development. The first machine passed its acceptance test on 30 March 1951 and a formal dedication ceremony took place in June. UNIVAC employed a binary-coded decimal, serial, architecture with a 12 decimal digit word length, 4 arithmetic registers, 43 machine instructions

6. Data Processing and the Birth of the Computer Industry

and an impressive 2.25 MHz clock speed. Primary storage was provided by a mercury delay line memory of 1,000 words capacity comprising 7 mercury tanks, each containing 100 individual delay lines or channels. This was supplemented with an auxiliary storage system based on magnetic tape. Like Lyons, EMCC had also realised the importance of high-speed input/output devices in processing large amounts of business data and had chosen magnetic tape for its cost-effectiveness. The tape unit developed for BINAC was redesigned to use nickel-plated bronze tape instead of plastic audio recording tape to improve durability. Up to ten of these new 'UNISERVO' tape drives could be connected simultaneously. Two additional devices were developed to handle magnetic tape data; a UNITYPER to convert manually typed input to tape and a UNIPRINTER to convert data stored on tape to printed output. To improve data flow between main memory and the slower input/output devices, UNIVAC was also equipped with two 60-word buffers, one each for input and output. These allowed simultaneous reading and writing of data without interrupting computation.

As with BINAC, EMCC's engineers placed a very high emphasis on reliability in the design of UNIVAC. The costly dual processor design of BINAC was dropped in favour of a single arithmetic unit that included extensive duplicate circuitry for checking and comparing results. This was augmented with an error detection system which checked the odd-even state of binary numbers passing through the circuitry at five locations. UNIVAC also featured a primitive form of conditional branch in which the machine had the ability to store the accumulator control counter value in memory, making it possible for the flow of a program to go to a subprogram and then return to the same location in the main program.

UNIVAC

UNIVAC was introduced to the market at a price of $950,000 for a basic system, a realistic figure based on the actual cost of manufacture but one that was many times higher than the purchase prices agreed by EMCC for all six machines in the company's order book. Remington Rand quickly sought to renegotiate these early contracts but failed to do so with the three machines on order to the US Government. Its private sector customers were easier to deal with. Following threats of protracted legal proceedings which could have held up delivery for several years, both the Prudential Insurance Company and the A C Nielsen Company wisely decided to cancel their orders.

UNIVAC was originally conceived by Eckert and Mauchly as a general-purpose computer which would be capable of satisfying as wide a range of computing needs as possible. However, Mauchly's skill in eliciting the needs of potential customers, in particular those of the Prudential Insurance and A C Nielsen companies, led to the evolution of UNIVAC into a machine more suited to the data processing requirements of large businesses than scientific computation. This made UNIVAC particularly appealing to private sector customers which helped to deliver an impressive sales total of 46 machines over the following 6 years. Nevertheless, UNIVAC had several design inadequacies, such as a lack of support for punched cards and poor printing performance, although these deficiencies were later remedied.

The development of UNIVAC had taken nearly 5 years and several million hard won dollars. It was an immensely ambitious project but the end result was an engineering triumph, hugely influential and a major milestone in the history of business computing.

Engineering Research Associates

Engineering Research Associates was formed in January 1946 by investment banker John E Parker and three former US Navy Reserve officers, Howard T Engstrom, William C Norris and Ralph I Meader. Engstrom, a mathematician, and Norris, an electrical engineer, had both worked for the US Navy's top-secret Communications Supplementary Activities - Washington (CSAW) agency during the war and had spotted an opportunity to commercialise some of the advanced cryptanalysis technology developed by the agency's equipment engineering group. They were soon joined by Meader, who had been in charge of the US Naval Computing Machine Laboratory (USNCML) in Dayton, Ohio, a research laboratory established in 1942 to oversee the design and manufacture of cryptanalysis machines supplied to CSAW and other government agencies by the National Cash Register Company. Parker was introduced to Meader through a mutual acquaintance and quickly realised that the business proposition presented by the three men might also provide a solution to the problem of what to do with another of his companies, Northwestern Aeronautical Corporation, a major wartime supplier of combat gliders to the US military, now that the war had ended. US Navy officials actively encouraged the formation of the new company, which they considered as a useful means of keeping the CSAW engineering group together following the end of the war. This endorsement provided the assurance Parker needed to invest.

6. Data Processing and the Birth of the Computer Industry

ERA set up operations in Northwestern Aeronautical's glider factory in St Paul, Minnesota. Around 40 ex-CSAW staff moved to St Paul to work for the new company. The Navy continued to demonstrate its support for the venture by placing two cost-plus-fixed-fee contracts for work to be defined later in terms of a series of tasks up to the limit of the value of the contract and USNCML personnel were relocated to St Paul to supervise the work. However, the larger of the two contracts had to go through Northwestern Aeronautical as ERA was not yet an established business.

The company's first tasks for the Navy involved the development of data storage devices and the construction of special-purpose electronic computers for cryptanalysis. Another early task was to conduct a review of the latest developments in computer components for the US Office of Naval Research. In gathering the information for this review, ERA employees were granted unrestricted access to the other US Government funded computer development projects and this information was soon to prove invaluable for the company's own general-purpose computer developments.

Task 13, assigned by the Navy in August 1947, was for the design of a general-purpose electronic computer. The proposed design was submitted in March 1948 and approval was given to begin construction of the machine, which was codenamed Atlas. Atlas employed a binary, parallel, stored-program architecture with a 24-bit word length and a 400 KHz clock speed. Storage was provided by a fast-access, directly addressable magnetic drum memory of 16,384 words capacity which rotated at 3,500 rpm and incorporated 200 read/write heads to improve access time. The design team, led by Arnold A Cohen, who had joined ERA from RCA's Tube Division in December 1946, had initially considered the use of electrostatic storage tubes but a corporate obsession with reliability resulted in the slower but familiar magnetic drum storage being chosen over the relatively unproven electrostatic technology. The completed machine was delivered to CSAW in December 1950.

Conscious of an unhealthy reliance on the US Navy as its main customer, ERA began efforts to diversify into the commercial sector and the success of the Atlas project prompted the company to request the Navy's permission to produce and sell an unclassified version for the scientific computing market. After some deliberation, this request was granted and the commercial version was announced in December 1951 as the ERA 1101, which is the number 13 in binary notation. However, the 1101 suffered from limited input/output facilities, inadequate documentation and no software. Therefore, despite having a technologically advanced design and proven reliability, none were sold commercially and only two 1101 machines were ever built, one for the Navy and the other for ERA's own use. The latter machine was subsequently used by the company as the basis of a computing service bureau but this venture proved unsuccessful and in November 1954 it was donated to the Georgia Institute of Technology.

By early 1950, development of the Atlas had been progressing sufficiently well that the Navy authorised the construction of a more ambitious model, codenamed Atlas II, which would incorporate both magnetic drum and electrostatic storage. The logical design of the new machine was again carried out by Arnold Cohen, with Jack Hill and Frank C Mullaney leading the engineering team. The binary parallel architecture of the previous model was retained but an increased clock speed of 500 KHz, a longer word length of 36 bits and additional machine instructions all made for a considerably more powerful machine. Memory was provided by 36 Williams electrostatic storage tubes of 1,024 bits each, giving a total of 1,024 words. This was augmented by a directly addressable 16,384 word capacity magnetic drum of similar design to the drum in the 1101. A comprehensive range of input and output equipment included magnetic tape units, paper tape and punched card devices.

The first Atlas II was delivered to the US Armed Forces Security Agency, the forerunner of the National Security Agency, in September 1953. During development, ERA once again sought government permission to produce an unclassified version for the commercial market. This was granted but the company was required to remove several machine instructions from the commercial version, which was announced as the ERA 1103 in February 1953 at a basic price of $895,000. In developing the 1103, ERA had addressed many of the issues that contributed to the commercial failure of its predecessor, the 1101. However, ineffective marketing, compounded by production delays and a lack of customer support, resulted in poorer sales than might otherwise have been expected and only 20 machines were sold.

ERA was an early example of a technology-driven business. It was geared towards delivering products that met the specification of an individual customer rather than addressing the needs of the wider commercial market and this lack of market focus hampered the company's attempts to diversify. Consequently, ERA had limited success in winning private sector business and a downturn in US defence funding in 1950 inevitably led to a period of financial difficulty. ERA's growing reputation in the computer industry also began to suffer through accusations in the press that its cosy relationship with the US Navy and the absence of competitive bidding on Navy contracts amounted to an unfair competitive advantage.

Cost-plus-fixed-fee contracts provide an excellent way of protecting contractors against the unexpected cost overruns that frequently occur in high-technology development projects but they also make it virtually impossible for the contractors to generate significant profits. With a seemingly unbreakable dependence on government cost-plus-fixed-fee contracts, Parker had been aware for some time of a need to raise additional capital in order to grow the business and had already begun sounding out other firms as prospective buyers for the company. Parker's search concluded towards the end of 1951 when he was approached by an executive from Remington Rand with a declaration

6. Data Processing and the Birth of the Computer Industry

of interest in acquiring ERA. However, the secrecy surrounding much of the company's work and a lack of security clearance for the Remington Rand executives created an unusual problem when valuing the business, leading to later speculation that the final figure may have been based simply on the number of engineers employed rather than the company's assets. Nevertheless, by December 1951, Parker was able to announce to the ERA board that he had accepted Remington Rand's offer for the company of approximately $1.7 million in Remington Rand common stock, although the transfer of ownership was delayed until May 1952 due to a requirement to obtain US Government approval for the deal following Remington Rand's earlier acquisition of Eckert-Mauchly Computer Corporation.

Merger with Sperry

Remington Rand was now the proud owner of three separate computer divisions, each with its own distinct line of products; Philadelphia-based EMCC with its industry-leading UNIVAC business model, St Paul, Minnesota based ERA with a range of scientific computers and special-purpose machines, and the original Remington Rand computer division in Norwalk, Connecticut, with its punched card calculators. The three product ranges appeared to complement each other perfectly but the company made little effort to coordinate or unify the activities of its computer divisions and relations between the divisional managers became increasingly hostile.

One positive decision was made early in 1953 to market the ERA and Remington Rand products under the better-known UNIVAC name. EMCC's UNIVAC was renamed the UNIVAC I, ERA models were rebadged as UNIVAC Scientific Computers and the latest versions of Remington Rand's 409 calculator became the UNIVAC 60 and UNIVAC 120. However, the internal problems resulted in the company losing considerable momentum and by early 1955 urgent action was needed if Remington Rand was to remain a key player in the rapidly expanding computer industry. Company Chairman James H Rand Jr. began looking for a suitable commercial partner and soon found a likely candidate in the Sperry Corporation.

The Sperry Corporation had been incorporated in the US in 1933 as a management and holding company for a number of small firms specialising in instrumentation for aviation and military applications. These included the Sperry Gyroscope Company and the Ford Instrument Company, two of the leading manufacturers of analogue computers for gunnery fire control applications. Sperry also had experience of digital computers, having built a prototype magnetic drum machine, the Sperry Electronic Digital Automatic Calculator or SPEEDAC, in 1953. The proposed partnership with Remington Rand looked an attractive proposition. Sperry would gain access to Remington Rand's civilian customer base and Remington Rand needed Sperry's ample reserves of capital for new product development. Consequently, the deal went ahead and the two firms merged in June 1955 to form the Sperry Rand Corporation.

In an attempt to force them to work more closely together following the merger, the three computer divisions were consolidated into a single entity and renamed the Remington Rand UNIVAC Division. ERA co-founder William Norris, who had become known for his tough, no-nonsense management style, was the logical choice to head up the new division and was appointed as its first General Manager in October 1955. However, these efforts were only partially successful and the continuing problems were exemplified in the development of two machines that came out of this period, the large-scale UNIVAC II and the medium-scale UNIVAC File-Computer.

In early 1954, Remington Rand had begun planning a replacement for the now 3-year-old UNIVAC I. Advances in random access memory technology had effectively rendered the serial access mercury delay line obsolete and, as the only large-scale computer on the market with a mercury delay line main memory, the UNIVAC I was now looking decidedly dated. The project was initially conceived as a joint effort between the Philadelphia and St Paul groups, with Philadelphia taking responsibility for design and St Paul for manufacturing. However, the Philadelphia group was already committed to another large-scale computer project for the US Atomic Energy Commission and senior management were concerned that the group would be seriously overstretched with two major development projects running simultaneously. Therefore, despite all the UNIVAC I experience residing in Philadelphia, the decision was taken to hand over both design and manufacturing responsibility to St Paul. Predictably, this resulted in considerable delays.

ERA veteran Jack Hill was given the difficult task of managing the project. The UNIVAC II was designed to be software compatible with its predecessor, therefore, the binary-coded decimal, serial, architecture and 12-digit word length were retained but advances in memory technology favoured the new magnetic core storage over electrostatic storage tube or mercury delay line technology. Accordingly, memory was provided by up to 10,000 words of magnetic core storage in blocks of 2,000 words. The instruction set was improved and extended with 11 additional machine instructions. The high emphasis placed on reliability in the design of the earlier machine was also carried over to the new model so that, where feasible, registers and other circuits were duplicated. As well as an increase in performance over the earlier model, due mainly to the use of faster access magnetic core memory, the new machine also had upgraded input/output facilities. Up to 16 magnetic tape drives could be connected simultaneously and buffer performance was also improved through the use of an additional transfer buffer of 9 words capacity to augment the 60 words of core storage each for input and output.

The UNIVAC II was announced in December 1955 at a purchase price of

6. Data Processing and the Birth of the Computer Industry

approximately $1M for a basic machine. However, the lengthy development delays resulted in the first deliveries not appearing until November 1957, by which time Sperry Rand's competitors had not only introduced similarly specified machines but were close to announcing the next generation of large-scale computers based on transistor technology. Consequently, the new model's impact was limited and, despite an expanding market, only 32 UNIVAC II machines were built, 14 less than the model it replaced.

Development of the UNIVAC File-Computer began in January 1955. Conceived as a joint effort between the ERA group in St Paul and the punched card calculator division in Norwalk, the machine was Remington Rand's response to the growing market for medium-scale magnetic drum computers. However, unlike other machines in this sector, the File-Computer was designed from the outset for on-line storage of large amounts of data.

Building on the St Paul division's industry-leading experience with magnetic drum computers, the File-Computer was based on the ERA Speed Tally System, a special-purpose computer for on-line inventory control built for the John Plain Mail Order Company of Chicago in 1954. It employed a binary-coded decimal, serial, architecture with a 12 decimal digit word length. Memory was provided by a high-speed magnetic drum of 1,070 words capacity operating at 12,000 rpm plus up to 10 additional drums of 15,000 words capacity operating at 1,750 rpm for backing storage. To facilitate data storage operations, the instruction set featured special machine instructions for searching the backing storage drums. A comprehensive range of input/output devices was available, including support for paper tape, punched cards and magnetic tape. Each device was equipped with its own magnetic core input/output buffers of either 12 or 120 digits and up to 31 devices could be connected simultaneously in any combination.

The UNIVAC File-Computer was announced in early 1956. Surprisingly, the initial version, known as the Model 0, was not fully stored-program. Programming was performed externally by means of a 48-step plugboard but with provision for storing additional instructions on the high-speed drum. Despite this severe limitation, around 30 Model 0 machines were built, with deliveries beginning in September 1957. A stored-program version, known as the Model 1, was developed during 1957 and incorporated 20 words of magnetic core memory for storage of instructions. Deliveries of the Model 1 began in June 1958 and most Model 0 machines were subsequently field upgraded to the stored-program version.

The machine's advanced data handling features made the UNIVAC File-Computer particularly well suited to the demands of business data processing and a notable success was its use by Eastern Airlines to implement a seat inventory application. However, its complex hierarchical storage system made the machine difficult to program. Also, at around $560,000 for a complete system, the File-Computer was expensive in comparison with rival products and was late in entering the market. Consequently, only around 190 File-Computers were built.

Sperry Rand's inability to capitalise on its ample resources caused the company to lose ground during a critical period in the formative years of the computer industry. Consequently, by the time that the UNIVAC II and File-Computer models were both announced around the beginning of 1956, the company's early market lead had been all but lost. Sperry Rand's closest competitor did not hesitate to seize the opportunity to catch up and overtake. That company was IBM.

The Inexorable Rise of IBM

In the years following the 1924 renaming of the Computing-Tabulating-Recording Company (C-T-R) as International Business Machines Corporation, the company had grown steadily in size and influence under Thomas Watson's strong leadership. In the immediate post-war period, the firm whose name was to become synonymous with the computer was enjoying a period of sustained success. Although smaller in overall size than Remington Rand, IBM's tighter business focus on tabulating and accounting machines had made it the leading manufacturer in this market sector. The company's avowed policy of renting rather than selling its equipment delivered a steady income stream which allowed it to comfortably ride out periods of economic uncertainty. Its sales force was the envy of the industry and would play a major role in the company's remarkable rise to a position of dominance in the computer industry within only three short years of entering the market with its first computer product.

IBM began introducing electromechanical calculating technology into its punched card products as early as 1931 and had built special-purpose relay calculators for the US Army during World War II but the company was slow to grasp the significance of digital computer technology. IBM's formidable Chairman, Thomas Watson, was initially sceptical that computers would have a significant impact on the data processing industry. However, for a company in IBM's position, it would have been foolhardy to ignore developments in what was obviously a closely related field. Therefore, the company remained tightly wedded to its punched card technology but hedged its bets by making a modest investment in the development of digital computer technology, the initial focus for which was the ill-fated collaboration with Howard Aiken at Harvard University.

Fortunately, IBM's early ambivalence towards computers did not extend to electronics. Electronic technology was gradually introduced into the company's punched card calculator product line through a development programme initiated in October 1943. In September 1946, IBM unveiled the first fruit

6. Data Processing and the Birth of the Computer Industry

of this new programme (and the first commercial product anywhere in the world to feature electronic calculation technology), the Type 603 Electronic Multiplying Punch. It was developed from an experimental multiplier built in 1942 by Ralph L Palmer and Byron E Phelps at IBM's Endicott Laboratories in New York. Boasting over 300 thermionic valves, the 603 was capable of multiplying two 6-digit decimal numbers and punching the 12-digit result at a speed of 100 cards per minute, an order of magnitude faster than its electromechanical predecessor.

An enthusiastic customer response to the Type 603 prompted IBM executives to immediately sanction the development of an improved version and production of the 603 was limited to 100 units. On his return from military service in February 1946, Palmer had been put in charge of a new electronic engineering laboratory at IBM's Poughkeepsie plant it was this group which was tasked with developing the successor to the 603.

The IBM Type 604 Electronic Calculating Punch was announced in July 1948. More versatile than the 603, it was capable of performing all four basic arithmetic operations on 5-digit numbers and could be programmed, via a plugboard and 8 internal registers, to perform sequences of instructions in up to 40 steps, all at a constant operating speed of 100 cards per minute. It also featured 'pluggable' subassembly units, an early form of modular construction, for ease of maintenance of the 1,250 miniature thermionic valves. The 604 was introduced to the market at a rental of $600 per month and was hugely successful, with over 5,600 built before production finally ended more than a decade later. It also paved the way for IBM's later range of business computers, as many 604 customers went on to upgrade to small computer systems.

A Scientific Calculator

IBM's links with the scientific computation community had broken down following the souring of the relationship with Aiken and Harvard. However, a growing awareness within the company of a potential market opportunity for large-scale scientific calculators prompted action to help re-establish these links, the mechanism for which would be the creation of a new research facility at a rival institution. In February 1945, IBM established the Thomas J Watson Scientific Computing Laboratory at Columbia University in New York and appointed Wallace Eckert as its first Director. Eckert, a prominent astronomer, was well qualified to lead this initiative, having pioneered the use of IBM punched card equipment for scientific computation in the 1930s.

Eckert's first assignment at the Watson Lab was to oversee the specification of a large-scale sequence-controlled calculator that would eclipse the achievements at Harvard and showcase IBM's technical prowess to the scientific world. Known as the Selective Sequence Electronic Calculator (SSEC), the machine was designed and built at IBM's Endicott Laboratories by a team led by ASCC veteran Frank Hamilton in the role of Chief Engineer and former Harvard mathematician Robert R Seeber as Chief Architect.

The SSEC was designed and built in less than two years, this impressive development schedule made possible by the judicious use of existing components and subsystems. However, the result was an inelegant hybrid of electromechanical and electronic technologies, incorporating over 21,000 electromagnetic relays in addition to its 12,500 thermionic valves.

The SSEC employed a binary-coded decimal architecture with a 20-digit word length. Fast electronic circuits were used in the arithmetic unit and to provide 8 high-speed arithmetic registers of 20 decimal digits each plus a 28-digit accumulator. These were based on the arithmetic circuits designed by Byron Phelps for the Type 603 Electronic Multiplying Punch. Relay-based circuitry provided 3,000 decimal digits of intermediate storage. This was implemented in 150 separate memory units operated in parallel to increase access speed. Punched tape was employed for storage of instructions and data, and in the implementation of up to 36 look-up tables for frequently used functions. A total of 66 tape readers and punches were installed, with the 7⅜-inch wide tape itself fabricated from uncut IBM punched card stock.

Although not a stored-program computer, the SSEC was capable of performing operations on stored instructions. Because instructions were represented internally in the same form as data, they could be loaded into memory, modified and then executed. Decisions on modifications could be based on intermediate results in what was essentially an extension of the conditional branch feature first implemented in the company's punched card calculators which triggered an alternative operation on sensing a change of sign in a specific register. Though somewhat limited, the SSEC's instruction modification capability was sufficiently innovative to allow IBM to file a US patent application in January 1949. This would place the company in a strong patent position when the stored-program concept was successfully realised by other groups later that same year.

The completed SSEC was installed towards the end of 1947 on the ground floor of a building adjoining the IBM Headquarters Building on New York's Madison Avenue where it was visible to the public walking past. The machine cabinets lined the walls of a room of 60 by 30 feet, its gargantuan proportions and myriad arrays of flashing lights attracting considerable attention and defining the image of an 'electronic brain' in the public imagination for years to come. A formal dedication ceremony took place in January 1948 in which Watson magnanimously dedicated the machine "to the use of science throughout the world".

6. Data Processing and the Birth of the Computer Industry

IBM Selective Sequence Electronic Calculator (Photo © IBM Corporation)

Despite the difficulty in programming the SSEC due to its cumbersome hierarchical memory, it was extensively used by both Columbia University and the US government for scientific research. One example of this work was Eckert's calculations of the moon's orbit which he published in 1954 and were later used by NASA for the Apollo moon landings. The SSEC remained in operation until July 1952 when it was dismantled to make way for IBM's first true stored-program computer, the Defense Calculator.

Listening to the Customer – The CPC

In the absence of commercially available digital computers, the ability of punched card tabulating machinery to perform rapid calculations had been attracting the attention of scientists and engineers. As early as 1928, the New Zealand born astronomer Leslie Comrie was experimenting with tabulating equipment for scientific computation at the H.M. Nautical Almanac Office in the UK. Comrie successfully used Hollerith punched card tabulating machinery to calculate the motions of the moon. His work directly influenced Wallace Eckert at Columbia University and through Eckert's links with IBM, first brought scientific computation to the attention of the company.

In 1947, engineers at the Northrop Aircraft Company in California began interconnecting IBM punched card calculators and tabulating machines in order that results could be transferred between machines electronically rather than manually via decks of punched cards. This led to a discussion between the company and IBM's Director of Engineering, John C McPherson, on the possibility of modifications to the standard punched card machinery to improve control and provide additional storage. IBM's willingness to keep Northrop, an important customer, happy resulted in the construction of a prototype machine at Endicott in 1948, which was based on a combination of a

Type 603 Electronic Multiplying Punch and a Type 405 Accounting Machine.

Following the publication of an article by Northrop's George Fenn describing the new machine, requests for similar configurations soon began to arrive from other IBM customers and this encouraged the development of a commercial product. Announced in May 1949 as the Card-Programmed Electronic Calculator or CPC, the production version consisted of three standard units linked together; a 604 Electronic Calculating Punch, 402 Accounting Machine and 521 Gang Summary Punch, and up to three 941 Auxiliary Storage Units. The enhanced sequencing capability provided by the replacement of the 603 with a 604 in the production version facilitated the implementation of fairly complex mathematical functions. With rental set at an affordable $1,500 per month, the CPC became popular with many of IBM's scientific and engineering customers, and nearly 700 systems were built before advances in low cost stored-program computers eventually rendered it obsolete.

The Defense Calculator

Following the outbreak of the Korean War in June 1950, IBM Chairman Thomas Watson was keen for his company to display its patriotic credentials by contributing in some way to the war effort. The promise shown by the company's research activities in high-speed memory and magnetic tape storage, and the popularity of the CPC with scientific and engineering customers, suggested that IBM could make a useful contribution by building special-purpose large-scale electronic computers that would serve the rapidly increasing computational needs of the US defence industry.

To gauge the commercial viability of this idea, visits were made to a number of prospective customers by IBM's Director of Product Planning James W Birkenstock and Applied Science Divisional Director Cuthbert C Hurd. This resulted in 30 letters of intent and also confirmed the view that a single general-purpose design could satisfy the majority of customer requirements. However, with no likelihood of a development contract from a specific customer to help offset development costs, it also meant that IBM would have to bear the full cost of development. Fortunately, the project was championed at senior management level by Watson's son, Thomas J Watson Jr., who had risen to the position of Executive Vice President and had advocated IBM's post-war research focus on electronics. Watson Jr. saw the project as an opportunity to ease the company into the commercial computer business and pushed through the decision to proceed in December 1950.

Following approval of a $3 million budget, work started at the IBM Poughkeepsie Laboratory in February 1951. The new machine, which had been codenamed the Defense Calculator, was to be a stored-program computer, its logical design strongly influenced by the work of John von Neumann at the Institute for Advanced Study. In charge of development was Jerrier A Haddad

6. Data Processing and the Birth of the Computer Industry

who, assisted by Nathaniel Rochester, led a team which peaked at 155 staff. The machine employed a binary parallel architecture with a 36-bit word length, the capability to address and process half-length words, 3 arithmetic registers and a clock speed of 1 MHz. Following the IAS example, electrostatic storage tubes were chosen as the storage technology for the new machine. The Williams Tube technology was licensed from Manchester University and further developed by Arthur L Samuel for use in the 701. A main memory of 2,048 36-bit words was provided via 72 IBM-manufactured Williams Tubes of 1,024 bits each. This was supplemented with two magnetic drums of 4,096 words storage capacity each.

IBM's engineers compensated for the somewhat conservative design of the Defense Calculator by employing innovative packaging. For ease of installation, the machine was configured as 11 separate units, each small enough to fit through a standard doorway. Pluggable subassembly units, first used in the 604 calculating punch, were also employed for ease of maintenance of the 4,000 miniature thermionic valves.

Input and output was provided by two Model 726 magnetic tape units, each fitted with two tape drives, a card reader, a card punch and a printer. The use of magnetic tape in the Defense Calculator was a bold move for a company whose business was built on punched card technology. Magnetic tape storage had been under development by Ralph Palmer's group at Poughkeepsie since late 1949 as part of an experimental system known as the Tape Processing Machine or TPM. IBM's magnetic tape technology differed from Remington Rand's pioneering UNISERVO magnetic tape technology in that it used conventional plastic audio recording tape, supplied by the 3M Company, rather than unwieldy metal tape. However, with a read/write performance of 7,500 characters per second, IBM's first production tape unit was considerably slower than Remington Rand's UNISERVO I unit which could read and write data at up to 12,800 characters per second. Furthermore, the IBM drive was unable to read data in both directions, which made data sorting operations difficult.

Rental for the new machine was initially set at $11,900 per month, a significantly higher figure than the estimates that Birkenstock and Hurd had used to obtain the letters of intent from prospective customers. Despite this higher price, a total of 19 machines were built, the first of which was retained by the company as a demonstrator and replaced the SSEC at IBM's World Headquarters in New York in December 1952. A formal dedication ceremony took place in April 1953 where the new computer was introduced as the 701 Electronic Data Processing Machine, named after the model number of its control unit. The Electronic Data Processing Machine (EDPM) label was coined by James Birkenstock in an effort to avoid the confusing term 'computer', with its connotations of human calculators, whilst reflecting the office equipment nature of IBM's core business.

The Story of the Computer

The announcement of the 701 on 27 March 1953 marked IBM's official entry into the computer business. The 701's performance compared favourably with that of the ERA 1103, which had been announced the previous month. However, its success was short lived, as the machine proved hopelessly unreliable in the field. The poor reliability of the electrostatic storage tube memory and the absence of any error detection capability in the machine's design resulted in critical programs having to be run twice and the 701 was quietly withdrawn as soon as a replacement model was announced in May 1954.

A Computer for Business

Having committed to the development of commercial machines for scientific computation, IBM now turned its attention to the business data processing market. The loss of one of its oldest customers, the US Census Bureau, to Remington Rand provided the motivation for a project to develop a large-scale stored-program computer suitable for business use and growing pressure from the sales department for a new machine that could be offered to customers for business applications strengthened the case considerably. By early 1953, the order was given to proceed with the development of a new computer which would be called the IBM 702 Electronic Data Processing Machine.

The 702 was specified by Stephen W Dunwell and based on the experimental TPM machine that Dunwell had also specified three years earlier. Studies were conducted to help determine customer requirements and refine the specification. The resulting machine employed a binary-coded decimal, serial, architecture with 3 arithmetic registers and a 1 MHz clock speed. Business-oriented features included a variable word length and the ability to handle character-based data represented as 6-bit alphanumeric characters. As the project shared development team members with the 701, it was inevitable that components would migrate between projects. Therefore, in common with the 701, the new machine employed electrostatic main memory plus a magnetic drum for intermediate storage and magnetic tape for bulk storage. Storage was provided by 84 Williams Tubes of 1,000 bits each, giving a memory capacity of 10,000 alphanumeric characters and two 512 character accumulators, and up to 30 magnetic drums of 60,000 characters storage capacity each.

An improved magnetic tape unit was developed for the new machine which could operate at twice the speed of the tape unit used in the 701 and was also capable of reading data in both directions. Flexible input/output arrangements on the 702 permitted the connection of up to ten of the new Model 727 magnetic tape units (each fitted with up to ten tape drives), card readers, card punches and printers. Another interesting feature was the early use of magnetic core memory in the buffers of the input and output devices.

The 702 was announced September 1953 but deliveries did not begin until February 1955, more than three years after the introduction of Remington

6. Data Processing and the Birth of the Computer Industry

Rand's UNIVAC I. With several major computer development projects running simultaneously, IBM had overstretched itself and was unable to deliver the 702 within the expected timescale. During this period, it also became clear that the IBM machine compared poorly to the UNIVAC I. Despite the addition of parity error detection to the electrostatic storage tube memory, the 702 suffered from the same reliability problems as the 701. It was also relatively slow, due to the variable word length and inadequate buffering of data between main memory and the input/output devices which severely limited the machine's performance. Consequently, the 702 was withdrawn as soon as a replacement model was announced in October 1954 and only 14 were built.

An Unforeseen Success – The Magnetic Drum Calculator

While the Poughkeepsie Laboratory was developing the 701 and 702 large-scale machines, the calculator group at IBM's Endicott Laboratories had also been busy developing their own stored-program computer. In 1948, a project was initiated under Frank Hamilton at Endicott to develop a small computer for scientific applications. The initial plan was to extend the 604 Electronic Calculating Punch using technology developed for the SSEC but by early 1949 the design had evolved to include a magnetic drum storage device and a stored-program architecture.

The project team began by developing their own magnetic drum storage device but progress was slow and IBM senior management, who had earmarked magnetic drum technology for use in another project, initiated discussions with Engineering Research Associates to supply drum devices based on ERA's more mature drum technology. These discussions led to a proposal from ERA to design a complete magnetic drum computer for IBM.

IBM management recognised the need for a stored-program successor to the CPC but were faced with a difficult decision on whether to proceed with the Endicott machine, now called the Magnetic Drum Calculator, the ERA proposal or one of two prototype medium-scale machines which were under development at Poughkeepsie. ERA was authorised to proceed with a design study for a magnetic drum computer with punched card input. However, Cuthbert Hurd, who as head of the Applied Science Division, was ultimately responsible for marketing IBM's scientific computing products, favoured the Magnetic Drum Calculator on the basis of ease of programming and this was a major factor in the decision in November 1952 to select the Endicott machine for production.

The Magnetic Drum Calculator employed a biquinary-coded decimal, serial, architecture with a 10 decimal digit word length and a modest 125 KHz clock speed. It featured a fast access magnetic drum, designed by Al Brown at Endicott, which rotated at a high speed of 12,500 rpm and had a storage capacity of 1,000 or 2,000 words depending on the number of read/write heads fitted. The machine also featured a look-up table for frequently used functions and a rudimentary hardware interrupt system for machine errors. Interrupts are an important feature of most modern computers and provide a means of signalling the processor following a specific event to allow a response to the event to be triggered under program control. The interrupt on the Magnetic Drum Calculator removed the need for a manual restart following detection of a machine error by permitting an instruction to be automatically loaded from a specific memory location.

The Magnetic Drum Calculator was announced in July 1953 as the IBM Type 650 Magnetic Drum Data Processing Machine at a rental of $3,250 per month for a model with 1,000 words of drum storage or $3,750 per month for the 2,000 word model. The first machine was delivered in December 1954. An Auxiliary Unit, the Type 653, announced in May 1955, provided 60 words of high-speed magnetic core storage to improve performance and could also be programmed to operate as a buffer memory for magnetic tape data.

IBM Type 650 Magnetic Drum Data Processing Machine (Photo © IBM Corporation)

Although the 650 had been designed and initially marketed for scientific and engineering applications, its decimal architecture was more typical of a business data processing model and its punched card calculator origins had also bestowed excellent compatibility with existing punched card equipment. Consequently, the 650 became popular for business applications and many of IBM's punched card customers chose to make the relatively painless upgrade to a 650 rather than the huge leap to a large-scale 702. In a shrewd marketing move, IBM also encouraged universities to adopt the 650 by offering a 60% discount, or in some cases supplying machines free of charge, in return for agreeing to run courses in data processing or scientific computing. In all nearly

6. Data Processing and the Birth of the Computer Industry

2,000 were built before production ended in 1962, making the 650 the world's most popular computer during this period.

The First Disk Drive

A contributing factor to the sustained success of the 650 Magnetic Drum Calculator was the development of an early magnetic disk drive, the IBM Type 350 Disk File. The project to develop the 350 Disk File was started in September 1952 by Arthur J Critchlow under the direction of Laboratory Manager Reynold B Johnson at IBM's recently established Research and Engineering Laboratory in San Jose, California.

Critchlow's remit was to seek a better method of storing and accessing data for punched card calculating equipment. All manner of storage technologies were considered until the idea of a continuously rotating magnetic disk was arrived at through the discovery of an August 1952 journal article by the prolific inventor Jacob Rabinow of the National Bureau of Standards which described a 'Notched-Disk Memory'. As it stood, Rabinow's complicated multi-disk device was not an ideal solution as the design, which featured notches or segments removed from each disk to permit the read/write head to pass from one disk to the next, restricted the speed of rotation. Nevertheless, the team could envisage its potential. A configuration based on a disk-shaped magnetic surface would have a much larger surface area than an equivalent sized drum but would be much more challenging mechanically due to the need for a more complex mechanism for positioning the read/write heads.

The magnetic disk configuration was adopted and over the following two years the San Jose project team refined the design and eliminated the problems associated with the head positioning. A novel air bearing system was developed with the help of consultant Al Hoagland, who had worked as a postgraduate student on the CALDIC magnetic drum memory at the University of California Berkeley. This used compressed air fed through small jets in the read/write heads to maintain the critical 0.001 inch gap between the heads and the disk surface. The final design incorporated 50 24-inch diameter aluminium disks stacked on a cantilevered spindle rotating at 1,200 rpm with data recorded in 100 concentric tracks on each side of each disk. This arrangement provided 5 million 7-bit characters of data storage, a considerably higher capacity than any similar storage device of the time. However, at 600 milliseconds, access time was much slower than a typical fixed-head magnetic drum. This was due to the use of only two pairs of read/write heads rather than a pair for each track and a positioning mechanism which relied on a servo-controlled arm to move the heads to the appropriate disk in the stack.

The 350 Disk File was first used in the IBM 305 RAMAC (Random-Access Memory Accounting Machine), a small business computer which was announced in September 1956. It was also available on the 650 RAMAC, an enhanced version of the 650 Magnetic Drum Calculator announced at the same time. Rental for the 350 Disk File was set at $650 per month.

IBM was not the first manufacturer to develop magnetic disk storage technology. The Technitrol Engineering Company had patented a magnetic disk storage device in September 1952 which was developed as part of its 'Reservisor' computerised airline seat reservation system. The UK's Elliott Brothers had also built a magnetic disk device for the Elliott 401 prototype computer which was completed in January 1953. Nevertheless, the 350 Disk File was the first commercially successful implementation of magnetic disk technology and its design proved highly influential. The 350 was the forerunner of the modern hard disk drive and its introduction on 14 September 1956 was a major milestone in information storage technology.

Large-Scale Progress

By early 1954, IBM's hopes that its large-scale 701 and 702 machines would provide strong competition for Remington Rand's UNIVAC I and 1103 models had begun to fade. Both machines had failed to set the market alight due to questionable reliability, indifferent performance and inadequate input/output. However, with a strong service-based culture, IBM took great pride in its ability to respond rapidly to customer needs. In April 1954, general sales manager T Vincent Learson was appointed to the new post of Director of Electronic Data Processing Machines and tasked with coordinating activities to better meet these needs. Learson set about this task by assigning top priority to a product development programme that would finally deliver IBM's first competitive large-scale computers, the 704 and 705.

Planning for an improved version of the 701, designated the 701A, had begun in early 1953 and was led by rising star Gene M Amdahl. Amdahl had joined IBM from the University of Wisconsin in June 1952, having built an electronic computer as part of his PhD research. Initially, the only significant change planned was the replacement of the troublesome electrostatic memory with magnetic core storage, a relatively straightforward engineering task as IBM's work on the development of magnetic core storage technology was already well advanced through the company's role as a contractor for Project SAGE, a US government funded project to develop a network of powerful computers for an air defence application. However, input from John Backus, an IBM programmer who had been developing programming techniques for the 701, led to the inclusion of several important new features and the final design was deemed sufficiently different from the 701 that it was given a new model number, the Type 704.

The 704's new features included 3 index registers, hardware-based floating-point arithmetic and additional machine instructions to facilitate access to the new hardware. Developed for the Manchester Mk 1 and first used commercially

6. Data Processing and the Birth of the Computer Industry

in the Ferranti Mark 1, index registers greatly simplified a computer's ability to perform operations on arrays of numbers, a common requirement in most scientific computing applications. Floating-point hardware, which automatically keeps track of the decimal point location when performing calculations using floating-point arithmetic, was first implemented by Konrad Zuse in the electromechanical Z3 in 1941. It had recently become available as an option on the medium-scale ElectroData Datatron but the 704 was the first commercially available large-scale machine to incorporate this feature as standard.

The performance of the new model was almost double that of the 701 despite the same 1 MHz clock speed due to a faster magnetic drum and shorter access time of the magnetic core storage. Main memory capacity was quadrupled to 8,192 words of magnetic core storage and considerable effort also went into providing software support for the new machine, most notable of which was the seminal high-level programming language FORTRAN. John Backus also led the IBM team that created FORTRAN for the 704 in November 1954 as part of his work on easier and more cost-effective methods of programming computers. FORTRAN compilers were supplied as standard with the 704 from April 1957.

The Type 704 Electronic Data Processing Machine was announced in May 1954. Deliveries began in January 1956 but production delays with the magnetic core memory meant that early examples had to be supplied with electrostatic storage tube memory and subsequently upgraded to magnetic core memory in the field. Nevertheless, IBM's second attempt to develop a large-scale machine for the scientific computing market had hit the mark. The improved reliability, powerful new features and supporting software ensured the popularity of the new model amongst customers, despite a high rental cost of around $30,000 per month for a typical installation. The 704 became the first commercially successful large-scale scientific computer, with a total of 123 machines built over a 6-year lifespan. Its success eclipsed that of its nearest rival, the UNIVAC 1103, prompting Remington Rand to introduce similarly enhanced versions of the 1103 with magnetic core memory, the 1103A in 1955, and floating-point hardware, the 1103AF in 1956.

The development of a replacement for the 702 business model, designated the 705, was led by Werner Buchholz. As with the 704, development effort initially concentrated on the use of magnetic core storage to replace the unreliable electrostatic storage tube memory but the designers also took the opportunity to reconfigure the memory into groups of 5 characters to further improve memory access speed. Other improvements over the 702 included a doubling of memory capacity to 20,000 alphanumeric characters, a revised instruction set and extended input/output facilities. Close attention was also paid to the design of the input/output equipment, in particular the card reader and record storage unit which both incorporated a small amount of magnetic core memory for buffering of data.

Announced in October 1954 at a rental of $14,000 per month for a basic system, deliveries of the 705 began in early 1956. IBM finally had a product to rival the UNIVAC I and Sperry Rand's failure to deliver a suitable successor until 1958 provided the perfect opportunity for IBM's formidable sales and marketing operation to establish the 705 as the new leader in the large-scale business sector of the marketplace. However, inadequate input/output performance remained an issue until the Type 777 Tape Record Coordinator became available with the 705 Model II later in 1956 and the announcement of the 705 Model III with internal buffering in 1957. Nevertheless, the 705 models proved very successful for IBM and a total of 175 machines had been produced by the time it was withdrawn in April 1960.

Strong demand for the 704 and 705 combined with the phenomenal success of the 650 which had begun in late 1954, gave IBM a considerable market lead over Sperry Rand and by the end of 1956 the company had secured an unassailable 75% share of the US computer market.

A Third Party

Although Remington Rand and IBM came to dominate the early years of the computer industry, it was by no means a two-horse race. The third US office equipment manufacturer with an influential role in the development of the computer during this period was the Burroughs Corporation. While Burroughs was the smallest of the three, its computers were highly regarded for the degree of technological innovation shown in their design and the near-fanatical loyalty they inspired amongst users.

The company originated in 1886 when the American Arithmometer Company was formed in St Louis, Missouri, by inventor William Seward Burroughs and three businessmen, Thomas Metcalfe, Richard M Scruggs and William Pye, to manufacture mechanical calculators to Burroughs' designs. Following a relocation to Detroit in 1904, the company was renamed the Burroughs Adding Machine Company in honour of its founder who had died from tuberculosis in 1898. A series of acquisitions followed and by the 1920s Burroughs was the leading supplier of adding and bookkeeping machines to the banking industry. The product line was gradually expanded to include other items of office equipment, such as typewriters, and in 1953 the company's name was changed to the Burroughs Corporation to reflect the changing nature of its products.

Early Computer Projects

In 1948, Burroughs was in a comfortable financial position. Revenues had exceeded $100M for the first time and the company was also enjoying success in new territories. Having directed a successful programme of retooling and

6. Data Processing and the Birth of the Computer Industry

expansion since assuming control of the company two years earlier, Burroughs President John S Coleman now began planning for the future. Recognising the potential of electronic computers in the office equipment market, Coleman hired the Moore School of Electrical Engineering's controversial Supervisor of Research, Irven Travis, and gave him a remit to establish an electronic digital computer division. Permanent facilities for electronic research and development were established in the Philadelphia area in 1949, with particular emphasis placed on the development of electronic circuits for high-speed calculation. This growing electronics R&D capability was augmented in 1951 through the acquisition of the Control Instrument Corporation, a small New York based supplier of electronic fire control equipment to the US Navy.

By February 1951, the newly established Burroughs Research Center in Paoli had built a working prototype of a stored-program scientific computer, the Burroughs Laboratory Computer. It employed a binary-coded decimal, serial, architecture with a 10 decimal digit word length, a small magnetic drum memory of 800 words capacity supplied by ERA and a modest clock speed of 125 KHz. An interesting feature was the adoption of standard components from the company's pulse control equipment for the arithmetic unit. This machine was renamed the Unitized Digital Electronic Calculator or UDEC and further developed and upgraded over the following two years to include 100 words of magnetic core storage plus an increase in magnetic drum storage capacity to 5,300 words. In December 1953, the UDEC was installed at Wayne University (now Wayne State University) in Detroit. A second prototype, the UDEC II, which was equipped with 1,000 words of magnetic core memory and a 10,000 word magnetic drum, was completed in October 1955 for use by the company's Electronic Instruments Division.

The Burroughs Research Center also conducted a series of design studies for a large-scale business data processing machine over the period 1950-55 under the project title Burroughs Electronic Accounting Machine or BEAM. The last of these designs, the BEAM IV, was progressed to the prototype stage but events led to the project being cancelled before the machine could be completed. Nevertheless, the project spawned the development of a new electronic device, the beam switching tube, a type of decade counter packaged as a thermionic valve, and this was to prove a useful component for the company's only commercial computer of this period, the E101 Desk Size Electronic Computer.

The E101

Having built a prototype medium-scale computer and initiated the design of a large-scale machine, Burroughs now turned its attention to the development of a small-scale model that could fill the gap between manual desk calculators and stored-program computers. The project to develop such a machine was initiated in 1953 and resulted in the introduction of Burroughs first commercial computer product, the E101 Desk Size Electronic Computer.

The Story of the Computer

The E101 was a compact, sequence-controlled electronic calculator. It featured a binary-coded decimal, serial, architecture with a 12-digit word length and was programmable via removable pinboards. These were functionally similar to plugboards only without the unwieldy mass of connecting cables, which made them more compact and simpler to use. Up to eight pinboards could be used simultaneously, each of which contained 16 locations and up to 3 pins per instruction, giving a total of 128 program steps in an instruction sequence. Pinboards could also be swapped over during a run to further increase the number of steps. Programs were 'stored' on card templates which were fitted over the pinboard and marked by the programmer to indicate which holes the pins should go into.

Besides the use of removable pinboards, the E101 also differed from rival products in that it featured a small magnetic drum memory with a capacity of 100 words (expandable to 220 words) for the storage of data. The use of a magnetic drum for data storage significantly reduced the requirement for bulky thermionic valves, only 160 of which were needed, and made for a very compact design that allowed the machine to be packaged as a desk-shaped unit. Input and output was provided by a Keyboard Printer unit, essentially a modified Burroughs bookkeeping machine, which incorporated an 11-column keyboard for input and a semi-gang printer with formatting capability for output. An optional paper tape reader was also available.

Burroughs E101 (Photo © Burroughs Corporation)

The E101 was announced in May 1954 at a purchase price of $32,500 or a rental of $850 per month and deliveries began in November 1955. However, despite an innovative design, high specification and ease of programming, the E101 failed to capture an appreciable share of the sequence-controlled electronic calculator market due to its incompatibility with punched card equipment. Consequently, only 127 examples of the E101 and its successor, the E102, were built. Burroughs carefully planned entry into the electronic computer business was in danger of failing almost as soon as it had begun.

ElectroData Corporation

One of the early pioneers of magnetic drum computers was Consolidated Engineering Corporation (CEC), a scientific instrument manufacturer based in Pasadena, California. CEC had built a prototype general-purpose magnetic drum computer, the Model 30-201, which had been developed in-house over the period 1951-52. The impetus behind this work came from the company's Assistant Director of Research, Clifford Berry of Atanasoff-Berry Computer fame. The logical design of the machine was carried out by the Norwegian cryptologist Ernst S Selmer, who had been employed by CEC as a consultant on the recommendation of John von Neumann.

To make the machine as flexible as possible, the Model 30-201's designers had cleverly incorporated features common to both business-oriented models and scientific computers. The basic design was that of a typical business data processing machine of the time, with a binary-coded decimal, serial, architecture, a 10-digit word length and a modest clock speed of 142 KHz. However, a four-digit index register was provided and there was also a floating-point hardware option available, the first of its kind on a commercial computer, predating the floating-point hardware on the IBM 704 by almost two years. Memory was provided by a 4,000 word capacity magnetic drum device of similar design to the University of California Berkeley CALDIC magnetic drum but the CEC engineers had also implemented a high-speed section of 80 words capacity. The drum rotated at a conventional 3,750 rpm but high-speed access was effected on one section through the use of 4 recirculating loops of 20 words each, created between pairs of read/write heads and repeated several times at equal intervals around the circumference of the drum. These were in addition to the arithmetic registers which were implemented electronically using flip-flops.

In 1953, the CEC computer division was spun out as the ElectroData Corporation to commercialise the company's computer technology and the production version of the Model 30-201 was introduced as the ElectroData Datatron. First deliveries of the Datatron took place in June 1954. The purchase price of a basic system was $225,000, which was considerably cheaper than rival machines of similar performance. The Datatron was moderately successful, with 16 machines built, and this encouraged ElectroData to produce an upgraded version with minor improvements, the Datatron 205, which was announced in July 1954. This new model proved a resounding commercial success for ElectroData, with more than 100 machines built.

A key factor in the popularity of the Datatron 205 was the impressive range of input/output equipment available, in particular the Cardatron punched card unit. The Cardatron was developed to provide a buffered punched card facility for the Datatron. It incorporated a small magnetic drum memory to buffer the flow of data between the input/output devices and the Datatron processor, and a plugboard controller which allowed for basic off-line processing to be performed on punched cards to improve throughput. Up to 10 magnetic tape devices plus a combination of paper tape devices and printers could be connected to a single Datatron machine.

Although Burroughs had been quick to recognise the potential of electronic computers, the company's progress with the development of stored-program machines was painfully slow due to the diverting of precious R&D resources onto military projects. Major projects undertaken by Burroughs for the US military during this period included a contract from the US Army in 1951 to design and construct a magnetic core memory upgrade for the ENIAC and the development of a special-purpose computer for the US Air Force, the AN/FST-2 Coordination Data Transmitting Set which was completed in 1955.

By 1956, Burroughs found itself in a similar position to that of Remington Rand a few years earlier; it only had one computer product on the market, the small-scale E101 calculator, and was being left behind by the competition. An acquisition was the only way to recover lost ground. Fortunately, ElectroData already had a medium-scale stored-program machine in production that Burroughs could start selling, the Datatron 205, and was working on the design of the next generation model. In June 1956 Burroughs purchased the ElectroData Corporation in exchange for $20.5 million in Burroughs stock, a huge sum in comparison to what Remington Rand had paid for both EMCC and ERA but one that reflected the increased size and prospects of the computer industry in the few years that had passed since those earlier deals.

The Burroughs 220

Following the acquisition of ElectroData, a decision was taken to cancel the BEAM IV large-scale computer development in favour of ElectroData's next generation model which became known as the Burroughs 220. Burroughs had gained valuable experience in magnetic core storage technology through the contract to build a core memory for ENIAC. Therefore, following the industry trend, the B220 was conceived as a magnetic core replacement for the successful Datatron 205 model.

The designers of the B220 stuck to the design philosophy of earlier ElectroData models by incorporating both business and scientific features. The binary-coded decimal, serial, architecture and 10-digit word length were retained but the new machine featured a heavily revised and extended instruction set. In order to maintain compatibility with earlier Datatron models, an early form of

6. Data Processing and the Birth of the Computer Industry

software emulator was developed which translated Datatron 205 instructions into code that could be run on the B220. The Datatron's magnetic drum storage was replaced by a magnetic core memory of 2,000 words which could be expanded in blocks of 1,000 words up to a maximum of 10,000 words. The four-digit index register was also retained and floating-point hardware was now included as standard.

An extensive range of input/output equipment was also available for the B220, which included the DataFile, a novel magnetic tape storage device that used 50 parallel loops of magnetic tape to significantly improve access time and increase storage capacity to a maximum of 4,880,000 words. Up to 10 magnetic tape devices, in any combination of DataFiles and conventional magnetic tape units, could be connected simultaneously, giving the B220 a massive potential data storage capacity.

The B220 was announced in early 1957 at a price of $320,000 for a basic system and deliveries began in October 1958. Although considerably more expensive than both the model it replaced and the market leading IBM 650, the magnetic core B220 delivered superior performance and also incorporated many of the features usually associated with large-scale machines. This made the B220 particularly attractive to those customers seeking to upgrade from a magnetic drum computer but who were unable to justify the cost of a large-scale system. However, the B220's technical superiority was tempered by it being one of the last thermionic valve machines to be introduced. The next generation of production machines based on transistors was imminent and the B220's days were numbered almost as soon as deliveries began. Consequently, only around 50 were built and its development, which took place towards the end of the valve era, also contributed to Burroughs late entry into the transistor market.

Across the Pond

Although the computer industry became dominated by US manufacturers from the mid 1950s onwards, successful indigenous computer industries had also began to flourish in other countries during this period, notably in the UK, Germany and France. The most successful of these home-grown manufacturers were able to compete with the US firms both in their own and in overseas markets. After the Americans, the British were the next most eager nation to embrace the commercial market for computers and a number of firms were active in establishing a viable computer industry in the UK. The ingenuity that was evident in the early British computer projects can also be seen in many of the machines developed by UK manufacturers during this period.

Elliott Brothers

Predating both Ferranti and J Lyons & Company in its involvement with the development of electronic digital computers in the UK was Elliott Brothers (London) Limited, a long established firm of instrument makers which could trace its origins back to 1800 and included none other than Charles Babbage among its early customers. Unlike all of its UK competitors, Elliott's computer technology was developed entirely in-house, a remarkable achievement for a relatively small company.

Elliott Brothers had moved into analogue computation early in the 20th Century with the introduction of the Dumaresq in 1904. The company also manufactured the Admiralty Fire Control Table which was extensively used by the Royal Navy during World War II. In 1947, the company became involved with the development of electronic digital computers as part of an Admiralty contract to develop an advanced digital fire control system for the Royal Navy. This project required a special-purpose real-time computer to be developed for on-line control. The prototype was completed by 1950 and was known as the Type 152 Naval Gunnery Control Computer. It incorporated Williams Tubes for storage and a parallel arithmetic unit to improve calculation speed. Another interesting feature was the pioneering use of modular electronic circuits, which were developed by Charles E Owen to simplify the maintenance and repair of the computer when at sea.

Although the digital fire control system was ultimately rejected by the Navy in favour of one based on analogue technology, the valuable experience gained on the project prompted Elliott Brothers to develop a general-purpose computer for internal use. The machine, known as NICHOLAS, was operational by December 1952 and incorporated a novel magnetostrictive delay line memory device developed by physicist William S (Bill) Elliott. The magnetostrictive delay line used tensioned nickel wires instead of mercury to carry acoustic pulses which were propagated electromagnetically but otherwise functioned in a similar manner to a mercury delay line.

In 1949 the British Government established the National Research Development Corporation (NRDC) to encourage the commercialisation of public sector research. High on the organisation's list of priorities was electronic computers. It was an NRDC contract that had supported Ferranti in the development of the Mark 1* and the Corporation would also play a leading role in bringing Elliott's innovative computer technology to market. In September 1950, the NRDC placed a contract with Elliott Brothers for a study on the application of standardised printed circuits to general-purpose computers. This was followed in April 1952 by another contract to build a small prototype machine that would serve as a demonstrator for the company's modular circuit techniques.

The prototype computer was completed in January 1953 and was designated the Elliott 401. In addition to modular electronics and a magnetostrictive

6. Data Processing and the Birth of the Computer Industry

delay line memory, this innovative machine also featured one of the first examples of a magnetic disk storage device. The success of the prototype led to the development of a series of production machines, the first of which, the Elliott 402, was introduced in 1955. The Elliott 405, a business data processing model which was introduced in 1956, was marketed by NCR in the US and international markets and a total of 30 were sold.

Despite the success of the Elliott 401 prototype, all was not well at Elliott Brothers. Following cutbacks in Admiralty support and the departure of John F Coales, the Director of the company's Research Laboratories, morale began to deteriorate. In September 1953, several key members of the Elliott 401 design team, including Bill Elliott and Charles Owen, resigned and moved to rival firm Ferranti. The mathematician Christopher Strachey, who had been responsible for evaluating the progress of the Elliott 401 project, persuaded the NRDC to continue its support for the team's modular circuit technology at Ferranti through the development of a more ambitious computer, the Ferranti Packaged Computer or FPC1. Strachey's design specification for the new machine also required the development of a radically different logical design, known as a general register set architecture, which was based on general-purpose registers that could function as either accumulators or index registers. This innovation greatly simplified programming and proved highly influential in subsequent computer designs.

The production version of the FPC1, the Ferranti Pegasus, was introduced in March 1956 at a basic price of around £40,000 and was a modest commercial success for the company, with a total of 26 machines delivered. Pegasus is often credited with introducing modular construction techniques to the computer industry but some of this credit must also be shared with its older relative, the Elliott 402, and with IBM for its pluggable units.

English Electric

Another UK firm with an active involvement in the commercial development of the computer was the English Electric Company. English Electric was formed in 1918 through the merger of a number of firms in the UK heavy electrical industry. The company became interested in electronic computers in 1949 as a result of the then Chief Executive, Sir George Nelson, learning of the National Physical Laboratory's plans to develop a computer through his seat on the NPL governing council. This led to the company being selected as the industrial contractor to assist with the completion of the Pilot Model ACE.

A group of engineers and technicians was seconded from English Electric to work with the ACE project team and this collaboration resulted in the company adopting the Pilot Model ACE design for its first commercial offering, the Digital Electronic Universal Computing Engine or DEUCE. Introduced in 1955 at a purchase price of £55,000, the medium-scale English Electric DEUCE proved to be a reasonably popular model, despite the retention of the NPL prototype's unfashionable mercury delay line main memory, and more than 30 were sold in the UK and overseas markets before production ended in 1964.

BTM

The final firm with an important role in the early stages of the UK computer industry was the British Tabulating Machine Company. BTM was founded in 1907 and the following year acquired an exclusive license to manufacture and distribute Herman Hollerith's punched card tabulating machinery in the UK and the rest of the British Commonwealth except Canada. In October 1949, IBM terminated this non-competition agreement and replaced it with a less favourable reseller agreement that only permitted BTM to market and sell IBM products. When it became clear within a few years that computers were the future of business data processing, IBM's decision prompted BTM to seek out an alternative source of computer technology.

A series of meetings were held with the Burroughs Corporation in January and February of 1952 to discuss a possible joint venture but these came to nothing. In the same year, John Womersley, who, as Superintendent of the Mathematics Division at NPL before joining BTM, had been instrumental in initiating the Pilot Model ACE development, began taking an interest in the work of Andrew Booth at the University of London's Birkbeck College Computation Laboratory. Booth had built several versions of a small magnetic drum computer, the APE(X)C, for various research groups. At Womersley's suggestion, an agreement was reached that BTM could manufacture Booth's machines.

By 1953, BTM had completed the prototype of a commercial version of the APE(X)C which was known as the Hollerith Electronic Computer or HEC. The production version, the HEC 2M, was marketed as the BTM 1200 from 1954. A business data processing model, the HEC 4 or BTM 1201, was introduced in 1956. With a purchase price of only £25,000 for a typical installation, around 125 of this model and its successor, the BTM 1202, were built, making it the most popular British computer of the period.

Further Reading

Bashe, C. J., Buchholz, W. and Rochester, N., The IBM Type 702: An Electronic Data Processing Machine for Business, Journal of the ACM 1 (4), 1954, 149-169.

Birkenstock, J. W., Pioneering: On the Frontier of Electronic Data Processing, a Personal Memoir, IEEE Annals of the History of Computing 22 (1), 2000, 4-47.

Booth, A. D., Computers in the University of London 1945-1962, in A History of Computing in the Twentieth Century, Academic Press, New York, New York, 1980.

6. Data Processing and the Birth of the Computer Industry

Caminer, D. T., LEO and its Applications: The Beginning of Business Computing, Computer Journal 40 (10), 1997, 585-597.

Clarke, S. L. H., The Elliott 400 series and before, The Radio and Electronic Engineer 45 (8), 1975, 415-421.

Cortada, J. W., Before the Computer: IBM, NCR, Burroughs, and Remington Rand and the Industry They Created, 1865-1956, Princeton University Press, Princeton, New Jersey, 1993.

De Barr, A. E., Millership, R., Dorey, P. F., Robbins, R. C. and Atkinson, P. D., Digital Storage using Ferromagnetic Materials, Proc. 1952 ACM National Meeting (Pittsburgh), 1952, 197-202.

Eckert, J. P., A Survey of Digital Computer Memory Systems, IEEE Annals of the History of Computing 20 (4), 1998, 15-28.

Elliott, W. S., Development of Computer Components and Systems, Proc. 1952 ACM National Meeting (Toronto), 1952, 68-72.

Gray, G. T. and Smith, R. Q., Before the B5000: Burroughs Computers, 1951-1963, IEEE Annals of the History of Computing 25 (2), 2003, 50-61.

Gray, G. T. and Smith, R. Q., Sperry Rand's First-Generation Computers, 1955-1960: Hardware and Software, IEEE Annals of the History of Computing 26 (4), 2004, 20-34.

Hamilton, F. E. and Kubie, E .C., The IBM Magnetic Drum Calculator Type 650, Journal of the ACM 1 (1), 1954, 13-20.

Hurd, C. C., Computer Development at IBM, in A History of Computing in the Twentieth Century, Academic Press, New York, New York, 1980.

Koss, A. M., Programming at Burroughs and Philco in the 1950s, IEEE Annals of the History of Computing 25 (4), 2003, 40-50.

Lavington, S. H., Early British Computers: The Story of Vintage Computers and the People Who Built Them, Manchester University Press, Manchester, 1980.

McPherson, J. C., Hamilton, F. E. and Seeber, R. R., A Large-Scale, General-Purpose Electronic Digital Calculator – The SSEC, IEEE Annals of the History of Computing 4 (4), 1982, 313-326.

Norberg, A. L., New Engineering Companies and the Evolution of the United States Computer Industry, Business and Economic History 22 (1), 1993, 181-193.

Phelps, B. E., Early Electronic Computer Developments at IBM, IEEE Annals of the History of Computing 2 (3), 1980, 253-267.

Pinkerton, J. M. M., Hemy, D. and Lenaerts, E. H., The Influence of the Cambridge Mathematical Laboratory on the LEO Project, IEEE Annals of the History of Computing 14 (4), 1992, 41-48.

Pugh, E. W., Building IBM: Shaping an Industry and Its Technology, MIT Press, Cambridge, Massachusetts, 1995.

Pugh, E. W. and Aspray, W., Creating the Computer Industry, IEEE Annals of the History of Computing 18 (2), 1996, 7-17.

Pugh, E. W., Memories that Shaped an Industry: Decisions Leading to IBM System/360, MIT Press, Cambridge, Massachusetts, 2000.

Rajchman, J., Early Research on Computers at RCA, in A History of Computing in the Twentieth Century, Academic Press, New York, New York, 1980.

Rees, M., The Computing Program of the Office of Naval Research, 1946-1953, Communications of the ACM 30 (10), 1987, 831-848.

Rosen, S., Electronic Computers: A Historical Survey, ACM Computing Surveys 1 (1), 1969, 7-36.

Stern, N., From ENIAC to UNIVAC: An Appraisal of the Eckert-Mauchly Computers, Digital Press, Bedford, Massachusetts, 1981.

Tomash, E., The Start of an ERA: Engineering Research Associates, Inc., 1946-1955, in A History of Computing in the Twentieth Century, Academic Press, New York, New York, 1980.

Tweedale, G., A Manchester Computer Pioneer: Ferranti in Retrospect, IEEE Annals of the History of Computing 15 (3), 1993, 37-43.

Weik, M. H., A Survey of Domestic Electronic Digital Computing Systems, Ballistic Research Laboratories Report No. 971, December 1955.

Wilson, J. F., Scientists, Engineers and Market Forces: A Study of Ferranti and Computers, 1949-1993, Proc. Int. Symposium on Computers in Europe: Past, Present and Future, 1998.

PART THREE
SEGMENTATION

7. Revving Up for Higher Performance

> *"Anyone can build a fast CPU. The trick is to build a fast system"*,
> Seymour R Cray.

The Need for Speed

Almost as soon as it was established, the commercial market for computers began to polarise into machines designed for either scientific or business applications. The leading computer firms, with their office equipment roots, naturally chose to devote most of their resources to addressing the requirements of the market sector with the largest potential, business data processing. Therefore, as the commercial development of the computer became increasingly driven by the needs of business users, the needs of the computer industry's earliest customers, scientists and engineers, were no longer being met. By 1953 some mainstream large-scale scientific models had begun to appear, such as the ERA 1103 and IBM 701, but these only served to whet the appetite of scientists and engineers hungry for more computing power to tackle progressively larger and increasingly complex computational problems.

The need for scientific computing power was perhaps greatest in the new field of atomic weapons research as a result of the escalating Cold War. The simulation of nuclear explosions involves the solution of large sets of highly complex non-linear partial differential equations. To ensure the accuracy of the model, calculations need to be made for a large number of time intervals and locations, and the optimisation of design parameters may require many iterations of these calculations. This demand for upgraded computing capabilities by atomic weapons designers would be further intensified by the introduction of an international moratorium on nuclear testing in October 1958.

Early high performance computers were similar in many ways to today's Formula One racing cars, in that the leading edge technology created to obtain the highest possible performance eventually filtered down to the mass-produced models. Manufacturers used them as a test-bed for new ideas and many of the features first developed for these highly specialised machines are still employed in the mainstream computers in common use today. However, the development of computers of the highest performance has always been a very expensive business and manufacturers in the early years of the computer industry lacked the deep pockets necessary for such an enterprise. Without the support of government agencies in both the USA and UK, the swift progress made in this field would not have been possible. As on so many other occasions in the history of the computer, the first government body to seize the initiative was the US Navy.

The Naval Ordnance Research Calculator

In early 1950, a young engineer named Byron L Havens was working at IBM's Watson Scientific Computing Laboratory at Columbia University on the development of a new type of high-speed electronic component, known as a microsecond delay circuit, for use in high performance computing applications. Havens had been one of three talented researchers recruited from the MIT Radiation Laboratory in January 1946 by Watson Laboratory Director Wallace J Eckert to work on the development of advanced electronic circuitry and devices. By July 1950, Havens' work had progressed to the point where he was able to propose the construction of a full-scale computer. The new machine would be an all-electronic stored-program successor to the SSEC. His objective was to create a machine which would not only eclipse the latest offerings from IBM's competitors but would be significantly faster than any other computer on the planet, in his own words "to build the most powerful and effective calculator which the state of the art would permit". This objective sparked an industry trend that would continue for many years and culminate in the emergence of a new class of machine, the supercomputer.

IBM senior management, aware that the US Government had begun to support large-scale computer development projects through agencies such as the National Bureau of Standards and the Office of Naval Research, set out to secure public funding for the project. Former Director of Engineering John C McPherson, now an IBM Vice President, soon captured the interest of the US Navy Bureau of Ordnance in using Havens' advanced electronic technology for ballistic calculations and in September 1950, IBM received a formal request from the US Naval Ordnance Laboratory in Maryland for a "high-speed electronic calculator". This led to the Navy issuing a potentially lucrative cost-plus-fixed-fee contract to procure the new machine. As a demonstration of its willingness to do business with the US government (and recognising the valuable experience that would be gained with such a project), IBM waived the opportunity to make a profit from the contract by setting the fixed fee element at a nominal $1.

The Naval Ordnance Research Calculator (NORC) was built at the Watson Scientific Computing Laboratory under the direction of Wallace Eckert. Work on the new machine began in October 1950. Byron Havens was appointed as project leader and was provided with an able assistant in William J Deerhake, an Assistant Professor of Electrical Engineering at Columbia University.

The Watson Laboratory had learned a valuable lesson with the programmer's

7. Revving Up for Higher Performance

nightmare that was the SSEC. Therefore, as well as high performance, NORC was also designed for ease of use. The design team chose to stick with a decimal architecture in order to simplify programming, rather than opt for the computationally efficient binary architecture that was rapidly becoming the industry standard for scientific machines and would soon be used in IBM's own Defense Calculator. The basic design of NORC was that of a conventional binary-coded decimal, serial, machine with a 1 MHz clock speed. However, the word length was a generous 16 decimal digits to facilitate higher precision calculations and the arithmetic unit boasted automatic floating-point hardware and a specially designed multiplier that operated in parallel to improve multiplication speed. The logical design also included two arithmetic registers and three address modifier registers, the latter permitting the modification of operand (input value or parameter) addresses without changing the stored instruction, a powerful feature for calculations involving arrays of numbers.

Main memory comprised 2,000 words of electrostatic storage, implemented using 264 Williams Tubes. This was augmented with eight ultra high-speed magnetic tape units. Developed at IBM's Poughkeepsie Laboratory, these tape units incorporated a novel vacuum column feature which facilitated a blistering read/write performance of 71,500 characters per second. Vacuum columns were used to create long u-shaped loops of tape on both sides of the read/write head which helped to maintain contact between the tape and head at high speed and also acted as a shock absorber, preventing tape breakage when starting and stopping quickly. Such was the performance of the NORC tape units that they could be used to store intermediate results. NORC was also equipped with two Endicott-developed high-speed printers which could operate at 150 lines per minute and IBM's first implementation of input/output buffers to improve throughput.

With the use of troublesome electrostatic storage technology for main memory and around 9,800 thermionic valves, a high-reliability design was a necessity to ensure consistent operation. Modular construction, via IBM's well-established system of pluggable subassembly units, was employed throughout for ease of maintenance and the designers also included error checking hardware, both within the arithmetic unit and on the electrostatic memory. An extensive maintenance and testing system was also developed for NORC which included the Test Assembly, a unique piece of auxiliary equipment that permitted dynamic testing of all pluggable subassembly units away from the machine, thereby minimising disruption to its operation.

The completed machine was publicly demonstrated and formally presented to the US Navy in a prestigious ceremony held at Columbia University on 2 December 1954. The dedication address was given by none other than John von Neumann, who heaped glowing praise on the machine and hinted at the technological race to come by highlighting the importance to the industry of building "the most powerful machine that is possible in this day with the present state of the art". NORC had been originally scheduled for delivery to the US Naval Ordnance Laboratory in Maryland but Navy officials decided to reassign it to a more experienced computing group based at the US Naval Proving Ground in Dahlgren, Virginia. The eminent physicist Edward Teller attempted to have NORC diverted to the University of California Radiation Laboratory in Livermore, California, arguing that the Livermore Laboratory's work on nuclear calculations was of greater national importance than the ballistic calculations which were the main focus of the work at Dahlgren, but the Navy held firm and NORC was installed at Dahlgren in June 1955 where it remained in operation until 1968. In what must have seemed a fitting revenge for Howard Aiken's earlier treatment of IBM, NORC replaced Aiken's Harvard Mark III calculator as the principal computing resource at Dahlgren.

Naval Ordnance Research Calculator (Photo © IBM Corporation)

Despite a modest 1 MHz clock speed, the extra performance features incorporated into the design of the arithmetic unit gave NORC the ability to perform a multiplication operation in only 31 microseconds. This was several times faster than both the ERA 1103 and IBM's own recently announced 704 model, earning NORC the title of the fastest computer in the world, a title that it held for over 5 years. IBM's design objectives had been largely achieved but at a project cost of approximately $2.5 million, double the estimated budget.

NORC was created as a one-off, cost-no-object computer to push the boundaries of the technology and was never intended for commercial production. However, its development took place in parallel with that of the IBM 701 which benefited greatly from NORC technology, in particular electrostatic memory (although unfortunately without the error checking feature) and Havens' microsecond delay circuit. The Poughkeepsie Laboratory's innovative vacuum column tape drive technology also found its way into IBM's standard line of magnetic tape units, beginning with the Model 726 in 1952, and the vacuum column feature was subsequently adopted by the rest of the industry.

The Livermore Automatic Research Computer

Having failed to persuade the Navy to give him NORC, University of California Radiation Laboratory co-founder Edward Teller remained determined to have his own high performance machine. Livermore already possessed several large-scale computers, including a UNIVAC I and two IBM 701 models, but had rapidly outgrown them. Teller convinced the US Atomic Energy Commission to put up the budget for the project and in December 1954 the Laboratory invited proposals for a "superspeed" computer for atomic weapons research which would be called the Livermore Automatic Research Computer (LARC).

By April 1955, two bids had been received, one from Remington Rand's Philadelphia division and the other from IBM. Both bids promised machines that would amply satisfy Livermore's performance requirements but Teller wanted the new machine as quickly as possible. As Remington Rand promised delivery within only 29 months compared with IBM's lengthier schedule of 42 months, it won the $2.85 million fixed-price contract.

The LARC project began in September 1955 and was directed by ENIAC veteran Herman Lukoff. Considerable input to the specification was also received from Livermore Laboratory scientists via regular technical review meetings. The emphasis was on a balanced design with proven, rather than state-of-the-art, components. To facilitate this, the input/output functions would be handled by a separate processor and a second arithmetic unit could also be added to create a powerful multiprocessor configuration. The high-speed arithmetic unit, known as a 'Computer', employed a biquinary-coded decimal, parallel, architecture with 12 digits per word and could handle both fixed and floating-point arithmetic in single or double precision. The confusingly named 'Processor' input/output unit was an independent stored-program computer which provided control for I/O and could also be programmed to deal with secondary computing tasks.

The logical design of LARC's high-speed arithmetic unit borrowed from the general register set architecture of the newly introduced Ferranti Pegasus by employing general-purpose registers which could function as either accumulators or index registers. LARC was equipped with a set of 26 of these registers as standard and could be expanded up to 99 in total. LARC also featured the industry's first implementation of instruction pipelining, a now common technique first conceived by Konrad Zuse in 1947 (under a Remington Rand research grant) in which multiple machine instructions are overlapped in execution to improve performance. In LARC, four instructions, each at a different stage in the execution process (fetch, decode, execute, write), could be operated on in parallel within a single clock cycle. Each 4-microsecond clock cycle was actually divided into eight 500-nanosecond slots, with alternate slots allocated to each arithmetic unit in a multiprocessor configuration. This instruction pipelining feature allowed LARC to perform considerably faster than its pedestrian 250 KHz clock speed might otherwise have suggested.

Following the industry trend towards solid-state components, LARC was built using transistors rather than thermionic valves in its arithmetic circuits. The higher switching speed of transistors made them particularly attractive to designers of high performance computers. However, rather than seeking out the latest developments in transistor technology in a quest for superior speed, LARC's designers remained faithful to their conservative design philosophy by opting for Philco Corporation's reliable and readily available germanium surface barrier transistor which was introduced in 1954 and had already been proven in several experimental computers.

In order to cope with the scientific community's demands for increased performance and larger data sets, it would become necessary to significantly increase the primary storage capacity of high performance computers over that of mainstream models. The designers of NORC had been able to make do with a conventional main memory size by taking advantage of the machine's ultra high-speed magnetic tape units to store intermediate results but as performance increased beyond the capability of auxiliary storage technology, designers of high performance machines soon had no option but to increase main memory capacity. LARC was the first computer to feature a significantly larger main memory. It was equipped with a minimum of 20,000 words of magnetic core storage, ten times that of NORC, which could be expanded in blocks of 10,000 words up to an unheard of maximum of 97,500 words. This was further augmented with high-speed auxiliary storage of 12 or 24 magnetic drum memory units of 250,000 words capacity each and up to 40 UNISERVO II magnetic tape units, giving a massive storage capacity of over 73 million decimal digits.

A data transfer bus connected single or twin arithmetic units and the input/output unit to the main memory. Access was interleaved to maximise throughput, with every 4 microsecond bus cycle divided into eight 500 nanosecond time slots. Additional input/output devices included 1 or 2 high-speed printers which could operate at up to 600 lines per minute, a high-speed punched card reader and a pair of graphical display devices known as Electronic Page Recorders.

The specially designed magnetic drum units each comprised a horizontally mounted drum 24.2 inches in diameter with a moving read/write head mechanism positioned above the drum, parallel to the axis of rotation. The

7. Revving Up for Higher Performance

drum surface was divided into 100 bands of 6 tracks, with each band further divided into 25 sectors of 100 words. Although the speed of rotation was only 880 rpm, a data transfer operation could begin at any sector boundary which reduced access time considerably. UNIVAC engineers developed an air bearing read/write head assembly which comprised 6 read/write heads floating on a cushion of air in order to reduce friction and maintain the correct gap between heads and recording surface. The input/output unit could be fitted with up to 5 independent drum controllers, 3 for reading and 2 for writing, and this permitted access to be interleaved between pairs of drum units which also increased speed and delivered a sustained data transfer rate of 330,000 decimal digits per second.

The Philadelphia design team had incorporated the usual array of UNIVAC error checking features, such as duplicate circuitry to detect single bit errors and parity checking for the memory units. Although transistors had been employed chiefly for their performance characteristics, their improved reliability over thermionic valves also helped to ensure that LARC would be more reliable than earlier designs. However, LARC lacked the modular construction of IBM's NORC, which made the machine more difficult to repair when problems did occur. Circuit boards were hardwired rather than pluggable, resulting in faulty parts having to be replaced at individual component level rather than by simply swapping out circuit boards.

Livermore Automatic Research Computer

During the design phase it had become clear that the $2.85 million budget set by the fixed-price contract would be woefully insufficient and the final development cost would be at least twice that amount. The US Navy helped to alleviate this situation by placing an order for a second machine for use in ship and nuclear reactor design. The first machine, a single arithmetic unit model fitted with 30,000 words of core storage, 12 magnetic drum memory units and 8 tape units, was completed and delivered to the University of California Radiation Laboratory, now renamed the Lawrence Radiation Laboratory, in April 1960. Embarrassingly, not only was this more than 2 years behind the delivery date specified in the contract, it was later even than the date specified in IBM's rejected bid. The similarly specified second machine was delivered about 6 months later to the US Navy's David Taylor Model Basin in West Bethesda, Maryland.

The overall cost of the LARC project was estimated at around $19 million but income from the two government contracts amounted to only $5.7 million. From the outset Sperry Rand had planned to recoup any development losses from commercial sales so LARC was announced as a commercial product, the UNIVAC-LARC, and priced at $6 million or a rental of $135,000 per month for a single processor system. Its decimal architecture allowed it to be marketed as suitable for both scientific and business data processing applications. However, insufficient orders were received and the machine was never put into production. Philadelphia's design strategy of employing only proven components meant that by the end of the 4-year plus development phase LARC was looking dated, despite its industry-leading performance, and customers were unwilling to pay a top dollar price for a machine that was not quite state-of-the-art. Nevertheless, the project was far from a total failure as Sperry Rand's next large-scale offering, the UNIVAC III, which was announced in May 1960, incorporated much that the engineers had learned from the LARC project, including transistorised circuitry and magnetic core memory technology. Furthermore, the moving-head magnetic drum technology from LARC was further refined and used in the company's FASTRAND magnetic drum storage units which became a common feature on later Sperry Rand computers.

Although embarrassingly late and a whopping $13 million over budget, LARC met or exceeded all its design goals. With a calculation speed of 4 microseconds for a floating-point single precision addition and 8 microseconds for a floating-point single precision multiplication, the machine comfortably surpassed NORC as the fastest computer in the world. LARC represented a huge leap forward in technology but its reign as the world's fastest computer would be short-lived. Bitter at the loss of the LARC contract, Sperry Rand's great rival IBM had been determined to regain its coveted position as the high performance computer supplier of choice to the US scientific community and was already close to completion of a project which would realise this goal.

7. Revving Up for Higher Performance

Project Stretch

The December 1954 call for proposals for the LARC machine had triggered a fierce debate within IBM on the way forward for the company's high performance computer developments. Opinions were divided into two camps; those who favoured the further development of an existing machine architecture or the advocates of a grander strategy to develop a completely new computer that would embody many new architectural concepts. Livermore's tight delivery schedule and the Poughkeepsie Laboratory's already stretched resources favoured the less ambitious option, therefore, Cuthbert Hurd's initial discussions with Livermore had centred on the development of a transistorised machine based on NORC. However, pressure continued to grow from Ralph Palmer and Stephen Dunwell for a development project that would yield substantial advances in solid-state device technologies. By April 1955, a compromise had been reached. The final version of the proposal to Livermore was for a machine based on an existing architecture to be delivered within 42 months, with the option of a more advanced machine within the same timescale but at additional cost.

The eventual loss of the LARC contract to Remington Rand in May 1955 was a major blow to IBM's high performance computer plans. Undaunted, Cuthbert Hurd set about finding another customer willing to pay for the development of a large-scale state-of-the-art machine. By August 1955, he had secured $1.1 million funding from the US National Security Agency for a design study and this allowed the new project to begin at Poughkeepsie, which was codenamed 'Stretch' to signify the company's intention to stretch the state-of-the-art in high performance computing technologies. However, the funding provided by the NSA was insufficient to support the construction of a machine and the search for a customer with deeper pockets continued. Livermore's sister laboratory in New Mexico, the Los Alamos Scientific Laboratory (LASL), had expressed an interest in acquiring an advanced large-scale computer. Fearful that it would also choose Remington Rand's LARC, IBM made a preliminary proposal to LASL in September 1955 which was received with interest. The ensuing discussions culminated the following January with the US Atomic Energy Commission issuing a formal request for proposals for a high-speed computer to be installed at LASL. Four bids were received but on this occasion the final decision favoured IBM. A $4.3 million fixed-price contract was issued in November 1956, with delivery scheduled for 42 months.

The importance of Project Stretch to IBM was underlined by the glittering array of engineering talent assembled to serve on the design team, including Werner Buchholz, Gerrit A Blaauw, James H Pomerene and Frederick P Brooks. A joint mathematical planning group, comprising a team of eight LASL scientists and a number of senior IBM engineers, was also formed to oversee the design of the new machine. Early in the project, a power struggle for overall control developed between veteran engineer Stephen Dunwell and relative newcomer Gene M Amdahl. The decision to give Dunwell overall responsibility for the project in December 1955 proved unacceptable to Amdahl, who promptly quit IBM for a job at the Ramo-Wooldridge Corporation. Amdahl would return to IBM in 1960 but was never again associated with Project Stretch.

The design objective called for the development of a machine with at least 100 times the performance of IBM's latest large-scale scientific model, the Type 704. From the outset the team were under no illusions that such an ambitious goal could be achieved through the application of new solid-state components alone and knew that the design would also have to embody several innovative architectural features. The architecture conceived by the Stretch design team to meet this objective has been described as aggressive uniprocessor parallelism. Like the LARC design team, the designers of Stretch were attracted by the notion of parallelism of operations within the central processor as a means of delivering higher performance. But the IBM team, unfettered by the conservative design philosophy of their rivals, had the freedom to take this idea much further.

This design strategy began with the arithmetic unit. It incorporated both a serial integer section, each of which could perform binary or decimal arithmetic on fields of variable length, plus a parallel floating-point section that operated on full-length words of 72 bits, with the last 8 bits used for error checking and correction. Further parallelism was achieved through the implementation of instruction pipelining and an extension to pipelining called instruction lookahead which was originally conceived by Konrad Zuse for his electromechanical Z4 and refined by Gene Amdahl and John Backus for use in Stretch. Stretch featured a 4-stage instruction pipeline that permitted 2 instructions and up to 4 operands to be fetched simultaneously, with 2 instructions also being decoded and 1 executed at the same time. A lookahead unit controlled the fetching and storing of data for up to 6 instructions ahead of the one being executed in order to minimise memory delays. By tightly coupling the operation of the instruction and lookahead units, up to 11 instructions could be in various stages of execution concurrently. The instruction and lookahead units operated on a clock speed of 3.33 MHz while the arithmetic unit operated on a slower clock speed of 1.67 MHz. The complexity of the arithmetic unit was mirrored by the highly complex instruction set which comprised 160 basic machine instructions grouped into five subsets, allowing a total of 735 different permutations.

The innovative architectural features of Stretch were matched by impressive storage capacity. Primary storage comprised up to 16 magnetic core storage units of 16,384 words capacity, giving a maximum capacity of 262,144 72-bit words of fast-access storage. Each unit was independently accessible so that memory operations could be performed in parallel. This was complemented by up to 32 IBM Type 353 Disk Storage Units of 2,097,152 words capacity each. Developed from IBM's first magnetic disk storage device, the Type 350, specifically for use in Stretch, each disk unit comprised a 39-disk stack, with a separate arm and read/write head for each side of each disk but positioned via a common servo control mechanism. A higher rotation speed of 1,744 rpm, coupled with the additional read/write heads, gave an access time of 180 milliseconds, which was almost 4 times faster than the earlier model. A separate input/output processor was also developed. Known as the Exchange, it was equipped with 8 input/output channels to deliver an overall data transfer rate of 6 million bits per second. Additional input/output equipment included IBM 729 IV magnetic tape units and an IBM 1403-2 high-speed printer which could operate at 600 lines per minute.

The performance requirements of Stretch dictated the use of the latest drift transistors rather than surface barrier transistors in the machine's logic circuitry, which contained more than 169,000 IBM-developed graded-base drift transistors. A new form of high-speed switching circuit, the current-switch emitter-coupled logic (ECL) circuit, was also developed for use in Stretch and later adopted as the industry standard for high-speed switching applications. The move to transistors also provided an opportunity for IBM to develop a new generation of modular circuit packaging known as Standard Modular System (SMS). SMS printed circuit boards were considerably smaller and cheaper to manufacture than the system of pluggable units that they replaced. This combination of transistorised circuitry and modular construction facilitated a compact design. Despite having almost 23,000 individual circuit boards of 42 different types, Stretch required only 2,000 sq ft of floor space, the same amount required for the IBM 704. This is particularly impressive considering that the total number of individual components in a full specification model would have been in excess of 19 million, which is over 6 times the number of components found in a modern jet airliner.

Among many notable firsts, Project Stretch pioneered the use of software simulation to assist with the logical design process. A computer simulation was written by John Cocke and Harwood G Kolsky in November 1957 for the IBM 704. Known as SIM-2, it simulated the timing of logical operations within the Stretch lookahead unit. Software was also used to optimise the complex wiring layout for the back panel of the machine's central processing unit.

Stretch was announced in April 1960 as the IBM 7030 Data Processing System. Priced at an eye-popping $13.5 million for a full specification model, it was marketed as suitable for both scientific and business data processing applications, the latter possible because of the machine's ability to perform decimal arithmetic and to deal with variable field lengths. The first production example, fitted with 6 magnetic core storage units and 2 disk storage units, was delivered to Los Alamos in April 1961, 11 months behind schedule.

With calculation speeds of 1.5 microseconds for a floating-point addition and 2.7

The Story of the Computer

microseconds for a floating-point multiplication, Stretch easily outperformed the less aggressively engineered LARC. The overall performance was around 3 times that of LARC and 60 times that of the IBM 704, an impressive achievement but one that fell well short of the design goal for the project. An unanticipated but necessary reduction in clock speed from an initial target of 10 MHz during development had severely compromised performance. A lack of consistency in operation of the new architectural features, due to the complex interactions between overlapping operations, also contributed to the failure of Stretch to meet its advertised performance specifications.

IBM 7030 Installation at the Atomic Weapons Research Establishment

In May 1961, a deeply embarrassed Thomas J Watson Jr. was forced to announce a price cut of all 7030s under negotiation to $7.8 million and the immediate withdrawal of the model from further sales. Nevertheless, a total of 9 machines were built, two of which went to overseas customers in the UK and France. This total also included one 'Harvest' derivative for the NSA, the requirements for which came out of the NSA design study conducted at the start of the project. Designed by James Pomerene and Paul S Herwitz, the IBM 7950 Harvest was essentially a 7030 with the addition of a stream co-processor unit and a high-performance automated tape library known as Tractor, which facilitated the high-speed processing of non-numerical data for cryptanalysis.

IBM is reported to have lost around $40 million on the Stretch project. As project director, Steve Dunwell received much of the blame for this failure and his career within the company nosedived as a result. Nevertheless, the transistorised circuit technology and architectural innovations that came out of Project Stretch formed the cornerstone of the company's large-scale computer product developments for several years and particularly influenced the design of the IBM 7090, IBM's most successful large-scale computer of the period.

7. Revving Up for Higher Performance

Announced in December 1958, the 7090 was taken from concept to production in only 17 months, this unusually fast development schedule made possible by the wholesale adoption of Stretch components and subsystems. The true contribution of the Stretch project to IBM's technological heritage was only acknowledged some years later in March 1966 when Dunwell was awarded the accolade of IBM Fellow, the highest recognition for technical achievement in the company.

The commercial failure of Project Stretch severely dented IBM's reputation as the industry's leading supplier of high performance computers. This damage might have been fatal were it not for the lack of any serious competition following Sperry Rand's equally disastrous experience with the UNIVAC-LARC. With Sperry Rand out of the picture, the next challenge to IBM's high performance computing crown would come from an unlikely source.

The Microsecond Engine

In May 1956, a report to the Advisory Committee on High Speed Calculating Machines of the UK Department of Scientific and Industrial Research (DSIR) highlighted the massive scale of the LARC and Stretch projects in the USA and identified a woeful lack of effort in developing high performance computers in the UK. This view was endorsed by the Computer Sub-Committee of the National Research Development Corporation in January 1957, which called for a review of the matter and also proposed the establishment of a national centre of excellence in the field. At around the same time, the growing high performance computation requirements of the Atomic Weapons Research Establishment at Aldermaston and its sister laboratory, the Atomic Energy Research Establishment at Harwell, prompted a joint proposal to the UK Atomic Energy Authority for a British high performance computer and suggested that a design competition be held to select a suitable manufacturer.

Like their counterparts in the USA, the British agencies recognised the need for government support to stimulate activity in this field and over the following months, the idea of a national high performance computer project began to take shape. However, in typical British fashion, they were unable to decide on who should be given the task of developing the computer; an existing company, university, government laboratory or perhaps some combination of the three.

One of the technical experts consulted by the NRDC was Tom Kilburn of Manchester University. Kilburn had recently completed the development of the megacycle machine or MEG and was busy formulating plans for a new project. Codenamed the microsecond engine or MUSE, the aim of the project was to produce a high performance computer using existing components within a 3 to 4 year timescale. Through his dealings with some of the potential users of high performance computers as part of the NRDC consultations, Kilburn was able to formulate a basic user requirement specification for his new machine. The main requirements were a processing speed of 1 million instructions per second (hence the name microsecond engine), a large main memory of 100,000 words and a wide selection of input/output devices. The MUSE project became one of the many options under consideration by the NRDC as the basis for the national project, however, the Corporation's deliberations showed no signs of ending so in the autumn of 1956 Kilburn decided to proceed independently, using internal resources which included royalty income from the Ferranti Mark 1.

Tom Kilburn assumed the role of principal architect for the new machine. His small project team included electronics engineer Dai Edwards, who was responsible for hardware development, and mathematician Tony Brooker for software. However, progress over the first two years was frustratingly slow due to the limited funding and manpower available to the project. Fortunately, Ferranti had been considering an involvement in the MUSE project for some time. The company needed a large-scale computer to extend its product range and had successfully worked in partnership with Manchester University twice before, on the development of its Mark 1 and Mercury models, both of which were based on Manchester prototypes. In January 1959, the MUSE project formally became a joint venture between Manchester University and Ferranti Ltd. The computer was renamed Atlas in keeping with the company's convention of naming products after Greek gods. Ferranti had estimated the development cost of the production version at £850,000. The NRDC, having finally decided that this should be the project to support, contributed £300,000 in the form of a loan and helped to broker an order for one of the production machines.

Kilburn's design for the Atlas was based on a binary parallel architecture with a 48-bit word length. A 48-bit floating-point arithmetic unit operated in parallel with a separate 24-bit integer unit which was designed primarily for address arithmetic. The logical design featured 8 defined registers and 120 half-word general-purpose registers, of which 30 were reserved for system use. A form of instruction pipelining was also implemented in which up to 3 instructions and 1 indexing operation could be performed in parallel. Unusually, Atlas was an asynchronous machine in that it had no system clock. Instead, a single pre-pulse triggered the action of the various elements of the machine and the self-timing of each action was used to complete it before proceeding onto the next element.

The user requirement for a large main memory, set against the modest project budget, prompted the Atlas design team to consider schemes for reducing the cost of fast storage. Their inspired solution was to create a novel virtual memory addressing system in which the conventional two-level storage arrangement of magnetic core main memory and drum-based backing store was made to appear to the user as a one-level storage system with an addressable capacity,

7. Revving Up for Higher Performance

or address space, of 1 million words. Memory was subdivided into blocks of 512 words known as pages. Instead of transferring only an individual block of data into memory when needed, the whole page would be transferred. Because of the sequential nature of data storage, the next block of data needed would be likely to be held within the same page, thus reducing the overall number of data transfer operations required. A page address register for each page of memory recorded the address of the block transferred. Subsequent requests for data compared the address requested with the contents of each page address register to determine if the data has already been transferred before issuing a page swap command. A 'drum transfer learning program' was developed by Frank H Sumner to optimise page swapping by accumulating statistics on page utilisation. To facilitate the development of this program, Sumner created a computer simulation of the Atlas one-level storage system on a Ferranti Mercury.

Virtual memory addressing permitted a large main memory to be implemented using mainly cheaper drum-based auxiliary storage instead of core storage with little performance penalty. Hence the Atlas prototype was equipped with only 16,384 words of magnetic core memory, in independently accessible units of 4,096 words, supplemented by four magnetic drum storage units of 24,576 words capacity each, giving a total memory capacity of 114,688 words. The high-speed magnetic drums rotated at 5,000 rpm and could transfer data at the impressive rate of one 512-word block every two milliseconds.

Atlas was also equipped with 8,192 words of high-speed read-only memory which held the control program and test routines. Developed in partnership with the Plessey Company, this was constructed using ferrite and copper rods in a woven wire mesh, the position of the rods within the mesh representing the data bits. The resultant 'hairbrush' memory unit boasted a 0.35 microsecond access time, which was several times faster than the highest speed magnetic core storage available. However, unlike modern read-only memories, which are electrically programmable, reprogramming it was a laborious task requiring the physical repositioning of thousands of tiny rods.

As with Stretch, Atlas was built primarily with drift transistors, however, the Atlas team did not have the luxury of custom transistors enjoyed by the IBM designers. Except for a small number of specially designed symmetrical surface barrier transistors used in the parallel adder, most of the transistors used were standard germanium junction transistors manufactured by Mullard for the transistor radio market. Atlas employed a total of 60,000 transistors, packaged via approximately 500 circuit boards.

In addition to the production engineering work necessary to make the research prototype suitable for manufacture, Ferranti engineers also contributed to the logical design of Atlas and to the development of the magnetic drum units. However, it was in the development of the system software that the company made its most important contribution. A team led by David J Howarth developed a sophisticated control program for Atlas, known as the 'Supervisor', which featured a multiprogramming capability (the ability to interleave processes so that the computer appears to run more than one job at the same time) and a job scheduler. The richness of features in the Atlas Supervisor have led to it being generally considered to be one of the first true operating systems.

The Atlas prototype first ran towards the end of 1962 and began limited operation in January 1963. The official inauguration took place on 7 December 1962 at Manchester University. The inaugural address was given by the former Director of the Atomic Energy Research Establishment, Sir John Cockcroft, himself an eminent nuclear physicist and Nobel laureate. Some historians have claimed that the UK's scientific computing capacity instantly doubled as soon as Atlas was switched on. This is an exaggeration but Atlas certainly made a major contribution.

Calculation speeds for the new machine were 1.59 microseconds for a fixed-point addition and 4.97 microseconds for a floating-point multiplication. These figures placed Atlas somewhere between LARC and Stretch in performance terms but the elegance and simplicity of the Atlas design made for a considerably cheaper machine to construct, which was reflected in the pricing of the production version. Introduced at a price of £1.5-2.5 million (equivalent to $4.2-7.0 million), depending on specification, Atlas was slightly cheaper than the UNIVAC-LARC and around half the original price of the IBM 7030. However, only two production machines were ever built, one for the University of London which was delivered in October 1963 and the other, the largest of the three with 49,152 words of magnetic core storage, for the Atlas Computer Laboratory at Chilton near Oxford in April 1964. The Treasury had attempted to force the UK Atomic Energy Authority to order an Atlas for the Atomic Weapons Research Establishment in 1960 but AWRE, already comfortable with IBM as a supplier, chose to rent an IBM 7030 instead. When teething troubles delayed commissioning of the production machines and it also came to light that inadequate peripheral equipment would compromise the machine's competitiveness, other prospective customers soon began to look elsewhere too.

The Atlas Computer Laboratory was an attempt to provide a national computing facility for science and engineering research. It was established by the National Institute for Research in Nuclear Science (NIRNS) in December 1961. The Laboratory provided a bureau service, based around the Laboratory's Atlas computer, in which university researchers, the AEA, the DSIR and various UK government research organisations were permitted free access on a strictly rationed basis.

Even before the Atlas prototype was completed, it was becoming obvious to

7. Revving Up for Higher Performance

Ferranti that the new machine would be unlikely to set the market alight. So, in a second attempt to derive some financial return from the Atlas project, the company decided to develop a cut-down version which would provide most of the performance and functionality but at a considerably lower price. In October 1961, a deal was struck with Maurice Wilkes' group at the Cambridge University Mathematical Laboratory in which the Cambridge group would be supplied with various Atlas components and modules at cost price in return for assistance in developing the new model. The Project Coordinator was former Elliott Brothers and Ferranti designer Bill Elliott, who had joined the Laboratory from IBM in early 1962. Based on Atlas but with less logic, no virtual memory system and no magnetic drum storage, the new machine was codenamed Titan. To compensate for the absence of a virtual memory capability, Titan was equipped with a larger main memory of up to 131,072 words of magnetic core storage. The new model was announced in August 1963 as the Ferranti Atlas 2. However, despite a lower price, only two Atlas 2 machines were ever sold, one to the UK Atomic Energy Authority in 1963 for use at Aldermaston and the other to the UK Ministry of Technology three years later for installation at its Computer Aided Design Centre in Cambridge.

With the Atlas project, Ferranti became the third manufacturer to experience failure in its efforts to create a commercial market for high performance computers. Nevertheless, Atlas is notable for having introduced several important new concepts, including virtual memory and a fully functional operating system, a spectacular achievement for such a small, under-resourced project.

With the announcement of the IBM 7094 in January 1962, IBM was now offering a relatively affordable mainstream model with similar performance to the Atlas 2. Ferranti, with its limited resources, was unable to compete but even the mighty IBM was about to be humbled by a younger company that would make the high performance computing market its own with the introduction of the most remarkable computer of its generation.

The CDC 6600

By the beginning of 1957, Engineering Research Associates co-founder William Norris had been General Manager of Sperry Rand's UNIVAC Division for over a year and was growing increasingly frustrated in this role. Mired in corporate bureaucracy and internal politics, Norris was offered an opportunity to recreate the freedom and excitement of the early days of ERA when he was approached by a number of similarly disaffected Sperry Rand employees, including Frank Mullaney and Willis K Drake, suggesting that if he were to quit Sperry Rand and set up his own company, they would follow him. Mullaney himself had been contacted a short time earlier by Arnold J Ryden, a management consultant and former ERA colleague, who had outlined an intriguing plan to set up a new business and this formed the basis of the ensuing discussions with Norris. Spurred on by the prospect of demotion in an impending restructuring of Sperry Rand, Norris finally agreed to the proposition and Control Data Corporation was incorporated in Minneapolis in July 1957.

The start-up financing was provided by the sale of 615,000 shares of stock at $1 per share. Norris invested $75,000 of his own money and persuaded a medical doctor friend to invest $25,000. The other founders and their friends and families invested smaller amounts and Drake managed to sell the remainder of the stock to the public. With a total of around 300 stockholders, Control Data became the computer industry's first example of a publicly financed company.

Norris's strong leadership qualities made him a natural fit for the role of President of the new company. Mullaney assumed the role of Director of Engineering, Ryden that of Finance Director and Drake as Director of Marketing. In all, about a dozen staff left Sperry Rand to join the new venture. The defection of so many employees provoked Sperry Rand into filing a lawsuit against Control Data in April 1958, claiming that the company's key executives, while employed at Sperry, had conspired to leave their employer and take its trade secrets with them. This was eventually settled out of court in early 1962.

Initial accommodation was a rented newspaper warehouse in downtown Minneapolis. To provide production facilities, the company acquired Cedar Engineering Incorporated, a Minneapolis-based instrument maker, in November 1957 for $428,000. This was to be the first of many acquisitions. Early projects included the development of an air traffic display device for the US Civil Aeronautics Administration, which was delivered in November 1959. However, orders for custom equipment of this type soon began to dry up and by early 1958 the company had begun work on its first mainstream computer product, the CDC 1604.

The CDC 1604 was designed by an exceptionally talented young engineer named Seymour R Cray. Cray's career had begun 7 years earlier in 1951 when he joined the fledgling ERA on completion of a bachelor's degree in electrical engineering and a master's in applied mathematics at the University of Minnesota. Cray was keen to make the move to Control Data at the time of the new company's formation but was working on the design of the AN/USQ-17 computer at that time. The AN/USQ-17 was the central component of the Navy Tactical Data System, a special-purpose computer system for shipboard use. Norris persuaded Cray to remain at Sperry Rand until the AN/USQ-17 design was completed in order that the US Navy, which was a potential Control Data customer, would not be upset. But Cray only managed to stay for a further two months before abruptly leaving to join the new company in September 1957, his departure forcing a redesign of the AN/USQ-17 hardware when Sperry Rand subsequently won the $50 million contract to supply a number of these

7. Revving Up for Higher Performance

systems for shipboard installation.

Cray commenced work for his new employer by building a small laboratory prototype computer to validate the modular circuit techniques he had pioneered in the NTDS project. Known as 'Little Character' due to its one character word length, the prototype was a 6-bit binary machine with 64 words of core memory. Norris believed that he had spotted an opportunity to open up a non-military market for scientific computers through the introduction of a lower cost equivalent to the UNIVAC 1103 and a successful demonstration of the Little Character prototype in early 1958 convinced him to give Cray the go ahead to design a full-scale machine which would be called the CDC 1604.

With a binary parallel architecture, floating-point arithmetic and a magnetic core memory, Cray's design for the 1604 was outwardly similar to that of the UNIVAC 1103AF. However, at 48 bits, the 1604 had a word length 12 bits longer than the UNIVAC model, which greatly improved the precision of calculations. It also had a much larger main memory capacity of up to 32,768 words of magnetic core storage and a clock speed two-thirds faster at 833 KHz. Furthermore, unlike the valve-based 1103, it was fully transistorised and Cray's modular circuitry resulted in an exceptionally compact design.

The CDC 1604 was announced in October 1959 at a price of $750,000 for a basic specification model fitted with 8,192 words of magnetic core storage, undercutting the price of a similarly specified UNIVAC 1103AF by more than $100,000. However, with calculation speeds of 7.2 microseconds for an integer addition and 36.0 microseconds for a floating-point multiplication, the CDC 1604 was approximately 6 times faster than the UNIVAC machine, a performance which approached that of an IBM 7090, a large-scale scientific model that sold for almost $3 million and was popular with the US defence community. Consequently, it was military rather than civilian customers that took an interest in Control Data's new model and the first machine was delivered to the US Naval Postgraduate School in Monterey, California, in January 1960. Other orders quickly followed and within a short time the company had sold more than 20.

The success of the CDC 1604 was sufficient to convince Norris that Control Data's main business focus should now be on high performance computers for the scientific market. However, the development of a new model large and powerful enough to attract the most demanding customers would require considerable financial resources, well beyond the reach of a young company like Control Data. Fortunately, the Lawrence Radiation Laboratory in Livermore, California had purchased a CDC 1604 and its scientists were much impressed by the machine's capabilities. Sidney Fernbach, Livermore's Director of Computation, was on the lookout for the Laboratory's next large-scale computer and approached Control Data to find out what it could offer in this respect. Fernbach listened to the company's plans for its new model, a Stretch-class machine that would be called the CDC 6600, and liked what he heard so much that he immediately placed an order for one. To help with the development costs, he arranged for the US Atomic Energy Commission to make an advance payment of $3.8 million. The discussions with Fernbach also provided valuable user input on the specification of the new machine.

The development of the CDC 6600 began in the summer of 1960. The design goal for the project was to create a machine with a performance 15-20 times that of the CDC 1604. The design team was led by Seymour Cray and included two of Cray's old ERA colleagues, James E Thornton and Lester T Davis. Thornton, who had worked closely with Cray on the design of both the UNIVAC AN/USQ-17 and CDC 1604, collaborated with Davis on the design of the central processor. Cray took responsibility for the design of the memory and input/output subsystems. With such an ambitious design goal for the new machine, there would be no attempt at maintaining backward compatibility with the CDC 1604. Only a clean sheet would suffice.

Seymour Cray's increasing importance in the management structure of the rapidly expanding Control Data Corporation brought additional responsibilities and these began to distract him from his computer design work. Consequently, progress over the first year or so of the CDC 6600 project was frustratingly slow and the development schedule began to slip. In an effort to remedy this situation, Cray asked Norris to let him abdicate his current role as Director of Engineering and set up a new development laboratory in his hometown of Chippewa Falls, Wisconsin, 80 long miles away from the company's headquarters in Minneapolis. Anxious not to lose his star designer, Norris agreed. The Control Data Chippewa Laboratory was founded in January 1962 under Laboratory Director James Thornton and the relocation was completed by July. The rural setting and close proximity of the Laboratory to his newly built home clearly suited Cray's solitary working methods, as the next few years would prove to be a particularly productive period in his long career.

The design strategy for the CDC 6600 was based on functional parallelism, a natural extension of the aggressive uniprocessor parallelism employed by IBM in the architecture of Stretch. The central processor featured a binary parallel architecture with a 60-bit word length and a 10 MHz clock speed, and was made up of 10 separate functional units that could operate concurrently. Where the CDC 6600 differed radically from Stretch, however, was that the design also incorporated 10 input/output processors. Known as Peripheral and Control Processors or PPUs, these were independent 12-bit computers, each fitted with 4,096 words of magnetic core storage and capable of concurrent operation via 12 input/output channels. The design for the PPU was lifted from the CDC 160, a small-scale computer that Cray had designed as the front end processor for the CDC 1604. The PPUs handled the slower I/O and supervisory functions, thereby leaving the central processor to continue high-speed calculations uninterrupted.

7. Revving Up for Higher Performance

To resolve the inevitable conflicts that resulted from the complex interactions of the functional units within the central processor, a 'Unit and Register Reservation Control' was developed. Also known as the Scoreboard, this system monitored the status of registers and functional units. If execution of an instruction was not possible immediately, due for example to data not being available in a register, the Scoreboard delayed issuing the instruction until ready. In managing this process, the Scoreboard was capable of out-of-order execution, another technique pioneered by Konrad Zuse in the Z4.

The Control Data engineers were keenly aware that memory access time is critical to machine performance and that larger memories increase access time, therefore, the main memory of the CDC 6600 was kept relatively small at 131,072 words in order to maintain speed. To facilitate a high level of interleaving, this was arranged in 32 independently accessible banks of 4,096 words each. The modest main memory was offset by the use of a very large magnetic disk storage unit of 13.2 million words capacity. This impressive unit comprised 72 disks grouped into 4 stacks of 18 disks each and vertically mounted on 2 spindles, each rotating at 1,200 rpm. Two read/write head mechanisms were located between each pair of disk stacks with a separate arm for each disk surface, positioned by a common hydraulic control mechanism, and 6 read/write heads on each arm.

In an early example of what is now called Reduced Instruction Set Computing or RISC, the instruction set for the central processor of the CDC 6600 was designed for simplicity and heavily optimised to reduce the number of machine instructions required to perform a task. Despite the architectural complexity of the machine, the designers were able to keep the number of instructions down to only 74, each of which was tightly bound to the processor hardware to ensure maximum performance. In comparison, both Stretch and Atlas had very large instruction sets which, including permutations, ran into hundreds of machine instructions.

In creating the electronic circuitry for the CDC 6600, the design team chose to abandon the conventional 'building block' approach of standardised circuit modules, with its heavy dependence on back panel wiring, in favour of complex, custom modules that offered cleaner electrical characteristics and faster performance. However, the faster switching speeds and greater density of components resulted in a considerably higher operating temperature which threatened to exceed the limits of germanium-based transistors. Fortunately, a new type of silicon planar transistor had recently become available from Fairchild Semiconductor which offered a higher operating temperature and faster switching speed. Approximately 400,000 of these transistors were used in the construction of the new machine. The characteristics of this new transistor also allowed the Control Data engineers to employ a simpler type of logic circuit known as Direct-Coupled Transistor Logic (DCTL).

The faster switching speeds of the electronic circuits in the CDC 6600 also gave rise to a problem with cable lengths due to propagation delay (the time required for an electrical signal to travel from one end of a wire to the other), so the machine was enclosed within an X-shaped cabinet with the most important circuitry located at the centre in order to minimise cable lengths. The 10 PPUs were also contained within one arm of this cabinet. One compensation for the propagation delay problem was that the engineers found that they were able to use lengths of cable instead of delay circuits to synchronise the timing of the circuitry. However, the compact configuration also served to further increase the density of components and severely restricted the flow of air necessary to dissipate the large amounts of heat generated by the CDC 6600's circuitry. With forced-air cooling out of the question, an alternative approach was called for. Fortunately, the team's mechanical designer, M Dean Roush, was a former refrigeration engineer. He devised a novel cooling system which conducted heat away from the circuit boards by means of copper pipes filled with Freon refrigerant. These were embedded in the framework supporting the modules and connected to refrigeration units located at the outer ends of each arm of the cabinet.

Another innovative feature of the CDC 6600 was the use of a cathode ray tube display on the operator control console. Dual 10-inch CRT display screens and an alphanumeric keyboard replaced the traditional array of indicator lights and toggle switches. The screens could be independently programmed to display either text or primitive bit-mapped graphics formed by patterns of dots.

CDC 6600 (Photo © Jitze Couperus)

The CDC 6600 was announced in August 1963 at a price of $6.9 million in basic

7. Revving Up for Higher Performance

configuration and the first machine was installed at the Lawrence Radiation Laboratory in April 1964, only a few months behind schedule despite the delays due to relocation of the design team midway through the project. The performance of the new model was stunningly ahead of anything else available at that time. Able to perform an integer addition in only 0.3 microseconds and a floating-point multiplication in 1 microsecond, it was more than 20 times faster than the CDC 1604 and around 3 times faster than Stretch. Not only had the design goal been achieved comfortably but the radical design of the CDC 6600 had also redefined the state of the art. A book by James Thornton on the design of the CDC 6600 became required reading for students of computer architecture following its publication in 1970. Running at 3 million instructions per second, the CDC 6600 remained the fastest computer in the world for much of the 1960s and was only superseded by the delivery of its replacement model, the CDC 7600, in January 1969.

With IBM's withdrawal of the 7030 in May 1961 and nothing else on the market coming remotely close, customers looking for the highest possible computing performance in 1964 had an easy choice. Early CDC 6600 customers included the US National Center for Atmospheric Research (NCAR) in Boulder, Colorado, the European Organisation for Nuclear Research (CERN) in Geneva, Switzerland, and New York University's Courant Institute of Mathematical Sciences. The Lawrence Radiation Laboratory ordered a further two machines in August 1965 and its sister laboratory at Los Alamos acquired the first of four CDC 6600s the following year. Strong sales of the 6600 model helped to take Control Data to third place overall in the computer industry by 1965 and more than 50 were eventually built. With the CDC 6600, Control Data had finally proven that high performance computing could be a commercial success.

An article in the influential publication Business Week in August 1963 noted that the development team for the CDC 6600 totalled only 34 people "including the night janitor". This compared with a team of more than 300 on Project Stretch and prompted a furious memo from IBM Chairman Thomas J Watson Jr. to his own staff, lambasting them for allowing IBM to lose its industry leadership position to such a small effort.

The CDC 6600 was a major landmark in the history of the computer. It established Control Data Corporation as a force to be reckoned with in the marketplace, cemented Seymour Cray's reputation as the industry's pre-eminent designer of high performance computers and heralded the age of the supercomputer.

Further Reading

Brennan, J. F., The IBM Watson Laboratory at Columbia University: A History, International Business Machines Corporation, 1971.

Buchholz, W., Planning a Computer System: Project Stretch, McGraw-Hill, New York, 1962.

Cocke, J. and Kolsky, H. G., The Virtual Memory in the STRETCH Computer, Proc. Eastern Joint Computer Conf. 16, 1959, 82-93.

Elzen, B. and MacKenzie, D., The Social Limits of Speed: The Development and Use of Supercomputers, IEEE Annals of the History of Computing 16 (1), 1994, 46-61.

Gray, G. T. and Smith, R. Q., Sperry Rand's Transistor Computers, IEEE Annals of the History of Computing 20 (3), 1998, 16-26.

The Rt. Hon. The Earl of Halsbury, Ten Years of Computer Development, Computer Journal 1 (4), 1959, 153-159.

Harlow, F. H. and Metropolis, N., Computing & Computers: Weapons Simulation Leads to the Computer Era, Los Alamos Science, 1983, 132-141.

Hendry, J., Prolonged negotiations: the British fast computer project and the early history of the British computer industry, Business History 26 (3), 1984, 280-306.

Howlett, J., The Atlas Computer Laboratory, IEEE Annals of the History of Computing 21 (1), 1999, 17-23.

IBM 7030 Data Processing System, IBM General Information Manual, International Business Machines Corporation, 1960.

Lavington, S. H., The Manchester Mark I and Atlas: A Historical Perspective, Communications of the ACM 21 (1), 1978, 4-12.

Lavington, S. H., Manchester Computer Architectures: 1948-1975, IEEE Annals of the History of Computing 15 (3), 1993, 44-54.

Thornton, J. E., The CDC 6600 Project, IEEE Annals of the History of Computing 2 (4), 1980, 338-348.

Thornton, J. E., Design of a Computer – The Control Data 6600, Scott, Foresman & Company, Glenview, Illinois, 1970.

UNIVAC-LARC System General Description, Remington Rand UNIVAC Division, Sperry Rand Corporation, U-1797, date unknown.

Weik, M. H., A Third Survey of Domestic Electronic Digital Computing Systems, Ballistic Research Laboratories Report No. 1115, March 1961.

Weik, M. H., A Fourth Survey of Domestic Electronic Digital Computing Systems, Ballistic Research Laboratories Report No. 1227, January 1964.

Worthy, J. C., Control Data Corporation: The Norris Era, IEEE Annals of the History of Computing 17 (1), 1995, 47-53.

8. Solid Progress – Transistors and Unified Architectures

"Gain a modest reputation for being unreliable and you will never be asked to do a thing", Paul Theroux.

The Quest for Reliability

In a memo to senior managers dated 2 October 1957, IBM Vice President W Wallace McDowell set out a new company policy to abandon the use of thermionic valves in favour of solid-state components for all future machine developments. IBM had invested heavily in thermionic valves and had used them extensively in its electronic and electromechanical products since the introduction of the Type 603 Electronic Multiplying Punch in 1946. So what had led to this costly decision?

The reliability of computer equipment is something that we take for granted nowadays but most early commercial computers had woefully poor reliability and required elaborate maintenance routines to keep them running. The main reason for such poor reliability was the use of the thermionic valve as a principal component in the electronic circuitry.

With their thin glass envelopes and myriad tiny wires, thermionic valves are delicate devices which are prone to failure in a number of ways. Failure can occur due to the filament, or cathode, being slowly 'poisoned' by atoms from other elements that are present as impurities in the tube, damaging the valve's ability to emit electrons. Leaks in the vacuum can also damage the cathode or cause a phenomenon known as plate-current runaway due to ionisation of free gas molecules. Failures can also occur due to breakage of the heater wire as a result of the thermal shock that occurs when the heater voltage is first applied.

The thermionic valves used by computer manufacturers were designed to last for 2,000 hours or more before failure but poor quality control during manufacture meant that defective valves were not uncommon and the higher demands placed on them by computer applications also tended to shorten their life. A large-scale computer would typically require several thousand valves and a failure in any one of these could result in the machine operating erroneously or failing completely. The result was that the first generation of commercial computers had reliability rates, usually expressed in terms of average error free running period or Mean Time Between Failure (MTBF), of only a few hours. For example, the UNIVAC I, with 5,200 valves, was reported by users as having an average error free running period ranging from 2.8 to 24 hours.

Engineers working on some of the early electronic computer development projects began devising methods for extending the life of thermionic valves. At the Moore School of Electrical Engineering, Presper Eckert devoted considerable effort to this issue when building ENIAC, which contained more than 17,000 valves. New valves were tested exhaustively before installation and ENIAC's electronic circuits were carefully designed so that components operated at well below their rated voltage and current capacities. The machine was also kept permanently switched on to minimise the risk of thermal shock. Despite these preventative measures, ENIAC is reported to have had an MTBF of only 12 hours.

In September 1948, IBM hired Arthur L Samuel, an electrical engineering professor from the University of Illinois, to oversee an internal development programme for thermionic valves. Samuel was a renowned expert in thermionic valve design and his appointment was widely seen as a signal to the electronic component manufacturers that they should begin taking IBM's thermionic valve requirements more seriously. At the Massachusetts Institute of Technology, engineers working on Whirlwind, an early digital computer designed for real-time operation, also took matters into their own hands by designing custom valves with an extended cathode life.

As the commercial market for computers took hold and production numbers increased, some electronic component manufacturers, led by Sylvania, began offering higher quality valves specifically for use in computer circuitry. High-reliability valves with a lifespan of greater than 5,000 hours were also developed at Bell Telephone Laboratories in the early 1950s for use in undersea telephone repeaters but these were very expensive, which tended to preclude their use in large numbers.

The progress made by computer engineers and the electronic component manufacturers certainly improved matters, however, thermionic valves remained fundamentally unreliable for computer applications. Improved reliability would mean reduced cost of ownership, a necessary feature if computers were to succeed in the lucrative business data processing market. Reliability was also a prerequisite for real-time control applications. For the computer industry to grow, the reliability of the equipment simply had to improve and the most important step towards reliable computers was the emergence of solid-state components.

Building Blocks – Semiconductor Devices

A semiconductor is a material with an electrical conductivity that is intermediate between that of an insulator and a conductor. Its conductivity can usually be

8. Solid Progress – Transistors and Unified Architectures

varied by exposing the material to heat or light, or through the introduction of chemical impurities in a process known as doping. This ability to vary electrical conductivity in a controlled way is what makes semiconductors a vital component in the construction of solid-state electronic devices. Examples of semiconductors include the elements germanium and silicon but the property is also present in certain naturally occurring crystals.

The first electronic device to employ semiconductors was developed by the German physicist Karl Ferdinand Braun at the end of the 19th century. Braun had discovered in 1874 that a galena crystal, a semiconductor material composed of lead sulphide, rectifies alternating current when probed with a metal wire. In early 1898, he became involved in work to extend the range of wireless telegraphy systems. As part of this work, Braun used the rectifying properties of the galena crystal to create what became known as a 'cat's whisker' diode, essentially a crystal mounted in a suitable holder with its upper surface exposed and a wire held over it in a frame that allowed the surface of the crystal to be probed to find the optimum position. However, the new device made no discernable improvement to wireless telegraphy systems and it would be several years before a practical application was found for it.

In August 1906, the American radio pioneer Greenleaf Whittier Pickard filed a US patent application for a device that used a cat's whisker crystal rectifier to separate a radio signal from its carrier wave. Pickard had based his device on Braun's work but chose fused silicon as the semiconductor after testing a large number of materials. This device, in various forms, was to serve as a critical component in early radio and radar receivers.

The term 'solid-state', a reference to the flow of electrons through a solid material rather than through a vacuum or a gas, was coined to describe these new semiconductor devices and the scientific understanding of semiconductors spawned a new branch of science, Solid-State Physics. Advances in quantum mechanics in the 1920s boosted progress in this field and by 1928 the Swiss physicist Felix Bloch was able to develop a theory for the conductivity of semiconductor materials. He was soon followed by the British theoretical physicist Alan H Wilson, who provided a scientific basis for semiconductor theory in two fundamental scientific papers published in 1931.

Semiconductor Diodes

In the early stages of World War II, Bell Telephone Laboratories became involved in research to increase the high frequency performance of radar receivers. As part of this work, Russell S Ohl, a Bell Labs electrochemist, began to experiment with crystal rectifiers. Although these devices had been largely superseded by the introduction of thermionic valves for radio receivers in the early 1920s, Ohl believed that they might have potential in high frequency applications.

In February 1940, Ohl made a remarkable observation that a cracked sample of silicon crystal which he was working with generated a small current when exposed to light. Further investigation over the next few weeks by Ohl and two colleagues, Walter H Brattain, a physicist, and Jack H Scaff, a metallurgist, revealed that the crack marked the dividing line between two regions of the crystal that contained subtly different amounts of chemical impurities and this had given them different electrical properties. One region was found to contain an excess of electrons, which they labelled negative or n-type, and the other a deficit of electrons, which they labelled positive or p-type. The action of light on the crystal caused electrons to migrate from one region to the other with such efficiency that the device could be used as a practical solar cell. But the device also had another unusual property. The junction between the two regions allowed current to flow across the crack in only one direction, making it, in effect, a semiconductor diode.

Following the end of the war, electronic component manufacturers began to introduce commercial versions of semiconductor diodes, the first of which was the Sylvania 1N34 germanium point-contact diode in 1946. The 40-year reign of the thermionic valve, which had begun with John Ambrose Fleming's valve diode in 1904, was coming to an end. Ohl's discovery of what he and Scaff termed the p-n junction would also lead to the invention of the most important electronic device in history.

The Transistor

The potential shown by Ohl's work encouraged Bell Labs to put more resources into semiconductor research and a solid-state physics group was formed in September 1945 under the joint leadership of physicist William Shockley and chemist Stanley O Morgan. Shockley, who had been engaged on solid-state research before the war, immediately began to consider the possibility of creating the solid-state equivalent of the triode thermionic valve, a three-terminal semiconductor device that could be used to amplify electronic signals. He proposed a device in which the conductivity of the p-n junction could be increased by applying a strong electrical field across it. However, when the device was built, it could not be made to operate as Shockley had predicted. Furthermore, a patent search revealed that a field-effect semiconductor device had been patented by the German physicist Julius E Lilienfeld fifteen years earlier.

A prolific inventor, Lilienfeld had conceived a solid-state device for amplifying radio signals in 1930 while working at the Ergon Research Laboratories in Malden, Massachusetts. His device was almost identical to Shockley's subsequent design except that it used a compound of copper and sulphur as the semiconductor material. Lilienfeld continued to work on solid-state devices for several years, resulting in two further patents which were granted in 1932 and 1933, but there is no evidence that any of these devices were ever built and

8. Solid Progress – Transistors and Unified Architectures

his semiconductor technology was never taken up commercially.

Following this initial disappointment, Walter Brattain and his theorist colleague John Bardeen began experimenting with alternative three-terminal configurations based on a point-contact structure. In December 1947 they observed that when a current was applied to the metal base of a device comprising a germanium block and two closely spaced contacts covered with gold foil, the current flowing between the contacts increased slightly, thus demonstrating the much sought after amplification effect.

When Shockley returned from a short sabbatical to find that Bardeen and Brattain had not only achieved the vital breakthrough in his absence but had done so using a structure that he himself had previously dismissed, he was furious. His reaction was to redouble his own efforts to create a three-terminal device that did not rely on a point-contact structure. After only a few weeks of intense activity, Shockley had invented a three-terminal device based on a combination of two p-n junctions. His design consisted of two n-type blocks of germanium separated by a thin layer of p-type germanium. Current flowing from an electrode at one end of the device, known as the emitter, to one at the other end, known as the collector, could be controlled by a current applied to the central layer or base of the device. With the electrons moving through the device rather than across its surface, Shockley's junction design would be more robust than one based on a point-contact structure but it would also be much more difficult to manufacture. A further two years of effort would be required before the Bell Labs chemists were able to grow individual crystals of germanium with the desired properties.

Following a successful demonstration of the Bardeen and Brattain prototype on Christmas Eve 1947, Bell Labs senior management immediately recognised the enormous commercial significance of this new invention and sought to ensure that they had watertight patent protection before disclosing it publicly. Four patent applications were filed, two covering the field-effect device and one each for the point-contact and junction devices. However, the earlier Lilienfeld patents resulted in some of the claims being rejected and only two of the applications were granted, one to Bardeen and Brattain for the point-contact device and the other to Shockley for the junction device. Bell Labs now needed a name for these new three-terminal semiconductor devices so an internal ballot was held in May 1948. Six names were proposed, four variations on the word 'Triode' and two new words, 'Iotatron' and 'Transistor'. The name that won was Transistor, an abbreviation of transconductance varistor coined by Bell Labs engineer John R Pierce.

The Story of the Computer

The First Transistor (Photo © Alcatel-Lucent USA, Inc.)

With patent applications filed and a suitable name chosen for the new device, a public announcement of the invention could now go ahead. A press conference was held in New York on 30 June 1948 at which demonstrations of a transistorised radio receiver and an audio amplifier were given. The transistor's many advantages over the thermionic valve were highlighted, in particular smaller size, considerably lower power consumption with no need to heat it up before use, increased ruggedness and the prospect of greatly improved reliability.

In order to evaluate the new device, a batch of point-contact transistors was manufactured in-house by Bell Labs. Labelled 'Type A', these were miniaturised versions of the original Bardeen and Brattain prototype conveniently packaged into a metal cartridge. Bell Labs engineers used these early samples to assess the suitability of the transistor as a replacement for thermionic valves in telephone switching circuitry but some were also made available to a select group of US government agencies and research establishments. In late 1948, as demand began to increase, the US electronics firm Raytheon was contracted to produce a commercial version of the Type A transistor which they called the CK 703.

As the research arm of a government-owned public utility, Bell Labs was expected to transfer technology into the private sector for commercial exploitation. It began offering licenses to manufacture transistors for an initial fee of $25,000, offset against future royalties. These included a cross-licensing

8. Solid Progress – Transistors and Unified Architectures

agreement in order that Bell Labs could learn from the further development of the transistor by its licensees. The license opportunity was first announced at a transistor symposium held at Bell Labs in September 1951 which had been organised at the request of the US military in order that their contractors and research staff could learn of the latest progress in the field. Only government agencies and defence contractors were invited to attend and the initial supply of transistors for experimental purposes was also restricted to military projects. Nevertheless, 26 companies initially signed up. A second symposium was held in April 1952 in which the restrictions were relaxed and this served to increase the number of licensees to 35.

The first practical use of transistors was in telephone switching equipment which was installed by the Bell Telephone System in Englewood, New Jersey in October 1952. Point-contact transistors were used to replace vacuum tubes in the oscillator circuits of a long distance direct-dial telephone exchange. Due to the limited power handling capability of early transistors, the earliest non-telephone applications were in hearing aids. The first of these to reach the market was the Sonotone Model 1010 which was introduced in December 1952. As well as featuring a single junction transistor, it also included two miniature thermionic valves. In acknowledgement of company founder Alexander Graham Bell's lifelong interest in improving the quality of life for the hard of hearing, Bell Labs waived the license fee for hearing aid applications.

In January 1956, as part of a Consent Decree between AT&T and the US Department of Justice, Bell Telephone Laboratories was ordered to put transistor technology into the public domain, providing specific patents related to transistors royalty free to anyone desiring to use them. Later that same year, the astonishing achievement of the Bell Labs solid-state physics group was formally recognised by the scientific community when Bardeen, Brattain and Shockley were jointly awarded the 1956 Nobel Prize for Physics.

Solid-State Alternatives

Computers used thermionic valves for both amplification and switching, and in very large numbers, so the benefits that could be gained from using transistors to replace valves in computer circuits were obvious. However, although transistors became commercially available towards the end of 1948, it would be another decade before they came into widespread use for computers. Point-contact transistors with consistent operating characteristics proved difficult to manufacture and their reliability was initially poorer than thermionic valves. Junction transistors, which became commercially available in late 1952, were more reliable but were unable to operate at sufficiently high frequencies to cope with clock speeds in the megahertz range. They were also particularly sensitive to changes in temperature. Nevertheless, there was an immediate recognition within the industry of the massive potential of solid-state circuitry. Therefore, while awaiting the next generation of transistors that promised higher operating frequency and improved reliability, computer designers began to investigate alternative solid-state components which could be used to build logic circuits.

The property of a diode to permit current to flow in only one direction can be applied to logic circuits. Logic gates that employed diodes as the principal component had been developed at Bell Telephone Laboratories in the late 1930s for use in automatic telephone switching applications. A US patent application for a 'Telephone System' filed by Arthur W Horton in September 1939 included a description of a diode-based OR gate and a similarly titled application by another Bell Labs researcher William H T Holden in October 1941 described a diode-based AND gate. The introduction of semiconductor diodes in 1946 opened up the possibility of constructing solid-state logic circuits. However, it is not possible to create a diode-based NOT gate and this limited the ability of computer designers to implement fully diode-based logic circuitry.

It is also possible to construct logic circuits using magnetic cores. The property of a magnetic core to have two stable states is analogous to a switch and hence can be used to create the basic circuits for logic gates. Munro K Haynes, a postgraduate electrical engineering student at the University of Illinois, first applied magnetic cores to implement logic as part of his PhD thesis, which was completed in August 1950. Haynes took up a job offer from IBM shortly after completing his PhD and subsequently described his ideas in a July 1953 US patent application entitled 'Magnetic Core Logical Circuits' which was granted in November 1954.

Another electromagnetic device, the magnetic amplifier, could also be used to replace thermionic valves in computer applications. Invented in 1916 by Ernst F W Alexanderson of General Electric as part of his work on radio transmitters, the magnetic amplifier is based on a combination of a saturable reactor and a rectifier. A saturable reactor is a device which uses the magnetic field in a magnetic core to control the inductance of a coil. When unsaturated, the magnetic core causes the coil to act as a high inductance which impedes the flow of electrical current. When the magnetic core is saturated, the inductance drops and current flows unimpeded. Magnetic amplifiers can be used to amplify current and can also be configured to operate as an electronic switch, which makes them suitable for both amplification and logic circuits. They are extremely robust and reliable but are limited to operating frequencies to below 1 MHz.

The Standards Eastern Automatic Computer

The first computer to employ solid-state logic was developed by the US National Bureau of Standards as an interim facility while awaiting delivery of

8. Solid Progress – Transistors and Unified Architectures

its first commercial machine, the UNIVAC I. Under the leadership of Samuel N Alexander, the NBS Electronic Computers Laboratory in Washington, D.C. had successfully developed several electronic components for computers and had also built input/output equipment for the IAS Computer. Having accumulated considerable in-house expertise in electronic computation, the NBS had taken on the role of contracting agent for the US Census Bureau in its dealings with the Eckert-Mauchly Computer Corporation. The contract for the initial phase of the UNIVAC I development had been signed in October 1946 but by early 1948 concern was growing that the new machine would not be completed in time for the 1950 US Census.

The Census Bureau was joined in its concern by two other US Government agencies that were each about to place a contract for a UNIVAC I via the NBS, the US Air Force and the US Army Map Service, and the three agencies began to discuss the need for an interim machine. The US Air Force seized the initiative and placed a contract with the NBS for a machine that could be quickly deployed to fill the gap until its own UNIVAC I was delivered and which would also be available for use by the Census Bureau, the US Army and the NBS itself.

Construction of the new machine began in May 1948. The Chief Engineer was Ralph Slutz, who had been one of the principal engineers on the IAS Electronic Computer project at Princeton before moving to the NBS. Initially known as the NBS Interim Computer, it was a binary serial machine with a 45-bit word length, 512 words of mercury delay line memory and a 1 MHz clock speed. The design was kept as simple as possible in order to minimise the project timescale. This philosophy carried through to the memory as, rather than devote a substantial proportion of development effort to developing a memory in-house, the team simply purchased a mercury delay line memory as a complete unit from the same contractor that supplied the memory for EDVAC.

The NBS team possessed a meticulous approach to engineering which had been instilled through working on standards and the reliability of thermionic valves was of particular concern. They were attracted to using solid-state components in the design of the new machine but transistors were not yet available commercially and the use of magnetic cores to implement logic had yet to be devised. Fortunately, semiconductor diodes were readily available so the NBS engineers decided to adopt these as the principal component for the logic circuits. A total of 10,500 germanium point-contact diodes were used to implement all the logic gates, clocking and pulse reshaping circuits. In the absence of power transistors, thermionic valves were still required for amplification but the total number of valves used, at 747, was considerably reduced.

The NBS Interim Computer was operational by April 1950 and was renamed the Standards Eastern Automatic Computer, or SEAC, to differentiate it from another computer under development at the NBS Institute for Numerical Analysis in Los Angeles, which naturally became known as the Standards Western Automatic Computer. A combination of solid-state logic circuitry and a rigorous preventative maintenance schedule ensured high reliability and the machine quickly became established as the principal computing platform at the Bureau's Washington headquarters. As one of the earliest stored-program computers in the USA, SEAC also attracted wider attention and was used by a number of groups on a broad range of computational problems over its 14-year lifespan.

Experimental Transistor Machines

The Manchester University Transistor Computer

The potential of solid-state devices in computer applications had also caught the attention of Tom Kilburn's pioneering team at Manchester University in the UK. In 1952, Richard L Grimsdale, a research student on the team who had been developing test routines for the Ferranti Mark 1, was about to begin the construction of a small valve-based computer when he learned that the first commercial samples of transistors had recently become available from a number of UK electronic component manufacturers. This presented an excellent opportunity to investigate the possibility of building a transistor-based computer. Grimsdale managed to obtain a small quantity of LS737 point-contact transistors from Standard Telephones & Cables in early 1953. Assisted by his colleague, Douglas Webb, and supported by other members of the Electrical Engineering staff, Grimsdale set to work building a simple stored-program machine which became known as the Manchester University Transistor Computer.

The Transistor Computer employed a binary serial architecture with a 48-bit word length that included 4 spare bits for timing. A magnetic drum unit cannibalised from the Manchester Mk 1 provided 64 words of main memory plus storage for the arithmetic and control registers. The registers were implemented as recirculating loops by positioning the read heads a short distance away from the write heads around the circumference of the drum. As the drum rotated at a leisurely 2,500 rpm, this dictated a slow clock speed of 125 KHz, which also suited the poor switching characteristics of the early point-contact transistors. A basic instruction set of 7 machine instructions was created to program the machine.

With transistors in very short supply in the UK, most of the logic circuitry was implemented using point-contact diodes and the first prototype incorporated only 92 transistors. Nevertheless, the machine was operational by November 1953 and executed its first program on the 16th of that month, earning it the

8. Solid Progress – Transistors and Unified Architectures

accolade of the world's first operational transistorised computer. The promise shown by this rudimentary prototype was sufficient to convince the team to proceed with the development of a full-scale machine.

For the full-scale prototype, the design was extended to include an index register, an eight word serial register and a multiplier. The hardware was rebuilt, this time requiring a total of 250 transistors and 1,357 diodes in order to implement the additional logic. The magnetic drum memory was retained but the speed of rotation was increased to 3,000 rpm.

The full-scale prototype was operational by April 1955. With a ponderous clock speed and magnetic drum memory, performance was never likely to be sparkling and this was reflected in calculation speeds of around 30 milliseconds for a 44-bit addition and 1 second for a division. Also, despite the use of solid-state components, the average error free running period was a disappointing 1.5 hours due to reliability problems with the early samples of transistors. Nevertheless, the machine consumed only 150 watts of power, less than that of a modern desktop computer and a tiny fraction of the 12 kilowatts power consumption of a typical medium-scale computer of the period.

Keen to replicate the successful transfer of technology into industry through their fruitful association with Ferranti, Kilburn's team secured the interest of another local firm, the Metropolitan-Vickers Electrical Company, which had collaborated with the University on the development of a differential analyzer some years earlier in 1935. During the development of the Transistor Computer, two Metropolitan-Vickers engineers were seconded to Manchester University to work with Grimsdale. This resulted in the design of the full-scale prototype being adopted by the company for their MV950 model, which was completed in 1956. The drum memory was retained but junction transistors were used instead of point-contact transistors to improve reliability.

The MV950 was put into limited production and a total of 6 were built. However, all 6 were earmarked for internal Metropolitan-Vickers use and the machine was never seriously marketed as a commercial product. A closely related model, the AEI 1010, was introduced commercially in 1960. Approximately 10 of these machines were sold before the company was acquired by GEC in 1967.

Bell Telephone Laboratories TRADIC

As the organisation responsible for the development of the transistor and having been active in digital computer development since the earliest days, Bell Telephone Laboratories was perfectly placed to play a leading role in the development of transistorised computers. Not surprisingly, the first ever application of transistors to computing took place at Bell Labs in 1949 when engineer Walter H MacWilliams incorporated early samples of point-contact transistors into a gating matrix circuit as part of a special-purpose computer for simulating naval gunnery systems. In 1950 another Bell Labs engineer,

James R Harris, successfully built a transistorised shift register and serial adder unit as part of a project for the US Army Signal Corps aimed at exploring the feasibility of transistors in a range of electronic applications.

Jean H Felker, a colleague of MacWilliams, was also experimenting with transistors and in 1951 constructed a transistorised regenerative amplifier which he realised could form the basis of a set of building blocks for a compact, low power digital computer. To demonstrate this capability, Felker then built a high-speed multiplier unit which could multiply two 16-bit binary numbers in only 272 microseconds and presented his results at the first Bell Labs transistor symposium in September 1951. The success of this demonstrator led to a formal proposal to the US Air Force for the development of a laboratory prototype to test the feasibility of a transistorised digital computer for aircraft bombing and navigational purposes. The Air Force, recognising the huge potential of transistors for military equipment, agreed to support the project and work began in early 1952.

The prototype was named the Transistorized Airborne Digital Computer or TRADIC. The development team was led by Felker, with support from James Harris. Felker and Harris adopted a two-phased approach to the development. To facilitate rapid progress, the architecture of the Phase 1 machine would be kept as simple as possible and a stored-program architecture would not be implemented until the second phase of the project.

The TRADIC Phase 1 machine was a sequence-controlled calculator with a 16-bit binary serial architecture and a 1 MHz clock, which was programmed via a replaceable plugboard with a capacity of up to 64 program steps plus the facility for a 64-step subroutine. Felker and Harris were strongly influenced by the design of SEAC. With the reliability of transistors still something of an unknown quantity, they decided to limit the number of transistors in their design by employing diode logic for the Phase 1 machine. Hence TRADIC Phase 1 incorporated 10,358 germanium diodes but only 684 Bell Labs point-contact transistors which were used for amplification. However, in the absence of power transistors, a small number of thermionic valves were still required in the power supply for the system clock.

The TRADIC Phase 1 machine was operational by January 1954, only a few weeks behind the first of the Manchester University transistor computers. Unlike the tiny Manchester prototype, TRADIC was a full-scale design but it was not a stored-program machine and it would be a further 2 years before the Bell Labs team would deliver a fully transistorised stored-program computer in the second phase of the project. Known as Leprechaun, the stored-program version of TRADIC used 5,500 junction and surface barrier transistors for the logic circuits plus 1,024 words of magnetic core memory.

Leprechaun became operational towards the end of 1956, eclipsing the earlier machine. Nevertheless, TRADIC Phase 1 was highly successful in proving that

8. Solid Progress – Transistors and Unified Architectures

transistors were the key to improved reliability. This was demonstrated in impressive fashion by subjecting the machine to continuous testing 24 hours per day, 7 days per week for a period of 2 years, during which only 8 transistors and 9 diodes had to be replaced.

The MIT Lincoln Laboratory TX-0

The MIT Lincoln Laboratory was created in 1951 as a Federally-funded research centre of the Massachusetts Institute of Technology. Located in Lexington, Massachusetts, the Laboratory's initial focus was the development of technology for air defence systems. The new organisation brought together Jay Forrester's Digital Computer Laboratory and a group from the MIT Research Laboratory of Electronics which had been involved in the development of microwave radar during World War II. Funding for projects came from the US military, the principal contributor being the US Air Force.

In 1953 the Lincoln Laboratory's Advanced Computer Development Group began to experiment with transistor circuitry. The group was led by William Papian, who had made his name as a member of Jay Forrester's team working on the development of magnetic core storage as part of the Whirlwind project. It also included two other prominent members of the Whirlwind team, Wesley A Clark and Kenneth H Olsen. After successfully building a shift register and then a small multiplier using transistors, the group proposed to construct an advanced high-speed transistorised computer, the Test-Experimental Computer Model 1 or TX-1, which would feature a 36-bit word length and a large capacity magnetic core memory. However, this proposal was deemed by Laboratory management to be overly ambitious for what was still a relatively small group. It was decided instead to split the project into two phases by first developing a much simpler prototype which could be used to evaluate the use of transistor circuitry for high-speed logic and would also provide the means to test a new magnetic core memory intended for the full-scale machine.

The small-scale prototype was referred to as the Test-Experimental Computer Model 0 or TX-0. Wesley Clark was responsible for the logical design and Ken Olsen for the circuit design and construction. The TX-0 featured the same binary parallel architecture proposed for the full-scale machine but with half the word length at 18 bits. To facilitate the testing of circuits, the clock speed could be varied up to a blazing 5 MHz maximum and a simple instruction set of only 4 instructions was created. As planned, the machine was initially fitted with the large magnetic core memory earmarked for the full-scale machine which had a capacity of 65,536 words.

In 1954, engineers working at the Philco Corporation in Philadelphia invented a new type of transistor known as a surface barrier transistor. Philco was a large well-established US electronics firm best known for its consumer radio and television products. Although one of the earliest licensees of transistor technology from Bell Labs, conservative Philco was hardly an obvious candidate for leading edge semiconductor developments. Nevertheless, the Corporation's scientists had developed a unique electrochemical etching process that allowed for an extremely thin germanium base layer, which in turn allowed the transistor to operate at very high frequencies. This was the breakthrough that computer designers had been waiting for and they soon began beating a path to Philco's door.

Papian's group immediately recognised the potential of Philco's newly announced surface barrier transistor for their new machine. However, the device was still at the experimental stage and had not yet been manufactured in sufficient quantities for the group's purposes. To remedy this, a subcontract was issued to the Philco Research Division in January 1955 for the development of the prototype surface barrier transistor into a device suitable for computer applications and to provide facilities for the testing of components. The result was the Philco L-5122 transistor, which was later commercialised as the 2N240. Approximately 3,500 of these new devices were used to implement the circuitry of the TX-0. However, due to the inability of early transistors to deal with high currents, several hundred thermionic valves were also required, principally in the driver circuits for the magnetic core memory and in the clock pulse amplifier circuits.

The TX-0 became operational in April 1956. Having been created primarily as a test facility, it was not designed to be physically compact nor low power, occupying over 200 square feet of floor space and consuming around 1,000 watts. However, it exhibited relatively high performance (83,000 addition operations per second) and good reliability and in 1957, following a period of highly successful operation, the decision was made to proceed with the development of the full-scale machine, which would now be called the TX-2 to reflect changes in specification from the original proposal. In the spring of 1958, the 65,536 word magnetic core memory was removed for installation in the TX-2 as planned and replaced with a smaller transistor-driven magnetic core memory of 4,096 words capacity. In July 1958, the TX-0 was transferred to the MIT Electrical Engineering Department on a long-term loan basis where it remained in use until 1976.

Missile Guidance Computers

Bell Labs' TRADIC and MIT's TX-0 had convincingly demonstrated the potential of transistorised computers to the US Air Force. Their small physical size and low power requirements made transistors particularly attractive for use in airborne computers and the huge improvement in reliability and higher performance that they brought also made them ideal for real-time control applications. Shortly after TRADIC became operational, the US Air Force

8. Solid Progress – Transistors and Unified Architectures

initiated a high priority weapon system development programme that would rely heavily on both airborne and control computers for its success. The aim of this programme was to develop an intercontinental ballistic missile (ICBM) but it would also provide the impetus to take transistorised computers out of the laboratory and into production.

Inertial Guidance

A ballistic missile is a missile which follows a parabolic, sub-orbital, ballistic trajectory. The missile travels at supersonic speed but is only powered during the initial phase of flight. Early missile guidance systems were based on a simple mechanical autopilot or a form of manual command in which the missile is directed to the target by an operator via radio control. However, both methods lacked precision and manual command was completely unsuitable for long-range missiles. The development of intercontinental ballistic missiles would require a more sophisticated form of navigation known as inertial guidance, the basic principles of which were established by German rocket scientists during World War II.

In inertial guidance, a set of three devices known as linear accelerometers which measure acceleration are mounted on a platform along three orthogonal axes. The platform is then suspended within a set of three gimbals that allow it to twist about any rotational axis and gyro-stabilised using a pair of electrically driven gyroscopes to maintain the correct orientation of the accelerometers in relation to the earth. The acceleration data from the accelerometers, suitably processed, can then be used to determine the missile's velocity, attitude and position relative to its starting point.

Inertial guidance systems can either be self-contained or combined with a ground-based tracking system in which the position of the missile is tracked via radar or radio interferometry and course corrections are transmitted to the missile from a ground-based control computer via radio signals. A self-contained system is preferable for missile guidance as it is less susceptible to interference and requires fewer personnel to operate. Known as an all-inertial guidance system, this type of system is capable of detecting deviations from the pre-programmed flight path and formulating midcourse corrections independently. However, an all-inertial system is considerably more difficult to engineer accurately and complex mechanisms are usually required to compensate for the effects of drift. It also places greater demands on the data processing, requiring more processing power, faster compute performance and 100% reliability.

German scientists successfully implemented an inertial guidance system in the V-2 rocket, which was first deployed to frightening effect against targets in France and Britain in September 1944. The system was of the radio-inertial type. Data processing was handled by an electronic analogue computer system known as Mischgerät, which translates into English as mixing computer. Developed by the electrical engineer Helmut Hoelzer, the Mischgerät system comprised a ground-based computer and a small on-board computer which communicated with each other by means of a radio link. Electronic analogue circuits were created to solve the differential equations used to control the motion of the rocket. As both computers were analogue, the system was able to operate in real-time.

The Atlas Missile Programme

The V-2 had shown the world how chillingly effective ballistic missiles could be and at the end of World War II, both the US and the Soviet Union began to investigate the strategic potential of ballistic missiles in the delivery of nuclear weapons. The programme to develop America's first operational ICBM, the Atlas missile, began with a request for proposals from the US Army Air Forces in October 1945. The USAAF, the forerunner of the US Air Force, envisaged a 10-year R&D programme to develop a family of four missiles with ranges of up to 5,000 miles and looked to the US aircraft industry to help realise this ambitious goal.

Consolidated Vultee Aircraft Corporation, popularly known as Convair, had been a major supplier of combat aircraft to the US forces during World War II and was keen to move into missile development, having recently developed a successful short-range rocket for the US Navy. Following the USAAF request, Convair submitted a proposal to evaluate two alternative designs for the longest ranged of the four types of missile envisaged; a jet-powered winged vehicle which flew at subsonic speed and a rocket-powered supersonic ballistic missile loosely based on Germany's V-2 rocket. Both designs proved to be of interest to the USAAF and the company was awarded a $1.4M study contract in April 1946.

The aim of the project, codenamed MX-774, was to evaluate the development of a missile which could carry a 5,000 lb payload over a range of up to 5,000 miles with a target accuracy of 5,000 feet. Early progress was sufficiently promising that the USAAF increased the value of the contract by 35%. However, a US Government economy drive initiated in December 1946 under President Truman forced a significant reduction in missile development funding which resulted in the USAAF first dropping the jet-powered winged design and eventually cancelling the entire project in July 1947, only a few months before it was due to be completed. Nevertheless, a small-scale test rocket was designed, the RTV-A-2 Hiroc, three examples of which were built and tested. Convair had also been working on an advanced ground-based radio tracking system known as Azusa, and this was deemed important enough for the USAAF to continue supporting it after the cancellation of the project.

In 1949 and 1950, various studies conducted by the RAND Corporation and

8. Solid Progress – Transistors and Unified Architectures

several US aeronautical firms highlighted the fact that recent technological advances greatly enhanced the feasibility of developing a long-range, rocket-powered guided missile capable of carrying heavy atomic warheads. This led to a revival of the research work at Convair on a relatively modest scale in January 1951 but with somewhat more ambitious goals, an 8,000 lb payload and a target accuracy of 1,500 feet. Initially codenamed MX-1593, the project was renamed Atlas in August 1951.

Over the next two years the scale of the research was gradually increased. In October 1953 the Air Force appointed an advisory committee of leading scientists and engineers to examine and evaluate the development of its missile programme. Codenamed the Teapot Committee, it was chaired by none other than John von Neumann. The Atlas programme was given a massive boost in February 1954 when the Teapot Committee, recognising the growing threat from the Soviet Union, recommended a crash programme to develop and deploy an ICBM within 6 years. This resulted in the Atlas programme being assigned the Air Force's highest development priority in May 1954, whereupon budgetary restraints were for all practical purposes lifted.

In an effort to guard against contractor failure and reduce timescales, separate 'associate' contractors were used for each of the missile's principal subsystems, with development carried out in parallel wherever possible. It was decided to adopt the more technically advanced all-inertial guidance system for Atlas, however, as a backup in case of difficulties a ground-based radio-inertial guidance system would also be developed.

The Atlas Guidance Computer

The contract to supply the ground-based radio-inertial guidance system for the Atlas programme was awarded to the General Electric Company in February 1955. The system that the company developed was known as the General Electric Radio Tracking System (GERTS). However, the Air Force policy of using separate contractors for critical components led to a separate contract being issued for the ground-based control computer for this system. The Burroughs Corporation was one of several firms to bid and won the contract.

The project began in April 1955 at the Burroughs Research Center in Paoli. The new computer was officially designated the AN/GSQ-33 but is better known as the Atlas Guidance Computer. It was designed by Isaac L Auerbach. Auerbach had been a key member of the team which developed the BINAC at Eckert-Mauchly Computer Corporation and was one of two former EMCC engineers hired by Director of Research Irven Travis in 1949 for the newly established Burroughs Research Division.

The Atlas Guidance Computer was a large-scale binary parallel machine with a 24-bit word length. To help deliver the high input/output performance necessary for real-time operation, Auerbach's design featured a Harvard architecture, where the instructions and data are stored separately and can be accessed simultaneously. The machine was equipped with 2,048 18-bit words of magnetic core storage for instructions and 256 24-bit words for data.

With reliability of prime concern, the US Air Force had originally specified a design incorporating redundant arithmetic units. Three identical units were specified, each of which would be capable of taking over on the failure of another. However, during contract negotiations, Auerbach succeeded in persuading the Air Force that a transistorised design with a single arithmetic unit and enhanced error correction would be reliable enough and much more cost-effective. Surprisingly, given the Air Force's enthusiastic support of transistor developments at Bell Labs and MIT, they remained concerned at the reliability of transistors and insisted that Burroughs also provide a backup design which employed thermionic valves.

The Atlas Guidance Computer was constructed using transistor logic circuitry originally developed at Bell Labs and featured Philco's first production quantities of the 2N240 surface barrier transistor. Having put such faith in transistors, there was an embarrassing moment for Burroughs when, in scaling up production, Philco began experiencing problems in maintaining transistor quality. A furious Auerbach was forced to remind Philco of the importance of the project to national security and the matter was quickly resolved.

The Atlas Guidance Computer was the first transistorised computer to go into production. In all, 18 machines were built at a total project cost of $37 million. The first of these was installed at the Cape Canaveral missile range in June 1957. His job done, Auerbach, who had been considering a career move for some time, left Burroughs the following month to set up his own computer consultancy firm, Auerbach Associates. Atlas missile launches began in September 1957 but test patterns were transmitted to the missile in place of actual guidance commands for the first few flights and the first computer-controlled launch took place on 19 July 1958.

The Atlas Airborne Digital Computer

Following humble beginnings as a supplier of gyroscopic correction equipment to the US Navy, the Arma Corporation of New York had built a solid reputation as a leading manufacturer of analogue fire control computers. It became a subsidiary of the American Bosch Corporation in 1948 and was merged to form the American Bosch Arma Corporation in 1954. In the early 1950s, under a series of study contracts funded by the US military, the company began to migrate from analogue to digital computer technology. Arma had also initiated an internal research programme on inertial guidance in 1951, which resulted in the development of a high precision accelerometer, and these two research strands would come together in the company's most successful product of this period, the Atlas Airborne Digital Computer.

8. Solid Progress – Transistors and Unified Architectures

In 1954 Arma submitted a bid for the contract to supply the all-inertial guidance system for the Atlas programme but lost out to a joint bid from MIT and the AC Spark Plug Company of Milwaukee, Wisconsin. However, the Arma proposal was not without merit and the US Air Force could afford to hedge its bets. In April 1955 it issued a contract to Arma for the development of a competing system.

The development of the Atlas Airborne Digital Computer was led by the Chinese-born electrical engineer Wen Tsing Chow. As with the ground-based control computer developed by Burroughs for the Atlas programme, the design requirements for the Arma computer included good reliability and high input/output performance. However, because the Arma computer was destined for installation within the Atlas missile itself, the list of design priorities also included robustness, small physical size and low weight. Chow rose to the challenge and created a computer that set new standards for reliability and compactness.

The Atlas Airborne Digital Computer featured a Harvard architecture and was fully solid-state in construction. The data memory was implemented using acoustic delay lines but with solid quartz as the transmission medium instead of liquid mercury. The targeting constants were stored in a novel non-destructive read-only memory which was implemented using 'burn-out' diode arrays. In this type of memory, a matrix of semiconductor diodes is used to modify a sequence of electrical pulses representing binary numbers. A high current is used to burn out individual diodes in each row to create the required binary numbers. Chow and colleague William H Henrich filed a US patent application describing their invention in December 1957. Granted in April 1962, this is the fundamental patent for what is now commonly known as programmable read-only memory or PROM.

The first prototypes of the Atlas Airborne Digital Computer were operational by late 1956. New packaging and miniaturisation techniques were developed by Arma engineers to create a system that fitted within a space of 8 cubic feet and a magnesium alloy chassis was employed to keep the weight low. The machine was designed for an MTBF of over 10,000 hours, a figure previously unheard of in the computer industry. The first flight test using the Arma all-inertial guidance system took place in March 1960 using a modified Atlas Model D rocket. Subsequent models of Atlas (E and F) incorporated the Arma all-inertial system and over 400 were eventually built at a cost of over $1M each.

Project Athena

Following the May 1954 decision to assign the highest development priority to the Atlas programme, concern began to grow that the prime contractor, Convair, might not be able to deliver. There was also a recognition that the

Atlas missile was somewhat limited in terms of range and payload, a view also endorsed by the Teapot Committee. In an effort to address these concerns, the US Air Force initiated a backup project in January 1955 to develop a larger and more powerful alternative ICBM known as Titan. Development took place in parallel with that of Atlas but the prime contractor was the Baltimore-based Glenn L Martin Aircraft Company.

The associate contractor for Titan's radio-inertial guidance system was Bell Telephone Laboratories. The system that Bell Labs developed was called the Bell Laboratories Command Guidance System, the basic principles for which were conceived by the renowned physicist Sidney Darlington in 1954. It combined radar tracking with inertial guidance, having evolved from earlier work by Western Electric to develop a shipboard radar using analogue computers. In March 1955 the contract to supply the ground-based control computer for this system was awarded to Remington Rand. The computer which Remington Rand developed was known as Athena and reliability was once again a major concern.

Engineers at Remington Rand's St Paul division had begun experimenting with solid-state components in late 1954. In an effort to find a suitable replacement for thermionic valves in future computer developments, Seymour Cray had initiated a project to investigate the feasibility of various alternative component technologies. With transistors not yet proven in computer applications and magnetic core logic looking promising, two experimental machines with identical instruction sets were constructed and assessed in an effort to determine which solid-state component technology to adopt. Working under the direction of Cray, James Thornton designed one of the machines using magnetic core logic, which was known as the Magnetic Switching Computer or Magstec. The other machine, known as the Transistor Test Computer or Transtec, was designed by Dolan Toth using Philco surface barrier transistors. On completion, Magstec was found to be the faster of the two experimental machines. However, by this time it was becoming clear that transistors would be the dominant solid-state device technology so a decision was taken to base the design of the Athena computer on transistor and diode logic.

Athena was remarkably similar in specification to the Burroughs Atlas Guidance Computer, being a large-scale binary parallel machine with a 24-bit word length and fixed-point arithmetic. It also employed a Harvard architecture, and was equipped with 256 24-bit words of magnetic core storage for data, supplemented by 8,192 17-bit words of magnetic drum storage for instructions. The arithmetic unit was constructed using 800 transistors and 4,000 diodes.

The first machine was completed in 1957. Although not fully solid-state, reliability in service comfortably exceeded the design specification, with an MTBF of over 1,000 hours, and a total of 23 Athena computers were built. Like Atlas, the Titan missile programme also moved eventually to an all-inertial

8. Solid Progress – Transistors and Unified Architectures

guidance system but the Athena computers were used in the Titan I from its first successful test launch in February 1959 until the missile was phased out in June 1965.

Despite having been rejected for Athena, James Thornton's magnetic switching technology was not discarded. It was adopted for a top-secret cryptanalysis machine designed by Remington Rand for the US National Security Agency under another project codenamed Bogart that began in July 1954. The prototype, which was known as the X-308, was completed in September 1956 and a total of four production machines were built.

Commercial Solid-State Computers

Philco's surface barrier transistor had proved beyond doubt that transistors were suitable for use in computer circuitry. This had not only been demonstrated in one-off experimental machines such as MIT's TX-0 but transistorised computers for missile guidance applications had also gone into production. Following Philco's lead, a number of electronics companies began to introduce new transistors specifically designed for computers and manufactured in quantity, one of the first of which was the Radio Receptor Company's RR156 computer transistor in 1955. The smaller, specialist companies were soon followed by the giant US electronics firms, such as General Electric with its 2N167 junction transistor in 1955 and RCA with its 2N404 junction transistor and 2N247 drift transistor, both of which were introduced in 1957. As production increased and unit costs fell, computer manufacturers now had no reason not to incorporate transistors into their mainstream computer products.

Transistor Development at IBM

IBM had begun experimenting with transistors seven years prior to Vice President McDowell's policy changing memo of October 1957. A group led by thermionic valve expert Arthur Samuel was formed at the company's Poughkeepsie Laboratory in 1950 to acquire expertise in solid-state technologies and an education programme was initiated to train IBM's engineers in solid-state physics. The company was also in the first wave of firms to license transistor technology from Bell Labs in September 1951.

In the summer of 1953, a group of novice IBM engineers attending one of the company's in-house training courses redesigned the circuits of the Type 604 Electronic Calculating Punch as a practical exercise, replacing the machine's 1,250 miniature thermionic valves with 2,200 transistors. Their design was sufficiently promising that a demonstrator was built and exhibited at the dedication of a new building at Poughkeepsie in October 1954. This led to the development of a production version, using IBM-manufactured germanium junction transistors, which was announced as the IBM 608 Transistor

The Story of the Computer

Calculator in April 1955 and began shipping in December 1957. Although fully transistorised and equipped with a small magnetic core memory of 40 words capacity, the 608 was not a stored-program machine, as it retained the earlier model's plugboard for programming. However, this machine was an important step towards transistorised computers and was the first commercially available data processing machine of any kind to employ solid-state construction.

A semiconductor device development group was also formed at Poughkeepsie under physicist Lloyd P Hunter, who had joined IBM in 1951. In 1954, this group developed a high performance graded-base drift transistor which was based on the work of the German theoretical physicist, Herbert Kroemer, who had proposed the structure of the drift transistor in a scientific journal paper the previous year. By 1958, IBM had commenced production of the drift transistor for use in Stretch and had also licensed the design to the electronic component manufacturer Texas Instruments as part of a cross-licensing and preferred supplier agreement. IBM had embraced transistors in a more fundamental way than other computer firms by becoming a device manufacturer.

The UNIVAC Solid-State

Despite IBM's achievements in transistor technology and the huge financial investment made by the company in solid-state devices, the first computer manufacturer to enter the market with a solid-state general-purpose computer was IBM's closest rival, Sperry Rand. The computer had its origins in a prototype machine built by Sperry Rand's Philadelphia division for the US Air Force Cambridge Research Center (AFCRC).

Development of the Cambridge Air Force Computer, as it was known, began in 1952 and was completed by 1956. With a biquinary-coded decimal architecture, 10 decimal digit word length and magnetic drum memory, it bore a striking similarity to the IBM 650 Magnetic Drum Data Processing Machine. However, solid-state logic circuitry and a higher capacity 4,000 word drum gave the Sperry Rand machine a considerable advantage over the 3 year old IBM model, making a commercial version the obvious next step for Sperry Rand.

The production prototype was completed in December 1957 and was known as the Universal Card Tabulator or UCT. It featured a larger, faster drum of 5,000 words capacity, which further improved the specification over the IBM model. The drum rotated at the impressively high speed of 17,667 rpm and featured a helium filled drum casing to eliminate air friction. As the prototype had been developed prior to the introduction of reliable transistors, the arithmetic unit contained only 700 transistors, with the bulk of the circuitry constructed from 23,000 germanium diodes and 3,000 specially designed magnetic amplifiers known as Ferractors.

Contrary to the name that it would eventually be given, the Solid-State was not an entirely solid-state design. It also contained 20 valves and the choice of a

8. Solid Progress – Transistors and Unified Architectures

magnetic drum rather than magnetic core storage for main memory, albeit an ultra high-speed one, further belied its pre-solid-state origins.

Unfortunately, the Solid-State was yet another victim of the internecine war that raged between the three computer divisions of Sperry Rand. The company had introduced its first medium-scale model, the St Paul division's UNIVAC File-Computer, in early 1956 but sales of this quirky and expensive machine were sluggish. In order to give the File-Computer more time to establish itself in the lucrative US market, a decision was taken by Sperry Rand senior management not to introduce the faster, more reliable Solid-State model in the US but to restrict sales to the European market where it would retain the prototype's UCT name. Costing around $450,000 for a typical system, the first production machine was delivered to the Dresdner Bank of Hamburg, West Germany, in October 1958.

Following pressure from potential customers, the UCT was eventually announced in the US as the UNIVAC Solid-State in December 1958, almost a year later than in Europe. Two versions were introduced, the Solid-State 80, with IBM-compatible 80 column punched cards, and the Solid-State 90 with UNIVAC-compatible 90 column cards. However, within less than a year IBM had introduced the IBM 1401, a transistorised replacement for the venerable IBM 650 which was considerably faster than the Solid-State due to its magnetic core main memory. Consequently, orders for the Solid-State soon began to dry up. Nevertheless, despite the damage done to sales through delaying introduction of the Solid-State in the largest market, over 500 examples of this important model were built.

The First Fully Solid-State Machines

With the availability of high frequency transistors and semiconductor diodes for logic circuits and magnetic core memory for storage, all the components necessary for the development of fully solid-state computers were in place by 1957. Given the lower cost of ownership that improved reliability brought, the first commercial machines to emerge were naturally targeted at the price sensitive business data processing market. However, with existing products based on valve technology, it took longer for the established computer manufacturers to make the move to transistors and this allowed newer entrants to take the lead. Those large US electronics firms that had flirted with computer technology at the beginning of the decade but had not yet committed were now perfectly placed to enter the market with transistorised offerings. The first to do so was the Philco Corporation.

The Transistor Automatic Computer

The US National Security Agency had been a keen supporter of the US computer industry from the outset. Though somewhat publicity-shy due to the nature of its business, the NSA had funded the development of the ERA Atlas II, the commercial version of which was the successful UNIVAC 1103, and would go on to commission the IBM design study that launched Project Stretch. Continually on the lookout for new components which could increase the performance or reliability of computers for use in cryptography, the NSA took an immediate interest in Philco's new surface barrier transistor. The agency was intrigued by the idea of a compact but powerful computer for individual use and high frequency transistors offered a means to build such a machine.

In June 1955, the NSA issued a contract to Philco to develop the proposed machine, which would be called SOLO to reflect the intention to create a computer for individual use. This was to be the Corporation's first computer development project and the lack of in-house experience made the adoption of an existing design a much more attractive option than having to develop a new machine from the ground up. Fortunately, the NSA had funded the development of the ERA Atlas II prototype and was able to provide the Philco engineers with access to its design details. Hence SOLO featured the same binary parallel architecture and 36-bit word length of the UNIVAC 1103. It also included a 4,096 word magnetic core memory, as featured in the recently introduced UNIVAC 1103A variant. Unlike the 1103A, however, the circuitry in the arithmetic unit of SOLO was fully transistorised and its calculation performance was more than twice that of the valve-based machine. It was also exceptionally compact for a medium-scale computer, occupying only 36 cubic feet of space, and consumed only 1,200 watts, a fraction of the 82,000 watts power consumption of the 1103. The commercial potential of this design seemed obvious.

The SOLO prototype was completed sometime during 1956. At around the same time Philco also began a $1.6M project to build a large-scale transistorised computer, codenamed CXPQ, for the US Navy's David Taylor Model Basin. The development of the CXPQ was directed by Morris Rubinoff, who was hired from the Moore School of Electrical Engineering to serve as Chief Engineer on the project. Rubinoff had already been providing advice to the Philco development team on a consultancy basis, during which time he had conceived a new type of logic circuit known as Direct-Coupled Transistor Logic (DCTL) that employed only transistors and resistors (and which Seymour Cray would later adopt for the high performance CDC 6600). However, the CXPQ project began to place greater demands on his time and in 1957 Rubinoff took a one-year leave of absence from the Moore School to work on the project full time. Earlier in his career Rubinoff had been involved in the development of the IAS Electronic Computer and the design of the CXPQ closely followed the binary parallel architecture and asynchronous operation of the influential Princeton machine but with a larger word length of 48 bits and magnetic core storage, rather than electrostatic storage tubes, for main memory. With the exception of a handful of valves in the magnetic tape and paper tape units, the CXPQ was fully transistorised, incorporating some 5,500 surface barrier transistors.

8. Solid Progress – Transistors and Unified Architectures

Towards the end of 1957, Philco executives took the decision to capitalise on the company's industry lead in transistor technology and enter the computer market with a commercial version of the CXPQ, which was announced as the Transistor Automatic Computer (TRANSAC) S-2000. To broaden its appeal in the marketplace, the specification was upgraded to include binary-coded decimal and alphanumeric numbering systems, optional floating-point hardware and an expandable main memory with up to 32,768 words of magnetic core storage. With a tried and tested medium-scale design already completed courtesy of the NSA, Philco was also able to simultaneously announce a medium-scale model, a commercial version of SOLO which would be known as the TRANSAC S-1000.

Rubinoff remained at Philco for a second year in order to complete the development of the S-2000 and the first machine, designated the TRANSAC S-2000 Model 210, was delivered to Philco's Western Development Laboratories in January 1960. The arithmetic unit incorporated more than 20,000 surface barrier transistors. However, in the 3 years that had elapsed since the project began, Philco had introduced the MADT transistor, a field-effect device with a higher and more consistent performance. This effectively rendered the company's surface barrier transistor obsolete and prompted a hasty redesign of the TRANSAC S-2000's circuitry to incorporate the MADT transistors and other improvements. Hence, only one Model 210 was built. The redesigned version was known as the Model 211. Able to perform a fixed-point addition in 2.1 microseconds, which was almost twice as fast as the Model 210, the Model 211 was proudly advertised by Philco as the "world's fastest all-transistor data processing system" and around 20 examples were built.

Introduced at around the same time as the UNIVAC Solid-State, it could be argued that the fully transistorised TRANSAC S-1000 was the first truly solid-state general-purpose computer to reach the market. However, only one example was built, destined for internal use, and none were ever sold commercially. Despite its technological lead, Philco was still seen as a manufacturer of consumer products and would struggle hard to establish a presence in the computer industry. Philco's early lead in transistor technology would also be short-lived, as two of the company's largest competitors in the US electronics industry, RCA and General Electric, were also hard at work developing transistorised computers of their own.

The RCA 501

Originally formed in 1919 as a joint venture between AT&T and General Electric, the Radio Corporation of America had grown to become the USA's third largest electronics company. Like its smaller rival Philco, RCA was best known for radio and television technology. However, unlike Philco, RCA had been active in computer-related research during World War II through the development of electronic fire control systems for the US military and had sustained this activity in the post war period through the development of the giant Typhoon electronic analogue computer for the US Navy, which was completed in 1950. RCA's corporate research facility in New Jersey, RCA Laboratories, had also provided technical assistance on the ENIAC project and was the location for the pioneering work of Jan Rajchman in the development of electrostatic and magnetic core storage technologies. This wealth of experience gave RCA a considerable advantage over Philco when the company came to develop its first general-purpose digital computer.

In 1952, RCA began the development of the Business Machine or BIZMAC, a large-scale data processing model developed at a cost of $4.5M for the US Army. BIZMAC featured an unusual architecture with a variable word length and both binary and decimal arithmetic. The system was designed to operate with a very large number of magnetic tape units connected simultaneously and included separate input/output processors to cope with the sorting and tape handling requirements. With logic circuitry constructed using a combination of diodes and miniature thermionic valves, the arithmetic unit was surprisingly compact for such a large machine, allowing it to be fitted within the desk-shaped console. RCA also took the opportunity to showcase the fruits of its pioneering research in electronic storage and BIZMAC was one of the first computers to feature a magnetic core main memory. However, as magnetic core storage was initially very expensive to manufacture, only 4,096 of the machine's 22,096 decimal digits of main memory were implemented using magnetic cores and the bulk of the storage was provided by a directly addressable magnetic drum.

The BIZMAC prototype was delivered to the US Army's Ordnance Tank and Automotive Command in Detroit in early 1956. RCA announced BIZMAC as a commercial product in November 1955, however, only 2 machines were sold, both of which were of basic specification, and the only full configuration example built was the US Army prototype. With RCA's colour television system having been adopted as the US standard in December 1953, Company President David Sarnoff anticipated a large increase in revenues from sales of colour television sets and planned to invest a substantial portion of this in commercial computer development. However, Sarnoff knew that if RCA was to succeed in the computer industry, it would have to do considerably better with its next offering. He hired a top-flight management consultant, John L Burns, to oversee the company's commercial computer activities and initiated the development of a new model, the RCA 501 Electronic Data Processing System.

As a leading manufacturer of thermionic valves, RCA had become very active in semiconductor research following the announcement of the transistor in 1948. This early flurry of activity resulted in a number of important transistor patents and the development of two transistors aimed at the computer industry, the 2N404 junction transistor and 2N247 drift transistor, both of which were introduced in 1957. Although the design of BIZMAC relied heavily on valves,

8. Solid Progress – Transistors and Unified Architectures

it did contain a small number of transistors within the peripheral units. When RCA decided to invest heavily in commercial computer development, the company was also able to capitalise on its transistor technology and the RCA 501 would be fully transistorised.

The RCA 501 was a binary serial machine with a variable word length and fixed-point arithmetic. Although it shared these features with BIZMAC, the 501 was an entirely new design with a strong emphasis on modularity and expandability. Aimed at the medium to large scale end of the business data processing sector, the RCA 501 featured a magnetic core memory of 16,384 characters which could be expanded up to an impressive maximum of 262,144 characters. This was supplemented by up to 12 magnetic drum units, each with a capacity of 1.5 million characters, and up to 63 magnetic tape units.

The RCA 501 was announced in May 1958 and the first production machine delivered to the US Air Reserve Records Center in Denver, Colorado, in October 1959. With a basic system costing around $580,000, the 501 was competitively priced and excellent expandability allowed the system to grow with the customer's needs. Following some judicious marketing, approximately 40 were sold. Encouraged by this success, RCA introduced two other models, the medium-scale RCA 301 and the large-scale RCA 601, both of which were announced in February 1960. That same year RCA also entered into an agreement with English Electric for the 501 to be manufactured under license and it was sold in the UK under the name KDP10.

The General Electric Model 210

The giant General Electric Company was among the first firms to license transistor technology from Bell Labs in September 1951 and, like its protégé RCA, contributed greatly to the development of semiconductor devices. GE had been similarly research-active in electronic computation at an early stage and had built several special-purpose machines for military customers. The company had also developed a general-purpose scientific model for the US Air Force, the Office of Air Research Automatic Computer (OARAC), which was delivered in 1953. However, despite the considerable resources available within GE for product development, OARAC was never put into production and remained a one-off prototype. During development, GE had made tentative overtures to insurance companies about a possible commercial version. When news of this reached IBM, the company threatened to stop using GE as a supplier. IBM was a major customer of General Electric and company President Ralph J Cordiner, averse to competing with IBM, strictly forbade entry into the civilian computer market. GE would eventually enter the market as a result of a need by the banking sector to automate cheque processing.

A dramatic increase in the use of personal cheques in post-war USA presented a growing problem for banks. By 1950 the increased workload in processing cheque transactions was causing many banks to close their doors early each day in order to cope with the extra work. For the Bank of America, which was at that time the largest bank in the world, the problem it faced was greater than most but it also had the means to do something about it. The California-based bank turned to a local contract research organisation, the Stanford Research Institute, to help solve this problem. In July 1950 it contracted SRI to undertake a feasibility study to investigate the use of electronic bookkeeping machinery as a means of automating cheque processing. The resulting report confirmed the feasibility of electronic equipment as a viable solution and this led to a second contract to SRI in November 1950 for further research to finalise the specification and execute the overall design of the proposed system.

As a contract research organisation, SRI was not well suited to building large, complex electronic systems and the expectation had always been that one of the established office equipment manufacturers would take on the responsibility for the detailed design and construction of the prototype machine. Following the successful completion of the second contract, the Bank of America approached several firms including NCR and Burroughs. Only Burroughs showed any interest but negotiations soon broke down following a disagreement with SRI over the likely cost of each system. Hence, SRI reluctantly accepted a third contract from the Bank of America in January 1952 to build and test a prototype, which would be called the Electronic Recording Machine or ERM.

The resulting device was far more than an electronic bookkeeping machine. Occupying 400 square feet of floor space and consuming 50 KW of power, it was a large-scale real-time electronic computer system with two magnetic drums and 12 magnetic tape units. However, the ERM was not a general-purpose machine nor was it a true stored-program computer, possessing a hard-wired program that required extensive rewiring if changes to the operation of the machine were necessary. At its heart lay a high-speed cheque reader and sorter unit that could automatically read the account number from cheques or deposit slips at a rate of up to 10 documents per second. The account numbers were used to retrieve the customer's account balance from magnetic drum storage and the amount, which was input manually by the operator via keyboard, then deducted or added, as appropriate, to create the new balance.

To reduce the construction load on SRI, many of the major components and subsystems were outsourced to other manufacturers. Circuit modules were supplied by Bendix and the magnetic drum units were supplied by the ElectroData Corporation. Peripheral equipment came from a number of manufacturers including Remington Rand and NCR. The increasing availability of point-contact transistors prompted the SRI engineers to consider using them for the logic circuitry of the ERM. However, following a thorough evaluation, transistors were rejected due to the poor quality of available samples. Instead, a total of 8,200 thermionic valves were used in combination with semiconductor diodes.

8. Solid Progress – Transistors and Unified Architectures

Outsourcing also freed up the SRI engineers to concentrate on the application of the technology to banking processes and this was where the Institute contributed most to the project. In order to process the cheques automatically, a system was devised in which account numbers were used instead of customer's names. A magnetic ink character recognition (MICR) encoding system was then developed by SRI engineer Kenneth R Eldredge which used special ink and characters to print the account number on each cheque so that it could be read automatically irrespective of its condition. Both systems were subsequently adopted as banking industry standards and are still in use today.

With a loosely defined specification typical of a research project, the construction phase became a process of continual improvement. In the spring of 1955, a decision was taken to freeze development and the 'completed' prototype was publicly demonstrated in a high-profile ceremony held at SRI in Menlo Park on 22 September 1955. To make the machine sound less intimidating, the Bank's public relations office renamed it ERMA by adding the word 'Accounting' onto the end of Electronic Recording Machine. Following the demonstration, the Bank was inundated with expressions of interest from companies keen on building the production machines and a total of 29 proposals were received. By late November 1955, this had been reduced to a shortlist of 4 companies; RCA, IBM, Texas Instruments and General Electric.

RCA proposed a system based on its soon to be announced BIZMAC. IBM took a different tack, choosing not to propose a system but offering instead to purchase the rights to the ERMA technology for $10M. Texas Instruments had no track record to speak of in computer technology but proposed to cooperate closely with SRI on the development of the production machine, a notion that appealed enormously to the SRI engineers and made TI the early favourite. The GE bid was led by Homer R Oldfield, manager of GE's Microwave Laboratory in Stanford, who saw the project as an opportunity to ease GE into the civilian computer business. Oldfield worked tirelessly to win the ERMA contract, submitting a bid which undercut the other bidders on price and included stiff penalties for failure to deliver. Because ERMA was a special-purpose computer, the bid was not considered to be a breach of Cordiner's policy forbidding entry into the civilian computer business but the computer was coyly referred to as a process control machine in company documents just to make sure.

Oldfield's efforts finally paid off in April 1956 when GE was awarded a $30M contract from the Bank of America to supply 32 ERMA machines. As a reward for his hard work, he was appointed as General Manager of a new GE department that would be established in Phoenix, Arizona to implement the ERMA development programme. However, his attempts to persuade key members of the SRI team to move to Phoenix to work for the new GE Computer Department failed so Oldfield was forced to look to other departments within the GE empire for suitable staff, which he was able to augment with an experienced computer designer from Burroughs and two senior engineers from the RCA BIZMAC project.

ERMA had been developed as a proof of concept demonstrator rather than a production prototype. This meant that the GE project team would have to put considerably more effort into the design phase than usual, so they planned to spend the first 6 months of the project working with SRI on finalising the design specification. It also provided a chance to make some changes, an opportunity that the Bank of America took full advantage of in specifying that the production version should have a stored-program architecture and solid-state circuitry. The GE team then proceeded to create a stored-program design which replaced the prototype's thermionic valves and magnetic drum storage with solid-state circuitry and a magnetic core memory.

The revised design was given the name GE 100 ERMA. It featured a binary-coded decimal, serial architecture with a 7-digit word length and 2,000 words of magnetic core storage. To cope with the additional sorting requirements of branches that dealt with large numbers of cheques from other banks, the design also included a separate input/output processor which received data from the cheque reader/sorter and stored it on magnetic tape in preparation for input to the main processor. However, this feature did not live up to expectations and was later dropped in favour of a doubling of magnetic core storage to 4,000 words.

GE's newly established Computer Department had limited in-house mechanical engineering capability so much of the input/output equipment development work was subcontracted to other manufacturers, including NCR, Pitney Bowes and Ampex. Remarkably, despite subcontractor delays and the setback with the input/output processor, the project was delivered on schedule and within budget, and the first of the 32 production machines was installed in the Bank of America's San José branch in December 1958.

Soon other banks began to show an interest in acquiring cheque processing systems. This was boosted further by a public unveiling of the first four ERMA installations in September 1959. The decision to move to a stored-program architecture and solid-state circuitry for the production version also provided GE with an opportunity to use the ERMA design as the basis for a general-purpose commercial model. The GE 100 ERMA was repackaged, given a wider range of input/output options and a larger memory, and announced as the GE 210 Data Processing System at a price of $225,000 for a basic specification machine. With Cordiner's policy now forgotten, the new model was marketed as suitable for a range of business data processing applications, its MICR technology promoted as an alternative to punched cards, but its special-purpose origins made the GE 210 really only suitable for banking applications. A total of 14 banks purchased GE 210 systems and around 60 machines were sold, making GE's unwitting entry into the civilian computer business a qualified success.

8. Solid Progress – Transistors and Unified Architectures

Shortly after delivery of the first production GE 100 ERMA machine, Oldfield left GE for a new job at Raytheon. Despite the success of the ERMA project, GE senior management remained uncomfortable with the company's entry into the civilian computer business. Nevertheless, they did sanction the development of a range of process control computers, the development of which took place in parallel with that of ERMA, and these became the basis of a series of business machines beginning with the medium-scale GE 225 in 1960. However, a continued lack of commitment to computer products stifled GE's potential in the marketplace, a situation exemplified by the decision of the Bank of America in 1966 to replace its ageing GE 100 ERMA installations with computer systems from none other than IBM.

IBM 7090 Data Processing System

Although IBM had not been directly involved in the Atlas missile programme, the company's first fully transistorised stored-program machine would come into existence as the result of a missile-related project. In the mid 1950s, advances in ballistic missile technology by the Soviet Union made the development of a Soviet ICBM a realistic prospect and opened up the threat of a Soviet missile attack against the US mainland. In early 1958, the US Air Force invited proposals for a missile detection radar system, codenamed the Ballistic Missile Early Warning System (BMEWS), that would provide long-range warning of a ballistic missile attack over the polar region of the northern hemisphere via a network of three radar stations located in Greenland, Alaska and the UK.

The contract to supply the real-time computer system for BMEWS was awarded to the Massachusetts-based electrical equipment manufacturer Sylvania Electric Products, which had recently begun developing real-time computers for the US Army. The requirement was for a pair of large-scale computers at each of the three sites running in parallel to ensure reliability. However, Sylvania decided against using its own relatively unproven computer hardware and chose instead to subcontract this element of the system.

IBM and Philco both submitted bids for the BMEWS computer hardware subcontract. The Philco bid was based on the TRANSAC S-2000, which had been announced a few months earlier but was still under development. IBM's bid was based on the IBM 709, the successor to the popular IBM 704 which had just gone into production. However, the Air Force had stipulated that the computer hardware for BMEWS should be fully transistorised and the 709 was a valve-based design. In order to secure the subcontract, IBM's Ralph Palmer proposed to supply Sylvania with the existing 709 model so that software development could begin immediately and to use the remainder of the project to develop a transistorised version, codenamed the 709TX, which would be software compatible. As a transistorised 709 had already been the subject of an in-house design study, Palmer was confident that his engineers could meet the tight development timescale required for delivery of the new machine.

Development of the 709TX began in June 1958 at IBM's Poughkeepsie Laboratory. The development team were able to draw upon the work to develop transistorised circuits as part of Project Stretch, which was under development at Poughkeepsie at the same time, and a large number of Stretch components and subsystems were adapted for use in the 709TX. They also took the opportunity to eliminate the antiquated magnetic drum auxiliary storage from the design and to include additional input/output channels. These changes aside, the new machine was architecturally and functionally identical to the 709, featuring the same binary parallel architecture, 36-bit word length and feature rich instruction set of the valve-based machine. However, it was now a fully transistorised design incorporating over 50,000 transistors. The combination of transistorised circuitry and faster magnetic core memory allowed the clock speed to be raised from 83 KHz to a respectable 459 KHz, giving a performance of more than 5 times that of the 709.

During development, a decision was taken to introduce the 709TX as a commercial product despite the inevitable damage this would cause to sales of the 709, which had only been on the market for a short time. The machine was re-designated the IBM 7090 Data Processing System and announced in December 1958 at a purchase price of $2,898,000 or a rental of $63,500 per month in typical configuration. The first two production machines were delivered to Sylvania on schedule in November 1959, however, the rush to complete them within the timescale resulted in reliability problems and considerable on-site work was required before this was remedied. With the reliability problems solved, the now obsolete IBM 709 was quietly withdrawn in April 1960.

The calculation speed of the 7090 was 4.36 microseconds for a fixed-point addition, which made the IBM model appreciably slower than Philco's 48-bit TRANSAC S-2000 despite having a shorter word length. However, software compatibility with the 709, which was itself backward compatible with the 704, made the 7090 very appealing to existing IBM customers and it became the industry's most successful large-scale computer of the period. The 7090 was also chosen as the central computer for SABRE (Semi-Automated Business Research Environment), an electronic airline reservation system developed by IBM for American Airlines, which was deployed in 1962. A total of over 300 of this model and the 7094 derivative were built before both models were withdrawn in July 1969. Despite being beaten to the transistorised computer market by several of its competitors, IBM had triumphed once again.

Unified Architectures

Having solved the reliability problem through the introduction of transistorised

8. Solid Progress – Transistors and Unified Architectures

models, the computer industry was now ready to turn its attention to another factor governing the successful adoption of computer systems. The leading manufacturers had product lines that offered a range of machines designed to cope with the majority of customer requirements. However, customers whose requirements encompassed both scientific computing and business data processing were forced to purchase two machines, neither of which was compatible with the other, or to compromise by choosing a model that suited only one type of application. Those customers looking for a truly general-purpose machine that could perform both types of calculation were poorly served. There was also growing criticism over the lack of compatibility between different models from the same manufacturer. Users whose needs had outgrown a particular class of machine often encountered problems with software compatibility when migrating to a larger class. As customers began to invest significant resources in developing software in-house, the thought of having to rewrite large numbers of programs so that they would run on a new platform was daunting. Therefore, the next challenge for the industry was to create a range of computers which unified the architectural features of both scientific and business machines and also offered software compatibility throughout the range.

Early Efforts

One manufacturer had already anticipated the need for both business data processing and scientific computing capabilities. ElectroData's Datatron, which was introduced in 1954, had shown that it was possible to incorporate features of both scientific and business machines within a single design. The Datatron was a typical business-oriented machine with a binary-coded decimal architecture but it also included a four-digit index register and a floating-point hardware option, two features that are necessary for scientific computing applications. When Burroughs acquired the ElectroData Corporation in 1956, this philosophy carried through to subsequent designs and Burroughs first transistorised large-scale business model, the B5000, not only featured floating-point hardware as standard but also incorporated both binary and decimal arithmetic.

The idea of combining the architectural features of scientific and business machines had also crossed the fertile mind of Presper Eckert. During planning for Sperry Rand's first transistorised large-scale business machine, the UNIVAC III, which was announced in 1960, Eckert suggested to senior management that the company should develop a single new model with a unified architecture, rather than one model for business, another for scientific computing and a third for real-time control applications. Perhaps unsurprisingly, given the constant feuding between the computer divisions of Sperry Rand, the St Paul division vehemently opposed the idea and Eckert's suggestion was dismissed. Instead, three new Sperry Rand models were developed in parallel; the UNIVAC III, the

The Story of the Computer

UNIVAC 1107 Thin-Film Memory Computer and the UNIVAC 490 Real-Time System.

Prior to the industry's standardisation on binary architectures for scientific computers, some scientific models were designed with decimal architectures and were thus naturally suited to handling character-based data of the kind required for business applications. The best example of this was the IBM 650, which was developed by IBM's Endicott Laboratories for scientific and engineering applications but became more popular for business data processing due to its biquinary-coded decimal architecture and excellent punched card handling capabilities.

The computer industry had been aware for some time of the need to provide new models with software compatibility with the machine that they replaced, especially for large-scale systems. In order to help users migrate to a new model, most firms provided extensive customer support but this was a costly activity which ate into company profits. Sperry Rand was one of the first to realise that customer support requirements would be reduced considerably if new machines could be made compatible with previous models and so the UNIVAC II was designed to be backward compatible with the UNIVAC I. IBM had a different reason for making the 7090 compatible with the 709 but had witnessed the favourable customer reaction when it did so. However, when the company attempted to reverse this trend in 1960 by replacing the valve-based IBM 705 with a transistorised model, the 7070, which was incompatible due to a move from variable to fixed word length, pressure from customers forced it to swiftly introduce another model, the 7080, which could run 705 programs.

Another area where Burroughs led the industry was in showing that it was possible to simulate an older model in software. The Burroughs 220, which was introduced in 1958, featured an early form of software simulator which automatically translated instructions written for the older Datatron 205 into code that could be run on the B220. Other firms soon seized upon this approach to provide their customers with the ability to run programs developed for rival machines and in June 1959, Sperry Rand announced the availability of a program for the UNIVAC Solid-State that could simulate an IBM 650. However, the performance of software simulators left much to be desired and software compatibility between different classes of machine had yet to be addressed.

The Compatibles/400 Family

The first serious attempt to introduce a line of compatible computers came not from one of the industry leaders but from the reluctant computer vendor General Electric. In a surprising turnaround, GE senior management took the decision in 1960 to approve an ambitious proposal to develop a family of compatible 24-bit general-purpose computers codenamed Mosaic. The intention was to

8. Solid Progress – Transistors and Unified Architectures

introduce a line of four models that shared a common architecture and would be aimed at both business and scientific applications. The line would span the range from medium-scale to large-scale, with each successive model having a performance 80% faster than its smaller sibling.

The GE Computer Department in Phoenix, now under the direction of Homer Oldfield's successor Clair Lasher, developed a sophisticated binary-coded decimal architecture for the Mosaic line which featured a 4 decimal digit word length and included additional machine instructions to support binary arithmetic plus 6 index registers and a variable-length accumulator. The new range was announced as the General Electric Compatibles/400 family and would comprise four models, the GE-425, GE-435, GE-455 and GE-465. Each would be fully transistorised, with an expandable magnetic core memory of up to 32,768 words and optional floating-point hardware.

The first two Compatibles/400 models, the medium-scale GE-425 and GE-435, were introduced simultaneously in 1963. Both were similarly specified, the only major difference being a clock speed of 196 KHz for the GE-425 and 370 KHz for the GE-435. A comprehensive set of input/output devices was also developed which included magnetic tape, punched card and punched tape units, a printer and a magnetic disk storage unit with an impressive capacity of 23.5 million 6-bit characters.

In contrast to the successful introduction of the two medium-scale models, development of the two large-scale Compatibles/400 models was progressing slowly and doubts began to creep in over whether their performance would be good enough to compete with the latest large-scale machines from rival firms. This led to a decision in February 1963 to halt development. Fortunately, GE's Heavy Military Electronic Equipment Department in Syracuse, New York had been working on a family of 36-bit binary computers derived from a model originally developed for missile guidance applications, so the company was able to quickly fill the gap left by the cancellation of the GE-455 and GE-465 machines with this new range, which would be called the GE-600 series. Both product lines would share the same input/output devices but they would not be software compatible.

Despite this setback, the two medium-scale models began to sell in reasonable numbers, particularly in Europe, and a small-scale model, the GE-415, was introduced the following year. Marketed as a suitable replacement for the popular IBM 1401 Data Processing System, which had become the market leader following its introduction in October 1959, the GE-415 was available with an optional IBM 1401 simulator and could also be used as a front-end I/O processor for larger GE models.

GE's decision to cancel the two large-scale Compatibles/400 models in favour of machines which were architecturally unrelated but had a longer word length may indeed have been the right one, as the GE-600 series models were reasonably successful in the face of stiff competition from IBM and Sperry Rand. However, the opportunity to become the first manufacturer to provide compatibility between different classes of machine had been thrown away, only to be picked up by GE's largest and most feared competitor in the computer business, IBM.

System/360

The immediate success of IBM's first transistorised stored-program model, the large-scale IBM 7090, was soon repeated at the lower end of the IBM product range following the introduction of a compatible replacement for the 1401, the IBM 1410, in September 1960. The importance of software compatibility was now manifestly clear to both the Poughkeepsie and Endicott divisions of IBM and would dominate the company's efforts to introduce its next series of computers.

Architectural incompatibility between the Poughkeepsie-developed high-end models and the smaller Endicott-developed machines meant that there was no smooth upgrade path for IBM customers. The separate responsibilities of IBM's two main engineering divisions had also become blurred through Endicott's development of the large-scale 7070 and Poughkeepsie's involvement in the development of the 1620, a small-scale scientific model with a binary-coded decimal architecture which had been conceived at Endicott but realised at Poughkeepsie. Consequently, rivalry between the two divisions began to intensify.

In January 1961, the Poughkeepsie Laboratory submitted a proposal to IBM's Corporate Management Committee for the development of a new family of five computers to replace the 7000 series, which had been a successful fixture in the large-scale sector since the introduction of the IBM 7090 in December 1958. Designated the 8000 series, the new family would span the range from small-scale business model all the way through to Stretch-class scientific machine. However, the proposed new line lacked a common architecture and so there would be no software compatibility between models. Endicott's future plans centred on the newly announced 1410 and the development of a revised version of the IBM 1620, a small-scale scientific model that had been well received on its introduction alongside the 1401 in October 1959. Both machines would overlap with the lower end of the 8000 series, thereby placing Endicott in direct competition with Poughkeepsie.

In May 1961 a decision was made to terminate work on the 8000 series, which was to be based on the SMS modular circuit technology developed under Project Stretch, in favour of new products that would utilise more advanced circuit technologies. However, this only served to exacerbate an already difficult situation. In an effort to broker a solution, IBM Vice President T Vincent Learson established a task group to examine the options and formulate

8. Solid Progress – Transistors and Unified Architectures

a corporate plan for future data processor products which would rationalise the company's increasingly disparate product range and perhaps even realise the dream of software compatibility between different classes of machine. The SPREAD (Systems, Programming, Review, Engineering and Development) Task Group comprised 13 senior IBM managers drawn from throughout the IBM empire and was co-chaired by John W Haanstra, Director of Development at Endicott, and Bob O Evans, who had recently been transferred from Endicott to head up planning and development at Poughkeepsie.

The final report of the SPREAD Task Group was issued in December 1961. It recommended the development of a cohesive product range comprising five compatible processors, each with an increase in performance of 3 to 5 times over its smaller sibling. Designated New Product Line, each processor would be designed for both business and scientific use and the range would also have the ability to combine processors to create larger systems. Full supporting software would also be developed, including a common operating system and a single high-level programming language that could serve both scientific and business needs.

On its presentation to IBM senior executives in January 1962, the SPREAD report received a mixed reaction. With development costs estimated at around $5 billion and no government agency to help pick up the tab on this occasion, the project would require an unprecedented financial investment. The new line would replace IBM's entire data processor product range and its development would necessitate a halt to all existing product development projects, including mid-life upgrades to existing models, thereby placing the company in a highly vulnerable position until the new line came on stream. One IBM executive summed up the project perfectly by telling Fortune magazine, "We call this project 'you bet your company'". However, Learson concluded that it offered the best way forward and firmly endorsed its implementation. The formal approval of the Corporate Management Committee duly followed in May 1962.

The scale of development of the New Product Line necessitated a combined effort by both Endicott and Poughkeepsie, with additional input from other IBM facilities including the IBM Hursley Laboratory in the UK. The principal architect was the gifted designer Gene M Amdahl, who had returned to IBM in September 1960 after several years working in the US defence industry. The project manager was another veteran of Project Stretch, Frederick P Brooks. Brooks was also responsible for the development of the operating system for the new line and his experiences would provide much of the material for his influential book on software project management, 'The Mythical Man-Month: Essays in Software Engineering', which was first published in 1975. Both men had learned valuable lessons from their involvement in Project Stretch and the new architecture would be much cleaner in design with a strong emphasis on scalability.

The new line was based on a 32-bit binary architecture, with 16 general-purpose registers, 4 optional floating-point registers and provision for both binary and decimal arithmetic. The design team chose to standardise on an 8-bit character size, rather than the 6-bit character that had been used previously, and this allowed a larger set of alphanumerical characters to be represented. The term 'byte', which had been coined by Werner Buchholz during Project Stretch, was adopted to describe the 8-bit unit and eventually became the standard term used throughout the industry. A system of base-register offset addressing was developed to allow each processor to address up to 16 million bytes, which would normally require 24 bits, using only 12-bit addresses. This was achieved by means of a two-part address format comprising a base address, or origin, and a displacement.

Full instruction set compatibility over such a wide range of computers was achieved through a technique known as microprogramming. Conceived by Maurice Wilkes at Cambridge University in 1951, microprogramming breaks down machine instructions into sequences of microinstructions which are held in a high speed read-only memory known as the control store. Microprogramming offers a more flexible approach than implementing an instruction set entirely in hardware, allowing a simpler logical design to be used and also permitting changes to be made much later in the design process.

Another important element of the New Product Line was the use of a new microminiaturised circuit packaging technology called Solid Logic Technology (SLT). The idea of microminiaturised circuits was first proposed by Geoffrey W A Dummer of the UK's Royal Radar Establishment in 1952. Dummer envisaged an integrated device in which a number of miniaturised electronic components were combined onto a single wafer of semiconductor material and connected to form a complete circuit. However, it took until 1958 before Dummer's monolithic integrated circuit concept was finally realised by engineers working independently at two different US electronics firms, Texas Instruments and Fairchild Semiconductor. Although integrated circuits became available commercially in 1961, the IBM engineers chose not adopt what was still felt to be an immature technology. Instead, they developed a hybrid technology which employed a combination of glass-encapsulated transistors and diodes with screen-printed resistors and connecting tracks on a ceramic wafer. Up to 36 of these half-inch square ceramic modules were then mounted onto a multi-layer printed circuit board to create an SLT logic card. The move to SLT resulted in a further increase in reliability of almost 100 times over the SMS technology that it replaced.

IBM's original intention had been to stagger the introduction of the New Product Line by introducing only two models initially, with the other models being introduced at subsequent intervals. However, this plan was hastily revised following Honeywell's announcement of its H-200 model in December 1963. The H-200 was a small-scale business model which was not only similar

8. Solid Progress – Transistors and Unified Architectures

in specification to the IBM 1401 but also featured the ability to run programs written for the IBM machine. Compatibility was achieved by means of a software package called Liberator that translated 1401 instructions into their H-200 equivalents. Soon IBM sales representatives were reporting losses of orders to Honeywell and IBM was forced to bring forward the introduction of the smallest model in the new line by a year to 1964.

The New Product Line was formally announced as the IBM System/360 series at a press conference held on 7 April 1964. An audience of 200 journalists gathered at Poughkeepsie to hear IBM Chairman Thomas J Watson Jr describe the occasion as the most important product announcement in the company's history. To emphasise its importance, simultaneous presentations were also given at 165 IBM sales offices throughout the USA. The name System/360 was intended to reflect the ability of the new line to cope with the full circle of computing applications. The entire line-up of six processors (Models 30, 40, 50, 60, 62 and 70) was announced, along with 44 peripherals, at prices ranging from \$133,000 for a basic configuration up to \$5.5 million for a large multiprocessor system. Each model differed not only in clock speed but also in memory capacity, which ranged from 8,192 to 524,288 bytes, and in the width of the input/output data bus, which varied from 8 to 64 bits. Floating-point hardware and decimal arithmetic were optional extras on the two smaller models but standard features on the four larger machines.

To help customers migrate to the new range, a 1401 'Compatibility Feature' was provided on the System/360 Model 30. Developed at Endicott, this mimicked the entire instruction set of the IBM 1401 and was implemented via microprogramming, which gave significantly better performance than software-based translation or simulation. Using this technique, which the IBM engineers called 'emulation', the Model 30 was able to execute 1401 instructions several times faster than the original machine. It was soon followed by the release of three additional emulator packages for high-end System/360 models that mimicked the IBM 7070, 7080 and 7090.

By the time deliveries of the first production System/360 machines began in April 1965, it had been decided to replace the Models 60 and 62 with a single higher specification machine designated the Model 65. Similarly, the Model 70 was also not shipped but replaced by the Model 75. Hardware delivery schedules were generally met but this was not the case with the supporting software. Development delays with the operating system, OS/360, and the high-level programming language, PL/1, resulted in both being delivered almost a year later than planned. Furthermore, problems with the size of OS/360 meant that it could not be used on machines with less than 32,768 bytes of main memory and IBM was forced to develop no less than three additional operating systems for low-end System/360 models.

IBM System/360 Model 65 (Photo © IBM Corporation)

Fortunately for IBM, the new line lived up to expectations in terms of sales. Within a month of the April 1964 product announcement, more than 1,000 orders had been received and, despite the problems with supporting software, over 14,000 System/360 models were eventually built. IBM's audacious gamble had paid off handsomely and the System/360 architecture would form the basis of the company's data processor products for the next two decades.

The System/360 Compatibles

System/360 was also a powerful catalyst for change in the computer industry. As sales increased, some firms began to consider an alternative strategy where, rather than attempt to compete head-on with IBM by introducing rival models with proprietary architectures, it might be wiser to produce machines that were compatible with System/360, but offered some improvement. The first company to take this approach was RCA.

Spectra 70

In December 1964, RCA announced Spectra 70, a new family of four compatible models. The four models, the 70/15, 70/25, 70/45 and 70/55, spanned a similar performance range to IBM's System/360 but offered a 15-20% better price/performance ratio and also featured the ability to run System/360 programs. To keep existing customers happy, the Spectra 70 range also included emulators for the RCA 301 and 501 models.

Using the company's expertise in semiconductor technology, the RCA engineers

8. Solid Progress – Transistors and Unified Architectures

had been able to keep manufacturing costs down through the extensive use of monolithic integrated circuits in the Spectra 70 range. Also, because they took the System/360 specification as their starting point, development costs were also relatively low, as much of the design work had already been done for them by IBM.

RCA had been working on the development of a new family of compatible machines for around a year when news broke of the System/360 announcement. Recognising the significance of IBM's new range, the company was able to quickly modify the design specification of its own range so that it used 8-bit characters and would therefore be compatible with the new IBM models. However, architectural differences between the two families meant that the ability to run programs written for System/360 did not extend to full instruction set compatibility and programs had to be recompiled on the RCA machines before they would run. Nevertheless, Spectra 70 offered a viable alternative to IBM customers looking for higher performance for their money and the range was moderately successful.

The UNIVAC 9000 Series

Sperry Rand followed suit in the spring of 1966 by introducing its own range of machines which were compatible with IBM's System/360 line, the UNIVAC 9000 Series. Unlike RCA, Sperry Rand did not attempt to match the full System/360 range. The company's large-scale UNIVAC 1100 Series of 36-bit machines were well established in the marketplace and there would have been little point in replacing them with incompatible models. Instead, it set its sights squarely on the lower end of the System/360 performance spectrum.

The UNIVAC 9000 Series initially comprised two models, the 9200 and 9300. Both were small-scale machines that featured integrated punched card handling equipment and were designed to replace the company's popular UNIVAC 1004 and 1005 punched card calculators. Both models implemented a subset of the System/360 instruction set, in line with the recently introduced System/360 Model 20, a small-scale punched card processor which was based on a cut-down version of the System/360 architecture. The 9200 and 9300 were followed by the introduction in January 1968 of a third model in the series, the UNIVAC 9400. This was a medium-scale machine which implemented the complete System/360 instruction set and also included support for multiprogramming and real-time data communications.

Once again, the strategy was to undercut IBM by reducing manufacturing costs. Like Spectra 70, the UNIVAC 9000 Series used monolithic integrated circuits but the Sperry Rand engineers went further than RCA by also employing plated-wire memory. Developed in 1957 at Bell Labs, plated-wire memory operated in a similar way to magnetic core memory but could be fabricated by machine, rather than hand assembled, which reduced the cost of manufacture considerably. As main memory was the most expensive component in computers of this period, the cost saving was dramatic and brought the purchase price of a basic specification UNIVAC 9200 down to only $39,000, which was a fraction of the price of the smallest System/360 model. This attractive combination of low price and System/360 instruction set compatibility produced strong sales and more than 3,000 UNIVAC 9000 Series machines were built.

Consolidation

The introduction of System/360 served to reinforce IBM's market dominance and by January 1965 the US computer industry was being referred to in the press as IBM and the Seven Dwarves, the dwarves comprising Burroughs, Control Data, General Electric, Honeywell, NCR, RCA, and Sperry Rand. Despite the increased competition from big new players such as GE and NCR, IBM was able to maintain its market share but the overall increase in market size had seen the company's revenues grow fivefold, from $563 million in 1956 to over $3.2 billion in 1965.

The transistor era was also a period of considerable change for the computer industry itself. Following the disappearance of many of the computer start-up and spin-out firms in the 1950s, the 1960s was characterised by numerous mergers and acquisitions. Ironically, the company which had led the commercialisation of transistorised computers was the first to succumb.

Farewell to a Pioneer

Following the introduction of the TRANSAC S-2000 in 1958, the Philco Corporation began to build on its technological lead in transistorised computers through the development of a range of industrial process control computers and mobile computer systems for the military market. In February 1960, the Corporation opened a new facility in Willow Grove, Pennsylvania, to house its growing computer design, manufacturing and sales operations. However, a fall in demand for its consumer products led to an alarming drop in Philco's profits and the company reported losses of $4.4 million in the first half of 1961. Clearly, Philco was no longer able to sustain the high cost of developing both computers and other new products.

Fortunately, Philco's plight had caught the attention of the Ford Motor Company. The cash rich giant of the US automotive industry was looking to diversify into new markets and saw Philco, with its strong technology development capability but lack of capital for investment, as an ideal acquisition opportunity. Ford moved rapidly to acquire the ailing firm and in December 1961 Philco became a wholly owned subsidiary of Ford known as the Philco-Ford Corporation.

The modest success of the TRANSAC S-2000 Model 211 had prompted the

8. Solid Progress – Transistors and Unified Architectures

development of an improved version. Announced in 1961 as the Philco 2000 Model 212, the new model was backward compatible with the Model 211 but the addition of instruction pipelining and a new 2-microsecond memory manufactured by Ampex increased performance by a factor of 4 over the older model. The new model easily outperformed IBM's latest large-scale offering, the IBM 7094. First deliveries of the Philco Model 212 were in February 1963. To complete the product range, a small-scale decimal machine, the confusingly named Philco 1000, was introduced which could also be used as an input/output processor for Philco's large-scale models and plans for a new medium-scale machine, the Philco 4000, were also drawn up.

Following the takeover, optimism within Philco's Computer Division was high. Ford installed a new team of senior executives but otherwise permitted the company to continue to operate autonomously. Philco had demonstrated that its computer products could be competitive but Ford, wary of competing with IBM, provided little investment, choosing instead to concentrate on growing its new subsidiary's promising aerospace business. A design for a multiprocessor version of the Model 212, designated the Model 213, was produced towards the end of 1964 but it never reached the market. Despite having led the industry in the introduction of large-scale transistorised computers, the Computer Division was closed down and Philco withdrew from the commercial computer business after only 8 years.

A British Contender

The increasing dominance of international markets by the US computer companies, led by IBM, was the trigger for an era of consolidation for UK computer firms. This would ultimately lead to the creation of a single national computer company and the largest non-US manufacturer of computers.

Two Rivals Unite

Anxious to stave off increasing competition from IBM in their home and overseas markets, the British Tabulating Machine Company and another long-established UK office equipment manufacturer, Powers-Samas Accounting Machines, took the decision to combine forces and announced in June 1958 that they would merge. The core business of both companies continued to be punched card equipment but BTM had been selling a line of small-scale magnetic drum computers in reasonable numbers since the introduction of the BTM 1200 in 1954 and Powers-Samas had recently introduced a sequence-controlled electronic calculator, the Programme Controlled Computer (PCC). Both companies recognised that computers were the future of office equipment, hence the new company was formally announced in January 1959 as International Computers & Tabulators Limited (ICT).

A few years before the merger, BTM had entered into an agreement with the General Electric Company (GEC) to jointly develop a successor to the BTM

1200 series, which GEC would manufacture and BTM would sell. Although this machine was still under development, pressure began to increase for the new company to introduce its first computer product to the market following IBM's announcement of the IBM 1401 in October 1959 and so the new machine was announced in May 1960 as the ICT 1301. Fully transistorised, with a binary-coded decimal architecture and a 12 decimal digit word length, the ICT 1301 was priced at around £100,000 in typical configuration. However, the premature announcement meant that it would not be ready for delivery for another 2 years and the company desperately needed a stop-gap model that it could begin selling immediately.

From the outset, ICT's Managing Director, Cecil Mead, had the strategic vision of consolidating all British computer interests together under one roof. To this end, a joint working party had been established with Ferranti Limited in the autumn of 1959. A business relationship with Ferranti would have provided ICT with an opportunity to sell Ferranti models under the ICT name but the discussions came to an end when Ferranti refused to accept the recommendations of the joint working party. ICT then entered into talks with the English Electric Company but these soon became complicated by English Electric's existing relationship with RCA. English Electric's subsequent rejection of a proposal by RCA for a three-way collaboration left the way clear for ICT and RCA to collaborate and in 1961, ICT entered into a technical agreement with RCA which provided scope for collaboration in future product development. As part of this agreement, ICT was also permitted to market RCA's medium-scale 301 model under the ICT name and it was sold as the ICT 1501. This would provide the necessary stop-gap model until the 1301 was ready.

Also in 1961, GEC decided to sell its computer operation to ICT, which brought the development of the ICT 1301 fully under ICT's control. In early 1962, ICT also acquired the computer interests of Electric and Musical Industries Ltd (EMI). EMI was primarily a recorded music business but its Research Laboratories had been involved in the development of radar and guided weapons technology during World War II and the company was also a pioneer in the field of television broadcasting. In 1959, it had introduced one of the first transistorised computers in the UK, the EMI EMIDEC 1100, a small-scale binary machine with a 36-bit word length which sold for £125,000 in typical configuration. This was followed by the introduction of a larger model, the EMIDEC 2400, the development of which had been supported financially by the National Research Development Corporation. Following the acquisition of EMI's Electronic Digital Computers Division, these models were renamed the ICT 1101 and ICT 2400.

Merger with Ferranti

In late 1962, Cecil Mead's vision moved another step closer to reality when

8. Solid Progress – Transistors and Unified Architectures

Ferranti finally decided that the time was right for a merger. Early success in the UK and overseas markets had given way to several years of losses for Ferranti's Computer Department, culminating in the costly commercial failure of the Atlas project. The company clearly needed a more effective sales and marketing organisation if it was to increase sales and this was something which ICT, with its extensive network of UK and overseas sales offices, could provide. In July 1963, it was announced that the Ferranti Computer Department, which had moved from its Moston factory to West Gorton in 1956, would merge with ICT to form Ferranti ICT. The merger was completed by December 1963. ICT paid £1.5 million in cash plus 1.9 million in shares of ICT stock, giving Ferranti a 10.6% stake in ICT. However, Ferranti's military computer development activity, which was based at a separate facility in Bracknell, was not part of the deal and was retained by the company.

The quickfire series of acquisitions and mergers had left the newly enlarged company with a motley collection of incompatible models, a situation made all the more obvious by the IBM System/360 announcement in April 1964. Ferranti ICT now needed to offer a range of compatible machines in order to remain competitive. The existing technical agreement with RCA allowed for the manufacture of RCA models under license and this prompted discussions with the US company on the possibility of licensing what would become the Spectra 70 range. The alternative was for Ferranti ICT to develop its own range of compatible machines based on an existing design but none of the in-house models had sufficient potential.

Looking further afield, one model that did look promising was the FP-6000, a 24-bit machine with strong multiprocessing capability which had been developed by Ferranti-Packard, the Canadian subsidiary of Ferranti. Fortunately, ICT executives had recognised the potential of this model during merger negotiations and had insisted that the design rights be included in the deal. The Ferranti ICT product planners now had a straight choice of licensing Spectra 70 from RCA or developing a new range based on the FP-6000. With doubts growing over the effectiveness of a strategy of IBM-compatibility in a European market that was not yet dominated by IBM and concerns over the lengthy time to market for RCA's new range, a decision was made to go with the FP-6000.

The Ferranti ICT engineers had to make a number of enhancements to the architecture of the medium-scale FP-6000 in order to turn it into a suitable platform for a full range of compatible models. Base-register offset addressing was added to increase the memory capacity from 32,768 24-bit words up to a maximum of 524,288 words. The input/output hardware was standardised to accept a larger selection of peripheral equipment and the operating system was extended to include additional functionality.

The new model range was announced as the Ferranti ICT 1900 Series at a

Business Efficiency Exhibition in London on 28 September 1964. It initially comprised 3 processors, each of which was available in either fixed or floating-point versions; the small-scale 1902 and 1903 models, medium-scale 1904 and 1905 models, and large-scale 1906 and 1907 models. The range also included an additional model, the 1909, which was a derivative of the 1905. Prices ranged from £105,000 to £700,000. The impressive progress made by the engineers in the development of the new range was demonstrated by the presence at the exhibition of working prototypes of the 1903 and 1904 models. Deliveries began in May 1965, several months ahead of the first IBM System/360 machines to reach the UK. The faster delivery and competitive pricing made up for the slightly poorer performance of Ferranti ICT 1900 models in comparison with their IBM equivalents and orders soon came flocking in. By 1968, over 1,000 Ferranti ICT 1900 Series machines had been installed, including substantial sales in overseas territories.

English Electric Computers

The merger of ICT with the Ferranti Computer Department in the summer of 1963 was preceded earlier that same year by the merger of another British computer company, Leo Computers Limited, with the computer division of the English Electric Company. In what was becoming a familiar story, both companies were experiencing difficulties in growing their computer businesses. English Electric had remained open to the prospect of partnering with another computer manufacturer since the breakdown of talks with ICT in 1961. Leo Computers, the subsidiary formed by J Lyons & Company in 1954 to commercialise the LEO II, had built up a loyal customer base of around 50 customers but was struggling to keep pace with the competition in a rapidly expanding market.

Both companies initially attempted to create a much larger consortium by also including a number of European computer manufacturers in their discussions but this proved too difficult. The deal was concluded by February 1963. Leo Computers was merged with the Data Processing and Control Systems Division of English Electric to form English Electric Leo Computers Limited and plans were formulated to rationalise the two separate product lines through the introduction of a new range of machines codenamed KLX.

In October 1964, with its core catering business now struggling, Lyons agreed to sell its 50% shareholding in the merged company to English Electric for £1.9 million. The company that had pioneered the use of computers for business data processing was finally bowing out of the computer industry after more than 15 years. Having assumed full control, English Electric was now free to absorb the military computer development capability of another of its subsidiaries, the Marconi Company, into the company and English Electric Leo Computers became English Electric Leo Marconi Computers (EELM). However, the recent introduction of the IBM System/360 and Ferranti ICT 1900 had forced

8. Solid Progress – Transistors and Unified Architectures

a rethink on the KLX project. The plan to introduce yet another new range to the market now seemed too risky. Instead, a decision was made to take advantage of a longstanding information sharing agreement with RCA, which English Electric had inherited when it acquired the Marconi Company in 1946, and manufacture the Spectra 70 under license.

The Spectra 70 series was renamed System 4, however, only the midrange Spectra 70/45 was adopted in its entirety. The EELM engineers were able to call upon the expertise in integrated circuit technology of recent acquisition Marconi to create two smaller models, the 4-10 and 4-30, and a powerful new model, the 4-70, was introduced at the top of the range. The operating system was also redesigned to exploit the advances in multiprogramming technology made by the Leo Computers team. The System 4 was announced in September 1965 at prices ranging from £100,000 to £600,000.

A few months prior to the System 4 announcement, another pioneering UK computer firm, Elliott Brothers, had suffered a serious financial blow when the British Aircraft Corporation's TSR-2 project was cancelled. TSR-2 was an ambitious UK government funded project to develop a highly advanced tactical strike and reconnaissance aircraft for the Royal Air Force. Elliott Brothers, which had changed its name to Elliott-Automation in 1957 to reflect a change in focus towards industrial process control machines and mobile computer systems for the military market, had secured the contract to supply onboard computer systems for the TSR-2, for use in the aircraft's automatic flight control and weapon aiming systems. The termination of this major contract in April 1965 led to financial difficulties for Elliott-Automation, forcing the company to explore the possibility of a merger, and discussions with English Electric Leo Marconi soon followed. These were strongly encouraged by the UK Ministry of Technology, which had made the continued survival of the British computer industry a top priority upon its formation in 1964. The merger of Elliott-Automation and English Electric Leo Marconi duly took place in June 1967 and the company's name was shortened to English Electric Computers Limited.

A National Computer Company

With the absorption of Elliott-Automation, only two British computer firms now remained, English Electric Computers and Ferranti ICT. The Ministry of Technology had first attempted to persuade English Electric and Ferranti ICT to merge their computer interests in November 1964 but both companies were working hard on delivering new product lines and neither felt the need for a merger. In September 1966 an enquiry commissioned by the Ministry had recommended against merging the two companies due to the incompatibility between the Ferranti ICT 1900 and EELM System 4 product ranges. Nevertheless the British government remained committed to establishing "a single national computer company able to maintain a position in the international computer industry" and in April 1967 it again attempted to persuade the two companies to merge. However, this time it came equipped with an offer of a grant of £25 million to help pay for the development of a new series of computers which would be compatible with both existing product ranges.

Ferranti ICT had been enjoying a relatively stable period of success since the introduction of the 1900 Series in September 1964. However, English Electric had fared less well with its System 4 range. Development costs had been higher than anticipated and delivery dates began to slip. English Electric now began to view a merger as an attractive option. Fortunately, Ferranti ICT's new Managing Director, Basil de Ferranti, could also see the benefits of an enlarged company and the merger finally went ahead. In July 1968, Ferranti ICT merged with English Electric Computers to form International Computers Limited (ICL). However, the public expenditure cuts that followed a devaluation of the pound in November 1967 now made the offer of a £25 million government grant politically unacceptable and only £13.5 million of the promised cash was provided.

It had taken 10 years but Cecil Mead's vision had finally been realised. Virtually the entire British computer industry had been combined into a single company large enough to compete with the industry giants. On its formation in 1968, ICL was the largest non-US manufacturer of computers with a workforce of over 34,000 employees.

The Dwarves Become a BUNCH

General Electric Sells Out

Having finally made up its mind that it should be in the computer business, General Electric went on a European buying spree. With limited success in its home market, GE sought to expand into international markets and selected the European region as its first target. In July 1964, GE purchased a 50% share in France's largest computer manufacturer, Compagnie des Machines Bull, for $43 million in the face of stiff opposition from the French government, who wanted the company to remain in French hands. The following month it acquired a controlling interest in the computer business of the Italian firm Ing. C Olivetti & Co. SpA through the formation of new company, Olivetti-General Electric, which was 75% owned by GE and 25% by Olivetti.

Attempts to rationalise the two European businesses were frustrated by French bureaucracy and little effort was made to integrate European product lines with those of the parent company. There were also problems with developing a much-needed successor to the large-scale GE-600 series. An ambitious plan to develop a new line of 32-bit machines at a cost of up to $1 billion was rejected by GE senior management in favour of further investment in the company's burgeoning nuclear energy and jet engines businesses. This

8. Solid Progress – Transistors and Unified Architectures

decision effectively ended the company's chances of securing the contract from the Bank of America to replace its ageing GE 100 ERMA installations, allowing IBM to step in.

In August 1967, GE attempted to revitalise its flagging computer business by recruiting John Haanstra from IBM. Haanstra was hired as a special adviser on advanced planning and strategy for GE's Information Systems Division, a role that his earlier experience as co-chair of the SPREAD Task Group made him particularly suited for. Haanstra made such an impact within GE that he was soon elected as a Vice President of the company and his appointment as General Manager of the Information Systems Division followed in April 1968. He began to strengthen links between the US and European businesses and initiated plans to rationalise GE's large-scale product developments. However, before these could be fully realised, Haanstra was killed in a plane crash in August 1969 and his vision of GE's bright future in computers died with him.

In May 1970, GE announced that its US computer business had been sold to Honeywell for $234 million. The Honeywell and GE computer divisions were then merged to form a new subsidiary, Honeywell Information Systems. Honeywell had originally entered the industry in April 1955 through the formation of the Datamatic Corporation, a joint venture with Raytheon to produce large-scale computers, which it subsequently purchased outright in 1957. In 1966, Honeywell acquired the Computer Control Company, a Massachusetts-based manufacturer of special-purpose computer systems for military applications which had recently introduced the DDP-116, a new low-cost 16-bit model aimed at the commercial market.

By 1969, Honeywell's new Series 16 range, which was based on the DDP-116, was selling well but the company's large-scale machines were less competitive. The purchase of GE's Information Systems Division in 1970 not only doubled Honeywell's annual revenues, lifting the company into second place behind IBM in the US computer industry, but also served to revitalise the company's large-scale range. Honeywell was able to take the latest model in the ageing GE-600 series, the GE-655, and give it an engineering and marketing makeover. The result was the Honeywell Series 6000, a new product line comprising 6 models, 3 of which were optimised for scientific applications and 3 for business data processing. Announced in February 1971, the Series 6000 machines sold well in the large-scale sector, with more than 130 orders taken in the first year alone.

RCA's Dilemma

At around the same time that GE was offloading its Information Systems Division to Honeywell, another of the Seven Dwarves was also experiencing major problems with its own computer business. RCA's strategy of selling machines which were software compatible with IBM's System/360 but cheaper or faster was becoming more and more difficult to sustain with every new IBM model that appeared. The situation worsened dramatically in June 1970 when IBM announced the first two models in its System/370 series, a new family of machines which incorporated significant advances in hardware, such as monolithic integrated circuits, but retained backward compatibility with System/360. The price/performance advantage of RCA's Spectra 70 was now wiped out.

In an effort to second-guess IBM, RCA had hired former IBM executive L Edwin Donegan in January 1969 to head up its computer sales operation. Although RCA had nothing new to offer against the System/370 range, Donegan believed that a price-reduced version of Spectra 70 might still be competitive and in September 1970 the company announced the RCA Series, a family of four models which looked new but were merely facelifted versions of the Spectra 70/45, 70/46, 70/60, and 70/61.

At first, it looked like Donegan's plan might actually work. However, only a week after the RCA Series announcement, IBM announced the System/370 Model 145, a medium-scale addition to the System/370 range with even better price/performance due to the use of an advanced new semiconductor memory technology instead of costly magnetic core storage. Donegan's plan had backfired and RCA customers began switching to IBM hardware in droves.

RCA was now on the horns of a dilemma. It needed to invest several hundred million dollars in order to deliver a competitive range of machines. Taking into account the accumulated losses of the last few years, this meant that the company's Computer Systems Division would be unlikely to break even until 1975 at the earliest. RCA's senior management decided that the only viable option was to sell the division and announced its decision in September 1971.

As a loss-making business with no significant new products in the pipeline, RCA's Computer Systems Division was not a particularly attractive acquisition prospect. However, the company had built up a sizeable customer base of around 500 customers and this was of interest to Sperry Rand. In November 1971, it was announced that the UNIVAC division of Sperry Rand would take over RCA's general-purpose computer customers for a purchase price of $70.5 million plus 15% of future revenues from existing RCA computers.

To make it easier for RCA customers to migrate to UNIVAC hardware, the Sperry Rand engineers modified RCA's Virtual Memory Operating System (VMOS) so that it could run on their new UNIVAC 90/70 model. This task was made somewhat easier by the fact that the UNIVAC 90/70 was itself based on the UNIVAC 9700, which was the last of Sperry Rand's System/360 compatible UNIVAC 9000 Series machines. Sperry Rand later introduced the UNIVAC 90/80, a machine based on an RCA design, at the top end of the Series 90 Family to provide an upgrade path for the Spectra 70/60, 70/61 and RCA Series 6 and 7 models.

8. Solid Progress – Transistors and Unified Architectures

With both General Electric and RCA now out of the computer business, the Seven Dwarves was replaced with a new moniker, the BUNCH (Burroughs, UNIVAC, NCR, Control Data and Honeywell).

Further Reading

Bashe, C. J., Buchholz, W., Hawkins, G. V., Ingram, J. J. and Rochester, N., The Architecture of IBM's Early Computers, IBM Journal of Research and Development 25 (5), 1981, 363-375.

Campbell-Kelly, M., ICL and the Evolution of the British Mainframe, The Computer Journal 38 (5), 1995, 400-412.

Fagen, M. D., A History of Engineering and Science in the Bell System: National Service in War and Peace (1925-1975), Bell Telephone Laboratories, New York, 1978.

Gandy, A., The Entry of Established Electronics Companies into the Early Computer Industry in the UK and USA, Business and Economic History 23 (1), 1994, 16-21.

Gray, G. T. and Smith, R. Q., Before the B5000: Burroughs Computers, 1951-1963, IEEE Annals of the History of Computing 25 (2), 2003, 50-61.

Gray, G. T. and Smith, R. Q., Sperry Rand's Transistor Computers, IEEE Annals of the History of Computing 20 (3), 1998, 16-26.

Gray, G. T. and Smith, R. Q., Sperry Rand's Third-Generation Computers 1964-1980, IEEE Annals of the History of Computing 23 (1), 2001, 3-16.

Gray, G. T. and Smith, R. Q., Sperry Rand's First-Generation Computers, 1955-1960: Hardware and Software, IEEE Annals of the History of Computing 26 (4), 2004, 20-34.

Harris, J. R., The Earliest Solid-State Digital Computers, IEEE Annals of the History of Computing 21 (4), 1999, 49-54.

Irvine, M. M., Early Digital Computers at Bell Telephone Laboratories, IEEE Annals of the History of Computing 23 (3), 2001, 22-42.

King, J. and Shelly, W. A., A Family History of Honeywell's Large-Scale Computer Systems, IEEE Annals of the History of Computing 19 (4), 1997, 42-46.

Lange, T., Helmut Hoelzer – Inventor of the Electronic Analog Computer, in The First Computers – History and Architectures, MIT Press, Cambridge, Massachusetts, 2000.

Lee, J. A. N., The Rise and Fall of the General Electric Computer Department, IEEE Annals of the History of Computing 17 (4), 1995, 24-45.

Logue, J. C., From Vacuum Tubes to Very Large Scale Integration: A Personal Memoir, IEEE Annals of the History of Computing 20 (3), 1998, 55-68.

Lonnquest, J. C. and Winkler, D. F., To Defend and Deter: The Legacy of the United States Cold War Missile Program, USACERL Special Report 97/01, Defense Publishing Service, 1996.

McKenney, J. L. and Fisher, A. W., Manufacturing the ERMA Banking System: Lessons from History, IEEE Annals of the History of Computing 15 (4), 1993, 7-26.

McKenzie, J. A., TX-0 Computer History, RLE Technical Report No. 627, Massachusetts Institute of Technology, 1999.

Melliar-Smith, C. M, Borrus, M. G., Haggan, D. E., Lowrey, T., San Giovanni Vincentelli, A. and Troutman, W. W., The Transistor: An Invention Becomes a Big Business, Proceedings of the IEEE 86 (1), 1998, 86-110.

Neufeld, J., The Development of Ballistic Missiles in the United States Air Force 1945-1960, US Air Force Office of Air Force History, Washington, D.C., 1990.

Oldfield, H. R., King of the Seven Dwarfs: General Electric's Ambiguous Challenge to the Computer Industry, IEEE Computer Society Press, Los Alamitos, California, 1996.

Rosen, S., Recollections of the Philco Transac S-2000, IEEE Annals of the History of Computing 26 (2), 2004, 34-47.

Ross, I. M., The Invention of the Transistor, Proceedings of the IEEE 86 (1), 1998, 7-28.

Slutz, R. J., Memories of the Bureau of Standards' SEAC, in A History of Computing in the Twentieth Century, Academic Press, New York, New York, 1980.

Tucker, S. G., Emulation of Large Systems, Communications of the ACM 8 (12), 1965, 753-761.

9. Small is Beautiful – The Minicomputer

"Where a computer like the ENIAC is equipped with 18,000 vacuum tubes and weighs 30 tons, computers in the future may have only 1,000 vacuum tubes and weigh only 1½ tons", Popular Mechanics, 1949.

The Advent of the Minicomputer

As the computer industry grew larger and its products became more commonplace in a widening variety of applications, a substantial market began to emerge for smaller scale machines in applications where the cost of a large 'mainframe' computer system could not be justified. These came to be called minicomputers.

The term minicomputer is not a technical term. It was coined in the mid 1960s by John Leng, UK head of operations for Digital Equipment Corporation, one of the leading manufacturers in the sector during this period. He sent back a sales report to DEC's head office in the USA that began with the phrase "Here is the latest minicomputer activity in the land of miniskirts as I drive around in my Mini Minor". The term soon became popular with marketing executives to describe the plethora of small-scale machines that were on offer from a growing band of manufacturers keen to grab a share of a rapidly expanding market. Consequently, there is no agreed definition of what constitutes a minicomputer. However, small-scale models existed almost from the start of the computer industry and their influence on it would be immense.

Early Small-Scale Computers

Since the earliest days of the industry, computer manufacturers had developed small-scale machines for use as front end processors to large installations. These machines acted as a buffer between the fast processor and the considerably slower input/output devices but they also provided off-line data pre-processing and formatting capability. The first computer of this type was the Cardatron which was introduced by the ElectroData Corporation in July 1954 as a $31,000 option for the company's medium-scale Datatron 205 model. The Cardatron featured a plugboard for sequence control and incorporated a small magnetic drum memory with a storage capacity of 140 10-digit words. However, it relied on its Datatron host for overall control and was incapable of functioning as a standalone computer. Later input/output processors would be sufficiently general-purpose that they could be used separately but the first true small-scale general-purpose machines would evolve from a requirement to provide scientists and engineers with an affordable upgrade path from desk calculators.

The Burroughs E101

The Burroughs Corporation, which would go on to acquire ElectroData in 1956, was the first manufacturer to attempt to create a market for small-scale general-purpose computers. Burroughs was arguably the most innovative of the big three US office equipment manufacturers that dominated the formative years of the computer industry but was the slowest of the three to enter the commercial market. In seeking a suitable market entry point, the company had identified a need for a compact, inexpensive computer that could bridge the gap between manual desk calculators and medium-scale machines. It set out to address this market gap with its first commercial model, the Burroughs E101 Desk Size Electronic Computer, which was announced in May 1954.

The E101 featured a small magnetic drum memory with a capacity of 100 12-digit words but this was for data storage only as the E101 was not a stored-program machine. Instruction sequences were stored on card templates and entered manually via pinboards. Although denied the advantages of a stored-program architecture, the machine was very easy to program and cleverly packaged, with all the hardware built into a cabinet the size and shape of a double-pedestal office desk which was mounted on casters so that it could be moved around easily. An optional paper tape reader allowed data to be input automatically.

Anticipating the later trend towards personal computing, Burroughs envisaged that the E101 would be used by small teams of engineers or scientists in a laboratory setting with a single person programming and operating it. Despite having a binary-coded decimal architecture better suited to business machines and no floating-point capability, the E101 did indeed prove attractive to engineers and scientists, finding favour with both the University of Rochester and Bell & Howell for optical design work. Five machines were also purchased by the US Army Corps of Engineers and several military contractors also used them. However, the E101 was too close in capability to the sequence-controlled punched card calculators then offered by Remington Rand and IBM, despite its higher specification and ease of programming. Moreover, it lacked the ability to handle punched cards, which hindered its use as a replacement for punched card calculators. Consequently, the machine was not a great success in the marketplace despite its low $32,500 price tag and total sales of the E101 and the E102, a revised version introduced in 1957, only reached 127 units.

The Personal Automatic Calculator

Burroughs lack of success with the E101 should have discouraged other

9. Small is Beautiful – The Minicomputer

computer manufacturers from introducing similar products but one such manufacturer, IBM, would do exactly that. Its genesis was a project conducted by the electrical engineer John J Lentz at IBM's Watson Scientific Computing Laboratory in New York.

Lentz had been recruited from the MIT Radiation Laboratory in January 1946, along with NORC designer Byron Havens, to work at the Watson Laboratory on the development of advanced electronic circuitry and devices. In 1948, having observed the tremendous difficulty in programming the giant SSEC calculator, Lentz set out to design a small computer system that could be operated without having to write complex programs for it.

Lentz's initial research involved the development of novel counting circuits in which decimal digits were represented by the amount of charge in a capacitor. He then designed a sequence-controlled calculator that utilised these inexpensive circuits in combination with a small magnetic drum memory. Named the Personal Automatic Calculator (PAC), it featured a decimal architecture with a 15-digit word length and 84 double word internal registers, each of which could be used as an accumulator. These were stored on the drum as recirculating loops. Numbers entered into the machine were automatically split so that the integer part of the number was stored in one half of the double word register and the fractional part in the other half. This allowed the position of the decimal point to be maintained during a calculation.

The PAC could be operated manually by entering instructions and data using the keyboard or automatically by recording instruction sequences on punched tape and replaying them. A plugboard allowed a further 200 steps to be preset and activated either manually or from punched tape under program control. A second punched tape unit provided up to 6,000 digits of data storage. A square root function was also provided.

The machine was built into a rectangular cabinet slightly larger than an office desk, with the punched tape devices accessible via a panel along the top and the plugboard mounted on the front. A separate desk was provided for an electric typewriter and a console unit which incorporated the input keyboard, various control switches and a small CRT device for displaying the contents of the registers in the form of two 16 by 9 arrays of spots.

In November 1953, with IBM's product engineering resources already at full stretch on the 650 and 702 models, development of the production version of the PAC was subcontracted to the ElectroData Corporation. The production prototype was completed the following year but introduction was postponed as IBM struggled to cope with demand for its larger models.

The Story of the Computer

IBM Type 610 Auto-Point Computer (Photo © IBM Corporation)

The PAC was finally announced in September 1957 as the IBM Type 610 Auto-Point Computer at a price of $55,000. IBM marketed it as "a desk-side computing tool" for the solution of engineering and scientific problems, the same market that Burroughs had targeted with the E101. The internal storage capacity of the IBM model was significantly larger than the Burroughs machine at 2,604 digits, however, it was considerably more expensive and over three years later. Other machines had also been introduced during the intervening period which provided a much higher specification for a similar price. Nevertheless, a total of 180 were built, a figure that probably owed more to the skills of the IBM sales force than the desirability of the machine itself.

Establishing the Market

For small-scale computers to differentiate themselves from punched card calculators in the marketplace, they would have to be stored-program machines. However, memory components remained very costly in the early 1950s and designing a stored-program computer which would be inexpensive to manufacture was a formidable engineering challenge. Two projects would independently rise to this challenge. Both involved individual American academics working in partnership with Southern California military contractors.

The Bendix G-15

The first commercially successful small-scale computer was the brainchild of the American computer pioneer Harry Huskey. Huskey had an impressive CV,

9. Small is Beautiful – The Minicomputer

having been involved in the latter stages of the ENIAC project while working as a mathematics instructor at the University of Pennsylvania and was the driving force behind the development of the NPL ACE Test Assembly during a one-year visiting appointment to the UK in 1947. He also led the development of the Standards Western Automatic Computer (SWAC) at the NBS Institute for Numerical Analysis during 1949 and 1950.

Huskey began working on the conceptual design for a small-scale computer in 1953 while based at Wayne University (now Wayne State University) in Detroit. He envisaged a stored-program machine with a similar architecture to that of the ACE Test Assembly but with a magnetic drum memory in place of the mercury delay lines to reduce the number of thermionic valves required. In 1954, Huskey obtained a position as Associate Professor of Mathematics and Engineering at the University of California Berkeley. The move to California placed him within easy reach of the numerous technology firms and military contractors clustered around the Los Angeles area, several of which were becoming interested in electronic computation. Sensing a commercial opportunity, Huskey prepared some block diagrams and a functional description of his proposed machine and approached a number of firms, two of which, the Bendix Aviation Corporation in Los Angeles and Librascope Incorporated in Glendale, responded positively. Huskey decided to go with Bendix.

Bendix Aviation had moved into electronic computation in 1952 when it acquired the rights to manufacture the MADDIDA digital differential analyser from the Northrop Aircraft Company. The Bendix version, the D-12, was priced at $55,000 and generated around a dozen sales. By early 1954, Bendix was seeking to develop a similarly priced general-purpose computer and Huskey's design concept was just what they were looking for. Huskey was hired as a consultant and the company's own engineers were put to work to develop the machine. The logical and circuit design was carried out by Robert M Beck, a physicist who had been a key member of the MADDIDA development team before transferring to Bendix from Northrop in 1952.

The completed machine was designated the Bendix G-15 All-Purpose Computer. It featured a binary serial architecture with a 29-bit word length. Memory was provided by a small magnetic drum device of 2,160 words capacity which rotated at 1,800 rpm. Four of the 24 tracks on the drum were fitted with extra read/write heads closely spaced around the circumference to provide recirculating loops of high-speed memory sufficient for three 2-word registers, two of which could function as accumulators, and a 1-word register. This approach dictated a modest 105 KHz clock speed but performance was bolstered by the use of a two-address instruction format which reduced access time by providing the location of the next instruction in memory.

The G-15's minimalist circuitry required only 450 thermionic valves and

The Story of the Computer

2,500 diodes and this allowed the entire machine to be packaged into a single cabinet the size and shape of a small domestic refrigerator. The cabinet also housed a high-speed punched tape reader and a tape punch unit. The efficient design ensured low heat output which reduced the need for installation in an air conditioned environment and a power consumption of only 3,500 watts allowed the machine to be powered from a standard electrical supply.

Bendix G-15 Brochure

The G-15 was introduced towards the end of 1955 at a bargain basement price of $44,800. This included the integrated punched tape reader, tape punch and an electric typewriter for input and output plus 5 weeks of training in the operation and maintenance of the machine. An extensive list of optional equipment included a magnetic tape unit, a coupler which permitted standard IBM punched card equipment to be connected, a graph plotter and a digital differential analyser for use in solving differential equations. Software support included INTERCOM, a command interpreter developed by Huskey to simplify programming. The G-15 was marketed by Bendix as suitable for industrial and scientific applications and it became popular with civil engineers for the 'cut and fill' calculations used in highway construction. Over 400 were built, a remarkable achievement for a company with no track record in the computer

9. Small is Beautiful – The Minicomputer

business.

Bendix had planned to introduce a transistorised version to capitalise on the success of the G-15 but a sharp fall in the cost of magnetic core memory served to render the machine's recirculating memory design obsolete. Rather than undertake a complete redesign of the G-15, the company decided instead to target the opposite end of the market with the G-20, a large-scale transistorised machine which was introduced in 1961 at a price of $290,000. However, this put Bendix into direct competition with established computer manufacturers such as IBM and Sperry Rand. Lacking the financial clout necessary to compete with these industry giants, Bendix struggled in the marketplace for a further two years until selling its computer division to Control Data Corporation in 1963.

The Royal Precision LGP-30

The firm Harry Huskey turned down in favour of Bendix would also play an important role in opening up the market for small-scale computers. Librascope was a successful military contractor which specialised in instrumentation and control and had developed electromechanical fire control systems for the US Navy during World War II. In 1952, the company introduced its first general-purpose computing product, the Librascope Equation Solver, a small-scale electronic analogue computer. The interest shown in this machine encouraged Librascope to expand its product line with a small-scale digital model and a suitable design was sought.

Huskey's decision to take his computer design elsewhere was not a serious setback for Librascope, as an evaluation exercise conducted by the company had identified an alternative design from physicist Stanley P Frankel. Frankel had been one of the first users of ENIAC, configuring the machine to perform design calculations for the hydrogen bomb in late 1945 as part of his work on the Manhattan Project at the Los Alamos Laboratory. Following the loss of his US Army security clearance in the McCarthy witch hunts of 1949 due to family links with the Communist Party, Frankel was invited to lead a newly created digital computing group at the California Institute of Technology (Caltech) in Pasadena, California.

At Caltech, Frankel set out to design and build a small stored-program computer using as few components as possible. After seeking advice from other groups, he decided to base his design around a magnetic drum memory. The machine was named the Minimal Automatic Computer or MINAC and featured a binary serial architecture with a 31-bit word length and a 120 KHz clock speed. Frankel used reject germanium diodes that he had obtained during a visit to the Hughes Aircraft Company for the majority of the electronic circuitry which reduced the number of thermionic valves required to around 100. The magnetic drum was designed and built by physics postgraduate student James Cass. Like the Bendix G-15, the arithmetic and control registers were implemented using recirculating loops on the magnetic drum.

Following completion of MINAC in 1954, Librascope licensed the design from Caltech. The company then hired Frankel as a consultant and Cass as an employee to assist with the development of the production version. The project was led by senior systems engineer Raymond Davis. Cass must have impressed his new employers, as he was given complete responsibility for the project when Davis left Librascope in 1955.

The production version, which was named the LGP-30, incorporated only minor changes to Frankel's design. It featured a magnetic drum rotating at 3,600 rpm with a 4,096 word capacity spread over 64 tracks plus an additional 3 tracks for the accumulator, instruction register and counter register. The minimalist design required only 113 miniature thermionic valves and 1,450 germanium diodes. These were packaged into a cabinet similar in size and shape to a chest freezer which was mounted on casters and fitted with a shelf on one end to support a Flexowriter electric typewriter and punched tape reader/punch. An oscilloscope built into the control panel displayed the contents of the registers. The minimalist design philosophy also extended to the instruction set which comprised only 16 machine instructions.

As a military contractor, Librascope had no experience of selling to the private sector. Therefore, rather than market and sell the computer itself, Librascope's parent company, the General Precision Equipment Corporation, decided to enter into partnership with the typewriter retailer Royal McBee Corporation of Port Chester, New York, to form the Royal Precision Corporation and the machine was sold under this name. The Royal Precision LGP-30 was introduced in November 1955 at a price of $47,000 for a basic system and deliveries began in September 1956. Optional equipment included a high-speed punched tape reader and tape punch unit. A generous educational discount which brought the basic price down to under $30,000 made the machine popular with universities and over 500 were sold.

A transistorised version, the LGP-21, was introduced in December 1962. The use of transistors and a magnetic disk memory allowed the size and cost to be further reduced and the new model was packaged as a desk-sized machine at the impressively low price of $16,250 in basic specification. However, despite having solid-state circuitry, the LGP-21 was actually slower than its predecessor due to a magnetic disk that rotated at a leisurely 1,180 rpm and a correspondingly lethargic 80 KHz clock speed. Consequently, sales were disappointing with only around 150 machines built. Clearly, the market that Librascope had helped to create now expected similar levels of performance from small-scale computers to that available with larger machines. This would only come through abandoning recirculating memory architectures in favour of fully solid-state designs.

9. Small is Beautiful – The Minicomputer

The First Solid-State Models

As the availability of transistors and magnetic core memory increased and prices began to fall, the use of such high-end technology in low-cost models became economically viable. IBM was well placed to take the lead in this respect, having invested heavily in both technologies, and it became the first manufacturer to introduce small-scale stored-program computers that were fully solid-state. During October 1959, the company announced not one but two new small-scale machines; a business data processing model designated the 1401 and a scientific model, the 1620. They were created to fill an embarrassingly wide gap in the company's product line between its punched card calculators, which rented at a few hundred dollars per month, and the medium-scale 650 Magnetic Drum Data Processing Machine, which rented at over $3,000 per month. This gap had already attracted the attention of Bendix and Librascope. Both were bit players in the computer industry but it would only be a matter of time before IBM's direct competitors targeted it.

The IBM 1401

The design of the IBM 1401 originated from a 7-week conference involving IBM engineers and product planners from France, West Germany and the US which was held in Sindelfingen, West Germany in the summer of 1955. The objective was to agree the specification for a new accounting machine capable of competing with rival offerings such as the Gamma 3, a sequence-controlled punched card calculator from Compagnie des Machines Bull that was proving more popular in European markets than IBM's Type 604 and Type 607 machines due to an ability to interconnect with tabulators and other punched card equipment.

Various design proposals presented during the conference were refined over the following months. After further discussions, a decision was made to proceed on the basis of a proposal from IBM's French laboratory for a plugboard controlled calculator with a variable word length and a small magnetic core memory for data storage. The accompanying input/output devices would be developed by IBM engineers in West Germany. However, as the corporate importance of the project began to grow, concerns were raised over the high production cost of the plugboard control panel and the lack of experience of IBM's European subsidiaries in major product development. A design engineer from IBM's Advanced Systems Development Department, Francis O Underwood, was asked to investigate production costs and concluded that a stored-program architecture would actually be cheaper to manufacture. This also became the trigger for the project being moved to the US.

Underwood's modified proposal was given formal approval in March 1958 and development began at IBM's Endicott Laboratories under the codename SPACE (Stored Program Accounting and Calculating Equipment). Underwood was appointed as chief architect. The logical design from the original proposal was retained so the machine featured a binary-coded decimal, serial, architecture with a variable word length and an 87 KHz clock speed. Three index registers were provided for handling tables and subroutines. The magnetic core memory of 1,400 alphanumeric characters could be expanded in increments of 2,000 up to a maximum of 16,000 characters.

The fully transistorised design employed low-cost Complementary Diode-Transistor Logic (CDTL) circuits and the SMS modular packaging technology developed under Project Stretch for ease of maintenance. The inherent reliability of the solid-state construction was enhanced by the extensive use of self-checking features. The hardware was contained within a cabinet comprising two cubes placed one on top of the other, with the control panel mounted on the front face of the upper cube. On models with additional magnetic core storage, this cabinet was expanded by a further two cubes. No special air conditioning requirements were necessary, however, unlike rival small-scale models the 1401 required a three-phase electrical supply.

The IBM 1401 Data Processing System was introduced on 5 October 1959 at a price of $70,500 for a basic machine fitted with 1,400 alphanumeric characters of magnetic core storage. Introduced along with the new model were a specially designed punched card unit which could read cards at a rate of 800 per minute and a high-speed printer. Up to 6 magnetic tape units could also be connected.

The 1401 offered IBM customers a painless migration path from punched card equipment to a fully computerised data processing system. Excellent expandability allowed a 1401 system to grow along with the customer's business needs. Furthermore, the machine could also be used as a front end processor for the large-scale IBM 7090. Despite the machine's lacklustre compute performance, these attributes resulted in sales of more than 10,000 units before the model was finally withdrawn in February 1971.

The spectacular success of the IBM 1401 prompted the introduction of similar machines from other manufacturers, such as the Honeywell H-200 which was announced in December 1963 and the General Electric GE-415 which was announced in May the following year. Both models were significantly faster in performance than the IBM 1401 and offered software compatibility with the IBM machine through the use of translation or simulation software.

The IBM 1620

In the spring of 1958, with project SPACE having been given the go ahead, a small team was assembled at IBM's Poughkeepsie Laboratory to study the small-scale scientific market. The 610 Auto-Point Computer had been acknowledged as a technological dead end with no successor planned, therefore, a completely

9. Small is Beautiful – The Minicomputer

new design was called for. With companies such as Bendix and Librascope already offering low-cost stored-program models in this market sector, the new IBM model would also have to be a stored-program machine.

In a light-hearted nod to the 1401 development, the new project was codenamed CADET. The objective was to produce a machine with at least half the performance of the medium-scale 650 at no more than half the price. A magnetic drum memory was initially considered but rejected on the basis that IBM had no technological or commercial advantage to bring in this area. Project manager Wayne D Winger had personal experience of the use of magnetic core storage, having previously served as engineering manager on the development of the IBM 705. A small-scale machine equipped with a magnetic core memory would clearly outshine the competition if the price could be held in check. Therefore, the decision was made to use the more expensive magnetic core technology for main memory and to compensate for this by keeping the cost of the other components to an absolute minimum. This was achieved by taking the radical step of eliminating most of the arithmetic logic circuitry in favour of look-up tables to perform arithmetic operations.

The CADET design featured a binary-coded decimal, serial, architecture with a variable word length, a two-address instruction format and a respectable 500 KHz clock speed. Eight 5-digit registers were provided for memory addressing and data flow. The magnetic core memory had a capacity of 20,000 'addressable positions', portions of which were reserved for the addition and multiplication tables. Subtraction was performed using the nine's complement method but division had to be accomplished using a software subroutine as the instruction set had no divide instruction.

Construction utilised very low-cost SDTRL logic circuits that eliminated the need for diodes through a clever combination of resistors and transistors. The SMS modular packaging technology from Project Stretch was again employed and this resulted in a highly compact design that allowed the machine to be packaged into a desk-sized enclosure similar to that of a Burroughs E-101. An electric typewriter which sat alongside the large console panel on top of the desk provided for basic input and output. An expansion unit provided an extra 20,000 or 40,000 digits of magnetic core storage. Following completion of the engineering prototype, CADET production was transferred to IBM's General Products Division in San Jose where it was given the nickname 'Can't Add, Doesn't Even Try'.

The new model was introduced as the IBM 1620 Data Processing System on 21 October 1959 at a price of $64,000 in basic specification. The design objective had been comfortably exceeded, with performance more than half that of the 650 at around a third of the price. More importantly, the 1620 was also four times faster than the Royal Precision LGP-30 at less than twice the price. The Model II, which was announced in December 1962, introduced further performance improvements including arithmetic logic circuitry for addition and subtraction plus index registers and a doubling of the clock speed to 1 MHz. Ease of programming due to its decimal arithmetic made the 1620 popular with educational institutions and around 2,000 were built before the model was withdrawn in November 1970.

The CDC 160

The market-leading performance of the new IBM 1620 would be short-lived, as a rival manufacturer best known for its high-performance scientific computers was about to enter the market. In February 1960, Control Data Corporation quietly announced the introduction of the CDC 160, a small-scale model with a performance that outstripped every other small computer on the market and even threatened that of some large-scale machines. The architect of this impressive and highly influential machine was the father of the supercomputer, Seymour Cray.

The development of the CDC 160 was triggered by the need to provide a front end processor for the company's large-scale CDC 1604 model. The design had its origins in a laboratory prototype known as 'Big Character', a 6-bit binary machine with 16,384 words of magnetic core memory which was built by Control Data engineers as a test-bed for the memory circuits of the 1604. Over what was reported to be a long weekend, Cray took the logical design for Big Character and reworked it to create a 12-bit production model with 4,096 words of magnetic core memory.

The CDC 160 featured a binary parallel architecture and a 2.5 MHz clock speed. The omission of multiplication or division circuitry and the use of diode logic helped to keep manufacturing costs low and Cray's modular packaging techniques resulted in a very compact design. As only 6 bits of a 12-bit word were available for memory addressing, relative and indirect addressing modes were included. These also provided additional flexibility for array processing. Having been conceived as a front end processor, input/output performance was impressive, with a data transfer rate of up to 70,000 words per second and the ability to connect with up to 30 devices simultaneously.

Like the IBM 610, the CDC 160 was packaged as a desk-sized machine with a console panel mounted on the desktop. Orange glowing numerals on the console displayed the contents of the 3 operational registers in octal digits. The desktop also accommodated an integrated punched tape reader and the left-hand pedestal housed a tape punch unit which was accessed through the pedestal door. Optional equipment included various magnetic tape and punched card units plus a high-speed printer. With a power consumption of only 700 watts, the CDC 160 could be powered from a standard electrical outlet and no special air conditioning was required.

The CDC 160 was introduced in February 1960 at a basic price of $60,000,

9. Small is Beautiful – The Minicomputer

undercutting the less powerful IBM 1620 by $4,000. With a compute performance of over 78,000 additions per second, the new model attracted considerable attention but the fixed memory capacity and software-implemented multiply and divide instructions were obvious deficiencies which made it much less suitable for scientific applications. These deficiencies were remedied in April 1961 with the introduction of a revised model, the CDC 160-A, which sacrificed compactness for expandability. Memory was doubled to 8,192 words by adding a second bank of 4,096 words. The 12-bit address space limitation was overcome by the inclusion of extra machine instructions for memory bank selection which operated in combination with four additional 3-bit registers to permit up to 8 banks of 4,096 words to be addressed. The additional 6 banks of memory were accommodated within a separate cabinet. An optional external arithmetic unit was also introduced that provided hardware-implemented multiplication and division in single or double precision plus double precision addition and subtraction.

CDC 160-A

Following the success of the CDC 1604, Control Data had set its sights on becoming the industry leader in large-scale scientific computers. Consequently, the small-scale sector was a low priority for the company and the CDC 160 was not aggressively marketed. Nevertheless, more than 300 CDC 160 and 160-A models were built, the total boosted by a contract from the National Cash Register Company for a programmable accounting machine which resulted in 50 machines being supplied to NCR and rebadged as the NCR 310. A larger scale variant was also developed for the National Aeronautics and Space Administration (NASA) which featured a 13-bit word length and a highly modular design that permitted up to 9 processors to be connected to a common memory unit. The production version of this machine was announced in February 1964 as the CDC 160G.

With its high-performance binary parallel architecture, strong input/output capability, solid-state construction and compact packaging, the CDC 160 was the first of a new breed of small-scale computers that eventually became known as minicomputers. However, the computer industry would be slow to recognise the commercial potential of such machines and would need further encouragement before fully embracing this new market opportunity.

The Influence of Industrial Process Control

Like gunnery fire control, the control of industrial process plant had traditionally been achieved using analogue computing mechanisms. However, manufacturers were quick to spot the potential for digital computers in industrial process control and the first article on the subject appeared in Scientific American in September 1948 but the poor reliability of the early valve-based machines made them unsuitable for real-time use. The advent of transistorised circuitry provided the requisite reliability and the suitability of digital computers for real-time control was soon proven with the installation of the first missile guidance computers in 1957.

The first commercially available computer to be specifically designed for industrial process control was the Ramo-Wooldridge RW-300 Digital Control Computer which was introduced in July 1957. Ramo-Wooldridge would later become the market leader in process control computers with a succession of fully solid-state machines but the company's first model employed a magnetic drum for main memory with only a small amount of magnetic core storage used for tape buffering. With a Mean Time Between Failure (MTBF) of only 500 hours, it would not have been sufficiently reliable for direct control applications. Consequently, the first wave of computerised process control applications retained much of the analogue control equipment and restricted the computer to a supervisory role. The earliest example of this was the installation of an RW-300 by the oil company Texaco for supervisory process control at its Port Arthur refinery in Texas which went on-line in March 1959.

Ramo-Wooldridge's pioneering efforts encouraged other manufacturers to develop computer systems for industrial process control. Not surprisingly, those firms which were already involved in both the process control equipment and computer markets, such as General Electric and Honeywell, took an early lead. The GE 312 Digital Control Computer System, which was introduced in November 1959, was a medium-scale transistorised model with a binary serial architecture and a 20-bit word length. It served as the central processor for

9. Small is Beautiful – The Minicomputer

GARDE (Gathers Alarms Records Displays Evaluates), a complete process monitoring and control system designed for electricity generating plants. Like the RW-300, the GE 312 also employed a magnetic drum memory, although magnetic core storage was available as an option. The basic price of the computer was $85,200 but a fully specified GARDE system could cost up to $500,000.

Honeywell's H-290 model, introduced in September 1960, included a small magnetic core memory as standard which was supplemented by a magnetic drum for additional capacity. This resulted in a MTBF of more than 1,000 hours, a much more acceptable figure for process control applications. However, the price remained high at around $170,000 for a typical configuration.

The Elliott 803

The first fully solid-state computer for process control was the Elliott 803 which was introduced by the British firm Elliott-Automation in early 1960. The 803 was also the first successful crossover model, achieving significant sales in both the process control and general-purpose computing markets.

The 803 was the third in a series of machines which began with the 801 prototype in 1957. The 801 was designed by Elliott-Automation chief design engineer John P Bunt, based on a new type of logic circuit developed by Bunt that employed magnetic cores for switching in order to reduce the number of costly transistors required. The production version, the Elliott 802, was introduced in November 1958 at a price of £17,000.

The 802 featured a binary serial architecture with a 33-bit word length and a 167 KHz clock speed. The accumulator, instruction register and program counter were implemented by means of a single magnetostrictive delay line, which reduced manufacturing costs but also impacted on performance due to the serial access characteristics of this type of device. A magnetic core main memory was provided with a standard capacity of 1,024 words which could be increased to a maximum of 4,096 words. The long word length allowed two 16-bit instructions to be stored, with the remaining bit reserved for use as a B-line flag that when set instructed the machine to modify the second instruction by adding to it the updated contents of the memory location specified by the first instruction. This made it possible for any location in memory to be used as an index register.

Another unusual feature was the provision of a loudspeaker to monitor the operation of the computer. This was driven by changes in the state of the most significant (highest) bit of the instruction register so that it produced a series of warbling tones which varied with the type of operation being performed.

The 802 was packaged into an L-shaped cabinet the size of a large desk, one leg of which contained the memory and arithmetic unit and the other containing the console and input/output circuitry. A separate cabinet housed the power supply which incorporated a huge nickel-cadmium battery to provide backup power in the event of a mains electricity failure. A punched tape reader and tape punch were also provided for input and output.

Despite the liberal use of solid-state components for logic circuitry and main memory, the 802 was not completely solid-state, as a number of thermionic valves were employed in the machine's power circuits due to a shortage of suitable power transistors during development of the prototype. Another limitation was the maximum memory capacity of 4,096 words as a result of having only 12 bits of a 16-bit instruction word available for memory addressing. Consequently, sales were poor and only 7 machines were delivered, none of which were destined for process control use.

Following a request from an American chemical company, the 802 design was overhauled to create a machine better suited to process control. The main changes were an increase in the word length to 39 bits, thereby providing an extra 3 bits to extend the memory address space, and the use of power transistors to replace valves. Floating-point arithmetic hardware was also included and main memory was increased to 4,096 words as standard, with an additional 4,096 words available as an option. The desk-sized cabinet was incapable of accommodating the additional circuitry and was replaced with a larger rectangular cabinet and separate desk for the console and punched tape equipment.

The new model was introduced in early 1960 as the Elliott 803 at a price of £29,000, the higher price reflecting the increased specification. However, overall performance was actually poorer than the 802 as a result of the increased access time on the delay line registers due to the longer word length. This embarrassing situation was remedied a few months later by the introduction of a revised version, the 803B, which incorporated individual magnetostrictive delay lines for the main operational registers.

With its 39-bit word length and floating-point hardware, the Elliott 803 was soon recognised as an affordable alternative to medium-scale scientific models such as the Ferranti Mercury and it became a popular choice for British technical colleges buying their first computers. Elliott-Automation had entered into a joint venture with the US firm NCR in 1957 to market and support various models of Elliott computers for business data processing under the name National-Elliott and the 803 was also sold under this name. A rebadged 803 also provided the central processor for the ISI 609, a data logging system manufactured by the US automation controls company Panellit. These efforts made the Elliott 803 the most successful British computer of the period with total combined sales of 211 units over a 6-year production time-span.

The Ferranti Argus

The arrival of fully solid-state models for process control opened up the

9. Small is Beautiful – The Minicomputer

possibility of using digital computers in direct control applications. The first computer to be used in this way was another British model, the Ferranti Argus Process-Control Computer System which was introduced in January 1961. In an early example of defence diversification, Ferranti took a machine which had been developed by its Automation Division in 1958 as the ground-based launch control computer for the Bristol Bloodhound Mk II surface-to-air missile and marketed it as suitable for industrial process control.

In common with other missile guidance computers of this period, the Argus featured a Harvard architecture, where instructions and data are stored separately and can be accessed simultaneously. The word length was 12 bits with 24-bit instructions and a 500 KHz clock speed. Conventional magnetic core storage was used for data but instructions were held in a type of read-only memory similar to the 'hairbrush' memory developed for the Ferranti Atlas which sensed the presence of ferrite pegs in pegboards, their positions representing the binary digits for each instruction. The machine also featured twin 6-bit arithmetic logic units that operated in parallel.

The idea of utilising the Argus for process control came from Alan Thompson of the British chemical company ICI. This resulted in a joint development project which culminated in the installation of an Argus at ICI's Fleetwood ammonia soda plant in 1962. This system went on-line in November 1962, making it the first example of direct digital process control. Over a two-year evaluation period, it achieved an overall reliability rate of 99.8%.

The Argus attracted only a handful of sales, potential customers being wary of the pegboard memory which enhanced reliability due to its non-volatile nature but made programming a particularly tedious task. Ferranti responded by producing a revised model, the Argus 100, that ditched the Harvard architecture in favour of a shared magnetic core memory for both data and instructions and also contained only one arithmetic logic unit. These changes reduced performance by a factor of six but the performance remained sufficient for process control purposes, a testament to the efficient design of the original model. Introduced in 1963 at a price of £20,000, the Argus 100 generated brisk sales in both the UK and in several European countries and around 100 examples were built.

The additional revenues from civilian sales of the Argus models would prove very useful when, in July 1964, Ferranti was forced to refund £4.25M to the British Government following an investigation by government auditors which revealed that the company had made much larger profits than originally agreed on the cost-plus-fixed-fee contract for the Bloodhound missile guidance programme.

The Programmed Data Processor

The Elliott 803 and Ferranti Argus 100 models were both reasonably successful in UK and European markets but neither captured an appreciable share of the much larger US market. Clearly, American customers preferred to buy from domestic manufacturers, which presented an opportunity for US computer firms to follow suit and develop inexpensive solid-state models for process control.

Rather than develop new machines, some US manufacturers attempted to market their existing small-scale general-purpose models as suitable for process control. Control Data began advertising the CDC 160-A as suitable for real-time control, emphasising the machine's excellent input/output performance and fast execution times. IBM took this a stage further, adapting its 1620 model for process control by adding an analogue-to-digital converter and real-time clock unit which simplified the collection and analysis of analogue data. It was even given a new name, the IBM 1710 Industrial Control System, when announced in March 1961 at a price of $111,000. However, none of these general-purpose models from established manufacturers would make any impact on the process control market, thus paving the way for two start-up firms to make this sector their own. The first of these was Digital Equipment Corporation.

Digital Equipment Corporation (DEC) was founded in August 1957 by two engineers from the MIT Lincoln Laboratory, Kenneth H Olsen and Harlan E Anderson. Olsen had worked for Jay Forrester on the Whirlwind project and was a founder member of the Lincoln Laboratory's Advanced Computer Development Group, performing the circuit design and supervising construction of the TX-0 and TX-2 experimental transistor computers. Anderson joined the group in June 1952 after obtaining his Master's degree from the University of Illinois, having been hired to work with Olsen on the development of a test computer for the Whirlwind magnetic core memory. In early 1957, Olsen was involved in a failed attempt to set up a joint venture with an electronics manufacturer to commercialise the Laboratory's electronic component technology. This experience gave him the idea to set up his own company.

The business plan for the new company was to manufacture and sell electronic test equipment and high-performance digital computers. Olsen had been developing a new generation of test equipment at MIT which was intended to replace valve-based kit produced by Burroughs from designs that came out of the Whirlwind project. The company's first products would be based on this work. There would be no need to worry about licensing of the intellectual property from MIT as the Lincoln Lab's US government funders were keen to see its technology transferred to industry. Once a firm financial footing had been established, a computer would be developed which would utilise the same circuits employed in the test equipment. Olsen and Anderson planned to target the industrial process control and scientific computing markets initially until they had developed the necessary input/output equipment to tackle the

9. Small is Beautiful – The Minicomputer

business data processing market.

Start-up financing was provided by an investment of $70,000 from American Research and Development Corporation (ARDC), a Boston-based venture capital firm, in return for a 70% share of the business. Olsen and Anderson each retained 10% with the remaining 10% reserved for a business partner that they would recruit later. The deal was concluded with little negotiation and was later considered to be a poor one for the two founders, however, ARDC's president, Georges F Doriot, also took on the role of business mentor to the inexperienced Olsen and his input would prove invaluable in helping to guide the company through its formative stages. The original choice of name for the company was Digital Computer Corporation but Doriot advised that the word computer should be dropped, as the impression in the US investment community at that time was that nobody made money in the computer business.

With the finance in place, suitable premises for the new company were rented on the second floor of an old civil war era textile mill in the town of Maynard, Massachusetts. Olsen and Anderson then hired a number of former colleagues from the Lincoln Lab and began developing their first products, a range of electronic circuit modules based on Olsen's TX-2 circuit designs. Their 5 MHz operating speed and ability to be plugged together to create advanced digital testing systems made the DEC modules unique in the marketplace. By early 1959, the company was making sufficient profits from them to begin the next phase of its operation, the development of a high-performance computer. Another former Lincoln Lab colleague, Benjamin M Gurley, was hired from MIT to design the new machine which would be called the Programmed Data Processor Model 1 or PDP-1. Once again, the word computer had been deliberately left out, in this case to avoid a recent moratorium on the purchase of computer systems for US government use.

The PDP-1 featured a binary parallel architecture with an 18-bit word length and 5 MHz clock speed. The logical design employed the 'one's complement' method of binary arithmetic which removed the need for subtraction circuitry. Main memory comprised 4,096 words of fast access 5 microsecond magnetic core storage which could be expanded in banks of 4,096 up to an impressive maximum of 65,536 words. As only 12 bits of an 18-bit instruction word were available for memory addressing, a system of memory bank selection was implemented using a 'bank switch' instruction and two 3-bit bank registers. The machine also supported multiple-step indirect addressing but an index register was purposely omitted to reduce costs.

A key strength of the PDP-1 design was its advanced input/output capability. This included a 16 level program interrupt mechanism and a high-speed data channel that operated at a data transfer rate of up to 200,000 words per second using a direct memory access (DMA) technique which bypassed the input/output register. A real-time clock was also provided for timing purposes.

Despite these sophisticated features, the instruction set comprised only 28 machine instructions.

The PDP-1 prototype was designed and built by Ben Gurley in less than 4 months using mainly standard DEC system modules. The completed machine comprised 544 modules encased in 4 tall cabinets fitted with standard 19-inch rack mounts. The prototype was exhibited at the Eastern Joint Computer Conference in Boston in December 1959 and announced for sale at a price of $120,000 for a basic specification model with 4,096 words of memory. Standard equipment included a punched tape reader, tape punch and an IBM Model B electric typewriter. The modular construction allowed the basic price to be kept down by offering many of the features as options, including the 16 level interrupt, high-speed data channel, real-time clock and the circuitry required for hardware multiplication and division.

Possessing a compute performance far in excess of much larger models and with the CDC 160 not yet announced, the PDP-1 generated considerable interest from scientists and engineers looking for high performance at a low price. The first production machine was sold to Bolt, Beranek and Newman (BBN), an acoustics consultancy firm based in Cambridge, Massachusetts, and delivered in November 1960. A total of 53 machines were built, 20 of which were purchased by International Telephone & Telegraph Corporation for use in message switching systems and rebadged as the ITT 7300 ADX.

DEC's first computer was a financial success, albeit a modest one. However, Olsen was cautious about over-committing the company or growing it too quickly and he turned down an opportunity to tender for a request from NASA for 100 machines. He also donated one machine to MIT in September 1961, a not altogether selfless act as it incurred tax benefits for the company and also made sound marketing sense by encouraging graduates from MIT who had used the machine to lobby their future employers to buy DEC products.

The PDP-1 was to be Gurley's only computer design. In December 1962, he left DEC for a position as Vice President of a new computer consultancy firm, Information International Incorporated. The following year, he was shot dead by a disaffected ex-employee of DEC while at home having dinner with his wife and children.

The PDP-4

Despite achieving sales in the scientific computing market, the PDP-1 failed to make an impression on the industrial process control market. The DEC model was around 30% more expensive than the Elliott 803 and Thompson Ramo Wooldridge TRW-230, both highly capable process control computers despite their inferior performance. Following discussions with the Foxboro Corporation and Corning Glass Works, two large potential customers with process control needs, the decision was taken to develop a new model based on

9. Small is Beautiful – The Minicomputer

the PDP-1 that could be manufactured more cheaply and sold for half the price.

Development of the new model, which would be called the PDP-4, was initiated in November 1961. It was designed by C Gordon Bell, a talented electrical engineer who had been recruited from MIT in May 1960 after designing a tape controller for the TX-0. A 12-bit design was initially considered after studying the CDC 160 but it was decided to stick with an 18-bit word length so that the memory subsystem and input/output equipment from the PDP-1 could be used.

Bell was able to reduce production costs by simplifying the logical design of the PDP-1 and implementing the majority of the circuitry using cheaper DEC 4000 Series system modules that operated at 500 KHz. Logic was eliminated by having a single 18-bit register function as both the accumulator and the input/output register and eliminating two 6-bit control registers used for program flags and sense switches. Input/output capability was simplified by reducing the number of operating modes supported and eliminating the real-time clock. These changes resulted in a reduction in the size of the instruction set to 16 machine instructions.

Bell also took the opportunity to make some design improvements. These included auto-index registers, memory locations that automatically incremented before use when used as indirect addresses, and a larger memory address space of 8,192 words which was the result of freeing up an extra bit in an instruction word for memory addressing.

The PDP-4 was introduced in November 1962 at a basic price of $65,500. The design goal of reducing the price to half that of the PDP-1 had been achieved but performance had also been cut to around 60% of the earlier model. Although the new model met the requirements of the process control market, it offered no significant advantages over the competition nor was it sufficiently powerful to interest scientific computing customers. Consequently, the machine was a market failure, with only 45 machines built.

The PDP-5

One of the few companies which purchased a PDP-4 for process control was Atomic Energy of Canada Limited (AECL). The machine was intended as the central computer for a nuclear reactor control system that also incorporated a large amount of analogue control equipment. In order to reduce the complexity of this system, AECL engineers identified a need for a special-purpose computer to serve as a front end processor for their PDP-4. Instead of a special-purpose machine, which would have had limited commercial potential, DEC offered to supply an inexpensive general-purpose computer that could be programmed to perform this task.

Development of the new model, which would be called the PDP-5, began in February 1963. The system design was carried out by Gordon Bell and Alan

The Story of the Computer

Kotok, who had joined DEC the previous summer as a new graduate from MIT. Kotok also began working on the logical design but this task was later completed by Edson D de Castro, an applications engineer who had been working on the front end equipment for the system, when Kotok was reassigned to another project.

The starting point for the new machine was a design proposal for a 12-bit digital controller that had been shelved during the discussions that led to the PDP-4. The standard 500 KHz system modules employed in the PDP-4 were again used to build the new machine which featured a binary parallel architecture with a 12-bit word length. With only 7 bits of an instruction word available for memory addressing, a system of indirect addressing was implemented which was based on a 128 word block size. The auto-index register feature developed for the PDP-4 was again employed to remove the need for a separate index register.

Production costs on the PDP-5 were kept low by replacing the radial wiring arrangement used for connecting input/output equipment in the two earlier models with an extendable bus system that allowed the equipment to be daisy-chained using 50-wire cables and connectors. An inexpensive analogue-to-digital converter was also implemented which utilised the accumulator and memory buffer register. With less circuitry than an 18-bit design, the hardware could be fitted into a single 19-inch rack cabinet.

A sharp drop in the price of transistors and core memory components during development of the PDP-5 provided a further reduction in production costs. When introduced in August 1963, it was pitched at the impressively low price of $27,000 for a system with 4,096 words of magnetic core memory and a Teletype ASR-33 teletypewriter. Software support included a FORTRAN compiler, symbolic assembler and a collection of utility programs and subroutines. Although intended mainly for process control, an extensive range of input/output equipment options and a respectable compute performance of 55,555 additions per second made the PDP-5 attractive for low-end scientific and engineering applications. A total of 116 were produced, giving DEC its first taste of success in the computer business and underlining the importance of price in this new market sector.

Scientific Data Systems

The other US start-up firm with a major role in establishing the minicomputer market was Scientific Data Systems. SDS was founded in September 1961 but its origins lie in an earlier venture by the company's founder, the logician Max Palevsky.

Palevsky was a mathematics and philosophy graduate who became interested in electronic computation in 1950 while working as a teaching assistant at UCLA. In 1951, he obtained a job with Bendix Aviation on the development

9. Small is Beautiful – The Minicomputer

of the company's digital differential analysers. Palevsky had an aptitude for logic design and was the principal designer of the D-12 analyser but he left Bendix in late 1956 when the company refused to let him develop his concept of a parallel digital differential analyser. He then contacted local electronics firms in an effort to find one that would allow him to build it. Packard Bell, a Los Angeles based military contractor and television set manufacturer, was sufficiently impressed by Palevsky's ideas that it established a new computer division in March 1957 and appointed him as company Vice-President and head of the new division.

Palevsky began in his new role by hiring a number of former colleagues from the Bendix Computer Division, including G-15 logic designer Robert Beck. He then secured some development funding from the US Army rocket team at Redstone Arsenal in Alabama and this allowed the development of the parallel analyser to proceed. The machine, which was named TRICE (Transistorised Real-Time Incremental Computer Expandable), was completed in late 1958. TRICE was capable of solving differential equations at the phenomenal speed of 100,000 iterations per second using a parallel architecture and solid-state circuitry that operated at 3 MHz. Modular construction allowed up to 108 computing modules, such as integrators, multipliers and summation circuits, to be interconnected. After filing a US patent application in January 1959 which covered the new technology embodied in the machine, a production version was introduced by Packard Bell which generated a handful of commercial sales.

The TRICE project had also provided the technology behind Packard Bell Computer's first commercial product, a high-speed analogue-to-digital converter which was based on circuits designed for the analyser by Palevsky and Beck with colleague George J Giel. Echoing the experience of DEC with its system modules, sales of this unit provided much needed revenues until the computer division was in a strong enough financial position to take on the development of higher value products.

The PB250

Feedback from customers provided Palevsky with the idea for his next product, an inexpensive general-purpose computer that could be programmed to suit a range of real-time applications. Palevsky wanted LGP-30 designer Stanley Frankel to design the new machine which would be called the PB250 but contractual issues prevented this. Instead, Frankel was employed as a consultant on the project and the machine was designed by Robert Beck under Palevsky's guidance.

The involvement of two of the key figures in the development of the Royal Precision LGP-30 and Bendix G-15, both of which employed recirculating memory architectures, clearly influenced the design of the new machine and the PB250 featured a magnetostrictive delay line memory supplied by

Ferranti in the UK. Production costs were kept low through the choice of a serial architecture with a relatively short 22-bit word length and by having a modest memory capacity of 2,320 words which could be expanded in blocks of 256 up to a maximum of 15,888 words. The clock speed was a respectable 2 MHz which ensured decent performance despite the recirculating memory design. A comprehensive instruction set included support for double precision arithmetic and variable length multiplication and division.

Many of the circuits in the PB250 were originally developed for the TRICE digital differential analyser. The resulting machine used only 400 germanium transistors and consumed a miniscule 110 watts of electrical power. This also ensured a highly compact design which featured a similarly shaped cabinet to that of the Bendix G-15 but small enough to fit on a desk.

The Packard Bell PB250 was introduced in December 1960 at a price of $40,500 for a basic system fitted with 2,320 words of memory and a Flexowriter electric typewriter. Options included various punched tape devices, a magnetic tape unit and a battery backup power supply. The machine was modestly successful, with sales boosted by an agreement with the Bailey Meter Company, an Ohio-based process control firm, to supply the PB250 for the control of boilers in power plants.

The SDS 900 Series

The modest sales of the PB250 hinted at the potential for low-cost computer systems in real-time applications. Palevsky was keen to build on this success by developing a new model with a magnetic core memory. However, Packard Bell had been experiencing financial difficulties with its television business and refused to provide the necessary investment to grow the computer division. Palevsky's plea for financial independence from the parent company also fell on deaf ears. Undaunted, Palevsky decided to set up his own firm and persuaded Beck and four of their colleagues to join him.

Palevsky's first task was to raise finance for the new venture which would be called Scientific Data Systems. He was introduced through a mutual acquaintance to Arthur J Rock of Davis & Rock, a newly established venture capital firm based in San Francisco. Rock agreed to invest $300,000 and also took an active role in the management of the new company, becoming Chairman of the Board of Directors. Palevsky was able to use personal contacts at the University of Chicago to secure a similar investment from members of the wealthy Rosenwald family, heirs to the Sears Roebuck empire. He also invested $60,000 of his own money and the other founders each invested $10,000, bringing the total funding package to approximately $1 million. This impressive figure was a far cry from the paltry $70,000 raised by DEC only four years earlier, indicating that the US investment community was finally starting to warm to the computer business.

9. Small is Beautiful – The Minicomputer

Palevsky set up shop in a rented building in Santa Monica in September 1961 and embarked on the development of the company's first products, a range of low-cost computers for real-time control and general-purpose use. The first two models were developed in parallel, the SDS 910 for on-line process control and the SDS 920 for scientific computing. The design team was led by Jack Mitchell and Emil Borgers, who had both joined SDS from Packard Bell, plus Henry Herold who had been recruited from the General Electric Computer Department. Both machines shared the same 24-bit binary architecture, 1.25 MHz clock speed and magnetic core memory with parity checking but the 920 added an extended instruction set, larger main memory as standard and circuitry for hardware multiplication and division. The component count was kept low by employing a hybrid parallel/serial logic design that combined parallel memory access with serial arithmetic.

A highly flexible input/output system was provided, with two separate channels available that supported several modes of operation, including DMA, and an optional program interrupt mechanism with up to 896 levels. Scientific computing features included an index register and indirect addressing. Other features, such as support for double precision arithmetic and variable length multiplication and division, were carried over from the PB250. Basic compute performance of both machines was 62,500 additions per second. Both models were also capable of floating-point arithmetic but this was implemented in software on the 910 to reduce costs. Another notable feature was the use of silicon transistors rather than germanium to increase speed and improve reliability, an industry first for non-military computers.

The experience of the SDS team was evident in the speed of development and the two new models were introduced together in July 1962, only 10 months after the company had been established, with first deliveries starting the following month. The 910 was priced at $48,000 with 2,048 words of core memory and the 920 at $98,000 with 4,096 words of core memory, undercutting DEC's similarly specified PDP-1 by almost 20%. Palevsky had maintained close links with the US rocket development community and the first customer was NASA which purchased a 910 model for its Goddard Space Flight Center in Greenbelt, Maryland. Having been spurned by DEC, NASA would become a major customer of SDS, buying around 40% of the estimated total of 312 910 and 920 models built.

After successfully launching the small-scale 910 and 920 models, the company initiated development of a third model in the series, the medium-scale SDS 930. The 24-bit architecture and 920 instruction set were retained to ensure software compatibility but the clock speed was raised to 5.7 MHz and other improvements made to the circuitry in order to increase performance. The magnetic core memory was expandable to 32,768 words, twice that of the other models, and input/output capability was also extended to provide additional flexibility.

The SDS 930 was introduced in December 1963 at a basic price of $108,000. It sold in similar numbers to the two smaller machines in the series, with a total of 173 machines built. Combined sales of the 900 series models resulted in SDS overtaking its closest rival DEC by the end of 1964 after only 3 years in business.

Building Blocks – Integrated Circuits

Crossover models such as the Elliott 803, DEC PDP-5 and SDS 910 allowed manufacturers to explore the general-purpose computing sector using inexpensive machines developed primarily for process control and this helped to open up the minicomputer market. Process control computers also influenced the design of later generations of small-scale machines and many of the features necessary for real-time control, such as program interrupt and direct memory access, would become standard features in minicomputers. However, the experiences of manufacturers such as DEC and SDS revealed that the nascent minicomputer market was much more sensitive to price than other sectors. Sales were not yet at a level where economies of scale would push down production costs as the number of units built increased. Fortunately, a new electronics packaging technology had recently become available which would reduce production costs significantly, microminiaturised or 'integrated' circuits.

The driving force for integrated circuits was the connectivity problem, also known as 'the tyranny of numbers'. As computers became more sophisticated and the number of electronic components increased, the amount of electrical wiring necessary to connect these components increased at an even greater rate, leading to serious reliability problems. Production costs also spiralled, as most wiring was carried out by hand. The use of modular circuit packaging technology, such as IBM's Standard Modular System and DEC's system modules, helped to alleviate the problem by eliminating the wiring of individual components through the use of printed circuit boards but the modules themselves were expensive to manufacture and many wires were still required to interconnect them.

Integrated circuits were first proposed in 1952 by the British electronics engineer Geoffrey W A Dummer. Dummer envisaged an integrated device in which a number of miniaturised electronic components were fabricated from a single block of semiconductor material that had been treated to produce several layers with different electrical properties. The components would be connected together to form a complete circuit by cutting out the layer of insulating material between them. He presented his ideas at a symposium on electronic components held in Washington DC in May 1952. However, Dummer's work at the UK's Royal Radar Establishment was concerned with

9. Small is Beautiful – The Minicomputer

component reliability and did not involve the fabrication of devices, which made it very difficult for him to reduce his concept to practice. Nevertheless, by September 1957, with the help of scientists at Plessey's Caswell Research Centre, he succeeded in producing a non-functioning model of a flip-flop integrated circuit using a solid block of silicon. Unfortunately, he was unable to convince his short-sighted Ministry of Supply bosses to support further development and the initiative would shift to the USA.

Over the following 18 months, Dummer's monolithic integrated circuit concept would be realised by two Americans working independently at different US electronics firms, electrical engineer Jack S Kilby at Texas Instruments (TI) and physicist Robert N Noyce at Fairchild Semiconductor. Kilby had worked for a number of years at the Centralab Division of Globe-Union Incorporated, an electronic component manufacturer and pioneer of printed circuit board technology, before moving to TI in May 1958. His experiences at Centralab, which involved the development of techniques to build passive electronic components directly into circuit boards, would provide the inspiration for his later work. Kilby is also known to have attended Dummer's 1952 presentation but it is not clear how much of an influence this was on him. Upon arrival at TI, Kilby was assigned to the Micro-Module programme, a major initiative led by RCA to develop electronic components that could be snapped together for the US Army Signal Corps. A few weeks later, he found himself in an empty laboratory, his colleagues having left for their two-week summer holiday which Kilby, as a new employee, was not yet eligible for. Unimpressed with the Micro-Module concept, Kilby used the time to develop an alternative approach, sketching out a method for fabricating different types of electronic component on a single wafer of silicon. He showed this to his boss, Willis Adcock, following his return from annual leave. Adcock could see the potential but was sceptical about the suitability of a semiconductor material for making resistors and capacitors so he asked Kilby to build a proof of concept demonstrator.

By September 1958, Kilby had succeeded in building a working prototype of a phase-shift oscillator circuit with 5 electronic components on a single bar of germanium. It was not a fully integrated device, as the components were connected together with tiny gold wires, but it proved that different electronic components could indeed be fabricated from a single block of semiconductor material. After demonstrating the prototype to senior TI executives, Kilby was given permission to take it further and a small team was assembled to assist him. The US Army Signal Corps remained committed to the Micro-Module programme, however, the US Air Force showed an interest and provided funding for further development. A US patent application was filed in February 1959 and Kilby's invention was publicly announced at a trade show the following month.

In January 1959, Robert Noyce was pondering what to put into the patent application for a new type of high-frequency transistor fabricated using a planar construction process developed by colleague Jean A Hoerni. Noyce had worked for transistor co-inventor William Shockley at Shockley Semiconductor Laboratory before leaving with Hoerni and 6 other colleagues to establish Fairchild Semiconductor in 1957. His role as Director of Research & Development in the new company involved identifying solutions to the problems of manufacturing electronic components in large numbers.

During discussions with the patent attorney, Noyce realised that Hoerni's planar construction process also lent itself to the creation of interconnections between components that had been fabricated from a single block of silicon. The components could be electrically isolated from each other within the semiconducting layer using a method developed by Kurt Lehovec of the Sprague Electric Company in which an insulating ring is created around each component using a reverse-biased p-n junction. Then the interconnections between components required for a specific circuit could be created by depositing a metal layer on top of the semiconducting layer and creating the geometric pattern for the circuit in this metal layer using a similar photolithography process to that used for the manufacture of printed circuit boards.

Unlike Kilby's invention, this was a complete realisation of the monolithic integrated circuit concept using a process which was much more suited to large volume manufacturing. Noyce filed his US patent application in July 1959, 5 months after Kilby, but it was the first to be granted. A patent interference action awarded priority to Kilby but this was later overturned following an appeal. By then the two companies had already settled on a cross-licensing agreement that included a net payment to Fairchild.

Integrated circuits became commercially available in March 1960 with the introduction by TI of the SN502 flip-flop at a price of $450. Within two years the price had fallen to below $100. The era of the 'silicon chip' had arrived.

Kilby was awarded the Nobel Prize in Physics in 2000 "for his part in the invention of the integrated circuit". Sadly, Noyce had died in 1990 and an award cannot be made posthumously. The citation made no mention of Geoffrey Dummer.

A Commercial Breakthrough

The challenge for the designers of early minicomputers had been to minimise the number of electronic components in order to lower production costs while keeping the specification intact. With the introduction of integrated circuits, computer designers now had the ability to dramatically reduce the component count without compromising machine performance or omitting any key features. By adopting integrated circuits, minicomputers could be both inexpensive and powerful.

9. Small is Beautiful – The Minicomputer

The two start-up companies that had helped to create the minicomputer industry, Massachusetts-based Digital Equipment Corporation, led by Ken Olsen and Georges Doriot, and California-based Scientific Data Systems, led by Max Palevsky and Arthur Rock, would continue to dominate it. However, DEC was uncharacteristically slow in incorporating the new integrated circuit technology and this allowed SDS to take an early lead.

The First Integrated Circuit Model

The introduction of the DEC PDP-5 in August 1963 at a price which undercut the SDS 910 by over 40% prompted the SDS marketing department to press for the development of a new entry level model that could compete with the PDP-5 at the low-cost end of the process control market. SDS engineers realised that integrated circuits could help to achieve the necessary cost savings and worked with the Signetics Corporation, a specialist electronics firm established in 1961 by a number of ex-Fairchild employees, to develop a suitable flip-flop integrated circuit. This device was then incorporated into the design for a new 12-bit model which would be called the SDS 92.

The SDS 92 incorporated a number of novel features, including two independent 12-bit accumulators, one of which could be used as an index register, and the ability to directly address up to 32,768 words of memory using a two-word instruction mode that permitted 15-bit addressing. The use of fast access 1.75 microsecond magnetic core memory guaranteed solid compute performance and a parallel transfer input/output mode provided a data transfer rate of up to 572,000 words per second.

Introduced in August 1964 at a price of $29,000 for a machine fitted with 2,048 words of memory, the SDS 92 was the first commercial computer of any size to incorporate monolithic integrated circuits. However, despite having superior performance to the DEC PDP-5 at a similar price, only 47 were sold. The likely reason for this is DEC's announcement only a few months later of a successor to the PDP-5 that would rewrite the rule book in terms of price-performance.

The DEC PDP-8

Sales of the PDP-5 hinted at what might be achieved by offering a minimal design model with good performance at a low price for use in task-specific environments such as process control and laboratory instrumentation applications. This prompted DEC to develop a successor to the PDP-5 that would have significantly higher performance at a similar or lower price. Integrated circuits were now available to help to realise this goal, however, DEC engineers considered them too new and expensive to implement. They decided instead to build the machine using the company's latest 'Flip Chip' modules which featured circuit boards populated with discrete components at a much higher packaging density than previous designs.

The new model, which was designated the PDP-8, was specified by Gordon Bell and designed by a team led by PDP-5 logic designer Edson de Castro. The binary parallel architecture, 12-bit word length, and minimal arithmetic circuitry of the PDP-5 were retained but performance was increased by using faster 1.6 microsecond magnetic core memory and moving the program counter from main memory to a separate register. An additional boost was provided by the Flip Chip modules themselves which featured silicon transistors and operated at twice the speed of the 500 KHz system modules used previously. The result was a machine with a compute performance of 312,500 additions per second, almost 6 times that of the PDP-5. Software compatibility with the earlier model would ensure a painless upgrade path for PDP-5 customers.

The PDP-8 was also designed to facilitate the use of automatic wire wrapping machinery for wiring the module mounting blocks. This eliminated the need for hand wiring during assembly, thereby reducing production costs significantly. The higher component density of the Flip Chip modules allowed the entire machine to be fitted into a half-sized cabinet that was small enough to be placed on a table top or easily integrated with other equipment using a 19-inch rack mount option.

DEC PDP-8

The PDP-8 was introduced in March 1965 at the impressively low price of $18,000 for a machine supplied with 4,096 words of memory and an ASR-33 teletypewriter. Sales were equally impressive, with a total of 1,450 sold over

9. Small is Beautiful – The Minicomputer

the 3 years that it was in production. The PDP-8 also provided a platform for an extended family of derivative models which included the PDP-8/S, a low-cost serial version that sold for under $10,000, and the PDP-8/E, a high-performance model introduced in 1970. DEC engineers had been slow in adopting integrated circuit technology due to their reliance on standardised modules in which the advantages of integrated circuits were much less pertinent. Following the introduction of a new series of Flip Chip modules with integrated circuit components, an integrated circuit version, the PDP-8/I, was introduced in December 1967. Each derivative boosted sales and extended the lifespan of the PDP-8 architecture, resulting in total combined sales of around 50,000 machines over a 15 year period.

Increasing Competition

The DEC PDP-8 was the breakthrough model for the minicomputer industry. Soon other firms began eying up this potentially lucrative market. IBM for once was ahead of the pack, pre-empting the introduction of the PDP-8 by a month with the announcement of the IBM 1130.

The IBM 1130 featured the now familiar binary parallel architecture but with a slightly unusual 16-bit word length and up to 8,192 words of 3.6 microsecond magnetic core memory. Other familiar features included indirect addressing and a multi-level program interrupt capability. The 1130 was packaged as a desk-sized machine in the style of the earlier IBM 1620 model but the logical design had much more in common with DEC's PDP-4, having originally been conceived as an 18-bit machine. The word length was changed to 16 bits during development to make it a better fit with the 8-bit character size and 32-bit word length of the recently introduced System/360 range. SLT circuit packaging technology, which had been developed for the System/360 as an alternative to integrated circuits, was also adopted.

The IBM 1130 was introduced in February 1965 at a price of $32,280 for a basic model fitted with 4,096 words of memory. Compute performance, at 120,000 additions per second, comfortably exceeded contemporaneous low-end models from DEC and SDS. However, the 1130 was quickly eclipsed by the arrival of the PDP-8 which offered more than double the performance at a much lower price. Later versions introduced in April 1967 were fitted with faster access 2.2 microsecond memory which increased performance by around 40% but still fell short of the market leading PDP-8. Nevertheless, the 1130 proved reasonably popular with IBM's loyal customer base, encouraged by free access to a large range of software programs developed by IBM for scientific and engineering applications.

IBM was by no means the only established computer manufacturer to embrace the burgeoning minicomputer market. Honeywell, which had enjoyed some success with its H-200 business model and a series of dedicated process control computers, increased its presence in the marketplace by acquiring the Computer Control Company (3C) in 1966.

Originally a military contractor, 3C moved into the computer business in June 1963 with the introduction of the DDP-24, a medium-scale model that became popular with the US aerospace industry. The machine was built using standard digital logic modules which 3C sold commercially in competition with similar products from DEC. In October 1964, 3C introduced a small-scale model, the DDP-116, a 16-bit binary parallel machine with minimal arithmetic and a magnetic core memory which sold for $20,000. This was reasonably successful, with more than 100 machines sold.

The subsequent adoption of integrated circuit technology for the company's logic modules led to the introduction of two integrated circuit models in 1966, the high-performance DDP-516 and a low-cost version, the DDP-416. Following Honeywell's acquisition of 3C, these two machines became the basis for the Honeywell Series 16 family. With Honeywell's extensive sales and marketing organisation behind them, several thousand Series 16 machines were sold and by 1970 the company's 3C Division was generating annual revenues of $100 million.

Data General

Despite the huge resources available to IBM and Honeywell, the most serious threat to DEC's market dominance would come from a start-up firm that had been created by former DEC employees as a result of the company's refusal to implement a new design based on an 8-bit character size.

The introduction of the IBM System/360 range in April 1964 not only demonstrated the benefits of software compatibility across a model range, it also introduced the 8-bit character size. The increase from 6 to 8 bits allowed a larger alphanumerical character set to be represented and would become standard throughout the industry, prompting an industry trend for word lengths that were multiples of 8 bits. In the minicomputer sector, this manifested itself as the adoption of a standard 16-bit word length which began with the DDP-116 in October 1964 and gained momentum with the introduction of the IBM 1130 in 1965 and Hewlett-Packard 2116A in 1966. SDS also embraced this trend with the introduction of the Sigma series in December 1966 but, with products that straddled the small and medium sectors, the company decided to include both a small-scale model with a 16-bit word length and a medium-scale 32-bit model.

In early 1967 DEC initiated a project codenamed PDP-X to create a new machine design that would embody the latest advances in computer architecture and provide a scalable platform for a wider range of models than previous designs. The project team was led by PDP-8 designer Edson de Castro and fellow

9. Small is Beautiful – The Minicomputer

hardware designer Larry Seligman. One of the first design decisions was to adopt an 8-bit character size.

DEC's policy of using predominantly standard circuit modules for the construction of its computers made for a more straightforward design process but resulted in additional complexity and expense in the hardware realisation of the design, with more wiring and interconnections required than would otherwise have been necessary. The PDP-X team proposed to reduce the number of modules required in future machines by designing much larger custom modules that combined four Flip Chip modules onto a single circuit board. These modules could then be used as the basis of a new family of compatible computers with 8, 16 and 32-bit word lengths that would replace DEC's entire product line.

Unsurprisingly, the PDP-X team's radical "bet your company" proposal to replace every computer in the range with a new model which was incompatible with the machine that it replaced met with fierce resistance from several areas within the organisation. In an effort to get their proposal accepted, the PDP-X project team approached Gordon Bell, who had left the company in 1966 for an academic post at Carnegie Mellon University, and asked him to endorse it, which he did. However, DEC management remained unconvinced and the proposal was rejected in the spring of 1968, prompting de Castro and two colleagues, Henry Burkhardt III and Richard G Sogge, to leave DEC and set up their own company.

Data General Corporation was founded in April 1968 by de Castro, Burkhardt and Sogge plus Herbert J Richman, a sales manager from Fairchild Semiconductor. With the help of lawyer Frederick R Adler, they raised $800,000 in venture capital and set up business in a former beauty parlour in Hudson, Massachusetts. Their plan was to develop an entry level machine that could be designed and brought to market quickly. While embodying some of the concepts proposed for PDP-X, this would have to be a much less ambitious project befitting a start-up firm with limited resources.

The hardware design for the new machine, which was named Nova, was carried out by de Castro and Sogge and the software by Burkhardt. It featured a binary parallel architecture with a 16-bit word length and 4 accumulators, 2 of which could be used as index registers. The original intention was to use an 8-bit word length to keep costs down but Burkhardt realised that the internal structure of the machine could be made to operate on a smaller word length than was presented to the programmer, so a 16-bit word length was chosen which was implemented internally using 4-bit components. A 1.5 MHz clock speed and 1.6 microsecond magnetic core memory helped to keep performance respectable at 178,500 additions per second despite the compromises that this approach entailed.

The Nova design team put considerable thought into how the machine would be manufactured and packaged. Liberal use of the latest medium-scale integrated (MSI) circuits facilitated a compact design that would be inexpensive to produce. The entire processor was contained on two 15-inch square circuit boards which plugged into a system bus using a 200-pin connector on the rear edge of each board. The system and input/output buses were also implemented using a printed circuit board backplane which eliminated most of the wiring and allowed the machine to be fitted into a 19-inch rack cabinet only 5¼ inches high. Up to 5 additional circuit boards could be accommodated within the frame, allowing extra memory or optional hardware to be added without the need for any cabling or wiring modifications.

The Data General Nova was introduced at the Fall Joint Computer Conference in San Francisco in December 1968. It was priced at the astonishingly low figure of $3,950 for a basic specification machine without memory or $7,950 with 4,096 words of magnetic core memory and an ASR-33 teletypewriter. Memory was expandable up to 20,480 words internally or 32,768 words using a separate expansion cabinet. For high reliability applications such as real-time control, the machine could also be fitted with a type of read-only memory which was programmed by the company using punched tapes supplied by the customer.

Data General Nova 1200

The Nova's low price and innovative design, aided by an audacious marketing campaign that billed it as "the best small computer in the world", made an immediate impact on the minicomputer market and more than 200 machines were sold within the first year. Buoyed with success, Data General went public in November 1969 after only 18 months in business, raising around $5 million. The injection of new capital allowed the company to invest heavily in product development. In April 1970 the company introduced a high-performance model, the aptly named Supernova, which replaced the Nova's 4-bit internal structure with a full 16-bit parallel design that operated in combination with ultrafast 800 nanosecond memory to boost performance to around 7 times that of the original machine. Later that same year Data General also introduced two further models, the Nova 800 and 1200, which replaced the original Nova with an improved design that featured a single-board processor and a choice of 800 or 1200 nanosecond memory. These variants boosted Nova sales up to an eventual total of 50,000, equalling DEC's impressive sales total for the entire PDP-8 family.

The DEC PDP-11

The announcement of the Data General Nova in December 1968 was a wake up call for DEC. Here was a competing product that took DEC's very own philosophy of providing high performance at low cost through the use of a minimal arithmetic design and the latest manufacturing techniques and used it to undercut DEC's lowest-priced PDP-8 model by a significant margin. It was also a 16-bit machine, providing compatibility with the emerging industry standard 8-bit character size and possessing all the other advantages of a larger word length.

DEC responded by initiating a crash programme in early 1969 to develop a new model that could compete directly with the Data General Nova. The idea of a family of compatible computers was carried over from the PDP-X project but scaled back to a 16-bit word length only, with no initial plans to replace DEC's established 12-bit or 36-bit model ranges. The project team toiled for several months to produce a functional design for the new machine which they then presented to Gordon Bell and professorial colleague William A Wulf at Carnegie Mellon University. One of Bell's most promising students, Harold McFarland, was invited to participate in the design review. McFarland had been working on some computer designs of his own, taking inspiration from a textbook that Bell had written with colleague Allen Newell on computer architectures, and was able to suggest some major improvements to the DEC design. These were so impressive that McFarland was hired by DEC to work with the project team and the design substantially altered to incorporate his ideas.

The new model was designated the PDP-11. In addition to a 16-bit word length, it featured a general register set architecture of the type pioneered in the Ferranti Pegasus. This was implemented using 8 general-purpose registers, each of which could function as an accumulator, index register or address pointer. Another innovative feature was the use of a single universal bus to connect both system and input/output components. Named UNIBUS, this was a bidirectional, asynchronous, 56-wire bus that supported up to 20 devices and allowed mapping of input/output devices into main memory so that they could be directly addressed without the need for separate input/output instructions, thereby allowing data transfer operations to bypass the processor. Also, because the processor connected with memory through the asynchronous bus, it was not tied to memory of a particular speed and the choice of memory was only limited by the bandwidth capability of the bus itself.

The engineering prototype of the PDP-11 was built using Flip Chip modules but these were gradually replaced with custom circuit boards to create a machine that fitted into a standard 19-inch rack cabinet 10½ inches high. The circuit boards incorporated SSI circuit technology using TTL integrated circuits.

The first model in the PDP-11 family, the PDP-11/20, was introduced in March 1970 at a price of $10,800 for a basic specification machine with 4,096 words of memory. A 3.5 MHz clock speed and 500 nanosecond magnetic core memory delivered a compute performance of 435,000 additions per second, well over twice that of the Data General Nova which now looked a much less attractive purchase against the PDP-11's more sophisticated architecture and richer instruction set. DEC's uncharacteristically vigorous marketing campaign for the new model also highlighted the advantages of UNIBUS. The new model was an instant success, with 150 orders received within the first week and the PDP-11 family went on to generate total sales of more than 170,000 units over a 25-year lifespan. It also created a market in plug compatible products from third party suppliers offering memory and input/output products that could plug into UNIBUS.

Winners and Losers

The spectacular sales of the PDP-11 propelled DEC into the big league and within a few years it had overtaken every other firm in the BUNCH to become the second largest computer manufacturer in the world. Although primarily a minicomputer manufacturer, it also competed successfully in the medium and large scale sectors with a line of 36-bit models that began with the introduction of the PDP-6 in 1964. However, DEC's oldest rival fared less well. Following Robert Beck's retirement in 1967, Max Palevsky decided to sell SDS. In a deal brokered by Arthur Rock, the company was sold to the giant US photocopier firm Xerox Corporation in May 1969 for $920 million in Xerox stock and renamed Xerox Data Systems (XDS). This was a spectacularly high valuation, given the company's modest 1% market share and $10 million annual profits, and one which probably said more about Rock's negotiating skills and Palevsky's salesmanship than Xerox's financial acumen. It may also have been a sign of Xerox's increasing desperation, having already been spurned by several other firms in its efforts to acquire a computer manufacturer as part of a new diversification strategy.

Palevsky remained a director for 3 years following the takeover but was no longer involved in the day-to-day running of the company. Without strong leadership, and with its largest customer NASA facing savage budget cutbacks in the wake of the moon landings, XDS failed to maintain its share of a rapidly expanding market. Efforts to target the business data processing sector instead failed and the company limped on for a few years until the decision was taken in July 1975 to close it down. In order for XDS to survive, Xerox had estimated that it would have had to invest $150-200 million to develop a new model range. However, with the US still recovering from the effects of a deep economic recession, such a large investment would have been too risky. Without it, the company's fate was sealed.

During this period, the minicomputer market grew to become the dominant market sector for the computer industry. New companies were attracted by the relatively low market entry costs and by 1970 there were around 50 separate firms manufacturing over 100 different models of minicomputer. However, profit margins were also low in this highly price-sensitive sector, prompting some minicomputer manufacturers to expand their product ranges upwards into the medium and large scale sectors where higher profits could be found. Minicomputer manufacturers would later adopt a 32-bit word length as standard, blurring the lines between minicomputers and medium-scale models with a series of higher performance machines that would become known as superminicomputers.

Further Reading

Bashe, C. J., Johnson, L. R., Palmer, J. H. and Pugh, E. W., IBM's Early Computers, MIT Press, Cambridge, Massachusetts, 1986.

Bell, G., Cady, R., McFarland, H., Delagi, B., O'Laughlin, J., Noonan, R. and Wulf, W., A New Architecture for Mini-Computers - The DEC PDP-11, Proceedings of the Spring Joint Computer Conference, 1970, 657-675.

Bell, C. G., Mudge, J. C. and McNamara, J. E., Computer Engineering: A DEC View of Hardware Systems Design, Digital Press, Bedford, Massachusetts, 1978.

Cook, R. L., Time-Sharing on the National-Elliott 802, Computer Journal 2 (4), 1960, 185-188.

Dummer, G. W. A., Electronic Components in Great Britain, Proceedings of the Symposium on Progress in Quality Electronic Components, 1952, 15-20.

Frankel, S. P., The Logical Design of a Simple General Purpose Computer, IRE Transactions on Electronic Computers EC-6 (1), 1957, 5-14.

Huskey, H. D., From ACE to the G-15, IEEE Annals of the History of Computing 6 (4), 1984, 350-371.

Kilby, J. S., Invention of the Integrated Circuit, IEEE Transactions on Electron Devices ED-23 (7), 1976, 648-654.

Koudela, J., The Past, Present, and Future of Minicomputers: A Scenario, Proceedings of the IEEE 61 (11), 1973, 1526-1534.

Leclerc, B., From Gamma 2 to Gamma E.T.: The Birth of Electronic Computing at Bull, IEEE Annals of the History of Computing 12 (1), 1990, 5-22.

Lentz, J. J., A New Approach to Small-Computer Programming and Control, IBM Journal of Research and Development 2 (1), 1958, 72-83.

Orden, A., Application of the Burroughs E101 Computer, Proceedings of the Eastern Computer Conference, 1954, 50-54.

Pearson, J. P., Digital at Work: Snapshots from the First Thirty Five Years, Digital Press, Burlington, Massachusetts, 1992.

Reid, T. R., The Chip: How Two Americans Invented the Microchip and Launched a Revolution, Random House, New York, 2001.

Stout, T. M. and Williams, T. J., Pioneering Work in the Field of Computer Process Control, IEEE Annals of the History of Computing 17 (1), 1995, 6-18.

PART FOUR
PERSONALISATION

10. The Big Picture – Computer Graphics

> *"A display connected to a digital computer gives us a chance to gain familiarity with concepts not realizable in the physical world. It is a looking glass into a mathematical wonderland."*, Ivan E Sutherland, 1965.

The Origins of Computer Graphics

A key attribute of today's computers is their ability to display and manipulate information in graphical form. We think of computer graphics as originating with the early video games and computer-generated imagery of the 1980s and 1990s, however, its development goes back to the very beginning of electronic computation. In a November 1945 memorandum outlining the purposes of what became the IAS electronic computer project, John von Neumann suggested that the output of the machine should be pictorial, or graphical, rather than numerical and displayed on an oscilloscope or photographed for permanent storage. As a mathematician, von Neumann's interest stemmed from the requirement to produce graphs of mathematical functions, an ability that many traditional analogue computing instruments possessed.

Following completion in 1952 the IAS machine was equipped with a graphical display device but it would not be the first. The earliest developments in computer graphics were driven by a different, though equally familiar, set of needs, those of the US military.

Project Whirlwind

Unlike other aspects of the history of the computer where advances were made as a result of contributions from different groups, all the earliest developments in computer graphics have a single source, the highly influential Whirlwind project at the Massachusetts Institute of Technology.

Project Whirlwind began in December 1944 as an analogue flight trainer/analyser development project for the US Navy and was originally codenamed ASCA (Airplane Stability and Control Analyzer). The huge increase in demand for trained flight crew during World War II had prompted major activity in both the US and UK to develop simulators that could mimic the performance of aircraft in order to reduce the time and cost of training. In the USA, Bell Telephone Laboratories had successfully developed an operational flight trainer for the US Navy's iconic PBM-3 flying boat. This featured an electronic analogue computer which operated in combination with electromechanical servo mechanisms through a mock-up of the aircraft's front fuselage and cockpit to provide a realistic experience for the trainee. The intention for the ASCA project was to employ similar technology to the Bell Labs trainer but to create a universal system that could simulate various types of aircraft.

The ASCA project was led by Jay Forrester, an assistant director of MIT's Servomechanisms Laboratory who would later make a name for himself as one of the pioneers of magnetic core storage. He was assisted by Robert R Everett, a young electrical engineer who had joined the staff of the Laboratory in 1943 after obtaining his Master's degree at MIT. As the preliminary design work progressed, Forrester began to encounter various problems that highlighted the limitations of analogue calculating machinery. He discussed these problems with MIT colleagues in September 1945, one of whom was Perry Crawford, whose 1942 Master's thesis introduced the magnetic drum memory concept. Crawford had kept abreast of recent advances in electronic digital computation and suggested that digital electronics might provide a solution.

The following month Crawford and Forrester attended an MIT-hosted conference on advanced computation techniques which included a presentation by John Brainerd and Presper Eckert on progress in electronic computation at the Moore School of Electrical Engineering. Intrigued by the presentation, Forrester then arranged a visit to the Moore School in order to see the ENIAC and EDVAC developments for himself. Further investigation convinced him that a move to digital technology would provide the flexibility needed for the flight trainer/analyser and that the resulting electronic digital computer could also provide a valuable resource for other projects, particularly if a stored-program design was employed. Forrester proposed these changes to the US Navy Bureau of Aeronautics and in March 1946 the ASCA contract was modified accordingly, along with a corresponding increase in its value to $2.4 million and an extended completion deadline of June 1950. The project was also given a new name, Whirlwind, in acknowledgement of the substantial changes to the specification.

The Whirlwind computer was operational by early 1949, although it was not fully completed until late the following year. The machine was designed primarily for speed and featured a binary parallel architecture with a relatively short 16-bit word length and a 1 MHz clock speed. Like many of the early high-performance computer development projects, Whirlwind was initially fitted with an electrostatic storage tube memory, using 32 CRTs to provide 256 words of main memory. It was this use of CRTs for storage which led to the earliest experiments in graphical display.

The CRTs in Whirlwind were of a different design to the standard television CRTs used in Williams Tubes, having been developed by the MIT Radiation

10. The Big Picture – Computer Graphics

Laboratory for radar applications. They contained a secondary set of lower voltage electron guns known as holding guns which flooded the entire screen with electrons in order to boost the spot pattern written by the main gun on the screen. When used in combination with a long persistence phosphor coating, the spot pattern was prevented from decaying and could be maintained for as long as necessary. This removed the requirement for a refresh cycle but increased the access time of the device to around 25 microseconds. Reliability also remained an issue.

The operation of individual storage tubes could be monitored by connecting an oscilloscope to the deflection circuits of a tube so that the spot pattern was reproduced on the oscilloscope screen. To test Whirlwind's electrostatic memory, the machine was programmed to generate a series of test patterns. These required the operator to manually inspect the oscilloscope screen to verify that a spot was present at the correct position but with 256 individual spot positions available, it was very difficult for the operator to identify a specific position. Everett solved this problem in early 1949 by inventing a pointing device that allowed the address of a particular spot to be identified. The device, which Everett called a light gun, was equipped with a photocell on the end that sensed the light given off by a spot when positioned over it. Spots were identified by comparing the timing of the selected spot as it appeared on the screen with the scan timing for the generation of the spot pattern. A specific spot would be correctly identified where the timing coincided.

The use of the light gun also enabled the operator to interact with the machine by selecting a spot which the computer could then be instructed to erase. This additional functionality was picked up by the laboratory's public relations staff who asked the Whirlwind engineers to create the letters MIT on the oscilloscope screen by filling the screen with spots and erasing selected regions. The resulting image, which was used for publicity purposes, was the first ever example of interactive computer graphics.

The Whirlwind team soon realised that they could also create simple graphs by using one of the storage tubes as an output device. The computer would be programmed to switch on the appropriate spots in the 16 by 16 array and the resulting image viewed on the oscilloscope screen. This technique of creating images using a two-dimensional array of spots on a screen later became known as raster graphics.

The display resolution was increased to a much more useful 1,024 spots following the replacement of the storage tubes with higher density CRTs as part of an ongoing effort to increase the size of Whirlwind's main memory. However, as interest in this new form of output began to grow, the MIT engineers realised that it would make more sense to have a dedicated display device. They developed a system that employed two high-speed 8-bit registers which were implemented using flip-flops and coupled to the horizontal and vertical inputs of an oscilloscope by means of digital-to-analogue converter circuits. Numbers placed in these registers were converted into voltages which were proportional to the value of the number. By operating the oscilloscope in x-y mode, the oscilloscope beam could thus be moved to any one of 16,384 discrete positions on the screen. Further circuitry provided the control pulses required to switch the oscilloscope beam on and off as necessary in order to build an image from multiple data. This allowed solid lines to be drawn from one position to another in a technique now known as vector graphics. With the beam switched on, a line could be drawn from one position to another and so on simply by changing the values of the numbers held in the high-speed registers. To begin a new line, the beam would be switched off, moved to a new start position and then switched back on. In this way complex line drawings could be built up.

MIT Whirlwind I (Photo © Massachusetts Institute of Technology)

The dedicated display equipment was completed towards the end of 1949. One of the earliest uses of the equipment was by MIT mathematicians who successfully used Whirlwind to perform the calculations for a series of mathematical equations and display their solution graphically by plotting the results in the form of dots on the oscilloscope screen. Another notable early use was made by Charles W Adams and John T (Jack) Gilmore, two of the eight programmers on the Whirlwind team. Adams and Gilmore created a program that used three differential equations to calculate the trajectory of a bouncing ball and display it on the oscilloscope screen. By adjusting the bounce frequency using a control knob, the ball could be made to pass through a gap as if it had gone down a hole in the floor. This is the earliest known example of an interactive computer game.

10. The Big Picture – Computer Graphics

Whirlwind's Legacy

Whirlwind is often cited in relation to the development of magnetic core memory or for its influence on the design of DEC minicomputers but it is the project's achievements in computer graphics that have had the most lasting impact. Summary progress reports were compiled quarterly and widely distributed throughout the US defence research community. Through these reports other computer development groups soon picked up on the idea of constructing graphical display equipment using electrostatic storage tubes or digital-to-analogue converters. Both methods utilised the CRT screen of an oscilloscope as the display device. Because CRTs were already commonly used as the display device in television and radar, extending their use in this way seemed perfectly natural.

Towards the end of 1950, a CRT display was added to the Standards Eastern Automatic Computer at the NBS Electronic Computers Laboratory in Washington, D.C. The equipment was used to simulate a radar display for aircraft guidance and control and included a joystick input device which allowed the operator to guide an aircraft to intercept an enemy plane.

The IAS electronic computer, which was completed in January 1952, was equipped with a slave CRT for monitoring the status of the 40 Williams Tubes that made up the machine's electrostatic storage tube memory. The IAS project team followed Whirlwind's example by using one of the Williams Tubes as an output device and viewing the resulting images on the slave CRT in order to create a simple raster graphics display with a 32 by 32 resolution, thereby fulfilling von Neumann's wish in his 1945 memorandum.

CRT displays were also fitted to the University of Illinois ILLIAC, which was completed in September 1952, and the NBS DYSEAC, an improved version of SEAC completed in April 1954. The ILLIAC display was of the raster type with a 256 by 256 resolution and capable of operating at a speed of 1,000 spots per second. DYSEAC featured a vector graphics display which could reportedly operate at 2,000 words per second.

Although the Whirlwind computer was completed by 1950, the project would continue in a different guise and with a new funder at the helm, the US Air Force. This would provide the first major application of computer graphics technology.

Project SAGE

In March 1950, a meeting was held at MIT with representatives of the Office of Naval Research to discuss the future of Project Whirlwind. The Whirlwind computer was now almost fully operational and ready to be put into service. However, the original aim of creating a universal flight trainer/analyser had been shelved in June 1948 in order to concentrate fully on the development of the computer itself and a new purpose had yet to be identified.

One of the attendees at the meeting was George E Valley, a professor in MIT's Physics Department. Valley was also chairman of an influential US Air Force committee which had been set up in 1949 to investigate US air defence capability following an increase in the perceived threat from Communist Bloc countries in the wake of the Berlin Blockade and Soviet nuclear bomb testing. He proposed that Whirlwind be applied to experiments in air defence. This was agreed and an interim study was initiated in February 1951 to examine the feasibility of a computer-based system for air defence command and control. Using data from an old radar system, experiments were conducted which showed that Whirlwind was able to track enemy aircraft and compute aircraft interception commands. The success of this study led to the creation of a new facility for air defence research, the MIT Lincoln Laboratory, and a new project, Project SAGE (Semi-Automatic Ground Environment).

The aim of Project SAGE was to develop a data processing system for air defence capable of gathering data from an extensive network of radar installations, processing it and using it to generate weapons guidance commands. A key feature would be the use of interactive graphics displays to present Air Force observers with a visual representation of the position of enemy aircraft in each geographical sector and allow them to select targets for interception. This ambitious aim necessitated the development of a computer with the highest possible inherent reliability in order to minimise downtime and permit round-the-clock operation. It would be based on Whirlwind but supplied by an established computer manufacturer due to the number of machines required and the need for continued maintenance of the equipment throughout the operational life of the system.

IBM was selected as prime contractor for the SAGE computer equipment in October 1952 from a shortlist of potential suppliers which included Raytheon and Remington Rand's ERA and UNIVAC divisions. The new computer was originally named Whirlwind II but later given the US military designation AN/FSQ-7. Development was a joint effort between the Whirlwind project team, now part of the MIT Lincoln Laboratory, and around 300 staff from IBM's Poughkeepsie Laboratory led by project manager John M Coombs. The design retained the binary parallel architecture of the Whirlwind prototype but the word length was doubled to 32 bits, the clock speed also doubled to 2 MHz and main memory increased to 8,192 words. Auxiliary storage was provided by 12 magnetic drum units of 12,288 words capacity each. Each AN/FSQ-7 computer was capable of supporting more than 100 display consoles.

Reliability was improved over Whirlwind by abandoning the electrostatic memory in favour of Forrester's new coincident-current magnetic core storage technology but a fully solid-state design was not yet possible as high-frequency

10. The Big Picture – Computer Graphics

transistors remained unproven so thermionic valves had to be used for much of the circuitry. These were specially designed with an extended cathode life, a common point of failure in standard valves. Other reliability features included duplex arithmetic units, one of which operated in active standby mode, parity error detection on both main memory and input/output buffers plus built-in test circuits that operated under program control to vary the supply voltage within acceptable limits in an effort to induce failure in weak or deteriorating components, a technique known as marginal checking.

Development of the SAGE display console was led by a group from IBM's Endicott Laboratories who had been chosen as a result of their experience in developing a CRT display for the AN/ASQ-38 bombing and navigation system used in the B-52 bomber. The console incorporated two display screens; a 19-inch 'Charactron' CRT which could display both vector graphics and alphanumeric characters, and a 5-inch 'Typotron' CRT for displaying supplemental alphanumeric data. Screen information was updated every 2 seconds. A light gun was provided for target selection purposes and, in an era where smoking in the workplace was common, the console also featured an integrated cigarette lighter and ashtray for the operator's comfort and convenience during long shifts. Having finalised the design of the display console, Endicott management decided that it was too far removed from their existing manufacturing activity so production was subcontracted to the Hazeltine Corporation.

The Charactron was a special-purpose CRT for the high-speed display of alphanumeric characters which was invented in 1949 by Joseph T McNaney of Convair. It contained an additional set of deflection coils and a masking plate with an 8 by 8 array of holes, each shaped for a different character in the set. Characters were formed by defocusing the electron beam slightly to create a larger spot and deflecting it onto the appropriate area of the masking plate so that the portion of the beam passing through would take the shape of the selected character. The second set of deflection coils located beyond the masking plate then allowed the selected character to be positioned at any point on the CRT screen. An aperture in the centre of the masking plate allowed the Charactron to also function as a conventional CRT by permitting a focused beam to pass through the plate unaltered, whereupon the second set of coils would be used to control the position of the resulting spot on the screen in the normal way. This arrangement allowed vector graphics to be combined with alphanumeric characters on the same screen to create annotated images.

The Typotron was another special-purpose CRT developed by the Hughes Aircraft Company. It was very similar in design to the Charactron but added a holding gun and long persistence phosphor coating so that images could be held on screen for as long as necessary without having to refresh them. Both types of CRT could display data at a rate of up to 20,000 characters per second but both also suffered from problems with the brightness of the image, as most of the energy from the electron beam was lost when it struck the masking plate. To counteract this, the ambient lighting in the rooms in which the SAGE display consoles were installed was carefully controlled and coloured filters were used on both the CRT screens and room lights to improve contrast.

SAGE Display Console (Photo © Royal Canadian Air Force)

The first production AN/FSQ-7 computer was completed in June 1956. Over the next few years, a total of 24 of these giant computers plus 3 AN/FSQ-8 variants were built by IBM for installation in 24 specially constructed Direction Centers and 3 Combat Centers located throughout North America. Each system contained more than 49,000 thermionic valves packaged into 70 cabinets with a combined weight of 225 tonnes and a power consumption of 3 megawatts. Software was equally impressive, comprising more than 100,000 instructions created as a result of a gargantuan development effort involving 800 programmers toiling under the watchful eye of the RAND Corporation.

SAGE was the largest computer system ever built and is probably also the most expensive. The total project cost remains classified but is estimated to have been at least $8 billion. The system became operational in July 1958 and remained in service until 1983, however, it was a white elephant, having been designed to detect and intercept enemy bombers when the threat had since shifted to ICBMs. Fortunately, the funding was not entirely wasted. IBM benefited enormously from SAGE, the project providing a massive boost to the company's revenues from computers during a critical period and giving it a head start over the competition in several new areas of technology, not least of which was magnetic core storage. The interactive technology from SAGE was subsequently adapted by IBM for use in air traffic control systems and the SABRE electronic airline reservation system. The project also contributed to the establishment of a US software industry, both directly through the creation of the System Development Corporation (SDC), an early software company spun out by the RAND Corporation in 1957, and indirectly through the valuable experience gained by the many hundreds of programmers who worked on it.

Early Commercial Display Devices

The move from military to civilian applications of graphics technology came with the introduction of display devices for adding graphical output capability to existing computer systems. An important attribute of these early devices was their ability to record the output for permanent storage. However, as CRTs could only store an image for as long as the device was powered up, engineers initially looked to other technologies to provide graphical output capability.

Pen Plotters

The ability to automatically plot the results from a computation in graphical form by moving a pen across paper is a common feature of many mechanical computing devices, beginning with the chart recorder used by Lord Kelvin in his tide predicting machine of 1873 and reaching its zenith in the twin pen flatbed plotter of the MIT Differential Analyzer. As analogue computation moved from mechanical to electronic technology, analogue computer manufacturers began offering electromechanically driven graph plotters. An early example was the flatbed plotter introduced by Electronic Associates Incorporated (EAI) in 1952 as an accessory for the company's PACE electronic analogue computer. This operated by converting the output voltage from the computer into a proportional movement of the pen along the vertical axis of the graph and moving the horizontal axis at a constant rate. Around the same time the F L Moseley Company of Pasadena, California introduced the Autograf X-Y Recorder, a type of electromechanical flatbed plotter intended for industrial process control applications. This operated in a similar manner to the EAI plotter but featured two inputs, allowing both axes to be controlled externally. From these devices, it was a relatively short step to the introduction of plotters that would accept digital input.

In September 1952, the Logistics Research Company introduced the first plotter designed for use with digital computers, the LOGRINC Digital Graph Plotter. Based in Redondo Beach, California, Logistics Research was founded by Glenn E Hagen, an engineer who had previously worked at the Northrop Aircraft Company on the development of the MADDIDA digital differential analyser. The company's main product was the ALWAC, a small magnetic drum computer, and the LOGRINC plotter was developed as an output device for this machine and as a third-party accessory for digital differential analysers. Equipped with two digital inputs, it was capable of plotting graphs at speeds of up to 20 steps per second in increments of 1/64 inch. A drum mechanism rotated the paper under a pen mounted on a carriage that traversed across the drum. This drum-based design was more compact than a flatbed plotter and permitted continuous plotting on a 12-inch wide roll of paper. The LOGRINC plotter was also used by Logistics Research as the basis for an automatic graph follower, where the pen was replaced by a photoelectric sensor which allowed the plotter to automatically follow the curve of a graph in order to digitise it.

The LOGRINC was followed by the introduction in April 1953 of a digital plotter from another Californian company, the Benson-Lehner Corporation, a Los Angeles based manufacturer of instrumentation cameras and analysers for chart recorders and cine film. The Incremental Plotter was a flatbed design with a capacity of 11 by 17 inches which was capable of plotting graphs at a rate of 10 steps per second in 0.02 inch increments. It could also be converted into an automatic graph follower by fitting a photoelectric line following device in place of the pen. Originally developed as an accessory for Computer Research Corporation's CRC 105 digital differential analyser, data input was via punched tape rather than by a direct connection to the computer. A tape punch was also included for the output of digitised data when operating as a graph follower and for reproducing punched tapes.

Although they could be used with general-purpose computers, these early models were created primarily to plot graphs from data generated by digital differential analysers. Later plotter designs such as the CalComp 565 Drum Plotter, which was introduced by the California Computer Products Company in 1959, added a mechanism to allow the pen to be lifted off or lowered onto the paper under computer control. This gave plotters the ability to generate complex vector graphics images in the same manner as a vector graphics CRT display.

The IBM 740 CRT Output Recorder

Digital plotters were capable of producing accurate, high-quality graphical output that could be stored permanently but they were painfully slow. A faster alternative would be to use a CRT display to generate the image then record it by photographing the screen with a camera. However, this was not as straightforward as it might seem, as CRT screens are notoriously difficult to photograph properly due to the problem of synchronising the camera shutter with the screen refresh cycle. Nevertheless, it was this approach which resulted in the development of the first commercially available CRT display device, the IBM 740 CRT Output Recorder.

The development of the IBM 740 began with a request from the RAND Corporation in 1953. One of the main areas of research conducted at RAND

10. The Big Picture – Computer Graphics

during the early 1950s was cognitive learning. Seeking an environment in which to test their ideas that would be of relevance to their US military backers, the RAND researchers chose a US Air Force radar direction centre. An experimental system was developed using RAND's recently acquired IBM 701 computer and a rudimentary plotter constructed using the printer mechanism from an IBM 407 tabulator which plotted marks on successive sheets of fanfold paper to represent blips on a radar display. These preliminary experiments led to RAND receiving a contract from the US Air Defense Command to develop a training simulator for radar equipment.

To provide a more realistic display for the simulator, RAND engineers conceived a system whereby a CRT display coupled to their 701 computer would be used to simulate a changing radar screen, with successive images photographed under computer control. The processed photographs would then be scanned electronically to determine the positions of the blips and the output used to drive actual radar display screens.

Development of the CRT display was subcontracted to IBM and directed by Bob O Evans at IBM's Endicott Laboratories. The synchronised camera and computer-controlled film transport system were developed by the Mitchell Camera Corporation and RCA was contracted to develop the film scanner and interface to the radar display screens. The CRT display incorporated two screens which operated in tandem; a 21-inch CRT for visual display and a higher accuracy 7-inch CRT for recording the images via the camera. Both displayed vector graphics at a resolution of 1024 by 1024 and a rate of up to 8,000 points per second.

On most vector graphics displays, the image would require constant refreshing as it would quickly decay. The usual solution was to create a 'display list' of vectors in computer memory which would be processed in a continuous loop in order to maintain the image on the screen, a technique that placed a heavy demand on the machine's input/output hardware. However, maintaining the image on screen was less of an issue if the image was to be recorded so the IBM engineers selected a CRT with a long persistence phosphor coating for the display screen which could maintain an image for up to 20 seconds before a refresh was required.

RAND eventually moved to a system which fed data directly into the radar displays. However, the IBM team had created a display device that would also be suitable for general-purpose graphics use and it was decided to offer it as a commercial product. The CRT display was introduced as the IBM 740 CRT Output Recorder in October 1954 at a rental of $2,850 per month. For customers who did not require photographic recording, the 21-inch screen was later made available separately as the IBM 780 CRT Display.

The IBM 740 was suitable for use with the 701, 704 and 709 computers. However, it was purely an output device and had no interactive capability.

The Story of the Computer

Unlike Whirlwind, which could operate in real-time, or SAGE, which was capable of supporting multiple interactive displays, these IBM models were designed for batch processing and were unable to support interactive operation. Nevertheless, the 740 was much faster than a plotter and was successfully used by a number of IBM 700 series customers in a range of technical applications requiring graphical display and recording capability.

Building Blocks – Timesharing

Much of the potential of computer graphics lay in giving the user the ability to interact with the computer. However, most general-purpose computers of this period operated by batch processing, where programs stored on punched cards or tape would be queued in batches for execution one at a time and the results output to cards or a printer for later collection by the user. The only interaction possible was by the operator when setting up a batch run. Some small-scale computers, such as the Bendix G-15, did support interactive operation which opened up the possibility of having a dedicated computer system for interactive computer graphics but this would have been an expensive solution that most customers could not afford. A key breakthrough in the use of large-scale general-purpose computers for interactive computer graphics was the development of timesharing.

Timesharing is a technique that allows a computer to support multiple interactive users simultaneously in such a way that they each appear to have exclusive use of the machine. It is based on the premise that the processor has the spare capacity when awaiting input from one user to run another user's program. Implementation of timesharing is normally through a combination of hardware and software. The hardware is necessary in order to manage the input/output activity from several devices simultaneously and to provide a memory protection mechanism and interrupt priority structure that permits user programs to be moved in and out of main memory efficiently. The software supervises this activity and provides a suitable interface to allow users to interact with the machine.

The SAGE AN/FSQ-7 was the first computer designed to support multiple users concurrently. It was capable of handling more than 100 users on a single processor using a real-time clock and sequence control program to synchronise operations. Each display console was updated at fixed 2 second intervals, allowing operation to be interleaved between multiple consoles. However, SAGE was not a true timesharing system as every user was interacting with the same program.

The earliest known reference to supporting multiple users on a single computer system was made by the American software engineer Robert W Bemer in an article in Automatic Control magazine in March 1957. Bemer envisioned a

10. The Big Picture – Computer Graphics

future where businesses would no longer purchase their own computers but would rent time on a remote computer that could "service a multitude of users", an idea that later took shape in the computer bureau services of the 1960s. Around the same time as Bemer's article was published, the mathematician John McCarthy was contemplating methods for increasing the productivity of a computer system while engaged on Artificial Intelligence research at Dartmouth College in Hanover, New Hampshire. He was given an opportunity to take these ideas further in the autumn of 1957 when he was awarded a Sloan Research Fellowship and transferred to the MIT Computation Center, a computing facility established by MIT the previous year in partnership with another 25 New England colleges and universities.

At MIT, McCarthy began conducting experiments with the Computation Center's IBM 704 mainframe computer. The 704 had no hardware interrupt but it did have a feature called transfer trapping, a special hardware mode for debugging programs in which branch instructions caused a 'trap' to occur by transferring program control to memory location 1 and placing the target address of the branch instruction into the address part of location 0. McCarthy was able to create a makeshift interrupt mechanism by installing a relay in the 704 so that the machine could be externally switched to transfer trapping mode and adding a set of relays to the machine's Friden Flexowriter teletypewriter to provide a small amount of input/output buffer memory. The interrupt was triggered as soon as anything was placed in this buffer, whereupon a supervisory program was initiated to monitor the contents of the buffer and swap the batch program with other software that allowed interactive operation as soon as it was full. The computer would then return to the batch program once the interaction had ceased. In this way, a user could interact with the computer to debug a program without having to cancel a batch run.

McCarthy wrote his ideas up in January 1959 in an internal memo entitled 'A Time Sharing Operator Program for our Projected IBM 709'. Shortly afterwards, he handed the project over to a colleague, Herbert M Teager, in order to concentrate on his main research area of Artificial Intelligence. Teager then secured funding from the US National Science Foundation for a more ambitious project to develop a large-scale timeshared computing facility which would allow remote access from various locations around the MIT campus. However, the protracted timescale for this project prompted another MIT group to develop an interim solution. This group was led by physicist Fernando J Corbató who had made extensive use of Whirlwind earlier in his career and had been inspired by the ability of the user to interact with the machine.

Corbató's group were able to make use of the Computation Center's newly acquired IBM 709 computer to create a prototype timesharing system that they named the Compatible Time-Sharing System (CTSS). The 709 had been modified by IBM to include a real-time clock that operated in conjunction with an optional input/output interrupt feature to provide an interval timer and this allowed multiple interactive sessions to be interleaved. Each user was given a fixed amount of processor time before the interval timer triggered an interrupt and returned the machine to the supervisor program which would then select the next user to be granted exclusive access to the processor. A successful demonstration of CTSS was given in November 1961, with the system supporting three Flexowriter terminals and a batch program simultaneously.

McCarthy and Corbató had proven that timesharing was possible but their work had also highlighted the limitations of the available hardware. Further progress would only come through closer cooperation with computer manufacturers to develop the additional hardware features necessary for efficient timesharing operation. This was the approach taken by another timesharing project that took place in parallel with CTSS at the Massachusetts consultancy firm of Bolt, Beranek and Newman.

In November 1960, BBN became the proud owner of the first production DEC PDP-1 which had been purchased to replace the company's ageing Royal Precision LGP-30. With a design optimised for real-time applications and a list of factory-fitted options that included a multi-level program interrupt and real-time clock, the PDP-1 was naturally better suited to timesharing than the IBM 700 series machines used at MIT. Although BBN's main area of expertise was acoustics, the company was also active in the emerging field of Artificial Intelligence and occasionally employed John McCarthy as a consultant. During one of his consulting visits McCarthy discussed timesharing with two of the people who had been instrumental in acquiring the PDP-1, electrical engineer Edward Fredkin and psychologist Joseph C R Licklider. Recognising the eminent suitability of the PDP-1 hardware to timesharing, Fredkin proposed a project to adapt the machine for timesharing operation.

Under McCarthy's guidance, a timesharing system was developed at BBN for the PDP-1. The system was based on the use of a fast access magnetic drum storage unit supplied by Vermont Research Corporation which allowed programs to be swapped in or out of core memory within a single revolution of the drum. It also featured a special hardware mode that triggered a specific type of interrupt upon any attempt to execute an input/output instruction or halt the program. The BBN project team, which included Fredkin, Licklider and software engineer Sheldon Boilen, worked closely with PDP-1 designer Ben Gurley at DEC on the necessary hardware modifications. The system was operational by September 1962 and was capable of supporting up to 5 interactive users using additional electric typewriter terminals and associated control logic.

In September 1961 MIT's Research Laboratory for Electronics (RLE) took delivery of the PDP-1 that DEC's Ken Olsen had donated to his former employer for educational purposes. Soon after the machine arrived, Jack B Dennis, an

10. The Big Picture – Computer Graphics

academic in MIT's Department of Electrical Engineering, proposed a student project to develop a timesharing system for it. He discussed his plans with John McCarthy who told him of the BBN project and suggested that both groups coordinate their efforts. Dennis and his students followed McCarthy's advice by adopting the same magnetic drum unit as the BBN system but chose to develop their own software. However, both groups jointly paid for DEC to develop the magnetic drum unit into a product, the MIT contribution coming from funding obtained from the US Office of Aerospace Research. The RLE timesharing system was operational in May 1963 and was capable of supporting 3 electric typewriter terminals and a CRT display terminal.

The MIT Computation Center's CTSS project and the RLE timesharing project were amalgamated into Project MAC (Project on Mathematics and Computation), an MIT-wide timesharing initiative led by the Italian-born electrical engineer Robert M Fano that began in July 1963 with $2 million funding from the US Department of Defense Advanced Research Projects Agency (ARPA). The project was championed by Joseph Licklider, who had moved to ARPA from BBN in October 1962 to head up the Agency's newly formed Information Processing Techniques Office. The Project MAC timesharing system was based on a modified IBM 7094 computer running a more advanced version of CTSS, with the PDP-1 now serving as a satellite system for computer graphics display.

A common problem with early timesharing systems was poor performance due to the requirement to continually swap user programs between main memory and a relatively slow backing storage device such as a magnetic drum or disk. This was alleviated by incorporating multiprogramming, a related technique which allows a computer to retain more than one program in main memory at the same time and run them concurrently by exploiting opportunities to transfer control from one to another while engaged on less compute-intensive tasks such as input and output. Memory protection hardware ensures that each program occupies a separate location in memory and cannot encroach upon the space taken up by another program.

Multiprogramming was conceived by the British computer scientist Christopher Strachey while working at the National Research Development Corporation in 1959. Strachey filed a UK patent application on multiprogramming entitled 'Improvements in or relating to electrical digital computers' in February 1959 which was granted in May 1963. This was followed by a US patent application 12 months later which was granted in December 1965.

Strachey's multiprogramming concept was first implemented in the groundbreaking Supervisor control program for the Ferranti Atlas and was also licensed to IBM for use in Project Stretch. As it could be used to improve the efficiency of batch processing operations, multiprogramming was rapidly adopted by mainstream computer manufacturers for use in their latest business data processing models. Ferranti became the first manufacturer to offer multiprogramming commercially with the Ferranti Orion which was announced in December 1959, although production problems delayed deliveries until early 1963. Other manufacturers soon followed suit with the introduction of multiprogramming enabled models such as the Honeywell H-800 in 1960, the Leo Computers LEO III in 1961 and the Ferranti-Packard FP-6000 in 1962.

Despite the swift uptake of multiprogramming, computer manufacturers initially showed little interest in timesharing. Batch processing remained the standard mode of operation for computer systems in business applications and the major suppliers of these systems saw no reason to change. IBM passed up on an opportunity to incorporate timesharing capability into its System/360 line during development in 1962-63 but later remedied this with the introduction of the System/360 Model 67 in August 1965. However, development of a timesharing operating system for this model was problematic and the software was never officially released. Apathy amongst the established manufacturers left the way clear for younger firms such as DEC and SDS to introduce computers with timesharing capability to the commercial market. The first to do so was DEC with the PDP-6, a large-scale model which was introduced in October 1964.

Computer-Aided Design

By 1960 many of the building blocks for interactive computer graphics were in place but the only graphics equipment available commercially were passive output devices such as plotters and CRT recorders. SAGE had demonstrated the power of interactive graphics for military purposes but what the industry needed was a mainstream application which could drive the development of the technology for civilian use. That application was Computer-Aided Design (CAD). CAD systems allow engineering drawings, maps and charts to be created interactively on a computer. Interactivity can accelerate the design process and improve problem solving. Their development arose from a growing interest in the use of computers for engineering design and manufacturing, particularly from the aerospace and automotive industries.

The earliest developments in this field were in the application of computers to the control of machine tools, the origins of which go back to the work of John T Parsons and Frank L Stulen at the Parsons Corporation of Michigan in 1946. Parsons and Stulen employed IBM punched card equipment to calculate the precise shape of templates used in the manufacture of helicopter rotor blades. By 1949, they had succeeded in connecting the punched card equipment directly to a milling machine. This work attracted the attention of the US Air Force which awarded the company a $200,000 contract for the

10. The Big Picture – Computer Graphics

development of a computer-controlled machine tool. However, the accuracy was found to be unacceptable without some form of feedback system, so they turned to MIT's Servomechanisms Laboratory for help. This led to the development of the world's first numerically controlled (NC) machine tool, a standard vertical milling machine fitted with servomotors on each axis and controlled by a special-purpose digital controller, which was demonstrated at MIT in September 1952.

As machine tool manufacturers took up the development of NC technology and commercial models began to reach the market, it became apparent that much of the labour saved through using these machines was cancelled out by the lengthy time required to prepare the punched tapes that controlled them. This prompted the development of programming languages for NC machine tools which allowed engineers to specify a machining operation in simple geometric terms and automatically converted this information into the complex sequence of numerical data that controlled the movement of the machine axes. The introduction of NC programming languages such as MIT's Automatically Programmed Tool (APT) and General Electric's Program for Numerical Tooling (PRONTO) opened up the market for NC machine tools, although the equipment remained an expensive option for all but the highest value manufacturing applications until the advent of low-cost minicomputers which could be adapted for use as NC controllers.

Having successfully applied computers to manufacturing, the researchers turned their attention to the use of computers in the engineering design process. The standard method of representing the design of a manufactured object, the engineering drawing, was highly suited to representation through vector graphics but the initial focus was on using computers for design calculations and it would be some time before the idea of computer-based draughting took hold.

At MIT, the mathematician Douglas T Ross was engaged on work aimed at extending the APT NC programming language so that it could also be used to describe the geometry of the component being machined and display it in graphical form. Experiments were conducted using Whirlwind's CRT display as an output device. However, Ross had no plans to employ interactive graphics, remaining firmly focused on a language-based interface between the user and the computer. His work formed the basis of the Automated Engineering Design (AED) project, a US Air Force funded initiative which began at MIT in 1959 and was later merged into Project MAC. AED further extended Ross's language-based concept into the realm of generalised problem modelling. While the project was important academically and provided a solid theoretical foundation for later developments, it was overshadowed by two industry-led development projects and another research project at MIT which were conducted in parallel and between them created the basic technology for Computer-Aided Design.

Design Augmented by Computers

In November 1956, an article in Fortune magazine by the geneticist George R Price entitled 'How to Speed Up Invention' described a hypothetical "design machine", a computerised system for engineering design that allowed the user to manipulate geometric shapes interactively on a display screen. This visionary article was to provide the inspiration for a groundbreaking project at the research and development arm of the giant US automotive firm General Motors.

In 1958, a company-wide exercise conducted by General Motors Research Laboratories (GMR) to identify problem areas in vehicle design and manufacture highlighted the production of engineering drawings as a particularly time-consuming process. The researchers in GMR's Data Processing group, led by Donald Hart, recalled Price's 1956 article and began to conceptualise a computer system for the design of vehicle bodies. Existing engineering drawings and design sketches were an important element of the design process so these would have to be incorporated into the system. By doing so, the researchers hoped that it would be possible to use the information extracted to automatically generate new drawings and the control tapes for NC machine tools.

A small project team was assembled and work began in early 1959 to explore the feasibility of this idea. The project was initially named 'Digital Design' but this was later changed to 'Design Augmented by Computers' or DAC-1 to avoid any confusion in an age when digits still meant fingers. GMR's IBM 704 computer, which was equipped with an IBM 780 CRT display, was used as a development platform. To provide an interactive capability, the project team created a rudimentary input device using a set of five rotary switches which they connected to the computer through the input/output circuitry of the system's line printer. A program was written to continually monitor the settings of these switches, each of which had 10 positions, thus providing 5 decimal digits that could be changed by the user as often as necessary in order to control program execution.

Drawings and sketches were input to the system using an improvised document scanner constructed with the assistance of GMR's Instrumentation Department. The drawing or sketch to be digitised was first traced onto clear acetate film and the tracing mounted on the face of the IBM 780 display screen. A light-sensitive photomultiplier tube was connected to the computer and positioned in a hooded enclosure in front of the display screen. Using this equipment, the presence or absence of a line could be detected by creating a spot of light at a specific location on the display screen and examining the output from the photomultiplier. Light from the spot would trigger a circuit connected to the photomultiplier unless the spot was beneath a line, in which case the line would obscure the spot and insufficient light would be detected. By scanning

10. The Big Picture – Computer Graphics

the spot over the screen and recording the position of these 'dropouts', the entire drawing could be digitised. However, rather than laboriously scanning the spot in a raster pattern, the team developed a time-saving technique which tracked lines by scanning in a circle around a detected point until the direction of the line was found and then following it.

The project team utilised the new FORTRAN programming language to create software which produced a 3-dimensional mathematical model of the vehicle body part from a series of design sketches through a process of surface fitting and interpolation. Cross sections and boundary curves of the resulting model were displayed on the IBM 780 screen and the designer could interact with the system to alter the shape of the model by changing various parameters. Full-size plots could also be generated using a modified vertical milling machine fitted with a pen and drawing table.

As the project progressed, the limitations of using standard computer equipment became more apparent and in August 1959 discussions were initiated with IBM on the acquisition of a more powerful computer and custom graphics hardware to improve the interactive capabilities of the system. Following a successful feasibility demonstration to GMR senior management in July 1960, a joint development contract was drawn up with IBM. Codenamed Project GEM (Graphic Expression Machine), the new system would be based on an IBM 7090 computer fitted with two additional data channels and a special memory protection mode originally developed for the CTSS timesharing project. The specialised equipment necessary for interactive graphics, document scanning and recording would be designed and built by IBM, with the system software developed by GMR under the direction of project leader Edwin L Jacks.

The IBM engineers were able to call upon their Project SAGE experience for the development of the interactive graphics display but the document scanning and recording equipment would prove to be far more challenging. Scratches on the 35 mm photographic film used to record drawings were a particular problem, causing erroneous lines to be detected during digitisation. Despite a subsequent reduction in the specification of the scanner, the project overran the planned 18-month timescale by almost a year and the hardware was not delivered to GM until April 1963. Designated the IBM 7960 Special Image Processing System, it comprised three units; a hardware interface, a graphics display console and a combined scanner/recorder which IBM referred to as an image processor.

The IBM 7960 display console featured a 10-inch square CRT screen which could display vector graphics at an impressive resolution of 4096 by 4096. A pointing device and a keypad with 36 programmable function keys and independently controllable indicator lights were provided for interactive operation. The console also incorporated an alphanumeric keyboard and card reader. The pointing device operated in a similar manner to a light gun but was implemented using an alternative technique in which an electrically conductive coating on the surface of the screen is used to create a voltage gradient across the screen and the voltage on the tip of the device measured at the point where it touches the screen. As this voltage will be proportional to the distance across the screen, the position of the device along that axis can be determined. By continually switching the voltage gradient between the horizontal and vertical axes, the position of the device at any location on the screen can be found.

Considerable software development effort was devoted to producing a multiprogramming operating system for the IBM 7090 which could support an interactive session and batch processing simultaneously. The GMR team also created two special programming languages, NOMAD and MAYBE, to ease the task of graphics programming and provide closer control over the custom input/output hardware. A language was also developed to allow users of the system to describe the geometry of an object in advance of an interactive session.

The DAC-1 system was operational by November 1963 but development continued until 1967. Although it was extensively used within GM, the company had no immediate plans to commercialise it, preferring instead to retain the system as a platform for a range of in-house research projects in areas such as image processing and mathematical modelling. IBM, however, had always intended to derive commercial benefit from the project and had sought to ensure that there were no restrictions in the joint development contract to prevent it from doing so. Following delivery of the graphics hardware to GMR, the company initiated Project Alpine to productise it. This resulted in the introduction of one of the earliest commercially available interactive graphics displays, the IBM 2250 Display Unit.

The IBM 2250 featured a 21-inch CRT which could display vector graphics at a resolution of 1024 by 1024 and a refresh rate of up to 40 frames per second. The DAC-1 display console had suffered from unacceptable flickering when displaying images containing more than a thousand or so vectors due to the relatively slow display rate and the need to continually interrogate the display list stored in the main memory of the 7090 computer. To avoid this situation, the 2250 incorporated fast access magnetic core memory of 4,096 or 8,192 bytes for use as a display buffer. An optional electronic character generator was also available which simplified the display of alphanumeric characters by automatically generating the sequence of analogue signals required to draw the characters in the form of strokes on the CRT screen. This device supported a standard set of 63 characters in two different sizes.

Input devices on the IBM 2250 included an alphanumeric keyboard, a keypad with 32 programmable function keys and an optional light pen pointing device. For the pointing device, the IBM engineers reverted to the light gun approach despite the greater speed and flexibility of the voltage measuring system used

10. The Big Picture – Computer Graphics

on the DAC-1. The reason for this is unclear but is likely to have been to reduce manufacturing costs.

The IBM 2250 Display Unit was introduced along with the 2280 Film Recorder and 2281 Film Scanner at the Fall Joint Computer Conference in Detroit in October 1964. Priced at $100,000, it was announced as part of the recently introduced System/360 computer range. Connection with the host computer was through a high-speed data channel which was a standard feature on all System/360 models. As one of the earliest interactive displays and the first commercial model to employ vector graphics, the 2250 was reasonably popular despite its high price tag, providing a platform for the development of numerous interactive graphics applications and facilitating the creation of a fledgling CAD software industry.

The Electronic Drafting Machine

The next significant development in Computer-Aided Design came through a project undertaken by a US military contractor to create a computer-based system for optical design. Itek Corporation had been established by Richard Leghorn, a former US Air Force reconnaissance expert, in Lexington, Massachusetts in 1957 to manufacture specialised photographic equipment for aerial reconnaissance. The project was the brainchild of Norman H Taylor who had been a member of the Whirlwind development team at MIT before moving to Itek.

In 1960, Taylor met former colleague Jack Gilmore at a computer conference. Gilmore mentioned a project called Scope Writer which he had been working on at the Lincoln Laboratory before leaving MIT in October 1959 to set up Charles W Adams Associates, a software consulting firm he co-founded with colleague Charles Adams. Scope Writer was an early attempt to create a word processor using the raster graphics display and light pen of the Laboratory's TX-0 computer to manipulate alphanumeric characters directly on the screen. The software could also handle simple diagrams.

Taylor recognised that the Scope Writer technology described by Gilmore could be further developed to provide the functionality required for an in-house system which would assist the Itek engineers in the preparation of drawings for optical systems. He convinced Itek management to take on this development, arguing that the resulting system would also have considerable potential as a commercial product. As Itek planned to move into computer technology but had not yet done so, the project was quickly approved and work began in August 1960.

Taylor and Itek colleague Earle W Pughe assumed responsibility for the hardware and Adams Associates was contracted to develop the software. The platform for the system, which became known as the Electronic Drafting Machine (EDM), would be DEC's newly introduced PDP-1, a machine Gilmore already knew well, having provided consultancy to DEC on the design of the instruction set during the machine's development. Itek took delivery of the second production PDP-1 in June 1961.

With its advanced input/output capability, optional graphics display and light pen, the PDP-1 was a good choice of hardware platform for a CAD system. However, the display was of the raster graphics type with a fixed resolution and relatively slow display rate. Vector graphics would be better suited to the complex line drawings that the EDM system would have to create, so Taylor and Pugh worked closely with DEC's Ben Gurley to develop a vector display for the PDP-1. The 16-inch CRT was replaced with a larger 25-inch screen to display vector graphics at a resolution of 1024 by 1024. In order to avoid the flickering problem experienced with these displays when displaying a large number of vectors, a magnetic disk memory unit was added to the PDP-1 to serve as a display buffer. Manufactured by the Telex Corporation, the disk rotated at 1,800 rpm with storage for up to 20,000 vectors and was able to sustain a refresh rate of 30 frames per second. For input, the light pen was retained and a keypad added with 15 function keys for operator selection of drawing functions. A CalComp drum plotter was also connected for the output of drawings onto paper.

The EDM software allowed an operator to create engineering drawings interactively on the screen using basic geometric constructs such as lines, circles and arcs or by sketching freehand. A tracking cross indicated the location of the light pen on the screen, allowing objects to be selected and dragged around, rotated, resized or copied. Dimensions could be added automatically and the system was also capable of generating a mirror image of selected regions of a drawing. Commonly used parts could be stored on magnetic tape and retrieved as necessary. These features significantly reduced the time and effort required to produce drawings containing repetitive elements, such as circuit diagrams.

Only 15 function keys were available on the keypad as these had been implemented using a bank of 18 toggle switches provided on the front panel of the PDP-1 for inputting a test word. To supplement the keypad, a set of virtual function keys known as 'light buttons' was also provided using a technique pioneered in the Scope Writer project. This took the form of several rows of dots displayed along the bottom edge of the screen which could be selected using the light pen. A plastic card overlay with holes for each of the dots and labels describing their function was placed over the screen to guide the operator in selecting the correct button.

Following completion of the EDM prototype towards the end of 1962, work began on the development of a production version codenamed 'Digigraphic' which Itek planned to sell for $400,000 to $500,000. Funding for product development was boosted by the early sale of a vector display unit and EDM software to the US Air Force Cambridge Research Laboratory for use with an

10. The Big Picture – Computer Graphics

experimental system that employed two PDP-1 computers operating in tandem, one of which was equipped with a colour graphics display. However, Itek had been experiencing financial difficulties for some time and it was decided to sell off the non-core parts of the business, including the EDM technology. A deal was brokered with high performance computing company Control Data which acquired the rights to the EDM technology in June 1963 and created a new division named Digigraphic Laboratories in Burlington, Massachusetts to develop and sell CAD systems based on it.

Digigraphic Laboratories contracted Adams Associates to port the EDM software to the CDC 3200, a new medium-scale model with a 24-bit word length and real-time capability. The magnetic disk was replaced with a Control Data magnetic drum unit which rotated at a lethargic speed of 30 rpm but was capable of transferring a complete frame within a single revolution of the drum, thus matching the 30 frames per second refresh rate of the prototype. The display console featured a 22-inch CRT vector graphics display with a massively increased resolution of 4096 by 4096, a 25 button keypad and a light pen. The sparkling performance of the CDC 3200 allowed a single computer to support up to three display consoles.

The system was introduced in 1965 as the CDC Digigraphic System 270 at a price of around $400,000 for a typical configuration including the CDC 3200 computer. Gilmore later estimated that it possessed only about 60% of the functionality of modern draughting systems, mainly due to the limitations of the early graphics hardware. Nevertheless, the Digigraphic System was the first CAD system to reach the market and, despite a high price and limited functionality, it became popular within the US aerospace industry where customers included Lockheed and Martin Marietta. It also paved the way for later developments in CAD systems, many of which were made by those same aerospace companies that had purchased a Digigraphic System.

Sketchpad

The third major contribution to the early development of CAD technology was all the more remarkable in that it was largely the work of a single person, Ivan E Sutherland, while a postgraduate student at MIT.

Ivan Sutherland was born in Hastings, Nebraska in 1938. The son of a civil engineer, he chose to study electrical engineering, obtaining his Bachelor's degree from Carnegie Mellon University and a Master's degree from the California Institute of Technology before moving to the MIT Lincoln Laboratory in 1960 to pursue a PhD. The research environment at the Lincoln Lab, with its focus on making computers easier to use through interactive graphics technology, was to prove an abiding influence on Sutherland. In April 1961 he approached Wesley Clark, the architect of the TX-0 and TX-2 experimental transistor computers, with a request for access to the TX-2 machine in order to investigate computer drawing techniques. Sutherland's request was granted and over the summer of 1961 he developed his ideas into a suitable topic for a PhD thesis. He also found a worthy supervisor to guide him, Claude Shannon, whose 1937 Master's degree thesis had introduced the concept of switching circuits as a tool for solving logical problems. In autumn 1961, Sutherland started work on the project which he called Sketchpad.

The TX-2 was an excellent platform for a computer drawing research project. It was large and powerful with a generous word length of 36 bits and a vast 69,632 word magnetic core memory, 4,096 words of which were implemented using fast transistor-driven circuitry to improve access time. The machine was also specifically designed to facilitate interactive operation and was capable of supporting two display consoles simultaneously. The consoles were originally developed by PDP-1 designer Ben Gurley and Charles E Woodward in 1958 for the TX-0 and featured a 12½-inch CRT screen which could display raster graphics at a resolution of 1024 by 1024 and a rate of up to 100,000 spots per second. Input devices included a light pen, a keypad with 37 pushbuttons and 4 control knobs fitted with digital shaft encoders which converted rotational movement into a 9-bit binary value.

Sutherland began by developing software that would track the position of the light pen and create a cursor on the raster display of the TX-2 to provide a visual indication of where the pen was pointing on the screen. From this he was able to create simple line drawing routines which allowed an operator to draw lines interactively on the screen using the light pen and pushbuttons. Following advice from Shannon, a circle drawing capability was then added and by early 1962 a basic drawing system began to take shape. However, editing drawings was much less straightforward, requiring the machine to laboriously search the entire display list for the drawing elements that required to be altered.

Taking his cue from the work of mathematician Douglas Ross in the AED project, Sutherland then devised a 'ring structure' in which drawing elements were stored in a database by means of a string of pointers that linked each element to all the other drawing elements relating to it in an endless loop. This greatly simplified the process of identifying the relevant drawing elements for editing. It also facilitated the application of geometric constraints to groups of drawing elements so that they could be made to obey certain rules when being manipulated. The use of a ring list structure gave Sketchpad a powerful capability which went far beyond that of Itek's EDM system.

Sketchpad was completed towards the end of 1962 and described in a paper presented by Sutherland at the Spring Joint Computer Conference in Detroit in May 1963. MIT also made a short film showing Sutherland demonstrating Sketchpad which was widely distributed throughout the US computer graphics research community. After obtaining his PhD, Sutherland joined the US Army where he was given the rank of First Lieutenant and assigned to the National

10. The Big Picture – Computer Graphics

Security Agency as an electrical engineer. His work on Sketchpad was taken up by Timothy E Johnson, a research assistant in MIT's Mechanical Engineering Department, as part of his Master's degree thesis.

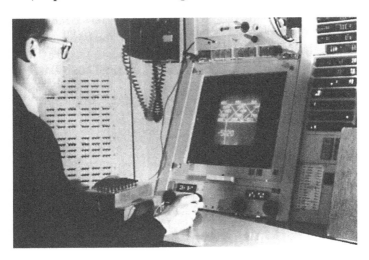

Ivan Sutherland demonstrating Sketchpad on the TX-2 (Photo © Massachusetts Institute of Technology)

Johnson extended Sketchpad into three dimensions. By providing the operator with the ability to view an object from different directions, 3-D objects could be constructed in the form of wireframe images on the screen. The software computed four views, a perspective view and three orthogonal views (elevation, end elevation and plan), which were displayed simultaneously by dividing the screen up into quadrants. Editing a drawing in one view automatically updated the other three views. To aid the tricky process of drawing in 3-dimensional space, objects could be rotated about any axis using one of the TX-2's control knobs. The system also incorporated a hidden line removal algorithm, developed by doctoral student Lawrence G (Larry) Roberts as part of his PhD thesis, which made perspective view wireframe images much easier to interpret by automatically removing the lines from those parts of the object that would be obscured by the parts of the object closest to the viewer.

Johnson's version of Sketchpad was operational by June 1963 and was known as Sketchpad III. The system remained a research tool and was not commercialised but it was highly influential. Although it was not the first CAD system capable of producing fully dimensioned engineering drawings, nor was it the first to handle 3-dimensional objects, Sketchpad was the first to combine both. However, what really set Sketchpad apart was Sutherland's use of a ring list structure and geometric constraints, which paved the way for later developments in parametric modelling, and Johnson's multiple-view display which would become the standard user interface for 3-D modelling.

Building Blocks – Pointing Devices

By the early 1960s, the light pen was firmly established as the industry standard pointing device for interactive computer graphics but demanding new applications such as CAD were beginning to reveal its limitations. Users would become tired when holding the light pen up against the display screen for long periods of time. Parts of the screen were also obscured by both the pen and the user's hand. This prompted efforts to develop alternative pointing devices that were more comfortable to use and did not obscure the screen.

Digitising Tablets

A pen is a natural pointing device but it would be much more comfortable to use over long periods of time if the target surface was horizontal instead of vertical and separate from the display screen. This was the reasoning behind the development of the digitising tablet.

A digitising tablet comprises a pen or stylus and a rectangular platen which is usually placed on a horizontal surface such as a desk. Electronics inside the platen enable it to detect the x-y position of the stylus when pressed against it. The stylus can then be moved around the surface of the platen to control the position of a cursor on the display screen of the computer.

The technology behind digitising tablets came out of research into handwriting recognition conducted at Bell Telephone Laboratories in 1957. The research was prompted by a desire to computerise the billing process for long distance telephone calls. Switchboard operators made a record of long distance calls on paper 'toll tickets', each of which contained an average of 25 handwritten numeric characters. With around two billion tickets generated each year, the time and effort required to manually transcribe this mountain of data onto punched cards would have been herculean. In an attempt to find a method of automating this process, BTL electrical engineer Thomas L Dimond, assisted by colleague Louis A Kamentsky, developed an experimental digitising device which he called a stylus translator or 'Stylator'.

The Stylator could read and recognise Arabic numerals as they were being written using a set of 7 metal conducting rods embedded in the plastic writing surface of the device. As the stylus of the device passed over a conducting rod, an electrical circuit was made. By constraining the position of the numbers using two dots around which they were written, the order in which the stylus passed over each of the conducting rods would be different for each number. A series of logic gates converted the output from flip-flops triggered by the stylus making electrical contact with the rods into an electrical signal representing the number written, allowing the device to operate in real-time with minimal load on the host computer. Dimond described the Stylator in a paper presented at

10. The Big Picture – Computer Graphics

the Eastern Joint Computer Conference in Washington DC in December 1957 but the device was never commercialised and it would be several more years before digitising tablet technology reached the market.

The Stylator was developed as a special-purpose device but the principle of operation had wider applications. The general-purpose digitising tablet grew out of research into man-machine graphical communication conducted at the RAND Corporation in Santa Monica, California in the early 1960s. The work was funded by the Advanced Research Projects Agency (ARPA), a government body established by the US Department of Defense in 1958 to fund the development of new technology for use by the US military. In studying traditional methods of communication, RAND researchers began to focus on developing a computer interface that mimicked a pen and paper. The result was the RAND Tablet.

The RAND Tablet was developed by RAND engineers Malcolm R Davis and Thomas O Ellis. Initial experiments were conducted using a flatbed plotter which had been modified so that the digital input operated in reverse, with the user manually moving the pen and the computer monitoring the output from the plotter. However, the movement of the pen was insufficiently smooth as the user had to push against the friction in the drive mechanism. Davis and Ellis then embarked on the development of a new device which became known as the RAND Tablet.

The RAND Tablet featured a platen with a 10-inch square writing surface and a stylus with a pressure-sensitive switch in the tip which was actuated when the stylus was pressed against the platen. The writing surface was manufactured from plastic sheet sandwiched between two thin layers of copper. Each outer surface was then etched using a printed circuit board manufacturing process to create a set of 1024 parallel lines in the copper, with one side oriented perpendicular to the other so that the lines combined to form a grid pattern. The upper surface was then sprayed with an epoxy coating to provide electrical isolation from the stylus and prevent wear. Electronics connected to the metal tip of the stylus allowed it to pick up the magnetic field from the electrified copper lines when placed in close proximity to them through a phenomenon known as capacitive coupling. By driving each copper line in the grid with a different sequence of electrical pulses, the location of the stylus on the writing surface could be determined to a resolution of 1024 by 1024. A data acquisition speed of 4,500 points per second ensured that the tablet was able to faithfully capture rapid movement of the stylus.

The prototype tablet was completed in September 1963. A larger version with a 36-inch square writing surface was also constructed, around a dozen copies of which were manufactured in-house by RAND at a cost of $18,000 each and supplied to other research groups engaged on ARPA projects. Applications included handwriting recognition and digital mapping systems. The smaller model was also commercialised by the Data Equipment Company, a division of BBN based in Santa Ana, California. Introduced in 1964 as the Grafacon Model 1010A Digitizing Tablet, it generated sales in specialist application areas but failed to catch on as a general-purpose pointing device. Similar devices followed over the next few years but digitising tablets would be eclipsed by another pointing device, the computer mouse.

The First Computer Mouse

The early 1960s witnessed a proliferation of research projects on the topic of man-machine communication. At the Stanford Research Institute in Menlo Park, California, electrical engineer Douglas C Engelbart had a grand vision of using computers to augment human intellect. In 1959, he obtained funding from the US Air Force Office of Scientific Research for a study into developing the means to increase human intellectual effectiveness. The result of this study was a weighty 134-page report entitled 'Augmenting Human Intellect: A Conceptual Framework' which was published in October 1962.

Engelbart's report was brought to the attention of timesharing pioneer Joseph Licklider, who had recently taken up his new post as head of ARPA's Information Processing Techniques Office (IPTO). Licklider shared Engelbart's vision, having published a seminal paper speculating on the possibility of man-computer symbiosis in 1960. He encouraged Engelbart to apply for ARPA funding which Engelbart did and in February 1963, he used the funding to set up a small research facility at SRI which he called the Augmentation Research Center (ARC).

To provide a suitable platform for his research, Engelbart installed an interactive computer system which was based on a CDC 160-A equipped with a magnetic drum storage unit and a CRT display capable of displaying 16 rows of 63 alphanumeric characters. A small team of researchers was assembled and work began on the creation of a system to facilitate collaboration between workers through interactive computing. The ARC researchers experimented with various pointing devices that would allow a user to interact with the system, including a light pen, joystick, Grafacon digitizing tablet and a fourth device which Engelbart conceived by analysing the characteristics of the other devices. This device, which Engelbart called an X-Y position indicator, was designed to be placed on a flat surface and moved around by the user to control the position of a cursor on the screen. The movement of the device was transferred to two potentiometers through the rotation of a pair of small wheels on the underside with their axes of rotation set perpendicular to each other. A ball bearing at one corner of the underside provided a third point of support. The device was connected to the computer via analogue-to-digital converter circuits which converted the potentiometer voltages into digital values representing the x and y positions of the cursor. A pushbutton on the top of the device allowed the user to send a signal to the computer for operations

10. The Big Picture – Computer Graphics

such as selecting text.

Engelbart's rough sketches were transformed into a working prototype by SRI's chief engineer William K (Bill) English. The prototype was encased in a small wooden box with the pushbutton located at one corner of the top face and was completed in 1964. One of the researchers nicknamed it the 'mouse', as the cable connecting the device to the computer resembled a tail, and the name stuck.

Douglas Engelbart's Mouse Prototype (Photo © SRI International)

A NASA-funded study, using volunteers who were timed to determine how quickly they could perform various tasks, showed that the mouse had the lowest error rate of all the pointing devices tested and was the most satisfying to use, although it required more practice to master than the other devices. Despite this, NASA showed little interest in the mouse, probably due to its unsuitability for use in zero gravity environments, and it would be several years before the device was taken up by other groups. Nevertheless, a US patent application was filed in June 1967 which was granted in November 1970. However, SRI's mean-spirited patent attorney refused to acknowledge English's contribution and Engelbart was named as the sole inventor.

When the ARC researchers moved to a larger timesharing computer system equipped with multiple terminals in 1968, a production version of the mouse was designed with a moulded plastic case and three pushbuttons and a small batch manufactured for SRI by Computer Displays Incorporated of Waltham, Massachusetts. Computer Displays also supplied the mouse as an option for one of its own products, the ARDS graphics display terminal, making it the first company to offer the device commercially. The two rotating wheels were eventually replaced by the more familiar ball in a design conceived by engineer

Ronald E Rider of the Xerox Corporation in 1972. Rider's improved design was based on the trackball, another pointing device which was originally developed by Thomas Cranston and Fred Longstaff of Ferranti's Toronto-based subsidiary Ferranti Canada in early 1952 for the DATAR (Digital Automated Tracking and Resolving) computer system. Rider took the trackball mechanism and simply turned it upside down to create his ball mouse. As Ferranti Canada had not patented the trackball, this allowed Xerox to obtain a patent on the ball mouse which was granted in September 1974.

Reducing the Cost of Computer Graphics

The introduction of commercial off-the-shelf display equipment and CAD systems by companies such as IBM and Control Data brought interactive computer graphics into the mainstream. However, the price of these products was so high that only governments or very large companies could afford them. In order for computer graphics technology to proliferate, the price of the hardware had to fall.

One of the first computer manufacturers to recognise the potential of computer graphics was DEC. The prototype of the PDP-1, DEC's first computer, featured an integrated raster graphics display with a 16-inch CRT screen and light pen built directly into the console. This became a $15,600 optional extra, the Type 30 Precision CRT Display, on the production version when introduced in 1960. The display was housed in a separate desk-shaped cabinet and controlled by the PDP-1 through the 10 most significant bits of the input/output register and accumulator. For an additional $4,900 it could also be supplied with an electronic character generator, the Type 33 Digital Symbol Generator, which created alphanumeric characters on the display screen using a 5 by 7 array of spots. Despite the extra cost that these options placed on the already not insignificant $120,000 price tag of a PDP-1, the DEC system remained the most affordable interactive graphics platform for several years and was only beaten by the introduction of another DEC model, the PDP-5, in August 1963 which brought the price of a basic interactive graphics system down to around $60,000. The following year DEC also introduced a vector graphics display, the Type 340 Precision Incremental Display, at a price of $28,600.

The relative affordability of DEC systems made them popular development platforms for some of the earliest computer graphics applications. As well as providing a platform for the development of the RLE timesharing system, MIT's PDP-1 was also used to create one of the first interactive computer games, 'Spacewar!'. Spacewar! was developed in 1962 by a team of postgraduate students led by Stephen R Russell and J Martin Graetz. Inspired by the science fiction novels of E E 'Doc' Smith, they created a two-player game in which each player controlled a spaceship and could fire a stream of missiles at their opponent's craft. The spaceships were displayed on the CRT screen as solid shapes, one of which was a wedge and the other a needle, against a background of random dots representing a star field. Each player controlled the movement of their spaceship using 4 of the test word switches on the front panel of the PDP-1 but this arrangement proved far from ideal and two rudimentary game controllers were later constructed to reduce the risk of a player flicking the wrong switch and halting the machine.

Spacewar! was publicly unveiled at an MIT open day in May 1962. As the team of students who developed it moved on, they took the game with them and it became popular as a test program for DEC computers fitted with a Type 30 display. It also proved a major influence on the first generation of video games.

The Direct-View Storage Tube

DEC graphics displays, although inexpensive, were dedicated devices which could only be used with a DEC computer. For users of other computer systems in the mid 1960s, the only alternative was the IBM 2250 Display Unit. In order to ensure a flicker-free image, the 2250 incorporated a relatively large amount of magnetic core memory for use as a display buffer and this accounted for much of the $100,000 price. Fortunately, a new type of CRT known as the Direct-View Bistable Storage Tube (DVBST) would remove the need for expensive memory and facilitate a low-cost approach.

The DVBST was a special type of CRT which incorporated a holding gun and long persistence phosphor coating in order to retain the image on the screen without having to continually refresh it. Although functionally similar to the CRTs used by MIT in the electrostatic memory of Whirlwind, the DVBST was developed independently by Robert H Anderson of the Oregon-based electronic instrumentation company Tektronix based on earlier work conducted at US Naval Research Laboratory in 1947. It added a fine metal grid to the inside surface of the CRT screen that collected the energy from the main electron gun and controlled the rate at which this was transferred to the phosphor screen by the flood of electrons from the holding gun. Using this method, the DVBST was capable of retaining an image for at least 15 minutes without degradation. It also had a 'write-through' capability which allowed transient features, such as a cursor, to be superimposed on an image without otherwise affecting it.

The DVBST was developed by Tektronix for use in a new range of storage oscilloscopes, the first of which was introduced in 1963. Its potential for use as a graphics display device was recognised by Robert H Stotz, an electrical engineer in MIT's Electronic Systems Laboratory (ESL). Stotz was a member of a small team known as the Display Group which had recently completed the development of an interactive graphics display aimed at extending the capability of the CTSS timesharing system in order to support Douglas Ross's CAD research. The ESL Display Console was based on a customised DEC Type

330 display fitted with two electronic character generators; a Straza Symbol Generator which generated spot patterns representing the complete Teletype character set of 64 symbols using a 15 by 16 array of spots, and an MIT-designed unit for generating non-standard characters.

The ESL Display Console was completed in 1963. Following the start of Project MAC, project leader Robert Fano challenged the MIT Display Group to develop a low-cost graphics display terminal capable of communicating with a timeshared computer system over a standard Teletype interface and which could be manufactured commercially for $5,000. The result was the Advanced Remote Display Station (ARDS).

Development of the ARDS prototype began in 1965 and was carried out by Stotz and Master's degree student Thomas B Cheek. An initial investigation of the range of display technologies available marked out DVBST as the having the greatest potential, as a storage tube would remove the requirement for a display buffer and could also reduce the need for the high-speed electronics which would normally be required to minimise flicker when displaying large amounts of data. The DVBST tube was not available as a separate component so a Tektronix Type 564 Storage Oscilloscope, which featured a 5-inch DVBST, was acquired and used as the display screen. Electronic circuits were designed to convert 10-bit vector data from a computer into voltages suitable for driving the horizontal, vertical and beam intensity inputs of the oscilloscope. An electronic character generator was developed by Cheek using a pair of counters to step the oscilloscope beam through the appropriate spot pattern in a 7 by 9 array for each character, with the patterns for a set of 94 characters stored in a small read-only memory. A standard computer keyboard was provided for text input. For graphics input, an electronic circuit generated a cursor on the display screen, the position of which was controlled by a computer mouse similar to the device developed by Douglas Engelbart at SRI.

Following completion of the ARDS prototype in early 1967, Stotz approached Tektronix in an effort to persuade the company to take on the development of a production version. Tektronix acknowledged the interest shown in its DVBST technology for computer applications by introducing the Type 601 Storage Display Unit, a stripped down version of the storage oscilloscope which was better suited to the display of graphical images, but the company showed no interest in entering the computer graphics market. Nevertheless, Stotz and Cheek remained convinced of the commercial potential of the ARDS and decided to go into business for themselves, establishing a new company, Computer Displays Incorporated in Waltham, Massachusetts, to manufacture and sell it.

The production version of the ARDS was introduced by Computer Displays in late 1968 at a price of $14,500. This was considerably higher than the $5,000 target price originally specified by Robert Fano but still significantly cheaper

10. The Big Picture – Computer Graphics

than competing products. It was based on the new Tektronix Type 611 Storage Display Unit which featured an 11-inch DVBST to give an addressable resolution of 1081 by 1415. The display screen was mounted in portrait orientation on top of a base unit which housed the electronics and a detachable 58-key keyboard. Two options were provided for graphical input, a 3-button mouse (the first available commercially) and a joystick. A standard Teletype interface was provided for communication with a timesharing computer system at a maximum data transfer rate of 1,200 bits per second.

The introduction of the ARDS spawned a similar product from another Massachusetts company, Computek, which was founded in 1968 by Donald R Haring and Michael L Dertouzos. Both were former MIT researchers who, like Stotz and Cheek, had spotted the potential of the DVBST for low-cost graphics display equipment. Computek's first product, the Model 400/20 terminal, which was introduced in 1969, was also based on the Tektronix Type 611 display. Priced at $12,000, it was slightly cheaper that the ARDS and it also supported the drawing of continuous curves.

The example shown by Computer Displays and Computek in introducing graphics terminals based on the DVBST prompted Tektronix to revise its opinion of the computer graphics market. In 1970, the company announced its first product specifically designed for computer graphics, the Type T4002 Graphic Computer Terminal. The T4002 was fitted with an 11-inch DVBST in landscape orientation to display vector graphics at a resolution of 1024 by 760. It was also equipped with a built-in electronic character generator which could generate up to 5,000 characters per second, with text input via a 71-key keyboard. Graphical input was provided by a set of 5 cursor control buttons or an optional joystick. A novel text editing feature employed a 1-centimetre high strip along the bottom edge of the screen for displaying a line of up to 84 characters which, unlike the rest of the screen, was continuously refreshed. This operated in conjunction with a small amount of local memory configured for use as a line buffer.

As the manufacturer of the principal component used in rival products, Tektronix had little difficulty in undercutting its competitors and the T4002 was competitively priced at $8,800. The company's decision to enter the computer graphics market would also prove to be a good one, as Tektronix would go on to dominate the industry for the next decade with a succession of low-cost graphics display terminals based on its DVBST technology.

Standalone Graphics Systems

The new generation of graphics display terminals based on storage tube technology offered organisations in possession of a timesharing computer system low-cost entry into the world of computer graphics. However, storage tube terminals were not well suited to interactive graphics as they did not support selective erasing. In order to change part of an image, the entire screen would have to be erased and redrawn, a process that could take several seconds. Performance was also a major issue with interactive graphics on a timesharing computer but buying a dedicated computer for graphics applications was an expensive luxury that most organisations could not afford. Fortunately, the plummeting price of minicomputers soon facilitated the introduction of affordable standalone computer graphics systems which coupled a graphics display console with a dedicated minicomputer.

Early Research

The first project to create a standalone graphics system was initiated by CAD pioneer Ivan Sutherland shortly after he moved from MIT to the National Security Agency in 1963. Sutherland and three NSA colleagues worked with DEC design engineer William H Long to specify an interactive graphics display system that would provide a flexible platform for research into man-machine communication. The system was then constructed by DEC and delivered to the NSA in August 1964.

Known as the DEC General Purpose Experimental Display System, it comprised a PDP-4C minicomputer coupled to a DEC Type 340 vector display. The display was modified to enable it to directly access the memory of the PDP-4C which was fitted with the maximum 8,192 words in order to provide additional storage space for the display list. If more memory or floating-point arithmetic were required, the system could also be connected to a large-scale CDC 1604A computer. To ensure maximum flexibility, the system was fitted with all manner of graphical input devices including a light pen, a trackball, a Grafacon Digitizing Tablet, 3 control knobs and 2 keypads fitted with a total of 54 pushbuttons. Each of the pushbuttons incorporated an independently controllable indicator light for visual feedback. Various audio output devices such as bells and buzzers were also fitted. This flexible approach also extended to alphanumeric characters, which were drawn using software subroutines rather than an electronic character generator, as this would have limited the system to a specific character set.

The DEC General Purpose Experimental Display System was a versatile system and was extensively used by the NSA for a wide range of research projects. Several additional examples were built but the emphasis on flexibility severely compromised performance. Fortunately, the system remained an in-house research tool and no attempt was made to introduce it to the commercial market.

Another early project to create a standalone system for interactive graphics took place at Bell Telephone Laboratories in 1965. The GRAPHIC 1 remote graphical display console was developed to explore the use of local computing

10. The Big Picture – Computer Graphics

power for interactive graphics in a timeshared computing environment, where communication between the central computer and remote terminals would usually be over slow data transfer rate connections such as Teletype interfaces. Of similar design to the NSA system, it comprised a DEC PDP-5 minicomputer coupled to a modified Type 340 vector display. However, it differed in having a separate display buffer which was implemented using an Ampex RVQ magnetic core storage unit fitted with 4,096 words of 36-bit memory. This unit also allowed the system to be connected to an IBM 7094 mainframe computer, with the display buffer able to function as shared memory between the two computers. Like the NSA system, GRAPHIC 1 also featured a wide range of input devices including a light pen, trackball, 6 control knobs, 32 pushbuttons and 4 switches, one of which was foot operated.

With GRAPHIC 1, Bell Labs successfully demonstrated the effectiveness of local computing power for interactive computer graphics. The advantages of a separate display buffer in allowing the display to be continually refreshed independently of the PDP-5 were also evident but the system was an inelegant hybrid of 12-bit minicomputer, 18-bit graphics display and 36-bit memory unit. Better hardware integration would be necessary before such systems could be introduced commercially.

The Programmed Buffered Display

Those first experimental standalone graphics systems were hampered by the hardware limitations of the first generation minicomputers upon which they were based. To compensate, they could be connected to large-scale computers but this cancelled out the advantages of having a standalone system in the first place and made commercialisation a deeply unattractive proposition. As minicomputers increased in specification, manufacturers began to develop much more capable standalone graphics systems for the commercial market.

The first manufacturer to enter the market with a standalone graphics system was DEC, which was able to use the knowledge it had gained as supplier of the equipment for both of the early experimental systems to leap ahead of the competition. In October 1966, DEC introduced the Type 338 Programmed Buffered Display. The Type 338 comprised a PDP-8 minicomputer and a dedicated vector graphics display similar in specification to the Type 340 display used in both experimental systems. No separate display buffer was included but the PDP-8 could be fitted with up to 32,768 words of memory which the display was able to access directly through the minicomputer's high-speed data channel.

In order to cope with a display refresh rate of 30 frames per second, the display list was stored in consecutive memory locations and a special 'display jump' instruction provided which allowed program control to be transferred to any part of the display list. This operated in combination with an extra register called the Display Address Counter, which pointed at the starting address of the display list and incremented a counter each time it was accessed, and a 'pop' instruction which returned control to the main program without the need to specify a return address location. Additional instructions allowed the programmer to create a list of pointers in the form of a pushdown stack which was stored in the first 4,096 words of main memory. These features enabled powerful multilevel and recursive control of the display list.

The Type 338 included a 12-pushbutton keypad as standard but the light pen was an optional extra, as was the electronic character generator. Other hardware options included automatic search logic, which improved the speed at which the display list could be searched for certain instructions, and the confusingly named zoom mode, which allowed the entire 75 by 75 inch addressable display area to be compressed and displayed on the screen at the flick of a switch. The system was also capable of supporting up to 7 additional 'slave' displays, each of which could be fitted with a light pen and programmed to display different information. For particularly demanding applications, interfaces were also available for connecting the Type 338 to a wide range of large-scale computer systems, including IBM and CDC models. By offering these features as options, the basic price of the Type 338 could be kept relatively low at $55,000.

The Type 338 was capable of displaying vector graphics at up to 300,000 points or 700 characters per second. However, the PDP-8 had no floating-point hardware available so floating-point arithmetic was performed in software using a set of standard routines supplied with the machine. This would have constrained the performance somewhat, as floating-point arithmetic is generally preferable for geometric calculations in the majority of graphics applications. Nevertheless, the DEC Type 338 Programmed Buffered Display was the first standalone graphics system to reach the market and its introduction prompted several other manufacturers to follow suit.

The IBM 1130/2250 Graphic Data Processing System

With a product line that included an interactive graphics display unit, the 2250, and a powerful minicomputer, the 1130, IBM was ideally placed to follow DEC's example by introducing a standalone graphics system. However, as the 2250 was designed primarily for use with the System/360 range, which employed different input/output hardware and a longer word length than the 1130, certain changes to the 2250 were necessary to couple it to its new host.

The 2250 was connected to the computer through the storage access channel, a high-speed data channel fitted to the 1130 which was normally used to attach up to two internal disk storage devices. This allowed the display to directly access the main memory of the computer in order to extend the amount of storage space available for the display buffer. The display communicated asynchronously with the 1130 using a method of 'cycle stealing', where memory

10. The Big Picture – Computer Graphics

cycles were periodically grabbed by the display by means of a higher interrupt priority in order to access the computer's main memory.

IBM 1130/2250 Graphic Data Processing System (Photo © IBM Corporation)

The IBM 1130/2250 Graphic Data Processing System was introduced in August 1967 at a purchase price of around $200,000 for a typical single disk configuration. Standard software included a full set of subroutines for graphics programming. However, like the DEC PDP-8, the 1130 had no floating-point hardware so performance would have been quite limited for all but the most basic graphics applications. An optional synchronous data adapter was available which allowed the system to be connected to a System/360 mainframe computer but, with the smallest System/360 model costing over $130,000, this would have been an expensive upgrade for a system which was already very costly in comparison to DEC's Type 338.

The Imlac PDS-1 Programmable Display System

An opportunity to dramatically lower the cost of standalone graphics systems came with the increasing availability of entire families of integrated circuits employing transistor-transistor logic (TTL). These off-the-shelf components simplified the design and construction of digital systems, allowing electronic engineers with limited experience of computer development to create workmanlike designs for small computers which could also be manufactured and sold cheaply. This was a major factor behind the influx of new firms entering the minicomputer market in the late 1960s. It was also the approach taken by the Imlac Corporation, a company established in 1968 in Waltham, Massachusetts, in the development of its first product, the PDS-1 Programmable Display System.

The PDS-1 incorporated a 16-bit minicomputer which was designed and manufactured in-house using Texas Instruments 7400 series digital logic integrated circuits and Dataram Corporation magnetic core memory modules. Loosely based on the DEC PDP-8, it featured a binary parallel architecture with 8 index registers and indirect addressing for up to 32,768 words of 2.0 microsecond memory. This was coupled to a vector graphics display fitted with a 14-inch CRT screen in portrait orientation to provide 1024 by 1024 resolution at a refresh rate of 40 frames per second. The high refresh rate was possible through the addition of a second processor that offloaded the task of refreshing the screen from the computer. The Imlac engineers took the display list processing techniques pioneered in the DEC Type 338 a stage further by creating a special-purpose processor with the logic circuitry necessary to decode display instructions directly and convert them into the correct values for the circuits used to control the position and intensification of the CRT beam. This display list processor contained its own registers but shared the main memory of the computer, using cycle stealing and a higher interrupt priority to access it.

The Imlac PDS-1 was introduced in 1970 at the impressively low price of $9,000 for an entry level system with 4,096 words of memory and a light pen. The extensive use of integrated circuits facilitated a compact design with all the electronics neatly housed inside the pedestal base of the display unit. With a list price which was only marginally higher than a Tektronix T4002 display terminal, the PDS-1 attracted considerable interest and early customers included NASA, McDonnell Douglas Corporation and CERN, the European Organisation for Nuclear Research.

Imlac provided an interactive graphics software package with the PDS-1 which supported basic editing of graphics and text but there was initially very little application software available for the machine. This led many users to develop their own software. One example of this was the creation of an emulator program by engineers at NASA's Ames Research Center which allowed a PDS-1 to replace an IBM 2250 Display Unit. Another notable example, which was also developed at the Ames Research Center, was Maze War, one of the earliest 3-D computer games.

Maze War was created in the summer of 1973 by Stephen Colley, Greg Thompson and Howard Palmer, three high school students who were working at the Ames Research Center under a school study programme. The game required the player to steer an eyeball-shaped character through a maze which was simultaneously displayed in both perspective view, using a 3-D wireframe representation, and plan view. To make the game more involving, a multi-player version was also created by linking two PDS-1 systems together using the serial interface provided by Imlac for connecting printers. Maze War was later ported to other graphics-enabled hardware platforms where it proved useful as both a demonstration program and a research tool. It also spawned several commercial versions.

10. The Big Picture – Computer Graphics

An upgraded version of the PDS-1, the PDS-1D, was introduced in February 1972 at a price of $9,970. The new version featured faster 1.8 microsecond memory to increase performance by 10% and a larger 15-inch screen and 8,194 words of memory were also fitted as standard. Imlac later produced a more powerful model, the PDS-4, which was introduced in 1974 at a price of $17,300. By 1977, sales totals for both models had reached several hundred units in what was still a very small market, making Imlac the undisputed market leader in refreshed graphics displays during this period.

Display Processors

As computer graphics entered the mainstream, a new demand began to emerge for increased performance in high-end interactive graphics applications. Minicomputers were not well suited to the type of spatial coordinate transformation calculations required for moving objects around the screen, such as translation, rotation and scaling. Coordinate transformations are normally accomplished using matrix equations. These equations become even more complex for 3-D graphics. With each vector requiring dozens of individual arithmetic operations to be performed for every screen refresh cycle, the calculation rate for a complex moving image can easily run into millions of operations per second. Only the very fastest scientific computers of the day, such as the CDC 6600, could deliver this kind of compute performance but such systems were well beyond the financial reach of most organisations. If general-purpose computers were not the answer, the alternative would be to develop special-purpose processors that were specifically designed to handle coordinate transformations.

The Adage Graphics Terminal

Graphics systems with special-purpose processing hardware were first introduced by Adage, a Boston-based manufacturer of specialised computer systems for signal processing. The company's main product was the Ambilog 200 General Purpose Hybrid Computer which was introduced in 1964. It incorporated various types of analogue and hybrid computing modules that operated in parallel and were controlled by a 30-bit digital processor. An optional CRT display was also available for graphical output. In 1966, Adage engineers Thomas G Hagan and Robert Treiber published a paper describing various techniques for signal processing using the Ambilog computer, one of which was spatial coordinate transformation. This provided the idea for the development of a standalone graphics display system, the Adage Graphics Terminal.

The Adage Graphics Terminal took the 30-bit processor from the Ambilog, added up to 32,768 words of 2.0 microsecond magnetic core memory and coupled it to a vector graphics display. The coordinate transformation hardware comprised a parallel array of multiplying D-to-A converters which were configured to solve matrix equations. The number of elements in the array depended on the type of graphics required, with a 3 by 3 array necessary for 3-D graphics and a 2 by 2 array for 2-D graphics. Coordinate values for a vector were converted into analogue signals and fed into the elements in each column of the array whereupon they were multiplied by a digital coefficient representing the coordinate transform. The outputs from each row of the array were then summed and converted back into a digital value to give the transformed coordinates or fed directly into the analogue beam control circuitry of the CRT display.

In order to reduce the storage requirements for the display list, images could be broken down into sub-images and the coordinate transformation hardware used to perform repeated translation and scaling on the sub-image instead of having to store coordinates for every vector in the image. The hardware also supported variable intensity depth cueing, a technique where objects which are further away in the 3-D image are displayed with reduced CRT beam intensity in order to enhance the appearance of depth.

Three models of Adage Graphics Terminal were introduced simultaneously in 1968, the AGT-10 for 2-D graphics and the AGT-30 and AGT-50 (both of which were fitted with a larger array) for 3-D graphics. The 3-D models were capable of displaying moving images containing up to 5,000 vectors at 40 frames per second. Graphical input options included a light pen, joystick, digitising tablet and 6 control knobs. With prices starting at only $60,000, the Adage models soon became established as the display equipment of choice for 3-D graphics research.

The Line Drawing System

The Adage Graphics Terminal raised the bar for computer graphics performance. It also proved that there was indeed a commercial market for high-end display systems and opened the door for other firms to strive for even higher performance in order to create systems for customers for whom even Adage displays were too slow. The first company to seize this opportunity was led by CAD and standalone graphics system pioneer, Ivan Sutherland.

Sutherland had replaced Joseph Licklider as head of ARPA's Information Processing Techniques Office when Licklider returned to MIT in 1964. He then spent 2 years as an associate professor at Harvard before moving to the University of Utah in 1968 at the invitation of computer science department head David C Evans, who Sutherland knew from his time at ARPA. Evans also had an impressive track record, having served as project manager for the development of the first commercially successful small-scale computer, the Bendix G-15, and as co-leader of Project Genie, the early timesharing system

10. The Big Picture – Computer Graphics

which was developed at the University of California Berkeley, in 1963-64. Evans, a devout Mormon, wanted to establish a computer graphics research centre at Utah and Sutherland was the ideal person to help him. However, his offer to Sutherland of a professorial position came with an unusual caveat, that they also make efforts to commercialise their work.

In May 1968, the two men founded Evans & Sutherland Computer Corporation, setting up business on a research park adjacent to the University of Utah campus in Salt Lake City. Start up finance included funding from a group of investors associated with the Rockefeller family. The company's first product was a high-end graphics display system named the Line Drawing System Model 1 (LDS-1) which was developed with input from researchers at BBN who were also the first to place an order for one. The logic design for the system was the work of Charles L Seitz, a doctoral student at MIT whose area of expertise was in asynchronous logic. After obtaining his PhD, Seitz was also invited to join the computer science department at Utah where he continued to provide design services to E&S.

Employing a similar approach to the Adage Graphics Terminal, the LDS-1 incorporated a matrix multiplier unit which featured a 4 by 4 array of multiplying elements to perform translation, rotation and scaling of 2-D and 3-D coordinates in hardware. It also included a novel 'clipping divider', developed by Sutherland and colleague Robert F Sproull during his time at Harvard University, which automatically differentiated the parts of an image which are visible in the field of view and should be displayed from those which are not and can be ignored. Both units were controlled by a central processing unit with an 18-bit word length and an instruction set geared towards graphics operations. All three units were designed to operate asynchronously and arranged in a pipeline so that each could operate on a different part of the image at the same time.

The electronics were housed within two large cabinets and connected by a 50-foot cable to a moveable vector graphics display fitted with a 10-inch square CRT screen. Although eminently capable of operating as a standalone system, the LDS-1 was also designed for use with DEC's large-scale PDP-10 computer. Connection was via the PDP-10's memory port which allowed the LDS-1 to interface directly with the main memory of the DEC machine through the now common method of cycle stealing, aided by a word length which was exactly half that of the 36-bit PDP-10.

Deliveries of the LDS-1 began in August 1969. With a display performance of 100,000 lines per second, the LDS-1 was marginally faster than the Adage Graphics Terminal but it was also considerably more expensive, as a fully configured LDS-1 cost $250,000. Besides BBN, other LDS-1 customers included the US Navy and Princeton University's computer graphics laboratory. However, only a handful of systems were sold before it was superseded by the

LDS-2 in 1971. Nevertheless, the pipelined architecture first featured in the LDS-1 would become the industry standard for graphics hardware design and the Evans & Sutherland name would be synonymous with high performance graphics throughout the next two decades.

Further Reading

Anderson, R. H., A Simplified Direct-Viewing Bistable Storage Tube, IEEE Transactions on Electron Devices ED-14 (12), 1967, 838-844.

Astrahan, M. M. and Jacobs, J. F., History of the Design of the SAGE Computer – The AN/FSQ-7, IEEE Annals of the History of Computing 5 (4), 1983, 340-349.

Ball, N. A., Foster, H. Q., Long, W. H., Sutherland, I. E. and Wigington, R. L., A Shared Memory Computer Display System, IEEE Transactions on Electronic Computers EC-15 (5), 1966, 750-756.

Dimond, T. L., Devices for Reading Handwritten Characters, Proceedings of the Eastern Joint Computer Conference, 1957, 232-237.

English, W. K., Engelbart, D. C. and Huddart, B., Computer-Aided Display Control, Final Report, Contract NAS 1-3988, Stanford Research Institute, Menlo Park, California, 1965.

Everett, R. R., Zraket, C. A. and Benington H. D., SAGE – A Data-Processing System for Air Defense, IEEE Annals of the History of Computing 5 (4), 1983, 330-339.

Forrester, J. W. and Everett, R. R., The Whirlwind Computer Project, IEEE Transactions on Aerospace and Electronic Systems 26 (5), 1990, 903-910.

Hagan, T. G., Nixon, R. J. and Schaefer, L. J., The Adage Graphics Terminal, Proceedings of the Fall Joint Computer Conference, 1968, 747-755.

Jacks, E. L., A Laboratory for the Study of Graphical Man-Machine Communication, Proceedings of the Fall Joint Computer Conference, 1964, 343-350.

Krull, F. N., The Origin of Computer Graphics within General Motors, IEEE Annals of the History of Computing 16 (3), 1994, 40-56.

Lee, J. A. N., McCarthy, J. and Licklider, J. C. R., The Beginnings at MIT, IEEE Annals of the History of Computing 14 (1), 1992, 18-30.

McCarthy, J., Boilen, S., Fredkin, E. and Licklider, J. C. R., A Time-Sharing Debugging System for a Small Computer, Proceedings of the Spring Joint Computer Conference, 1963, 51-57.

Newman, W. M. and Sproull, R. F., Principles of Interactive Computer Graphics, McGraw-Hill, New York, 1979.

Ninke, W. H., GRAPHIC 1 – A Remote Graphical Display Console System, Proceedings of the Fall Joint Computer Conference, 1965, 839-846.

Potts, J., Computer Graphics – Whence and Hence, Computers & Graphics 1 (2/3), 1975, 137-156.

Rapkin, M. D. and Abu-Gheida, O. M., Stand-Alone/Remote Graphics System, Proceedings of the Fall Joint Computer Conference, 1968, 731-746.

Redmond, K. C. and Smith, T. M., Project Whirlwind: History of a Pioneer Computer, Digital Press, Bedford, Massachusetts, 1980.

Ross, D. T., A Personal View of the Personal Work Station: Some Firsts in the Fifties, Proceedings of the ACM Conference on the History of Personal Workstations, 1986, 19-48.

Sutherland, I. E., Sketchpad: A Man-Machine Graphical Communication System, Proceedings of the Spring Joint Computer Conference 23, 1963, 329-346.

Swets, J. A., The ABC's of BBN: From Acoustics to Behavioural Sciences to Computers, IEEE Annals of the History of Computing 27 (2), 2005, 15-29.

Ware, W. H., RAND and the Information Evolution: A History in Essays and Vignettes, RAND Corporation, Santa Monica, California, 2008.

11. The Microcomputer Revolution

"There is no reason anyone would want a computer in their home",
Kenneth H Olsen, 1977.

Creating a Mass Market

The minicomputer had shown how the commercial market for computers could be dramatically expanded by reducing the price of the equipment to the point where their use could be justified in hitherto unjustifiable applications such as process control. The real-time capability of minicomputers also opened up exciting new possibilities for interacting with computers in a much more direct way, particularly when combined with advanced input/output technologies such as graphics displays and pointing devices. Interactive applications, such as Computer-Aided Design, were early indicators of how productivity soared when users were given their own personal interactive graphics terminal or standalone computer on which to work. By 1970, low-cost minicomputer models such as the Data General Nova and DEC PDP-11 were selling in their tens of thousands, hinting at the possibility of a mass market for computers. However, creating a mass market would require an even greater reduction in production costs than that seen following the adoption of integrated circuits. The opportunity to do this came in 1971 with the introduction of the microprocessor.

Building Blocks – Microprocessors and Semiconductor Memory

Microprocessors are electronic devices which combine all the logic circuitry of a computer's central processing unit onto a single integrated circuit. By miniaturising and packaging an entire processor onto a tiny silicon chip, huge savings in both space and cost are possible. Reliability is also improved through the elimination of the numerous wires and solder joints that are required to interconnect the components in a 'discrete' processor design.

The same integrated circuit technology which led to the development of the microprocessor also slashed the cost of manufacturing transistors, as large numbers of miniaturised transistors could be fabricated onto each chip of silicon wafer in a single process. This opened up the possibility of building entire computer memories using flip-flops, a technique that had previously been uneconomical due to the vast number of individual transistors required

Fuelling the Demand

The development of microprocessors was driven primarily by the needs of the calculating machine industry. Electronic desktop calculators first appeared in 1961 with the introduction of the Sumlock ANITA Mk VIII which was developed by the Bell Punch Company, a British manufacturer of key-driven mechanical calculators. Development was led by Norbert A Kitz, an experienced digital circuit designer who had worked for computer pioneer Andrew Booth while a Master's student at the University of London's Birkbeck College in the late 1940s and had also been a member of the design team for the English Electric DEUCE computer.

The ANITA Mk VIII was introduced in October 1961 at a price of £335 (approximately $940 at 1961 exchange rates), which compared favourably with similarly specified mechanical models. Adverts for the new model emphasised the speed and quietness of operation. However, the circuitry was not solid-state as it was based on the Dekatron tube, a type of thermionic valve designed to operate as a ring counter. Consequently, reliability was poorer than the mechanical models that it replaced, with most examples requiring repair after only a year of operation. Despite this, the ANITA was a popular model, with sales reaching almost 10,000 units per year while it remained the only electronic desktop calculator on the market until the introduction of the EC-130 by US teleprinter manufacturer Friden in June 1963.

The Friden EC-130 was the first desktop calculator to feature solid-state circuitry. It was soon followed by solid-state models from the Sharp Corporation in Japan and Industria Macchine Elettroniche (IME) in Italy. The adoption of solid-state electronics allowed manufacturers to increase the functionality of desktop calculators and the first programmable model also appeared in 1963, the bizarrely named Mathatronics Mathatron.

Mathatronics Incorporated was established in Waltham, Massachusetts in February 1962 by William M Kahn, Roy W Reach and David Shapiro. Kahn, an experienced computer design engineer who had worked at Honeywell and Raytheon, believed that many of the features found in digital computers could also be applied to desktop calculators and had been working in his spare time to develop the ideas behind such a machine. The result of these efforts was the first programmable desktop calculator, the Mathatronics Mathatron 8-48, which was announced in November 1963.

The Mathatron differed from conventional four-function decimal calculators in that it incorporated a small magnetic core memory which allowed the user to record and replay keystroke sequences of up to 48 steps. It also included 8

11. The Microcomputer Revolution

storage registers for storing intermediate results. Parenthesis keys permitted mathematical expressions to be input to the machine in the same format as they were written down. Memory was expandable up to a maximum of 480 steps and 48 registers, and an optional punched tape unit and printer were also available. These features gave the Mathatron a capability that went far beyond that of a desktop calculator and early marketing literature described the machine, not entirely inaccurately, as a digital computer.

Weighing almost 40 kilograms and with its complex circuitry containing over 1,000 individual transistors, the Mathatron was costly to manufacture and this was reflected in a correspondingly hefty list price of $5,000. Nevertheless, with the cheapest minicomputers in 1963 costing 5 times as much, the Mathatron addressed a gap in the market and was highly successful. Soon other manufacturers began to offer programmable models but these too were uniformly bulky and expensive. A move to integrated circuits, which began in February 1967 with the introduction of the Sharp Compet CS-31A, helped to reduce both the physical size and the cost of desktop calculators. It also led to the idea of an entire calculator on a single chip.

Early Efforts

The first integrated circuit devices introduced in 1960 contained only a handful of electronic components but component density increased steadily as semiconductor fabrication techniques improved and by 1968 this had risen to several hundred components, acquiring the term large-scale integration (LSI). As the number of components on a chip continued to rise, the idea of extending the integrated circuit concept to an entire calculator or computer processor began to occupy the thoughts of a number of people within the electronics industry.

One of the earliest attempts to develop a microprocessor was made by Gilbert P Hyatt, an electrical engineer who had worked on special-purpose computer systems at Hughes Aircraft Company and Teledyne. In 1968, Hyatt founded Micro Computer Incorporated (MCI) in Los Angeles, California with the idea of developing low-cost computers for industrial and business applications using custom integrated circuits. The company's first product was the Contourama IV, a dedicated computer for machine tool numerical control applications. This was used as a platform for further products including an illumination control system for use in the production of printed circuit boards.

Hyatt filed a US patent application in November 1969 describing the architecture and logic design of the numerical control system, including the bold claim that it could be integrated onto a single chip. However, this claim was never realised. MCI went out of business in September 1971 following a disagreement between Hyatt and his investors over Hyatt's refusal to assign his intellectual property rights to the company. Hyatt was deeply suspicious of the investors' motives so, rather than attempt to resurrect the company, he went into business for himself as an independent inventor and consultant to the aerospace industry. Although he would continue to pursue his patent application, he would never again have the means to further develop his technology.

The next significant advance in the development of microprocessors was made by another Californian start-up firm, Four-Phase Systems, and was the brainchild of company founder Lee L Boysel. Boysel was among the first wave of semiconductor design engineers to acquire expertise in the new metal-oxide-silicon (MOS) integrated circuit fabrication process that had facilitated the move towards large-scale integration. While working at Fairchild Semiconductor he had designed an 8-bit arithmetic logic unit chip, the Fairchild 3800, which demonstrated the potential of the new MOS fabrication technology. In October 1968, Boysel left Fairchild along with several members of his group in order to establish Four-Phase Systems with the aim of exploiting MOS technology for computer equipment.

In April 1969, Four-Phase Systems completed the development of the AL1 processor chip. The AL1 comprised an 8-bit arithmetic unit and eight 8-bit registers coupled to a 24-bit wide system bus. It was intended to be used in sets of three to create a 24-bit processor, along with a further 9 memory and input/output controller chips, although it was possible to use it in a single-chip configuration, supported by a smaller number of memory and input/output chips.

The AL1 was designed using four-phase logic, a methodology invented in 1966 at the Autonetics division of North American Aviation which delivered significant improvements in efficiency and component density over conventional methods when designing logic circuits using MOS technology. The resulting device contained more than 1,000 logic gates on an area of silicon wafer only 0.13 by 0.12 inches in size. It was encased in a standard 40-pin dual in-line package (DIP), a small rectangular housing with a row of 20 connecting pins on each of the two longer sides, and manufactured by Cartesian, another start-up firm established by former Fairchild Semiconductor employees.

The AL1 was the central component of a new dedicated minicomputer, the Four-Phase Systems Model IV/70, which was introduced in the autumn of 1970 for use as a communications processor in large IBM System/360 installations. Four-Phase Systems steadfastly refused requests to sell the AL1 chip separately, remaining tightly focused on developing and selling dedicated minicomputers within what turned out to be a highly lucrative niche market. The firm also made no attempt to patent it, as the idea was considered commonplace. However, Boysel and his colleagues did publish several papers describing the technology, including a tutorial article in the April 1970 issue of Computer Design magazine, which almost certainly influenced later microprocessor

11. The Microcomputer Revolution

developments. Several hundred Model IV/70 systems were sold, heralding a period of sustained success for Four-Phase Systems which culminated in the company's acquisition by Motorola in 1981 for $253 million.

Single-Chip Calculators

The Four-Phase Systems AL1 was a major step forward in the development of the microprocessor but it was expensive to manufacture and required the addition of several equally expensive chips to support it. The earliest single-chip solutions were developed in parallel by two small companies working independently, one based in Texas and the other on the opposite side of the Atlantic in Scotland.

The first of these two companies, Mostek Corporation, was established in Carrollton, Texas in June 1969 by a group of engineers from Texas Instruments led by L J Sevin and Louay E Sharif. Like the founders of Four-Phase Systems, the engineers recognised the enormous potential of the new MOS fabrication process. However, despite having a pioneering role in the development of integrated circuits, TI was a deeply conservative company and was reluctant to fully embrace MOS technology. This led Sevin, Sharif and six of their colleagues to leave TI and set up their own business specialising in the design of custom MOS devices. Start-up finance was provided by $250,000 in venture capital plus a $3 million investment by the Sprague Electric Company, which also provided access to its state-of-the-art semiconductor fabrication facilities in Worcester, Massachusetts.

In May 1970 Mostek secured a contract from the Nippon Calculating Machine Company, a leading Japanese manufacturer of electronic calculators which were sold under the brand name Busicom. The brief was to develop a single-chip solution for a new range of calculators that could compete with the recently introduced Pocketronic handheld model from arch rival Canon. The Pocketronic had been created in partnership with Texas Instruments and was based on technology developed at TI by a team led by integrated circuit co-inventor Jack Kilby. Priced at only $395, it employed a minimalist design which replaced the costly digital display with a thermal printer and reduced most of the circuitry onto a set of three chips supplied by TI. Nippon knew that further savings in manufacturing costs could be made if all of the calculator logic was reduced onto a single integrated circuit and agreed to place an order with Mostek for a minimum of 60,000 chips at $30 each if they could have a working device ready by mid November.

The next few months would involve frenetic activity by the Mostek engineers in order to meet Nippon's tight development timescale. This they succeeded in doing and the resulting device was designated the Mostek MK6010. It contained 360 logic gates and 160 flip-flops implemented using 2,100 transistors on a 0.18 inch square of silicon wafer which was encased in a 40-pin DIP package.

The MK6010 went into production in January 1971 and was first employed in an updated version of the Busicom Junior, a compact four-function desktop calculator, replacing 22 separate integrated circuits in the previous model. It was also used in the Busicom LE-120A, a small handheld model introduced in early 1971 which is generally considered to be the first true pocket calculator. The use of the new Mostek chip in this model allowed Nippon to offer a much more elaborate design featuring the latest LED display technology at a price which matched the Canon Pocketronic.

Another single-chip device for calculators was developed during the same period by Pico Electronics Limited. Pico Electronics was founded in the town of Glenrothes, Scotland, in 1970 by four engineers from General Instrument, a US electronics manufacturer which had established a microelectronics facility in the same town the year before. The new company was created with the blessing of General Instrument, which provided the necessary start-up finance in return for exclusive manufacturing rights to the single-chip devices that the new company was planning to develop.

While working at General Instrument, the Pico founders had designed a set of five integrated circuits for the Swedish calculator company Facit. Some members of the team had also worked on the development of MOS chips for the Sumlock ANITA 1000 calculator at another Glenrothes facility operated by Elliott-Automation. They were able to use this experience to design a single-chip MOS device which they named the PICO1.

The PICO1 was of a more advanced specification than the Mostek device in that it supported floating-point arithmetic. This was made possible through the inclusion of three numerical storage registers, a decimal point register and a small amount of read-only memory for storing the sequences of individual steps necessary to carry out each floating-point arithmetic operation. This inclusion of on-chip memory was an industry first, predating its use in later general-purpose microprocessors.

The prototype of the PICO1 chip was completed towards the end of 1970 and a UK patent application was filed in March 1971, followed by a US application four months later. The chip was manufactured and marketed by General Instrument under the name GI 250. It was first used in the Royal Digital III, a budget handheld model introduced by the Monroe Calculating Machine Company in late 1971 at a price of $139. This dramatically low price was achieved by limiting the display to only four digits and replacing the usual pushbutton keypad with a stylus which was used to contact metal pads on the circuit board through a set of apertures in the casing.

The Royal Digital III and similarly low-priced models which followed from other manufacturers turned calculators into a consumer product. Their existence would not have been possible without the development of single-chip calculator devices.

11. The Microcomputer Revolution

The First Microcontroller

The single-chip approach taken by Mostek and Pico Electronics for calculators was also adopted by a team at Texas Instruments for a different application, industrial control. This resulted in the development of the first microcontroller chip. Unlike microprocessors, which are normally designed to operate with external memory devices and other support chips, microcontrollers are self-contained processing systems with main memory incorporated onto the same chip. The project began in December 1970 and was led by electrical engineers Gary W Boone and Michael J Cochran. It arose out of a dawning realisation that the needs of many of TI's integrated circuit customers were so similar that they could be met by a single-chip device rather than designing a new set of custom chips for every contract.

Using experience gained in designing calculator chip sets, the team completed the development of the new chip, which was codenamed the TMS-100, in July 1971. A US patent application was filed the same month. The logic design was based on that of an 8-digit four-function calculator with 3 storage registers but the chip also incorporated a 3520-bit read-only memory for program storage, programmable input/output decoders and a real-time clock, all of which made it possible to use the chip for simple industrial control applications.

The new chip was announced as the TMS1802NC in September 1971 at a price of $150 which dropped to less than $20 when ordered in large quantities. The TMS1802NC incorporated approximately 5,000 transistors on a 0.23 inch square of silicon wafer encased in a standard 24-pin DIP package. TI's marketing literature emphasised the capability of the new device as a single-chip solution for calculator manufacturers but also highlighted its potential for use in non-calculator applications such as meters, cash registers and terminals.

The TMS1802NC was primarily a calculator chip and as such was not ideally suited to control applications. Nevertheless, it introduced the microcontroller concept to the semiconductor industry which prompted the development of dedicated microcontroller chips based on binary parallel computer architectures. These later microcontrollers found a market as 'embedded' systems in domestic products and were directly responsible for the emergence of a global market for consumer electronics products.

The Birth of an Industry Giant

The company most closely associated with the development of the microprocessor is the Intel Corporation of Santa Clara, California. Intel began life as yet another start-up firm established by former Fairchild Semiconductor employees, one of the so-called Fairchildren. It was founded in July 1968 by integrated circuit co-inventor Robert Noyce and his chemical engineer colleague Gordon E Moore under the name NM Electronics. Fairchild Semiconductor had been experiencing a period of instability which had seen several changes of Chief Executive in less than a year. Noyce and Moore believed that they had spotted a significant new market opportunity to supply semiconductor memory products to the computer industry. However, rather than wait for the dust to settle at Fairchild before pitching their idea, they decided instead to branch out on their own.

Start-up finance came from venture capitalist Arthur Rock, whose business acumen had helped grow minicomputer pioneer Scientific Data Systems into a billion dollar company. Rock had also played a key role in setting up Fairchild Semiconductor some years earlier and this was where he had first met Noyce and Moore. Rock was impressed by their vision for the new venture and agreed to help them raise the necessary capital, providing that they show their commitment by putting in some of their own money. Noyce and Moore were able to raise an impressive $490,000 between them, to which Rock added $10,000 of his own money. Rock then made a number of telephone calls to potential investors, all of whom agreed to invest. Within a single day he had succeeded in securing a total of $2.5 million in convertible debentures. Rock also took on the role of Chairman of the Board in the fledgling company.

Within a few weeks, the name of the company was changed from NM Electronics to the more familiar Intel, a word coined by Moore as a shortened version of 'Integrated Electronics', and work began on the company's first products. Two products, both semiconductor memory chips, were developed in parallel using different fabrication processes in order to see which could be brought to market first. The first to be completed was the Intel 3101, a 64-bit high-speed random-access memory (RAM) chip employing bipolar semiconductor technology, which was introduced in late 1969. It was followed a few months later by the Intel 1101, a 256-bit RAM chip developed using the newer MOS process. Despite taking longer to develop, the MOS process had again shown that it was capable of producing higher component density and it was chosen as the favoured process for future Intel devices.

The Intel 3101 and 1101 were both static RAM (SRAM) devices, comprising arrays of 1-bit memory cells, each of which contained 4 transistors in a flip-flop circuit plus 2 additional transistors to control access. The latching action of the flip-flops meant that they held their state indefinitely without the need for a refresh cycle but, with 6 transistors required for each bit of storage, the chips were no cheaper to produce than magnetic core storage. High numbers of transistors also require a relatively large amount of power which could give rise to heat dissipation problems when packing large numbers of transistors into such a small area. Fortunately, a new type of memory technology known as dynamic RAM (DRAM) had the potential to significantly reduce the number of transistors per bit.

DRAM was invented by IBM researcher Robert H Dennard in 1967. Dennard dispensed with the flip-flop and used a capacitor as the basic storage element.

11. The Microcomputer Revolution

His memory cell contained only two components, the capacitor itself and a single transistor which controlled charging and reading. Continuous refreshing was necessary as the reading process was destructive and the capacitors could only retain their charge for a short period, hence the name dynamic, but the dramatic reduction in the number of transistors required per bit of storage from 6 to 1 held the prospect of a similar reduction in manufacturing costs.

The first successful implementation of DRAM technology was a 1,024-bit DRAM chip developed by Four-Phase Systems in 1969 for the AL1 processor but this was never made available as a separate product. Intel's first foray into DRAM technology came through collaboration with computer manufacturer Honeywell on the development of a 512-bit memory chip for its 3C Division's minicomputer line. The chip would be based on a 3-transistor memory cell invented by Honeywell engineer William M Regitz which possessed certain advantages over Dennard's 1-transistor cell despite the higher number of transistors used. The resulting device was the Intel 1102, a 1024-bit DRAM chip which was announced in February 1970. However, the method used for connecting the individual components within a memory cell was found to be difficult to fabricate which resulted in a very low 'yield' (the proportion of defect free chips produced from each silicon wafer processed). Fortunately, the Intel engineers recognised this early on in the development of the 1102 and initiated a parallel project to design an improved version of the chip using an alternative 3-transistor cell arrangement. The completed chip, which was named the Intel 1103, was more complex in design but cheaper to manufacture. Intel decided to drop the 1102 in favour of the 1103, much to the dismay of the Honeywell engineers. They also hired Regitz to oversee 1103 production and take charge of ongoing development.

As the first commercially available DRAM chip, the Intel 1103 offered computer manufacturers a cost-effective alternative to magnetic core memory. Following its introduction in October 1970, spectacular sales of the new chip helped to establish Intel as a major player in the semiconductor industry. Rapid adoption as the industry standard memory chip prompted several other semiconductor firms to secure the rights to manufacture the 1103 under 'second source' license agreements and by 1972 the Intel 1103 was the largest selling integrated circuit device in the world. The success of the 1103 also gave Intel the confidence to branch out into other semiconductor products, not least of which was the microprocessor.

The Commercialisation of the Microprocessor

In September 1969 Intel received a contract from the Nippon Calculating Machine Company to develop a chip set for a new family of Busicom programmable calculators. With no products on the market yet, the project was a welcome source of income for Intel, despite being outside the company's main area of expertise in semiconductor memory. Nippon design engineer

The Story of the Computer

Masatoshi Shima had specified a set of 12 chips with various functions which could be used in different combinations to create a range of calculator models. However, Shima's design was very complicated, with each chip containing an average of more than 2,000 transistors encased in various non-standard packages requiring up to 40 pins. Marcian E Hoff, the Intel engineer responsible for liaising with Nippon on the project, was concerned that this design would be too expensive to implement. Taking inspiration from the minimalist logic design of his DEC PDP-8 minicomputer, Hoff proposed a smaller chip set which would combine all the required functionality into only 4 chips, using a general-purpose binary programmable microprocessor as the principal element. The microprocessor would only require a 4-bit word length to handle the simple BCD arithmetic necessary for the calculators. It would be supported by a 2048-bit read-only memory (ROM) chip for program storage, a 320-bit DRAM chip for data storage and a 10-bit shift register chip for input/output. Standard 16-pin DIP packaging would be employed throughout to reduce manufacturing costs.

Hoff's proposal was presented to Nippon Calculating Machine executives in October 1969 and was swiftly accepted. Shortly afterwards, Hoff was reassigned to another project and his design was completed by Intel colleagues Stanley Mazor and Federico Faggin, with input from Shima on programming and architectural improvements. Faggin, a physicist, had previously worked at Fairchild Semiconductor, where he had developed a new fabrication process known as 'silicon gate technology' which allowed a further reduction in the size of MOS transistors and also increased their reliability. Faggin's process was adopted by Intel for its own devices and would be used to fabricate the new chip set. However, with the bulk of Intel's limited design resources deployed on the development of the 1102 and 1103 memory chips, progress was slow until Faggin joined the team in April 1970 and it would be early 1971 before a complete set of prototype devices was available for testing.

As Nippon had paid for the development of the new chip set, it had been granted exclusive rights to the technology. This prevented Intel from selling the chips to anyone else despite the obvious potential of the microprocessor chip as a general-purpose device. Fortunately for Intel, Nippon was beginning to encounter fierce competition in the calculator market and was now anxious to renegotiate the terms of the contract. Encouraged by Hoff and Faggin, who had both recognised the commercial potential of the chip set in industrial control applications, Intel sales executives took the opportunity to regain the rights to the technology in return for a concession on price. Intel was now free to introduce the first commercially available microprocessor to the market but co-founders Noyce and Moore remained unconvinced, expressing concern over the additional resources necessary to support a much more complex product than memory chips. Their reservations dissipated following the hiring of experienced marketing executive Edward L Gelbach from Texas Instruments

11. The Microcomputer Revolution

in August 1971, who pointed out that a microprocessor product would also drive sales of Intel's semiconductor memory devices, as microprocessor customers would also require memory.

Intel 4004 (Photo © Intel Corporation)

The microprocessor was launched in November 1971 with an advertisement in the US trade journal Electronic News under the bold heading "Announcing a new era of integrated electronics". The advertising campaign featured the entire chip set, which Intel labelled the MCS-4, but it was the 4004 microprocessor which attracted the most attention. Despite having only a 4-bit word length, the 4004 was a fully featured processor with a binary parallel architecture, 16 index registers, a 740 KHz clock speed and a comprehensive set of 45 machine instructions. It contained 2,250 transistors on an area of silicon wafer 0.15 by 0.11 inches in size. Priced at less than $100 for the complete set of 4 chips when ordered in large quantities, the 4004 soon found a ready market as a replacement for custom logic circuitry in control systems but the word length was too short to interest computer designers. This would be remedied in Intel's next microprocessor product.

Early in the development of the 4004, Intel received a contract from Computer Terminal Corporation (CTC) of San Antonio, Texas, to develop a 512-bit shift register memory chip for a new programmable display terminal, the Datapoint 2200. The terminal would feature an 8-bit processor implemented using standard TTL integrated circuit components. Guided by their experiences on the 4004 project, Hoff and Mazor suggested that Intel could also produce a microprocessor chip based on CTC's processor specification which would replace many of the discrete components, thus helping to reduce manufacturing costs. CTC agreed to this proposal on the basis that Intel could supply 100,000 chips at a price of $30 each.

In March 1970 Intel hired Harold Feeney from General Instrument to lead the design of the new 8-bit microprocessor. However, shortly after Feeney started work on the design, CTC began experiencing financial difficulties and the development of a microprocessor was no longer considered a priority. The project was put on hold while Feeney was reassigned to more pressing projects. Work resumed in January 1971 but it soon became clear to the CTC engineers that the performance of the microprocessor would not be a match for TTL circuits, with their fast bipolar junction transistors, and the expected cost saving was no longer so impressive due to a subsequent fall in the price of TTL chips. CTC decided to return to its original plan of a discrete implementation of the processor and requested termination of the contract. This was not necessarily a disastrous outcome for Intel as, by terminating the contract with CTC, it would regain the rights to the technology and could sell the 8-bit microprocessor to other customers, such as the Japanese company Seiko which had expressed an interest in using the device for a programmable desktop calculator. It could also be introduced as a standard product to complement the soon to be released 4004.

Following termination of the CTC contract, Hoff and Mazor took the opportunity to make some changes to the instruction set of the new microprocessor. Having been specified by CTC engineers for the Datapoint terminal, many of the instructions were related to character string handling. These were retained but register increment and decrement instructions were added and branch instructions revised to make the processor more versatile. The final set comprised 48 instructions, a similar number to the 4004 but with a different mix of functions that would make the chip more suitable for higher level applications.

The new microprocessor was introduced as the Intel 8008 in April 1972 at a list price of $120. It featured an 8-bit binary parallel architecture with a 500 KHz clock speed, 6 general-purpose registers and support for interrupt handling. A generous address space provided for up to 16,384 bytes of memory, 4 times that of the 4004. Unlike the 4004, the 8008 was designed to be used with standard chips for memory and other support functions. The 3,300 transistor device had been squeezed into a standard 18-pin DIP package, the largest that Intel was then capable of manufacturing. However, this left only 8 pins available for both the 14-bit wide address bus and the 8-bit data bus. The Intel engineers solved this problem by employing a time-division multiplexing technique to interleave signals under the control of a two-phase clock but this also increased the number of support chips required to a minimum of 20.

The Intel 8008 was first used in the Seiko S-500, a programmable desktop

11. The Microcomputer Revolution

scientific calculator introduced in early 1973. It was also adopted by other Intel customers, such as the terminal manufacturer Sycor, as an alternative to discrete logic circuitry in new versions of existing products. However, computer manufacturers were initially reluctant to take on the 8008 due to insufficient technical information and a lack of development tools which made it very difficult to incorporate such a complex device into their products. Intel responded by producing detailed user manuals and organising a series of technical seminars and workshops around the country. Two evaluation boards, the SIM4-01 for the 4004 and SIM8-01 for the 8008, were also introduced. These comprised a printed circuit board fitted with sockets for the microprocessor and all the support chips necessary to create a rudimentary computer system. The board plugged into an edge connector on the top of a box-shaped chassis which contained the necessary power supplies and a standard interface for a Teletype terminal. A basic suite of software tools was also supplied, the development of which had been subcontracted to a local software company. This included various test programs, a 'bootstrap' loader for loading programs into memory and an assembler for converting assembly language programs into machine code.

Intel's user manuals and evaluation boards proved highly popular, with sales revenues from them outstripping those from the microprocessors themselves for the first few years. However, the expected uptake of the 8008 by mainstream computer manufacturers failed to materialise and feedback from engineers evaluating it revealed that performance was a major concern. To address this, Faggin proposed the development of a new 8-bit microprocessor which would employ the latest N-channel silicon gate MOS process to increase the speed of the logic circuitry and various design improvements to increase other aspects of the performance. As an interim measure, a faster version of the 8008, designed the 8008-1, was introduced with an increased clock speed of 800 KHz and a similarly increased price tag of $180.

Intel's new high-performance microprocessor was named the 8080 and introduced in April 1974. The design was directed by Masatoshi Shima, who had moved to the USA to become an Intel employee in November 1972. The 8080 retained the same overall architecture as the 8008 but added DMA support, greater parallelism and a 16-bit program counter for improved memory addressing. It also featured an enhanced instruction set which added 30 instructions while retaining upward compatibility with the 8008. A 2 MHz clock speed helped deliver a tenfold performance increase over the 8008. The N-channel fabrication process had facilitated a 40% reduction in the size of the individual electronic components, allowing 6,000 transistors to be fitted onto an area of silicon wafer only slightly larger than that of the 8008. Packaging was also improved through the use of a 40-pin DIP which removed the need for multiplexing of the address and data buses and reduced the number of support chips required to 6.

The Competition Responds

The Intel 8080 was the breakthrough device for microprocessors. Despite a price tag of $360, sales were strong from the outset, allowing Intel to recoup its development costs within only 5 months. However, this success inevitably led to competition from other firms. Intel had filed a US patent application on the 4004 in January 1973 but it only covered how the device interacted with the external support chips. Like many others in the industry, Intel considered the idea of a complete processor on a single integrated circuit commonplace, which left the door open for competitors.

The first company to go into competition with Intel was the giant US electronics and communications firm Motorola which introduced its own 8-bit microprocessor, the MC6800, in August 1974 at the same price as the 8080. Although functionally similar to the Intel device, the MC6800 employed a leaner architecture with fewer registers which simplified programming and helped deliver a performance close to that of the 8080 despite a clock speed of only 1 MHz. The Motorola designers were also unencumbered by the requirement to maintain compatibility with earlier devices which made for a simpler and more elegant instruction set.

Intel and Motorola were able to gradually reduce the price of their microprocessor chips as production numbers increased but prices did not fall far enough for a group of eight Motorola design engineers led by Charles I (Chuck) Peddle. Frustrated at their management's refusal to sanction the development of a low-cost version of the MC6800, the engineers decided to leave Motorola and join MOS Technology Incorporated, a small semiconductor firm based in Norristown, Pennsylvania. MOS Technology had developed a method for increasing the yield from the silicon fabrication process and this was just what Peddle and his colleagues needed in order to realise their goal.

In September 1975, MOS Technology introduced the 6502 microprocessor at a price of $25, undercutting the similarly specified 8080 and MC6800 devices by more than $150. The 6502 was very similar to the MC6800 but Peddle and his team of ex-Motorola employees had used their intimate knowledge of the Motorola chip to make some changes which improved performance, including a 1-stage instruction pipeline that allowed the next instruction to be fetched during execution of the current instruction. Intel and Motorola both reacted quickly by dropping their prices to $69. With three fully-featured microprocessors now available at low cost, the computer industry finally had the means to slash production costs and build affordable computers in pursuit of a mass market.

11. The Microcomputer Revolution

Early Microprocessor-Based Computer Systems

The development of the Intel 4004 and Intel 8008 had arisen out of a need to reduce the chip count for a programmable calculator and a display terminal. System designers making use of these first generation microprocessors were constrained to a specific architecture which may not have been the optimum one for their application. The embryonic design and fabrication methods used to create them also dictated that they were basic processor designs with few performance features and limited memory addressing capability. Consequently, they were mainly used as a low-cost alternative to discrete logic circuitry in dedicated systems such as calculators, terminals and industrial controllers. However, a handful of intrepid firms did make use of the 8008 to create the earliest fully programmable microprocessor-based computer systems. Surprisingly, the first of these firms to do so was not based in California or Massachusetts but on the other side of the Atlantic in France.

Réalisation d'Études Électroniques (R2E) was established in a suburb of Paris in 1971 by Vietnam-born electrical engineer André Thi Truong. In June 1972, the company received a contract from INRA (Institut National de la Recherche Agronomique), the French national institute for agricultural research, to develop a low-cost programmable system for process control. INRA needed a number of these systems for use in crop experiments but its budget did not stretch to the 45,000 FF price of the DEC PDP-8. R2E project team leader, electronic engineer François Gernelle, proposed a system based on the recently introduced Intel 8008 microprocessor, having seen a preliminary data sheet on the device a few months earlier. However, Intel products were virtually unknown in France at that time and he had to work hard to convince both his R2E colleagues and the customer to proceed.

Gernelle based the design of the system on various minicomputers he had worked with. It featured a modular architecture with a series of printed circuit boards connected through 74-pin edge connectors to a specially designed backplane bus and housed in a standard 19-inch rack mountable instrument case. The 8008 microprocessor and real-time clock circuitry were contained on one board. A second board contained 1,024 bytes of RAM for storing data in the form of a stack which could also be accessed by the input/output devices. Other memory boards were also provided, each with a capacity of 2,048 bytes, using ROM, RAM or a combination of both. A choice of input/output boards were available with parallel or serial interfaces. The front panel featured three banks of toggle switches and several rows of indicator lamps. As it was designed for unattended operation, the system was supplied without a console but it could be fitted with an optional console board which allowed a teletype terminal or modem to be connected for installation or testing purposes.

The prototype system was completed and delivered to INRA in January 1973. In April 1973, R2E began marketing the system commercially as the Micral N at a price of 8,500 FF (equivalent to about $1,900 at 1973 exchange rates) for a basic specification machine. As an affordable alternative to minicomputers for less demanding process control applications, the Micral N proved reasonably popular in European markets and around 2,000 were sold. It also formed the basis for a series of microprocessor-based process control models introduced by R2E over the next few years, with total combined sales in the region of 90,000 units. The company was acquired by Groupe Bull in 1981.

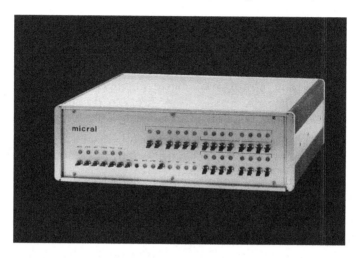

R2E Micral N (Photo © Rama)

A few months after the launch of the Micral N in France, an American engineer who had also been inspired by the potential of the Intel 8008 began taking his first steps towards realising his goal. Nat Wadsworth was working as a design engineer with General DataComm Industries, a telecommunications equipment company based in Danbury, Connecticut, when he and several of his colleagues attended one of Intel's seminars on the 8008. The engineers immediately recognised the potential of the new microprocessor for simplifying General DataComm's products but they were unable to convince their managers to let them explore this further so Wadsworth suggested that they design a small 8008-based computer in their own time which each of them could then build at home for personal use.

The computer was designed by Wadsworth and General DataComm colleague Robert Findley. Their initial approach was to create the circuits using a type of 'breadboard' construction normally used for building test circuits but they had underestimated the amount of work involved in designing an entire computer system and it soon became clear that there would be little additional effort required to commit the design to a set of printed circuit boards. The use of printed circuits would not only simplify the task of building the machine, eliminating much of the wiring needed to connect components, but would also

11. The Microcomputer Revolution

facilitate production in larger numbers should they later decide to introduce it commercially.

As the design work progressed, Wadsworth began to consider the commercial potential of the computer. This included the intriguing thought that a self-assembly version, which was the original intention of Wadsworth and his colleagues, might also interest electronics hobbyists if the price was set low enough. Wadsworth took the bold step of resigning from his job at General DataComm in order to devote himself full-time to the project and in August 1973 he set up a company, Scelbi Computer Consulting Incorporated, in Milford, Connecticut, to commercialise it. The name Scelbi was an acronym of Scientific, Electronic and Biological, the three application areas that he had in mind for the computer. Wadsworth was now able to work uninterrupted to complete the design of the computer and potential investors began showing an interest in the new venture but the project received a setback in November 1973 when, aged only 30, he suffered a heart attack. Following his recovery, Wadsworth was able to complete the design but the investors began drifting away so there would be precious little money available to bring it to market.

The new machine was completed in early 1974 and named the Scelbi-8H. Wadsworth had kept the overall design as simple as possible in order to reduce manufacturing costs and facilitate self-assembly. The appearance echoed Intel's SIM8-01 evaluation board system, with a box-shaped chassis fitted with edge connectors on the top for up to 8 printed circuit boards. The circuit boards were exposed on the top, sides and back but covered by a front panel with pushbuttons for interrupt, step and run plus 8 toggle switches for input and a rectangular aperture through which were visible several banks of LED indicator lamps mounted on the nearest board. The power supply was housed in a separate box which was connected by a cable to a socket on the rear panel of the chassis.

The Scelbi-8H was accompanied by an impressive selection of input and output options. These included a novel interface board which allowed a standard laboratory oscilloscope to be used as an alphanumeric display. Another board allowed an audio cassette tape recorder to be converted into a low-cost magnetic tape storage unit. An inexpensive alphanumeric keyboard was also available. Up to 4 memory boards of 1,024 bytes capacity each could be fitted. Software included an editor, assembler and an interpreter for the SCELBAL programming language, a derivative of BASIC which Wadsworth had created for the 8008. These options provided everything necessary to create a complete general-purpose computer system.

For a company with little money, mail order offered the cheapest route to market. An advert for the new machine was placed in QST, an amateur radio magazine, in March 1974. Wadsworth realised that many amateur radio enthusiasts were also keen electronics hobbyists who might be attracted by the idea of a computer construction project. The Scelbi-8H was priced at only $440 for a basic system supplied as a self-assembly kit. A fully assembled and tested machine fitted with 1,024 bytes of memory was also available for $565. The advert generated sufficient interest from potential customers that Wadsworth was able to hire a few employees, including computer co-designer Findlay, but the company received a major setback in May 1974 when Wadsworth had a second heart attack which necessitated a lengthy period of rest and recuperation. When Wadsworth returned to work, his priorities had changed and he was no longer so strongly focused on commercial success.

A second model, the Scelbi-8B, was introduced in April 1975 at a price of $849 for a fully assembled and tested machine fitted with 4,096 bytes of memory. The main difference from the 8H was the use of 4,096 byte memory boards to provide increased memory capacity up to the 8008 address limit of 16,384 bytes. Although a kit version was also available, this model was firmly targeted at technical and business users, hence the letter B designator, and was advertised as a low-cost alternative to minicomputers in simple control applications.

While in hospital recovering from his second heart attack, Wadsworth had started writing a book on 8008 machine code programming. Scelbi published the book in 1975 and it was an instant success, selling hundreds of copies per month at $20 each despite its primitive typographical appearance as a result of having been prepared using a Teletype terminal. Other books followed, many of which featured source code listings for Scelbi software products, each selling in similar numbers to the first. In contrast, monthly sales totals for the computers were in single figures and the company was making a loss on each one sold. With profits from the book sales keeping Scelbi afloat, Wadsworth decided to stop advertising the computers in order to concentrate fully on books and software.

Wadsworth's poor health and a lack of investment in the business clearly hampered Scelbi's chances of success. Total combined sales for both Scelbi models were only in the region of 200 units but they were highly influential as the first machines to demonstrate the potential of the microprocessor in the development of low-cost general-purpose computers. Wadsworth's visionary targeting of a separate hobby market also inspired others to do likewise and presaged the market segmentation that took place later on.

Hobby and Home Computers

R2E and Scelbi had proven that it was possible to create practical low-cost computer systems based on the Intel 8008 microprocessor, despite its architectural and performance limitations, but it was a commercial risk which few other firms were prepared to take. The second generation microprocessors

11. The Microcomputer Revolution

introduced by Intel, Motorola and MOS Technology in 1974 and 1975 swept away these limitations, offering computer designers the opportunity to create a new class of general-purpose machines which rivalled the performance and capability of a minicomputer at a fraction of the price. These became known as microcomputers.

As well as low-cost 'professional' models designed for technical or business applications, second generation microprocessors also facilitated the opening up of a market for hobby and home computers. Nat Wadsworth had shown that hobbyists were willing to accept hastily designed machines with a 'bare bones' specification and minimal software support as long as the price was right. Therefore, companies targeting this market could introduce products quickly and with minimal development costs. Supplying self-assembly kits reduced production costs and the mail order route to market also minimised marketing and distribution costs. These lower barriers to entry attracted a number of entrepreneurs into the computer industry and the companies that they established were among the first to experience success in the new microcomputer market.

The concept of a home computer was not entirely new. A few years earlier Honeywell had announced the Kitchen Computer, a minicomputer apparently intended for home use. It was based on the pedestal-mounted version of the H316, the entry level model from Honeywell's Series 16 family of industrial minicomputers. The Kitchen Computer was first advertised in the 1969 Neiman-Marcus Christmas Book, a catalogue of Christmas gifts produced annually by the US department store chain Neiman-Marcus, under the heading "If she can only cook as well as Honeywell can compute". A photograph showed the bright orange computer surrounded by baskets of vegetables in a typical kitchen setting with a woman wearing an apron leaning over it. The accompanying text explained that the computer could be used by the housewife to store her favourite recipes and plan menus. After learning to program it, for which a two-week programming course was helpfully included in the price, she could also use the machine to balance the family budget.

With a price tag of $10,600 and no keyboard or means of displaying text, the Honeywell Kitchen Computer was clearly not a serious attempt at establishing a market for home computers and there is no evidence that any were sold. Instead, it would appear to have been a publicity stunt dreamt up in collaboration with Neiman-Marcus, which had a tradition of featuring fantasy gifts in its Christmas catalogues to generate media attention. Nevertheless, the machine did help to inspire Gordon Bell of DEC to write a lengthy memo on what he called "the computer-in-the-home market". In a remarkable example of future gazing, Bell envisioned a home computer that could be used by the whole family for home study, business applications and games. Sadly, DEC management did not act on the memo and Chief Executive Ken Olsen would remain sceptical of a market for home computers for years to come.

The Story of the Computer

The MITS Altair

The first company to taste success in the hobby and home computer market was Micro Instrumentation and Telemetry Systems. MITS was founded in Albuquerque, New Mexico in December 1969 by H Edward Roberts, an electrical engineer and serial entrepreneur who worked in the Weapons Laboratory at Kirtland Air Force Base, along with two of his US Air Force colleagues, Forrest M Mims III and Robert Zaller plus college friend Stan Cagle. It was created as a subsidiary of Reliance Engineering, another small company which Roberts had formed the previous year, in order to take advantage of the established firm's credit rating.

The company's first products were transmitter modules and tracking light units for model rockets. These were sold by mail order through adverts placed in hobbyist magazines such as Model Rocketry. Mims, who was a keen model rocket enthusiast himself, also generated publicity for the products by writing magazine articles for Model Rocketry. However, sales were slow so in an effort to boost sales MITS also began selling the modules in self-assembly kit form at a reduced price. Sales increased but MITS had chosen to target a highly specialised market, as evidenced by the tiny 15,000 reader circulation of Model Rocketry magazine. For the company to survive, a much larger market would have to be found.

The combination of magazine articles and mail order kit sales gave Mims the idea to follow the example set by Southwest Technical Products Corporation, a Texas-based company which specialised in selling kits for the construction projects described in Popular Electronics, an electronics hobbyist magazine with an impressive circulation of 400,000 readers. Mims, who had decided to pursue a parallel career as a freelance technical writer, had already submitted an article to the magazine on light emitting diodes and the editor was receptive to the idea of a companion piece featuring an LED-based infrared voice communicator which Roberts and Mims had devised. Both articles appeared in the November 1970 issue of Popular Electronics but sales figures for the new kit were no better than their previous efforts despite the much larger readership.

Following months of poor sales, Roberts suggested that the company shift to making self-assembly kits for electronic calculators. Roberts had maintained a fascination with electronic computation since building a small relay calculator in 1959 while a student at the University of Miami. The market for electronic calculators was expanding rapidly as prices fell following the adoption of integrated circuit technology and the prospect of a calculator kit would be likely to generate significant interest from electronics hobbyists. However, Mims and Cagle disagreed, arguing that another change of direction was too risky at this stage. Roberts was nevertheless determined to press ahead with his plans so Mims and Cagle both decided that it was time to leave MITS. Roberts then

11. The Microcomputer Revolution

found another US Air Force colleague who was willing to invest in the business and used this money to buy them out.

The new MITS calculator kit was chosen as the cover story for the November 1971 issue of Popular Electronics, appearing under the headline "Electronic Desk Calculator You Can Build". The accompanying article on how to construct it was written by Roberts. Designated the 816, the 16-digit four-function desktop model was based on a standard set of 6 calculator chips manufactured by Electronic Arrays Incorporated. It featured an 8-digit fluorescent display which could be switched to display the remaining 8 digits of a 16-digit number using an additional key on the keypad. The kit was advertised at a price of $179 with a fully assembled version also available for $275, which was much cheaper than similarly specified models from established calculator manufacturers.

The MITS 816 calculator was an immediate success, generating thousands of orders per month. This prompted the company to introduce a string of new models over the next two years, including a programmable version, a series of inexpensive handheld models featuring LED displays and an advanced desktop model with scientific functions. Although the kits proved popular, the fully assembled calculators were selling in even greater numbers, so Roberts shifted emphasis away from hobbyists towards a wider market by placing advertisements in popular science magazines such as Scientific American. However, as the calculator market continued to expand and large office automation and semiconductor firms such as Texas Instruments entered it, a vicious price war ensued and MITS was unable to compete. By 1974, previously healthy profits had turned into a $200,000 debt.

Roberts needed an exciting new product which could remove the threat of bankruptcy and return the company to profit. He decided to attempt to leapfrog the calculator industry by introducing an affordable small computer based on Intel's new 8080 microprocessor. It would be targeted at the market MITS knew best, electronics hobbyists.

The computer was designed by Roberts with the assistance of senior MITS engineers William Yates and James Bybee. Development was funded by a $65,000 bank loan obtained on the basis of projected sales of 800 machines, a figure which Roberts privately admitted was wildly optimistic. Fortunately, the close relationship with Popular Electronics magazine had been unaffected by the company's ill fated foray into the realm of consumer electronics. The timing was also fortuitous, as the magazine's editorial director, Arthur Salsberg, was on the lookout for a suitable computer construction project following the publication of an article on the Mark-8 "personal minicomputer" in rival hobbyist magazine Radio-Electronics.

The Mark-8 was an early 8008-based computer designed by Jonathan A Titus, a doctoral student studying chemistry at Virginia Tech in Blacksburg, Virginia. Titus was also an electronics hobbyist with a keen interest in computers.

The Story of the Computer

As a research student, he had limited access to the computer systems at his university so he decided to build his own machine that he could use at home. Titus based his design on the circuit Intel had used for the SIM8-01 evaluation board, the diagram for which was reproduced in the 8008 user manual. Having successfully created a working prototype, Titus then contacted the two leading US electronics hobbyist magazines, Popular Electronics and Radio-Electronics, with the suggestion for an article on how to construct it. His suggestion was taken up by Radio-Electronics and the article appeared in the July 1974 issue.

Titus had no interest in commercialising the Mark-8 so there would be no kit or assembled version to buy. Experienced hobbyists seeking a challenge could obtain a booklet for $5 with further details on how to construct the machine. A set of 6 printed circuit boards was also available from Techniques Incorporated of New Jersey for $47.50, which eased the task of connecting components. Several hundred of these circuit board sets were sold but it is not known how many systems were actually built. Where the Mark-8 did make an impact, however, was in helping to raise interest in computers within the US electronics hobbyist community.

On hearing that MITS was developing a computer kit, Popular Electronics technical director Leslie Solomon contacted Roberts and offered the company a cover story in the top selling January issue if they could meet the tight publishing deadline. This spurred Roberts and his team to complete the prototype by October 1974. Roberts then arranged to fly to New York to demonstrate the computer at the offices of the magazine's publisher, Ziff Davis. The prototype was shipped to New York by rail but went missing in transit so Roberts was forced to improvise, describing the machine with the aid of schematic diagrams. For the cover photograph, Yates rigged up an empty case with working indicator lamps and despatched it to New York while a second prototype was hurriedly built.

Despite this setback, the MITS computer kit made the January 1975 issue of Popular Electronics, appearing on the front cover under the headline "Project Breakthrough! World's First Minicomputer Kit to Rival Commercial Models... Altair 8800". The Altair name was suggested by Popular Electronics staff editor John McVeigh as an imaginative alternative to the company's original name of PE-8, with Roberts adding the number 8800 in order to differentiate it from any future models. The accompanying article written by Roberts and Yates was published in two parts, with the second part appearing in the following month's issue.

A standard Altair 8800 system comprised three printed circuit boards; a processor board containing the Intel 8080 chip set and associated clock circuitry, a 1,024 byte memory board populated with 256 bytes of SRAM chips, and a third board containing the control circuitry for the front panel. With an eye to future add-on products, Roberts made expandability a key feature

11. The Microcomputer Revolution

of the Altair design. He obtained a batch of 100-pin edge connectors cheaply and used these to create a 100-line backplane bus. This allowed every signal of the 40-pin 8080 chip to be put on the bus with room left over for various front panel control lines and power lines at different voltages. The first production units were fitted with a single 4-slot 'motherboard' which left only one spare slot for expansion but later versions had up to 4 of these fitted, giving a total of 16 slots. A high capacity 8 Amp power supply was also included to accommodate the power requirements of such a large number of expansion boards. To give the computer a professional look and provide plenty of space for additional boards, it was enclosed in a high quality metal instrument case. The minicomputer-inspired front panel incorporated 25 toggle switches for input and 36 LED indicator lamps to display the status of the machine.

MITS Altair 8800

The MITS Altair 8800 was priced at $397 in kit form or $498 fully assembled. A board-only version without case or power supply was also available for $298. These incredibly low prices would not have been possible had MITS paid the full retail price of $360 each for the Intel 8080 processors but Roberts was able to use his experience of dealing with calculator chip suppliers to negotiate a discounted price of only $75 each for quantities of 1,000. Orders for the Altair soon came flooding in and by the end of March MITS had shipped more than 2,500 units despite a necessary increase in prices which pushed the price of an assembled machine up to $621.

Following the launch of the Altair 8800, the MITS engineers turned their attention to the development of a range of expansion boards. Over the next few months, various boards were introduced including a 4,096 byte memory board, a Teletype interface board and a cassette tape interface board. The popularity of the Altair 8800 also encouraged MITS to develop a second model, the Altair 680, which was introduced in November 1975. The 680 was designed to be much easier to construct than the 8800 and was based on a different microprocessor, the Motorola MC6800. The modular construction and 100-line bus were dropped in favour of a single-board design with integrated Teletype interface circuitry and ROM-based monitor software. However, the lack of an expansion capability made the 680 much less popular with hobbyists than the 8800 despite being significantly cheaper to buy. This was remedied by the introduction of the Altair 680b in April 1976 which added a 3-slot expansion board but the model remained less popular than its Intel-based predecessor. In June 1976, MITS introduced a revised version of the original model, the Altair 8800b, which featured the newer Intel 8080A processor, an 18-slot motherboard, a cooling fan and an upgraded power supply.

In order to cope with the huge demand for its computer products, MITS had grown rapidly from 20 employees at the launch of the Altair 8800 to 230 by late 1976. Roberts was becoming tired of his increasing management responsibilities and wanted to spend more time with his wife and five children. He also knew that larger competitors would soon be eyeing up the shiny new market that he had helped to create and he did not want to end up in another price war that might destroy his business. Roberts decided to sell MITS to a company with the financial strength to compete. He approached a number of potential buyers before settling on Pertec Computer Corporation, a Californian disk and tape drive manufacturer which was looking to expand its product line from computer peripherals to entire systems. In May 1977, MITS was sold to Pertec for $6 million in stock. Roberts used his $2 million share of the proceeds to buy a farm in Georgia. He later realised his lifelong ambition by qualifying as a doctor of medicine at the age of 47.

Pertec attempted to take the Altair range into technical and business markets by repackaging the machines, introducing two 'turnkey' models, the 8800bt and 680bt, for unattended operation in industrial control applications and a business model, the MITS 300/25 Business System. However, the repackaging failed to hide the low budget origins of the underlying Altair technology and they met with limited success in a more demanding marketplace. Within a year, Pertec replaced the Altair-based models with higher specification systems which were designed specifically for technical and business applications.

By the time production ceased in 1978, total combined sales for the Altair models had reached over 10,000 units. With its bare bones specification and minicomputer styling, the MITS Altair did little to push the boundaries of technology but it was highly influential in showing other manufacturers that there was a sizeable market for cheap 'personal' machines. The Altair also spawned an industry in the manufacture of third party expansion boards, with a host of small companies springing up to exploit gaps in the MITS

11. The Microcomputer Revolution

product range, mirroring a similar phenomenon which had occurred in the minicomputer industry following the introduction of DEC's UNIBUS in 1970. As the range of plug compatible boards for the Altair increased, other computer firms also began adopting the Altair expansion bus for their own machines. In 1977, it was formally recognised as an industry standard and renamed the S-100 bus, much to the annoyance of Roberts.

The Sol Terminal Computer

One of the small companies that began producing expansion boards for the MITS Altair was Processor Technology Corporation. Processor Technology was founded in Berkeley, California in April 1975 by computer hobbyist Robert M Marsh and his friend Gary Ingram, an electronics engineer. Marsh was a member of the Homebrew Computer Club, one of the earliest and most influential clubs for computer and electronics hobbyists that sprang up across the USA in the mid 1970s, and it was his membership of this club which provided the inspiration to set up the business.

The Homebrew Computer Club was formed in March 1975 by Gordon French, a model railway enthusiast, and political activist Fred Moore. The first meeting, which was held in the garage of French's house in Menlo Park, California, attracted 32 people including several professional electronics engineers and software developers. By the third meeting, the Club had grown to more than 100 members. At the first meeting, one of the attendees gave a report on a visit to MITS premises in Albuquerque where he had been shown the recently introduced Altair 8800. Another showed off his newly acquired Altair kit. From the ensuing discussions, Marsh could see that demand for the Altair would be huge. Moreover, the machine's excellent expandability and 'open' architecture could provide an opportunity for other firms to offer plug compatible products. He contacted Ingram and suggested that they go into business together to produce third party add-ons for the MITS Altair.

Marsh and Ingram were soon presented with the perfect idea for their first product, a 4,096 byte memory board for the Altair 8800. The MITS 4,096 byte board was an expensive purchase at $264 in kit form but a necessary one for Altair owners who wanted to run the Altair BASIC interpreter software, which required a minimum of 4 kilobytes of memory. However, Club members who had purchased the board reported that it was very unreliable. The MITS engineers had used dynamic memory in order to keep the power requirements of the board within specified limits for the Altair bus but quality control issues with the DRAM chips and a design flaw which caused an intermittent problem with the refresh cycle timing conspired to make the boards unreliable.

To avoid the refresh timing issue, Marsh designed a 4,096 byte memory board for the Altair using SRAM. The new board was introduced at the fourth Homebrew Computer Club meeting in June 1975, priced at $225 in kit form, with a generous discount of 20% for cash prepayment. Within a month, word had spread that the Processor Technology board was much more reliable than the MITS one and orders began flooding in, many of which included the cash advance. The money generated gave Marsh and Ingram the confidence to incorporate their partnership. Over the following months Processor Technology introduced several more boards for the Altair bus, including the first video display board for a microcomputer, the VDM-1 Video Display Module.

The VDM-1 turned a standard video monitor into an alphanumeric display device, displaying 16 rows of 64 characters at a rate of up to 2,000 characters per second. Marsh was a self-taught electronics engineer and a video display board would be a far more challenging project than previous products so he hired Homebrew Computer Club moderator Lee Felsenstein, who worked as a freelance computer consultant, to help design it. At the heart of Felsenstein's design was an electronic character generator which formed characters from a 7 by 9 array of dots. This operated with 1,024 bytes of dedicated video memory. The specific character and its position on the screen were determined by writing the appropriate character code into the desired location in the video memory. The board also supported scrolling and blanking.

The VDM-1 board was announced in the first issue of Byte magazine in September 1975 at a price of $199 in kit form, although production problems delayed deliveries until the following year. As the first of its kind, the board became a popular add-on for Altair owners who wanted a text output capability but whose budgets were unable to stretch to the $1,500 price of a Teletype terminal. It also led to Processor Technology developing a complete computer. As with the Altair, the catalyst for the project was Leslie Solomon of Popular Electronics magazine.

Solomon had been looking for an intelligent video terminal construction project for some time, having failed to persuade Ed Roberts of MITS to team up with Donald E Lancaster, another contributor to Popular Electronics who had written a seminal article on a simple video terminal called the TV Typewriter for the September 1973 issue of rival magazine Radio-Electronics. In November 1975, Solomon approached Processor Technology with an offer of a cover story if they could deliver a working prototype within a month. Given the unreasonably tight timescale, Felsenstein suggested creating a simple video terminal based on his VDM-1 board design but Marsh held out for a more capable system which would incorporate an Intel 8080 microprocessor. The resulting machine was named the Sol Terminal Computer. The name arose from a suggestion from Felsenstein that it should be marketed as having "the wisdom of Solomon" but may also have been a sly reference to Leslie Solomon.

The majority of the detailed design work for the Sol Terminal Computer was carried out by Felsenstein, with Marsh taking on the design of the power supply

11. The Microcomputer Revolution

and interface circuitry. The project manager was Homebrew Computer Club co-founder Gordon French who also performed the mechanical design work. The end result was a general-purpose computer based around a 2 MHz Intel 8080A with integrated video output circuitry, 1,024 bytes of dedicated video memory and an audio cassette tape interface, all of which were incorporated onto a single motherboard. The single-board design allowed the designers to eschew the boxy minicomputer styling of the Altair in favour of a sleek low-profile case with an integrated keyboard. To help convey an air of quality, the bespoke case was also fitted with wooden side panels made from off-cuts of walnut which Marsh had obtained cheaply for an unrealised digital clock project.

Despite Felsenstein's best efforts working long hours and weekends, the development of the prototype took two months but Solomon held true to his offer and the Sol Terminal Computer was featured on the cover of the July 1976 issue of Popular Electronics. In keeping with the original intention of the project, both the headline and the accompanying article by Marsh and Felsenstein described the machine as an intelligent terminal rather than a computer, glossing over the obvious standalone capabilities of the microprocessor-based design. However, when Processor Technology unveiled the production version at the Atlantic City Personal Computer Show the following month, they showed no such reticence and it was advertised under the provocative slogan "The First Complete Small Computer".

Two models were introduced, the Sol-10 at a price of $995 in kit form or $1,649 assembled, and the Sol-20 which squeezed 5 Altair expansion bus slots into the same low-profile case (by mounting them horizontally) and added an enhanced 85-key keyboard and higher capacity power supply for an additional $200. Both were fitted with 1,024 bytes of SRAM as standard. These prices were around 20% higher than the recently introduced MITS Altair 8800b, which also featured the Intel 8080A processor, but the Sol was a comprehensively specified system which did not require the purchase of several additional boards to make it fully functional. When compared with a similarly specified Altair system, the Sol Terminal Computer became the cheaper option.

The Sol Terminal Computer was also much easier to use than the Altair. A monitor program permanently stored in ROM allowed the user to perform basic functions by entering commands from the keyboard rather than through the toggle switches provided on the Altair. Three versions of the monitor program were available; a basic version which provided a set of simple commands for examining the contents of a memory address and executing a program at a specified address, a version which allowed more advanced terminal operations and a sophisticated version which provided a full standalone computer capability. These took the form of a novel 'Personality Module', a tiny interchangeable daughterboard which could be fitted through a slot in the rear of the case.

The Story of the Computer

Despite the limitations of the character-based graphics display, two collections of video games were produced for the Sol. To support the games, the ROM chip that held the character set could also be swapped for an alternative version which contained special characters. These helped to sell the machine to a new set of customers, computer gamers, and a total of 10,000 were produced. Processor Technology also marketed the Sol as suitable for technical and business applications but few software applications were available and there is no evidence of any significant sales in these markets.

As the first kit computer to include an integrated keyboard and video output circuitry as standard, the Sol Terminal Computer created the template for hobby and home computers for the next decade. However, Processor Technology Corporation failed to capitalise on the success of the Sol and the company ceased trading in May 1979, killed off by a fatal combination of management inexperience, lack of new products and increasing competition. One of these competitors would go on to dominate the industry, its breakthrough model directly influenced by the Sol Terminal Computer. The company was named Apple.

Apple Computer Company

Like Processor Technology, Apple was another Californian start-up created by members of the Homebrew Computer Club. The two founders, Steven P Jobs and Stephen G Wozniak, had first met in 1971 through a mutual school friend at Homestead High School in Cupertino. Though Wozniak was four years older than Jobs, both were keen electronics hobbyists with a passion for the music of Bob Dylan and they became firm friends despite having very different personalities.

Like many electronics hobbyists, Wozniak badly wanted a computer that he could use at home but he was unable to afford the $4,000 price of his favourite minicomputer, the Data General Nova. After attending the first meeting of the Homebrew Computer Club in March 1975, Wozniak realised that microprocessors offered the opportunity to design and build his own computer for just a few hundred dollars. Aged 24, Wozniak was already a highly experienced digital circuit designer, both through his day job working on the design of scientific calculators and in his spare time where he created various electronic devices for his own amusement. The son of an aerospace engineer, Wozniak had studied electrical engineering and computer science at the University of California Berkeley but took a year out in order to earn some money to fund the remainder of the course. After landing what he considered to be his dream job as an electronics engineer at Hewlett-Packard in Palo Alto, he decided not to go back.

Wozniak started his new project by familiarising himself with the workings

11. The Microcomputer Revolution

of the Altair's Intel 8080 processor but he was more impressed by the leaner architecture and neater instruction set of Motorola's rival MC6800 device. However, the high price of these chips in small quantities led him to seek out a cheaper alternative, which he found in September 1975 when MOS Technology introduced the 6502. Wozniak combined a 6502 microprocessor and some memory with an earlier circuit he had designed for a video terminal to create a single-board computer with an integrated video display. By early 1976, he had completed the design of the computer and had built a working prototype.

The shy Wozniak was keen to make an impression on his fellow Homebrew Club members so he began handing out photocopies of the schematic diagram for the computer at Club meetings. With its comprehensive specification and potentially low construction cost, Wozniak's design captured the interest of members who were looking for a more capable alternative to the MITS Altair but few of them had the electronics knowledge necessary to build it from a schematic. Jobs, who had more business acumen than Wozniak, spotted an opportunity for the pair to make some money by supplying printed circuit boards for Wozniak's computer. Jobs worked as a technician at Atari, a manufacturer of arcade video games based in Los Gatos. His plan was to pay a work colleague to produce the layout for the board, then have a batch of 50 manufactured at a cost of around $20 each and sell them for $40 each. He and Wozniak would set up a company as a vehicle for this enterprise which they would still have even if they failed to make any money from it.

Their first task was to raise some cash to finance the new venture. Neither had any savings to draw upon so they resorted to selling personal possessions, raising $1,300 between them by selling Wozniak's prized HP-65 programmable calculator and Jobs' battered Volkswagen camper van. There was also another hurdle to overcome before they could start selling products. As a company employee, the intellectual property rights to Wozniak's inventions legally belonged to his employer, irrespective of whether they were created during or outside of normal working hours. Wozniak would have to obtain Hewlett-Packard's permission to commercially exploit his computer design, running the risk that they might want to take it for themselves. He wrote to the company's legal department informing them of what he had created and asking if they wished to exploit it. Fortunately, none of the divisional managers within Hewlett-Packard were interested in it, as a product aimed at hobbyists was viewed as too far removed from the company's core technical and business markets, and Wozniak was issued with a legal release. Hewlett-Packard's Corvallis Division in Oregon would later develop a microcomputer aimed firmly at engineering and business professionals, the HP-85, which was introduced in January 1980.

Jobs and Wozniak established Apple Computer Company on April Fools' Day 1976 in partnership with Ronald G Wayne, a middle-aged work colleague of Jobs who had previous experience of setting up a business. The Apple name was suggested by Jobs as he had occasionally worked in an apple orchard in Oregon and the word also evoked the hippie counterculture that he and many electronics hobbyists were attracted to. The word 'Computer' was added in an attempt to differentiate it from the Beatles' record company, Apple Corps, but the similarity would cause problems later. Jobs felt that he needed someone who could arbitrate in any disagreements between himself and Wozniak, so Wayne was given a 10% shareholding, with Jobs and Wozniak splitting the remainder equally. In return, Wayne produced an operating manual for the new computer, using his skills as a draughtsman to create the schematic diagrams and a logo for the company showing Sir Isaac Newton sitting under an apple tree. He also drafted the partnership agreement.

The founding of the company was prompted by an order Jobs had secured from the Byte Shop, one of the first retail stores to specialise in selling computers which had recently opened in nearby Mountain View. The proprietor, Paul Terrell, had seen a demonstration of Wozniak's prototype computer at a Homebrew Club meeting and was sufficiently impressed to agree to take 50 for his new store. However, he wanted fully assembled and tested machines rather than bare circuit boards, for which Apple would be paid $500 each, cash on delivery. To fulfil this order, the cost of the components alone would be more than $10,000 but Jobs knew that the opportunity was too good to miss.

On the strength of the Byte Shop purchase order, Jobs was able to persuade a local electronic components supplier to provide the fledgling Apple with 30-day credit terms. He also obtained a $5,000 loan from the father of Wozniak's school friend and fellow Hewlett-Packard engineer, Allen J Baum, to use as working capital. The computers were assembled in the spare bedroom of Jobs' parents' house in Los Altos. In order to meet the tight delivery schedule dictated by the 30-day credit, Jobs recruited his stepsister Patty and various friends to help with the assembly, including Daniel Kottke, an old friend from his college days. Wozniak tested the completed boards using the Jobs family television set as a display monitor.

The fully assembled computer was named the Apple-1. It was supplied with 4,096 bytes of DRAM and an on-board power supply but without a mains transformer, case or keyboard. The absence of these components had come as a surprise to Terrell, who had expected complete systems, but he decided to accept the board-only machines at the agreed price. The integrated video terminal incorporated 1,024 bytes of dedicated memory to display 24 rows of 40 uppercase alphanumeric characters at a rate of 60 characters per second. This leisurely display rate was due to the use of inexpensive shift registers for video memory which only allowed one character at a time to be transferred to the video circuitry. A monitor program stored in 256 bytes of ROM scanned the keyboard interface and recognised a set of simple commands in hexadecimal format which the user could type to perform various functions, such as examine the contents of a memory address or execute a program at a specified address.

11. The Microcomputer Revolution

As the first batch of computers was being built, Wayne was having second thoughts. Unlike the penniless Jobs and Wozniak, he had personal assets that could be seized by creditors if the business failed. He was also growing increasingly concerned by the intensity of Jobs' ambition. Having experienced failure with his earlier business, he realised that he no longer had the stamina for what was likely to become a very stressful journey. Only 11 days after forming their partnership, Wayne decided to withdraw from the agreement, giving up his 10% shareholding in Apple for a payment of $800.

In addition to the 50 Apple-1 computers supplied to the Byte Shop, a similar number were sold direct to Homebrew Club members at the retail price of $666.66, a figure arrived at by adding a one third mark-up to the $500 wholesale price. An optional cassette tape interface, which included a BASIC interpreter written entirely by Wozniak despite having no previous software development experience, was also available for an additional $75. A brief review of the Apple-1 appeared in the July 1976 issue of Interface Age, a new magazine aimed at the computer hobbyist that had evolved from the newsletter of another hobbyist club called the Southern California Computer Society. Jobs also took out adverts for the machine in two other computer hobbyist magazines, Byte and Dr Dobb's Journal, inviting both direct sales and dealer enquiries but these generated few additional sales. The problem was that the Apple-1 was neither a kit nor a complete system. Moreover, with the announcement of the Sol Terminal Computer in the July issue of Popular Electronics, it now had serious competition.

In August 1976 another computer store proprietor, Stanley Veit, gave Apple some free space on his booth at the Atlantic City Personal Computer Show after turning down an offer from Jobs to acquire a 10% shareholding in the company for $10,000. The event was held over two days, during which Jobs and Daniel Kottke demonstrated several Apple-1 machines on the booth while Wozniak remained in his hotel room in an effort to complete the BASIC interpreter. Jobs worked tirelessly in buttonholing passing visitors to extol the virtues of the Apple-1 and the machine attracted considerable attention but no new sales were forthcoming. During the show, Jobs also took some time out to size up the competition. One machine that caught his eye was the Sol Terminal Computer. Jobs was impressed by the Sol's integrated keyboard, low-profile case and highly engineered appearance. Recalling Paul Terrell's expectation that the Apple-1 would be a complete system, he told Wozniak that they should adopt a similar approach for their next model which Wozniak was already working on.

The development of the new model would require much more cash than had been generated through faltering sales of the Apple-1 so Jobs began looking for a suitable investor. He contacted his old boss at Atari and was given an opportunity to make a pitch to the company's president, Joe Keenan. However, his barefoot hippie appearance, offhand manner and poor personal hygiene failed to create the right impression and his proposal was swiftly rejected. Shortly afterwards, Apple was approached by Commodore Business Machines, a leading US manufacturer of calculators and digital watches which was seeking an entry into the microcomputer market. Jobs offered to sell Apple for $100,000 in cash plus some Commodore stock and generous salaries for himself and Wozniak but this too was rejected.

Following a second failed attempt at capturing the interest of Atari, Jobs was given the name of Donald T Valentine, a prominent venture capitalist and former Fairchild Semiconductor marketing executive who had founded his own investment firm, Sequoia Capital, in 1972. On hearing Jobs' pitch, Valentine could see the potential of the Apple business but it was at too early a stage for him to invest. He was also concerned at Jobs' lack of marketing experience so he introduced Jobs to Armas C (Mike) Markkula, another former marketing executive for both Fairchild Semiconductor and Intel who had retired at the age of only 32 after making his fortune from stock options. Markkula maintained a keen interest in technology, having originally trained as an electrical engineer, and was on the lookout for a new project to devote some time to. Following a visit to Jobs' parents' house in which he was shown the prototype of the new model, Markkula agreed not only to invest in the business but also to become actively involved in the running of it.

Markkula's involvement triggered a restructuring of the business. In January 1977 Apple was incorporated and the original partnership acquired by the new company for $5,300, with Ronald Wayne receiving one third of this amount in order to avoid future legal complications. Markkula personally invested $92,000 in return for a 26% stake. He also provided additional financial support in the form of a guaranteed loan of up to $250,000. They rented some office space in nearby Cupertino and began the process of building the new business. Markkula assumed the role of business mentor to Jobs, helping him to create a long-term plan for the company. He also created the Apple marketing philosophy, a one-page document which listed the three tenets of empathy, focus and impute, the latter emphasising the importance of every aspect of the presentation of the product to the perception of product quality, including such mundane items as packaging. As Jobs lacked experience in running a business, Markkula hired Michael M Scott, an experienced executive and former head of manufacturing at National Semiconductor, as the company's first Chief Executive Officer. He also persuaded a reluctant Wozniak to quit his day job at Hewlett-Packard and commit to Apple full-time.

The Apple][

With the company now on a stable business footing, Jobs and Wozniak were finally able to realise their dream of launching a fully engineered small computer system. The new machine was named the Apple][. It retained the 1 MHz 6502 processor from the earlier model but the circuitry was redesigned to improve

11. The Microcomputer Revolution

performance and make the machine easier to manufacture. Wozniak was adept at reducing the number of chips necessary for a circuit in order to create the most efficient design possible which would meet the required specification and he used this skill to great effect in the design of the Apple][, raising the chip count only slightly while substantially increasing the specification.

The higher specification was most noticeable in the graphics capability of the new model. One of the most talked about third party expansion boards for the MITS Altair was the Cromemco Dazzler, a video display board introduced in April 1976 which supported colour graphics. Wozniak wanted to include this feature in the Apple][so he designed the video circuitry to support two colour display modes in addition to text mode; a block graphics mode with up to 15 colours at 40 by 48 resolution and a high resolution graphics mode with a resolution of 280 by 192 and up to 4 colours. As colour video monitors were relatively expensive, an optional RF (radio frequency) modulator allowed the computer to operate with a domestic colour television set.

Unlike the earlier model, the Apple][did not have dedicated video memory. Instead, a section of main memory was set aside for graphics use. The amount required depended on the graphics mode selected but could be as much as 8 kilobytes for high resolution graphics. The use of main memory for video rather than shift registers also remedied the slow display rate of the earlier model. However, the character set was again limited to uppercase only due to Wozniak's continued use of an inexpensive character generator chip which formed characters using a 5 by 7 array of dots rather than the 7 by 9 array used in the Sol Terminal Computer. Another compromise was in the number of displayable characters, which at 24 rows of 40 characters was dictated by the display resolution of a standard TV set.

In 1975 Wozniak had produced a circuit design for a hardware version of the classic video game 'Breakout' which Jobs had been working on at Atari. The circuit design was never used but Wozniak was proud of his work for Atari and thought that a software version of the game would provide an excellent demonstration of the Apple]['s colour graphics capabilities. However, for the game to run properly on the new computer, it would need a suitable input device and the ability to produce audio output. He decided to incorporate a game interface port, sound generation hardware and a built-in speaker into the Apple][specification, reasoning that these features would also encourage other video games to be developed for the new machine. The game interface port took the form of a 16-pin socket on the motherboard which allowed two 'paddle' controls or a joystick to be connected.

To allow users to access the graphics capability of the new model, Wozniak produced an enhanced version of his BASIC interpreter with additional graphics commands. This was now permanently stored in 8 kilobytes of ROM alongside an enhanced monitor program which included extra commands for tape control. Wozniak's cassette tape interface, which had proved a popular add-on for the Apple-1, was made a standard feature and the circuitry integrated onto the motherboard.

The Apple][also boasted expansion slots. Wozniak was unimpressed by the Altair bus despite its growing acceptance as the industry standard. Not only was the bus too closely tied to the Intel 8080 processor but it also employed an inefficient method for dealing with multiple boards which required the use of switches and duplicate circuitry on each board for manual selection of the unique address during installation. Wozniak and Allen Baum devised an alternative scheme in which the board address decoding circuitry was implemented on the motherboard. The saving in components and the use of shorter 50-pin edge connectors resulted in boards which were also much smaller than their Altair bus equivalents, allowing a generous total of 8 expansion slots to be fitted onto the Apple][motherboard.

The styling of the Apple][was inspired by the Sol Terminal Computer but alternative materials were employed to soften its appearance for the consumer market and set it apart from the industrial look of the Sol. The custom case was designed by Jerrold C Manock, a freelance product designer who had learned his trade working for Hewlett-Packard. It featured an integrated 53-key keyboard and a removable lid for access to the expansion slots. The use of injection moulded plastic construction instead of sheet metal panels allowed Manock to create a more sculpted shape with rounder corners and the beige colour was carefully chosen to suggest a domestic appliance rather than laboratory equipment. Jobs was closely involved in the design process, obsessing over which shade of beige should be used to the point where an entirely new shade had to be created before he was satisfied. He also insisted on a design that did not require a cooling fan, as he felt that the noise these made was distracting. This necessitated the development of a high efficiency power supply which would generate less heat than a standard unit. The power supply was designed by new employee Rod Holt, an experienced analogue circuit designer whom Jobs had headhunted from Atari. Holt utilised a switched-mode technique originally developed for use in the aerospace industry to create a highly efficient design which would operate without a cooling fan. However, the absence of a fan in the Apple][would lead to problems with overheating when expansion boards were fitted and later models were supplied with a modified case containing extra ventilation slots.

To complement the sleek design of the new product, a new Apple logo was also created. The logo was designed by graphic designer Rob Janoff of public relations agency Regis McKenna. Janoff replaced the image of Newton sitting under an apple tree with a much simpler silhouette of an apple with a bite taken out of it. The silhouette was filled with horizontal stripes of colour which alluded to the machine's colour graphics capabilities.

11. The Microcomputer Revolution

The Apple][was introduced in April 1977 at a price of $1,298 for a fully assembled machine or $598 in board-only form. It was supplied with 4,096 bytes of RAM as standard but spare sockets on the motherboard allowed the user to expand this up to a respectable maximum of 48 kilobytes simply by plugging in suitable memory chips. Marketing literature emphasised the Apple][´s status as the first colour microcomputer. Existing Apple-1 owners were offered a generous trade-in allowance on the new model and by the end of the year 570 had been sold, comfortably outstripping the sales total of 175 for the Apple-1. With Markkula's marketing know-how, sales increased to 7,600 units the following year, a figure which compared well with those for the Sol Terminal Computer. However, the microcomputer market was expanding rapidly and the Apple][was selling poorly in comparison to rival models from Commodore Business Machines and Radio Shack, two recent entrants to the microcomputer market who had both introduced similarly specified machines at a fraction of the price.

The 1977 Trinity

The first of these machines, the Commodore PET 2001, was introduced two months after the Apple][in June 1977. Commodore had been founded by Polish immigrant Jack Tramiel and investor C Powell Morgan in Toronto, Canada, in 1955. Best known for its calculators and digital watches, Commodore's move into the computer business came as a result of the company's acquisition of semiconductor firm MOS Technology in September 1976. Tramiel had been seeking to acquire a semiconductor fabrication capability in order to compete more effectively with Texas Instruments in the increasingly savage electronic calculator market. As MOS Technology supplied the chips for many of Commodore's calculators, they were the logical choice. MOS also happened to be struggling financially as a result of TI's increasing dominance so Tramiel was able to obtain the entire company for only $750,000 in cash plus 9.4% of Commodore stock. In doing so, Commodore also acquired a team of highly talented engineers including the lead designer of the 6502 microprocessor, Chuck Peddle.

A year earlier MOS Technology had introduced the KIM-1 demonstrator system to help sell their new 6502 microprocessor. Priced at only $245 for a system fitted with 1,024 bytes of RAM, an integrated hexadecimal keypad and a 6-digit LED display, the KIM-1 also appealed to hobbyists with the result that over 7,000 of these board-only systems were sold. Peddle realised that the KIM-1 sold well because it came fully assembled rather than in kit form but the machine had been designed primarily for evaluation purposes and was too limited for most applications. He proposed a more advanced model which would be aimed at the burgeoning home computer market but the company was now in the throes of its financial problems and his proposal was rejected. Following the acquisition, Peddle pitched his idea to Commodore executives,

The Story of the Computer

emphasising the competitive advantage that a fully fledged computer would have over the programmable calculators offered by rival firms. Peddle's case was strengthened by a request Commodore had recently received from one of its largest customers, the electronics retail store chain Radio Shack, to bid for the design and manufacture of an inexpensive computer for hobbyists. They agreed to let him establish his own independent development group in California with a remit to have a prototype machine ready in time for the Winter Consumer Electronics Show (CES) being held in Chicago the following January, where it would be demonstrated to Radio Shack.

The looming deadline prompted a decision to base the development on an existing piece of hardware. Peddle had met Apple's Jobs and Wozniak a few months earlier when they had requested customer support for the Apple-1's 6502 processor so he and his colleagues decided to approach them with a suggestion that Commodore acquire the rights to the Apple-1 design. This coincided with Apple's search for investment for the Apple][development and at first the two Apple founders were receptive to the idea. However, negotiations soon faltered over disagreements between Peddle and Wozniak on the technical specification for the new machine and between Tramiel and Jobs on the financial deal.

Without the Apple technology, Peddle and his group had no option but to press ahead with an in-house design. Fortunately, one of Peddle's MOS Technology colleagues had been developing a small 6502-based process control computer which contained most of the hardware necessary to create a working prototype. The group were able to use this as the basis for their prototype by adding a keyboard, video display circuitry, a monitor and a mocked-up case made from sculpted wood. This allowed them to meet the CES deadline but they knew that considerable additional development work would be necessary to turn it into a saleable product if Radio Shack chose to adopt it.

On the last day of the Show, the prototype was demonstrated to Radio Shack's Vice-President of Manufacturing, John V Roach. Roach was impressed by what the engineers had managed to achieve in such a short space of time and was unfazed by the hastily assembled prototype, as he had every confidence in Commodore's ability to complete the product development, but he was put off by Tramiel's attempt to secure a large calculator order on the back of the deal. He decided not to take up the Commodore design but to give the green light to an internal Radio Shack computer development project instead. Fortunately, Tramiel had also been impressed by the efforts of Peddle's group and agreed to let the product development go ahead. The computer would be launched as a Commodore product at the summer CES in 6 months time and Peddle would be given additional resources to help him meet this equally tight deadline.

The Commodore PET (Personal Electronic Transactor) 2001 was unveiled at CES in Chicago in June 1977. The name was chosen to fit with Commodore's

11. The Microcomputer Revolution

tradition of using three-letter acronyms for its products and to capitalise on the pet rock craze which was then sweeping the US. Keenly priced at $595, the PET featured a 1 MHz 6502 processor, 4,096 bytes of RAM, an integrated 73-key keyboard and onboard video terminal circuitry for displaying monochrome text at 25 rows of 40 characters formed using an 8 by 8 array of dots. Unlike the Apple][, the PET also featured an integrated cassette tape unit and a 9-inch CRT video monitor as standard which allowed it to be used straight out of the box. Although high resolution graphics was not supported, the extended character set included various lines and symbols which could be combined together to create simple graphics, a technique known as text semigraphics. An alternative character set, which included lowercase letters, could also be selected by a command entered from the keyboard or from within a program. Both were permanently stored in 18 kilobytes of ROM along with a full-screen editor and a BASIC interpreter supplied by a tiny software company by the name of Micro-Soft.

To keep costs down, the PET had no expansion slots but it did have a sophisticated IEEE-488 interface. Originally developed by Hewlett-Packard for controlling laboratory instruments, the IEEE-488 was an 8-bit parallel interface bus which allowed up to 15 devices to be connected simultaneously. The substantial case which also housed the monitor was manufactured from sheet steel by Commodore's office furniture division in Toronto, however, the 'Chiclet' style keyboard was a much flimsier affair supplied by Commodore's Japanese calculator division and attracted much criticism until it was replaced by a full-sized keyboard in later models.

Commodore marketed the PET as suitable for everything from video games to industrial process control, emphasising its status as "the only totally integrated, self-contained, personal computing system". Reviews were generally favourable and the machine also appeared on the front cover of the October 1977 issue of Popular Science magazine. The accompanying article on home computers also featured other models including the MITS Altair and Sol Terminal Computer but did not include the Apple][. Despite all this positivity, sales figures for the PET were initially unspectacular until the machine was launched in the UK and the rest of Europe in early 1978. A more focused marketing effort and new software packages led to the PET becoming very popular in the business and educational sectors in these territories, selling 30,000 units in the first year of sales in the UK alone plus a similar number in Germany. This contrasted with sales of only 4,000 in North America during the same period. Nevertheless, Commodore's success in overseas markets pushed total sales figures for the PET well beyond those of the Apple][during this period.

Two months after Commodore introduced the PET, Radio Shack announced its first microcomputer product, the TRS-80. The computer was the vision of Donald French, a buyer for the company. French had been inspired by Jonathan Titus's article on the Mark-8 computer in the July 1974 issue of

The Story of the Computer

Radio-Electronics magazine. When the MITS Altair kit became available the following year, he bought one in order to learn more. After studying the machine, French realised the potential for Radio Shack to enter the computer hobbyist market with its own microcomputer kit. He then produced a design specification which he took to John Roach. Roach was initially unconvinced but agreed to have the specification evaluated.

The evaluation was conducted by Steven W Leininger, an enthusiastic young engineer from the semiconductor firm National Semiconductor whom Roach had first met while visiting the company to discuss its SC/MP chip, a low-cost microprocessor which was under consideration for the Radio Shack computer. Leininger was hired on a consultancy basis for the evaluation but it soon became clear that he was exactly the right kind of person to lead the development of the new machine and he was taken on as a Radio Shack employee a month later in September 1976. One of Leininger's first tasks was to convince his new employers to opt for a fully assembled machine rather than a kit. Leininger had been working part-time in a Byte Shop store to earn some extra cash and had witnessed at first hand the difficulty many customers had in building their computer kits. He won this argument but it would be another four months before the third-party supplier bids were rejected and the project received the go ahead.

Leininger wanted a more capable processor than the SC/MP for the new machine. He looked at the Intel 8080 and Motorola MC6800 but settled instead on the recently introduced Zilog Z80, an 8-bit microprocessor designed by Federico Faggin and Masatoshi Shima following their departure from Intel. Faggin and Shima had made the Z80 binary compatible with the 8080 but had added a host of improvements including 80 new instructions, index registers and on-chip DRAM refresh circuitry. Leininger's design for the new computer coupled a Z80 processor operating at 1.78 MHz with 4 or 16 kilobytes of DRAM, video output circuitry, a cassette tape interface and a BASIC interpreter permanently stored in 4 kilobytes of ROM. Two display modes were supported, text at 16 rows of 64 uppercase characters or block graphics at a resolution of 128 by 48. A compact single-board design allowed the entire computer except for the power supply to be fitted into the generously proportioned plastic case of the 53-key keyboard, an idea later copied by Commodore for their highly successful VIC-20 and C64 models.

Leininger completed a working prototype of the new computer in February 1977. After demonstrating it to company president Charles D Tandy, Roach was given permission for a production run of 3,500 units, one computer for each Radio Shack store, on the basis that the stores could use the machine for inventory control purposes if they were unable to sell it. The new computer was introduced as the TRS-80 in August 1977 at a press conference in New York. The original aim had been to produce a machine that could be sold for $199 but this proved to be wildly optimistic and the eventual price was

11. The Microcomputer Revolution

$399.95 for the 4 kilobyte model. For an extra $200 customers also received a 12-inch video monitor, cassette tape unit and a software tape containing two video games, thus providing everything necessary to operate the computer for almost exactly the same price as a Commodore PET.

As French and Roach had anticipated, Radio Shack's customer base proved highly receptive to the new product and more than 10,000 orders were received within a month of the announcement. Deliveries began in September 1977 but the company struggled to cope with demand and reliability problems in some of the early production machines resulted in it being nicknamed the Trash-80, a moniker which remained popular with computer hobbyists long after these problems were resolved due to the cheap image of Radio Shack products. The model also suffered from a problem with electromagnetic radiation which caused interference with other electronic devices due to the poor screening offered by the plastic case, a problem also shared by the Apple][. Nevertheless, Radio Shack's ready-made distribution network of nationwide retail stores paid dividends and by the time it was discontinued in January 1981, the TRS-80 Model I had sold over 250,000 units.

As the first three microcomputer models to make a significant impact on the market, the Apple][, Commodore PET and Radio Shack TRS-80 were later referred to by Byte magazine as the 1977 Trinity. The Apple model was by far the most expensive of the three. Being a small company, Apple outsourced most of its manufacturing work which resulted in higher production costs but the machine's specification was only marginally superior and did not justify the significantly higher price. Now that there were two decent alternatives for under $600, computer hobbyists were no longer prepared to pay in excess of a thousand dollars for a home computer. Despite Markkula's clever marketing, Apple's sales figures reflected this situation. Fortunately, salvation for Apple would arrive in 1979 in the form of the first major business software application for microcomputers.

VisiCalc

In the spring of 1978 Daniel S Bricklin, an MBA student at Harvard Business School, was pondering how to simplify the manipulation of tables of financial data for his class work. Bricklin had previously studied computer science at MIT and had spent time working as a programmer for DEC before beginning his MBA course. His initial approach was to write BASIC programs to perform the calculations but he noticed that one of his classmates was having more success with a pocket calculator, obtaining the results instantly rather than having to rerun a program each time a number in the table changed. Bricklin was reminded of his work on the development of a word processing system at DEC, where interactive graphics had been employed to provide the user with the ability to edit text directly on the screen. He began to envisage a similar idea for numerical data, with the numbers presented in the form of a table of

The Story of the Computer

cells, or spreadsheet, on the screen. By changing a number in a cell, the entire table would update automatically.

Enthused by the commercial potential of his electronic spreadsheet concept, Bricklin created a rudimentary prototype using the School's DEC PDP-10 computer system which was equipped with interactive terminals. He then demonstrated it to his fellow students and tutors. One tutor warned him that the market was already awash with financial modelling software for mainframe computer systems, although none of these offered the same freedom to interact with the data in real time. Another introduced him to Daniel H Fylstra, a former student who ran Personal Software, a company which specialised in marketing and selling software for microcomputers. Fylstra had also been one of the founding editors of Byte magazine in 1975 and his continued involvement with the magazine allowed him to keep abreast of trends in the computer industry, including the untapped potential of microcomputers in business applications. Fylstra offered to publish the software if Bricklin would make a microcomputer the target platform.

Bricklin had little experience of microcomputers and no access to one so Fylstra lent him an Apple][which he used to recreate his electronic spreadsheet prototype using Apple BASIC. The Apple][was a capable platform for interactive applications, having been designed by Steve Wozniak to run video games. Unlike rival microcomputer models, it also possessed excellent data storage capability through an optional floppy disk drive unit which had been introduced in July 1978. However, the prototype software ran very slowly due to the inefficiency of the BASIC interpreter. Bricklin realised that it would have to be rewritten in machine code to produce the necessary performance for interactive use but this would take considerable time and effort, and his MBA studies would be likely to suffer as a result. Fortunately, his friend Robert M Frankston, a computer scientist at Interactive Data Corporation, had the requisite programming skills and was willing to help him develop the software. In January 1979, Bricklin and Frankston formed a company called Software Arts as a vehicle for the development of the electronic spreadsheet application. Frankston and Fylstra also coined an appropriate name for the new product, VisiCalc.

VisiCalc was released in October 1979 to widespread acclaim. With an introductory price of just $99.50, the software appealed to small business owners who wanted to keep track of their finances without having to employ a bookkeeper and more than 100,000 copies were sold in the first 18 months. As it was initially only available for the Apple][, this translated into direct sales for Apple and resulted in a tripling of sales of the machine in 1980. Computer retailers began bundling VisiCalc with all of the hardware necessary to get the most out of it, including an Apple][fitted with 32 kilobytes of memory, a floppy disk unit and a printer, which pushed the total price up to several thousand dollars. Nevertheless, the unique appeal of VisiCalc ensured that

11. The Microcomputer Revolution

sales of the Apple][remained strong. Other developers soon followed the example set by Software Arts in developing business software applications for the Apple][with the result that Apple's share of the microcomputer market, which was now expanding rapidly through the addition of business customers, steadily climbed above that of its two closest competitors, Commodore and Radio Shack.

Apple][running VisiCalc (Photo © Apple Computer, Inc.)

Despite its lowly hobbyist origins, the Apple][had become the microcomputer of choice for business users. Updated models introduced at regular intervals extended the lifespan of the Apple][platform and by the time production of the final model in the range ended in November 1993 total sales were in the order of 2 million units, its longevity testament to the excellence of Wozniak's design. The adoption of the Apple][by business users also paved the way for other computer manufacturers to introduce inexpensive microcomputers for business applications. Unfortunately for Apple, one of these manufacturers was the company which had dominated the computer industry for the past three decades.

The IBM PC

Hewlett-Packard's rejection of Wozniak's design for the Apple-1 in 1976 was typical of the computer industry's initial attitude towards microcomputers. Despite the early signs of a large untapped market for hobby and home computers, the established computer firms were highly sceptical. When they eventually did embrace the microcomputer, it would be for technical and business applications.

Some established computer manufacturers had been quick to adopt microprocessors for use in specialist products but failed to apply them to general-purpose models. The first to do so was DEC with the introduction of its Microprocessor Series (MPS) modules in March 1974. This consisted of a group of four modules and an optional operator's control panel which were designed for implementing low-cost digital controllers for process control and industrial automation. The processor module was based on the Intel 8008 but DEC appears to have been shy in admitting to the use of a third-party processor in one of its products, as there is no mention of Intel in any of the MPS documentation. Like many of the big computer firms, DEC was immensely proud of its engineering capability and may have felt that the use of a third-party processor design was a sign of weakness. It would be several years before the company would use a microprocessor again, in the DEC VT78 Video Data Processor which was introduced in May 1977, and the microprocessor chosen, the Intersil IMS 6100, was a single-chip implementation of DEC's own PDP-8 architecture.

Both DEC and IBM also dabbled with the concept of a personal computer. However, it was the normally more conservative IBM which would follow this through to a mainstream model. In doing so, IBM would turn the entire industry on its head.

Project Mercury

IBM's first foray into the world of personal computers began in 1973. The project was the vision of Paul J Friedl, an engineer in IBM's Palo Alto Scientific Center who specialised in real-time control applications. In December 1972 Friedl received an enquiry from IBM's General Systems Division (GSD) in Atlanta, Georgia regarding the development of a new product which would utilise the APL programming language. APL (A Programming Language) is an interactive language which was created by IBM in 1962 and is based on a mathematical notation conceived by the computer scientist Kenneth E Iverson. It is popular with scientists and engineers due to its powerful array manipulation capabilities. The GSD engineers had no idea what form this new product should take, other than it might be similar to a handheld calculator. Friedl agreed to investigate and promised to respond with a proposal within a month.

In January 1973, Friedl presented his proposal to IBM executives in Atlanta for the development of a self-contained portable computer based on the company's PALM processor which he named SCAMP (Special Computer, APL Machine Portable). As the computer would use mainly off-the-shelf components, Friedl

11. The Microcomputer Revolution

also proposed to build a prototype in only 6 months. His proposal was accepted and Friedl quickly assembled a project team of five software engineers from his own Center plus five hardware engineers from IBM's Los Gatos Advanced Systems Development Laboratory. Fortunately, Friedl's estimate of the development timescale proved accurate and the team were able to deliver a working prototype within the allotted 6-month timescale.

The PALM was a discrete processor which was implemented on a single circuit board using bipolar gate arrays. It was developed by IBM's Entry Level Systems (ELS) unit in Boca Raton, Florida for use in industrial control systems. Features included a 16-bit word length, 16 general-purpose registers and four interrupt levels. The SCAMP prototype was based on a 1.9 MHz PALM processor coupled with 64 kilobytes of RAM, a standard IBM keyboard and an integrated 5-inch CRT video monitor which could display 16 rows of 64 characters. Storage was provided by a standard audio cassette tape unit. An input/output board contained the keyboard and cassette tape interfaces plus an interface for an external printer. With the exception of the power supply, the components were all packaged into a case the size of a small suitcase.

An important feature of the new machine would be its ability to run APL, however, there was no APL interpreter available for the PALM processor and the SCAMP team would not have enough time to develop one. The solution to this dilemma was to make use of the microprogramming capability of the PALM processor to emulate the instruction set of the IBM 1130 minicomputer on the SCAMP and use the APL interpreter which had been developed for the 1130. The interpreter was permanently stored in 6 kilobytes of ROM.

Following a successful demonstration of the SCAMP prototype to IBM President John R Opel, a decision was taken to put the machine into production. The production version was developed by GSD's Rochester Laboratory in Minnesota, with Friedl acting as a consultant. The development was codenamed Project Mercury. For the production machine, the emulation was changed to a subset of the IBM System/370 instruction set so that a more up-to-date version of APL could be used and a BASIC interpreter was also provided. The tape unit was changed from an audio cassette to one which used the more robust QIC data cartridges and housed within the high quality metal alloy case. The external power supply of the prototype was replaced with an internally mounted switched-mode unit, the only major component which had to be specially designed.

The production version of the SCAMP was introduced in September 1975 as the IBM 5100 Portable Computer. Prices ranged from $8,975 for a model with 16 kilobytes of RAM and a BASIC interpreter up to a whopping $19,975 for a 64 kilobyte model with both BASIC and APL interpreters. Optional software included libraries of mathematical and statistical routines, and a computer-aided instruction course for learning APL programming.

The Story of the Computer

The IBM 5100 predated other self-contained personal computers, such as the Commodore PET 2001, by almost two years. It was also the first portable computer on the market, although weighing in at over 23 kilograms it would not have been the easiest machine to carry around. However, with a higher price than many minicomputers, it was clearly too expensive for hobbyists. In any case, with the market for home computers not yet established, IBM seems to have had no inkling of such uses for a personal computer and the 5100 was marketed to IBM's traditional customer base of scientific, engineering and business users.

Sales figures for the IBM 5100 were sufficiently encouraging to warrant the introduction of a second model, the IBM 5110 Computing System, in January 1978. Aimed squarely at business users, the 5110 retained the same specification as the earlier model but added various peripherals, including a printer and an auxiliary storage unit with a choice of single or dual 8-inch floppy disk drives, plus several business software applications. This was followed in February 1980 by a third model, the IBM 5120 Computing System, which sacrificed portability for a larger 9-inch video monitor and dual 8-inch floppy disk drives as standard. Both models proved reasonably popular despite prices which were only a fraction lower than those of the 5100.

The System/23 Datamaster

IBM finally dropped the PALM discrete processor in favour of a microprocessor with its fourth personal computer model, the ill-fated System/23 Datamaster. The Datamaster project began in February 1978 at IBM's ELS unit in Boca Raton and was led by Roger E Abernathy. The aim was to develop a new product for small businesses which combined the features of a dedicated word processor and a business data processing system. Dedicated word processors had first arrived in 1972 with the introduction of the AES-90 CRT Text Editor by the Canadian company Automatic Electronic Systems and had grown into a sizeable market in their own right. An article published in Business Week magazine in November 1977 reported that there were up to 400,000 word processing systems in the USA alone and the market was growing at a rate of 20% per year. As the name suggests, these were dedicated computers usually based on a minicomputer architecture with a video display, keyboard and software which allowed the interactive manipulation of text on the screen.

In a sign that the established computer manufacturers were finally dropping their 'not invented here' attitude towards third-party processors, the Datamaster was based on the Intel 8085 microprocessor. The 8085 was Intel's successor to the 8080, essentially a 5 Volt version of the 8080A with some minor performance enhancements and the addition of on-chip clock generator and system controller circuitry which reduced the number of support chips required. Early on in the project it became clear that the system would require more memory than the maximum 64 kilobytes permitted by the address limit

11. The Microcomputer Revolution

of the 8085, so the team had to create an external paging mechanism which allowed up to 256 kilobytes to be addressed. Problems were also experienced with the development of the BASIC interpreter which was to be the main means of interacting with the machine. Partway through the project, a corporate decision was made to standardise on the version of BASIC used in the new System/34 minicomputer in order to provide software compatibility between different IBM model ranges. This required considerable additional effort and resulted in the development schedule slipping by 18 months.

IBM's new microprocessor-based model was announced as the IBM 5322 System/23 Datamaster in 1979 but the slippage in the development schedule delayed introduction until July 1981. It featured a 12-inch video monitor which could display up to 24 rows of 80 characters, an 83-key keyboard and dual 8-inch floppy disk drives, all of which were housed within a substantial desktop case with an overall weight of 43 kilograms. Unlike its PALM-based predecessors, the Datamaster featured an expansion bus with two spare slots. Optional add-on boards were available for a high-speed connection to larger IBM computer systems and to allow a second monitor and keyboard to be attached. Prices ranged from $9,830 for the data processing version up to $12,030 for the word processing version. Both versions were supplied with a small dot matrix printer which printed text at 80 characters per second.

A by-product of the lengthy delay in introducing the Datamaster was the IBM 5120 Computing System, the development of which was led by Datamaster team member William L Sydnes. Sydnes took elements of both the 5110 and the Datamaster to produce a PALM-based computer which closely resembled the Datamaster in both physical appearance and specification, thus extending the lifespan of the 5100 series and providing a stop-gap product until the Datamaster was ready for release. Unfortunately, the 8-bit Datamaster performed poorly when compared with its 16-bit predecessors. It was also eclipsed by the introduction of IBM's first general-purpose personal computer model the following month.

Project Chess

By 1980, IBM was beginning to emerge from a long running anti-trust lawsuit with the US Department of Justice which had begun in January 1969. Freed from the restraints that this had imposed, IBM senior managers were becoming more receptive to new ideas within the company. Chief among them was IBM President John Opel, who wanted the company to be more imaginative and to have a strong presence in every area of the computer industry, including personal computing.

William C Lowe, the laboratory director of IBM's ELS unit in Boca Raton, had been studying the growing market in inexpensive microcomputers for some time and could see how Apple and Commodore were beginning to encroach upon IBM territory in the entry level business computing sector. In July 1980, he proposed to IBM's Corporate Management Committee (CMC) that the company should enter this market but he also warned them that the IBM culture was incapable of allowing a competitive product to be developed within a reasonable timeframe and they would have to either buy in the technology or acquire an existing business. Fortunately, Lowe's opinions were well respected within IBM and the Committee endorsed his proposal, instructing him to do whatever was necessary to circumvent this cultural quagmire and to return in a month with a prototype.

To create the prototype within such a short timescale, Lowe assembled an experienced task force of 13 engineers which included William Sydnes, who led the hardware team, Jack Sams, who led the software team, and interface expert David J Bradley, all of whom had been members of the Datamaster development team. The chief architect was Lewis C Eggebrecht, a 20-year IBM veteran with a keen interest in the computer hobbyist scene who was transferred in from IBM's manufacturing facility in Rochester, Minnesota. The task force worked hard to build a rough working prototype in time for Lowe's demonstration to the CMC in August 1980 and their efforts paid off when the Committee agreed to take it forward to the product development stage. The project, which was given the codename Project Chess, would be fast-tracked and set up as an Independent Business Unit (IBU), giving the task force a high degree of financial and technical freedom, but the pressure would remain relentless as they would have only 12 months in which to deliver the production version. This was in contrast to the 4-year product development timescale which would normally be required for a new IBM model. Now that his role in initiating the project was over, Lowe transferred responsibility for the day-to-day management of the project to Philip D (Don) Estridge, an experienced software development manager with an interest in microcomputers.

The task force had learned some valuable lessons from the disastrous Datamaster project. To improve their chances of meeting the tight development timescale, they proposed to depart even further from corporate convention and use third-party off-the-shelf components for software as well as hardware. Software partners would be recruited to work in parallel with the IBM team to ensure that third-party applications were available in time for the launch of the new product. Components and subsystems from the Datamaster would be used wherever possible but the machine, which was codenamed Acorn, would be based on a 16-bit processor to ensure adequate performance. It would also have an open architecture to encourage other firms to develop add-on boards and software for it.

One of the earliest decisions was to use an Intel processor, as many of the engineers on the team were already familiar with Intel support chips and development systems from the Datamaster project. The device chosen, the Intel 8088, was a low-cost variant of Intel's first 16-bit microprocessor, the

11. The Microcomputer Revolution

8086, the main difference being a reduction in the width of the external data bus from 16 to 8 bits which allowed the use of cheaper 8-bit support chips. The Acorn was based on an Intel 8088 processor running at 4.77 MHz with up to 256 kilobytes of RAM and 40 kilobytes of ROM. A socket was provided on the motherboard for an optional Intel 8087 maths co-processor which extended the capabilities of the 8088 by allowing double precision and floating-point arithmetic operations. Expandability was accommodated by an 8-bit expansion bus with 5 slots provided on the motherboard. Known as the ISA (Industry Standard Architecture) bus, it was based on the expansion bus from the Datamaster and used the same 62-pin edge connectors. Storage was provided by single or dual 5.25-inch floppy disk drives supplied by Tandon Corporation, each of which could store up to 160 kilobytes of data. For the hobby and home market, the machine also included a cassette tape interface, a built-in speaker and sound generation circuitry.

Considerable thought was put into the graphics capability of the new machine. Two separate video display boards were developed, a 'Color Graphics Adapter' (CGA) board for high-resolution colour graphics and text, and a 'Monochrome Display Adapter' (MDA) board for displaying monochrome text only but at a much higher quality for applications such as word processing. Both boards plugged into the ISA expansion bus. The CGA board incorporated 16 kilobytes of on-board video memory for displaying 4 colours at a resolution of 320 by 200 or 2 colours at a resolution of 640 by 200. It also supported text at up to 25 rows of 80 characters and 16 colours. The text-only MDA board was equipped with 4 kilobytes of on-board video memory for displaying text at 25 rows of 80 characters, each formed using a 9 by 14 array of dots to ensure high quality. It also incorporated a parallel printer interface. Both boards were designed by Mark E Dean, who also worked on the design of the ISA expansion bus, and were based on the Motorola MC6845 video controller chip.

The decision to offer a choice of display boards also impacted on the styling of the new machine, as different video monitors would be required for each board. The Datamaster's all-in-one case was abandoned in favour of a box-shaped system unit with a separate video monitor and keyboard. The keyboard itself was the same high quality 83-key unit used in the Datamaster but detached from the system unit and connected by a coiled cable. The system unit contained two bays for the floppy disk drives with apertures on the front panel which could be fitted with blanking plates if a drive was not installed.

By January 1981, the project team had grown from the original task force of 13 people to a large multidisciplinary team of 135. Contracts were put in place with external suppliers and several software partners had also been recruited to develop third-party applications. Unlike IBM's earlier personal computers, the only software permanently stored in ROM was a bootstrap loader and some low-level routines for hardware initialisation and input/output which IBM referred to as the BIOS (Basic Input/Output System). Programming

The Story of the Computer

languages and operating systems would be loaded from cassette tape or floppy disk. This would give the user more choice but it would also leave IBM at the mercy of its software development partners if they failed to meet the deadline for the launch of the new machine.

The Acorn was introduced in August 1981 as the IBM 5150 Personal Computer, priced at $1,565 for a basic specification model fitted with 16 kilobytes of RAM and a video display board. The addition of a floppy disk drive and monochrome video monitor pushed the price up to just over $3,000. Software available at launch included a choice of 3 operating systems (Microsoft PC-DOS, Digital Research CP/M or UCSD p-System), a BASIC interpreter, a compiler for the Pascal programming language and 7 applications with an emphasis on personal productivity, including the VisiCalc spreadsheet and a word processing program called EasyWriter.

IBM 5150 Personal Computer

From the outset, IBM had recognised that its existing sales network, though extensive, would not be enough to make the machine available to the broadest set of customers and that new distribution channels would be needed in order to penetrate the hobby and home computer markets. In a major break with tradition, agreements were made with the US department store chain Sears Roebuck & Co and with specialist computer retailer ComputerLand to sell the IBM Personal Computer through 49 Sears Business Centers and 190 ComputerLand outlets. This was combined with a high-profile advertising campaign which featured a lookalike of silent movie star Charlie Chaplin in his best known character of the little tramp, the character's 'everyman' image carefully chosen to appeal to potential customers who might otherwise have been intimidated by IBM's reputation as a company that only sold expensive

11. The Microcomputer Revolution

computers to big business. The resulting sales figures wildly exceeded expectations and by August 1982, 200,000 units had been produced.

Like the Apple][, the IBM Personal Computer was rather expensive for the hobby and home market and the majority of sales were for business use. Apple and Commodore had reacted to the news that IBM was about to enter the personal computer market by introducing new models targeted specifically at business users, Apple with the Apple III and Commodore with the CBM 8032. However, neither of these models proved effective competition for the IBM Personal Computer. Having dominated data processing for over 30 years, IBM had acquired an enviable reputation as a trusted supplier to businesses which Apple could only dream of. IBM also brought new standards of engineering excellence to the personal computer market and the build quality of its products was of a level previously unknown in this market sector. Consequently, IBM rapidly became the dominant player in personal computing, taking a 40% market share which continued until the emergence of PC compatibles in 1983.

Legacy

The IBM Personal Computer was not especially innovative. The open architecture, reliance on third-party technology and use of retail distribution channels may all have been new to IBM but not to the microcomputer sector. Even the term 'personal computer' was not new, having been first used in an advert for Hewlett-Packard's HP 9100A programmable scientific calculator in the October 1968 issue of the journal 'Science'. Nevertheless, the IBM PC was one of the most important products in the history of the computer. It legitimised the microcomputer in the eyes of the business community, kick-starting the process of achieving a mass market for computers. It also created a hardware platform for personal computers which is still being adhered to 30 years later.

When the editors of Time magazine came to choose their 'Man of the Year' for 1982, they chose the personal computer. The cover of the January 1983 issue featured an illustration of a typical personal computer under the headline "The Computer Moves In" and the accompanying article entitled "A New World Dawns" pondered the fact that 4 million personal computers had now been sold in the US alone. The PC era had finally arrived.

Further Reading

Ahl, D. H., Tandy Radio Shack Enters the Magic World of Computers, Creative Computing 10 (11), 1984, 292.

Antonoff, M., Gilbert Who?, Popular Science, February 1991, 70-73.

Aspray, W., The Intel 4004 Microprocessor: What Constituted Invention?,

IEEE Annals of the History of Computing 19 (3), 1997, 4-15.

Atkinson, P., The Curious Case of the Kitchen Computer: Products and Non-Products in Design History, Journal of Design History 23 (2), 2010, 163-179.

Bagnall, B., Commodore: A Company on the Edge, Variant Press, Winnipeg, Manitoba, 2010.

Bassett, R. K., To the Digital Age: Research Labs, Start-Up Companies, and the Rise of MOS Technology, Johns Hopkins University Press, Baltimore, Maryland, 2002.

Chposky, J. and Leonsis, T., Blue Magic: The People, the Power and the Politics Behind the IBM Personal Computer, Grafton Books, London, 1989.

Dennard, R. H., Evolution of the MOSFET Dynamic RAM – A Personal View, IEEE Transactions on Electron Devices ED-31 (11), 1984, 1549-1555.

Faggin, F., Hoff, M. E. Jr., Mazor, S. and Shima, M., The History of the 4004, IEEE Micro 16 (6), 1996, 10-20.

Freiberger, P. and Swaine, M., Fire in the Valley: The Making of the Personal Computer, McGraw Hill, New York, 1984.

Gernelle, F., La Naissance du Premier Micro-Ordinateur: Le Micral N, Proceedings of the second symposium on the history of computing (CNAM), 1990.

Grad, B., The Creation and the Demise of VisiCalc, IEEE Annals of the History of Computing 29 (3), 2007, 20-31.

Gray, S. B., The Early Days of Personal Computers, Creative Computing 10 (11), 1984, 6.

Jackson, T., Inside Intel, HarperCollins, London, 1997.

Littman, J., The First Portable Computer, PC World, October 1983, 294-300.

Malone, M. S., The Microprocessor: A Biography, Springer-Verlag, New York, 1995.

Mims, F. M., The Altair Story: Early Days at MITS, Creative Computing 10 (11), 1984, 17.

Noyce, R. N. and Hoff, M. E. Jr., A History of Microprocessor Development at Intel, IEEE Micro 1 (1), 1981, 8-21.

12. Bringing it all Together – The Graphics Workstation

"The best way to predict the future is to invent it", Alan Curtis Kay, 1971.

A Market Need

By the end of the 1970s, microcomputers were gradually becoming established in a number of professional applications. As well as business data processing, popular models such as the Apple][and Commodore PET were also being adopted by scientists and engineers as inexpensive replacements for dedicated minicomputers in applications ranging from data acquisition and analysis to computer-aided design. The Apple]['s expansion slots and the PET's IEEE-488 interface allowed the basic hardware specification of these machines to be augmented through a wide selection of add-on boards and peripherals. The availability of mathematically oriented programming languages, such as FORTRAN and Pascal, for both machines allowed scientists and engineers to create their own bespoke software. These same tools also allowed third-party software developers to produce commercial software packages for scientific and technical applications.

However, despite the growing availability of special-purpose hardware and software to extend their capabilities, these low-cost general-purpose machines with their puny 8-bit processors were less than ideal for the more demanding technical applications. A market need began to emerge for personal computer systems with significantly higher compute performance and extended display and networking capabilities.

A new kind of personal computer was developed to meet this need, one that embodied much of the leading edge technology in computer architecture, operating systems, interactive graphics and microprocessor design created over the previous decade. These 'super microcomputers' became known as graphics workstations and strong demand from the outset would make the graphics workstation market the fastest growing segment of the computer industry for much of the 1980s.

Early Efforts

The idea of a high performance personal computer system for scientific and technical applications predates the introduction of microcomputers by more than a decade. In May 1961, work began at the MIT Lincoln Laboratory in Lexington, Massachusetts, on a project to develop a small computer for laboratory instrument control and data analysis. The project was the brainchild of TX-0 logic designer Wesley Clark.

Three years earlier, Clark had been involved in a project for MIT's Communications Biophysics Laboratory (CBL) to develop instrumentation which could measure the neuroelectric activity in the human brain. Clark's solution employed digital electronics and hard-wired logic to process the signals but he also realised that a small stored-program computer equipped with a CRT display might be programmed to perform the same task and would be much more flexible than a dedicated instrument, allowing it to be used for a whole range of laboratory applications. To test this theory, Clark programmed the Laboratory's TX-0 experimental computer to perform the same functions, noting the relative ease with which the analysis and display of the data could be altered. Having satisfied himself of the feasibility of his idea, Clark set out to design a suitable machine which could be built for under $25,000. He called the machine LINC, an acronym of 'Laboratory Instrument Computer' which also referenced the machine's Lincoln Laboratory origins and alluded to the role of the computer as a link between the scientist and their instrumentation.

Clark based his design on DEC's new 4000 Series logic circuit modules, which he was already familiar with as they were based on the circuits from the Laboratory's TX-2 computer which Clark had also helped to create. The machine employed a binary parallel architecture with a 12-bit word length, 1,024 words of magnetic core memory and two small reel-to-reel magnetic tape units which were specially designed for LINC by Clark and colleague Thomas C Stockebrand. An output display was provided by a 5-inch CRT screen adapted from a standard Tektronix oscilloscope which could display raster graphics at a resolution of 256 by 256 or text using a somewhat coarse 4 by 6 array of spots to form the characters. User input devices included an alphanumeric keyboard and a set of 4 rotary control knobs. Analogue-to-digital converter circuits allowed the computer to be connected directly to laboratory instruments.

With the design finalised, Clark and postgraduate student Charles E Molnar constructed a prototype machine with the assistance of several Lincoln Lab colleagues. LINC was designed to be relatively compact to allow it to be moved around easily within a laboratory environment. A tall 19-inch rack mount cabinet housed the majority of the electronic circuitry and power supplies. This was connected by a set of 6 metre long umbilical cables to a tabletop console unit comprising four microcomputer-sized user interface modules, one each for the CRT display, twin tape units, console panel and another panel containing the control knobs and analogue input connectors. The keyboard was also a separate unit which connected to the console unit by a short cable. Software comprised the LINC Assembly Program (LAP), a screen editor and

12. Bringing it all Together – The Graphics Workstation

assembler with additional commands for controlling the display and tape units which was developed by programmer Mary Allen Wilkes using a software simulation of LINC running on the TX-2 computer.

The LINC prototype was completed by March 1962 and was publicly demonstrated at a life sciences conference organised by the US National Academy of Sciences in Washington DC the following month. The machine was then moved to a nearby US National Institutes of Health (NIH) laboratory where it was successfully used in an experiment to display the neuroelectric responses of a domestic cat called Jasper. This led to the NIH and NASA jointly sponsoring a $1.4 million evaluation programme which involved the construction of 16 additional LINC machines, 12 of which would be placed with selected scientists working at biomedical research laboratories across the country. In an effort to reduce costs and to educate the users on their new equipment, the machines would be supplied in kit form and assembled by the scientists themselves at two specially convened summer workshops to be held in July 1963.

The project team took the opportunity to make some improvements to the design of the LINC prototype before commencing the construction of the additional machines. However, with the exception of an option for an additional 1,024 words of memory, these were aimed at enhancing reliability and simplifying construction rather than increasing the specification. DEC staff also became closely involved at this stage, ensuring that sufficient quantities of 4000 Series modules were supplied in time for the workshops. Following the workshops, DEC began offering LINC commercially, selling a total of 21 examples at $45,000 each. LINC also heavily influenced the design of DEC's 12-bit minicomputer line, starting with the PDP-5 in August 1963.

In addition to the LINC evaluation programme, much more ambitious plans were also afoot to establish a new multi-institutional centre for research into biomedical computing which would be funded by the NIH and hosted by MIT. However, this failed to materialise so in 1964, Clark and several members of his team took up an offer to move to a new computer research laboratory which was being set up at Washington University in St Louis, Missouri. With sponsorship from ARPA's Information Processing Techniques Office, the team were able to continue their work on LINC, completing the development of the software and documentation in time for the final evaluation programme meeting in March 1965.

The introduction of the highly successful PDP-8 model in 1965 prompted DEC to undertake a major redesign of LINC which was led by Wesley Clark and the company's Richard Clayton. The new machine was based on the PDP-8 architecture with additional Flip Chip modules and an extended instruction set which allowed it to run LINC software. The separate electronics and console modules were integrated into a single tall cabinet and the keyboard was replaced by a Teletype ASR-33 teletypewriter which had been modified for the LINC character set. A set of 6 programmable relay contact outputs were also provided for instrument control. The new machine was introduced as the LINC-8 in March 1966, priced at $38,500 for a model fitted with 4,096 words of memory, and was moderately successful with sales of 143 units.

The third and final iteration of the LINC design was the PDP-12, a more powerful version of the LINC-8 based on the integrated circuit version of the PDP-8, the PDP-8/I. Designed by Richard Clayton, the PDP-12 closely resembled the earlier model but added a larger 11-inch CRT display and real-time clock as standard plus improvements to the tape units and analogue-to-digital converter. It was introduced in 1969 at the significantly lower price of $27,900. LINC had become the computer platform of choice for biomedical researchers and this was reflected in increased sales for the PDP-12. DEC also marketed it as suitable for other applications involving the processing of analogue signals but the LINC platform seem to have become too closely associated with biomedicine and the PDP-12 failed to capture the many other scientific and technical applications for which it would have been suited. Nevertheless, a respectable total of 725 PDP-12s were sold, increasing the overall number of LINC-based machines in existence to almost 1,000.

LINC is often hailed as the first personal computer. The machine was certainly significant in advancing the notion of a small, self-contained computer system for laboratory use first expressed in the Burroughs E101 and IBM Type 610 Auto-Point Computer a few years earlier. However, it also embodied the Lincoln Laboratory's philosophy of real-time interactive computing, with computer graphics as the key enabling technology. Therefore, a more fitting description of LINC would be the first graphics workstation.

In December 1975, another contender for the title of first graphics workstation was introduced. The Tektronix 4051 Graphics Display Computer was a novel hybrid of a graphics display terminal, a microprocessor and a magnetic tape drive. It featured a Motorola MC6800 processor with up to 32 Kilobytes of RAM and removable storage in the form of a magnetic tape cartridge unit with 300 Kilobytes storage capacity. As the industry leader in graphics display terminals, Tektronix endowed the machine with excellent graphics capability. Based on the company's 4000 Series Storage Display, the 11-inch DVBST storage tube displayed text at 35 rows of 72 alphanumeric characters or vector graphics at 1024 by 780 resolution. The integrated keyboard included a set of 10 user-definable keys plus additional keys for screen editing and controlling the tape drive. For controlling laboratory instruments, the 4051 was equipped with an IEEE-488 interface which allowed up to 15 devices to be connected simultaneously.

Software was no less impressive, with a BASIC interpreter permanently stored in ROM and an extensive library of optional software which included statistical,

12. Bringing it all Together – The Graphics Workstation

mathematical and electrical engineering software packages. The BASIC language had also been extended to include powerful graphics commands, matrix functions and string functions for text handling.

The Tektronix 4051 took its styling from the company's graphics terminals, thus introducing the familiar all-in-one desktop computer shape popularised a few years later by models such as the Commodore PET. The use of a storage tube removed the need for dedicated display memory and this allowed the machine to be keenly priced at $6,995. Optional extras included a joystick, printer and plotter. When compared with contemporary microcomputers such as the MITS Altair, the Tektronix 4051 was in a different league. However, it was clearly too expensive for the emerging hobbyist and home computer markets so Tektronix marketed it to their traditional science and engineering customer base where it sold in reasonable numbers. The company also attempted to target business applications requiring graphical output, such as financial analysis, but despite receiving a glowing review in Creative Computing magazine the machine failed to make a significant impact on the wider market. Nevertheless, the Tektronix 4051 was several years ahead of its time and has come to be regarded as a classic example of microcomputer design.

Building Blocks – Networking

One of the key attributes that set graphics workstations apart from their microcomputer contemporaries was an inherent ability to communicate with each other over a computer network. Networks allow computers to share resources and data, turning a standalone personal computer into a much more capable beast. Networking evolved from a need to connect remote terminals to large timesharing computer systems. It also shared concepts with telegraph and telephone networks, and the earliest documented example of remote operation of computing equipment was George Stibitz's dazzling September 1940 demonstration in which he used the US telegraph network to connect a Teletype terminal at Dartmouth College in New Hampshire to the Bell Labs Complex Number Calculator located 200 miles away in New York.

Modems

In 1949, a new device was invented to allow digital data to be transmitted efficiently over telephone lines. It was created by electrical engineer John V (Jack) Harrington and his group at the US Air Force Cambridge Research Laboratory (AFCRL) in Bedford, Massachusetts as part of the Digital Radar Relay, a project to develop a digital communication system for the remote reception and interpretation of radar signals.

Telephone lines are designed to transmit analogue signals within a narrow band of frequencies between about 300 and 2,200 Hz. Their electrical characteristics make it very difficult to transmit digital data due to the distortion of the electronic pulses which represent the binary data. These limitations were tolerable for teletypewriter equipment, which transmitted data at low speeds of typically 110 bits per second (bps), but became unacceptable for higher speed data transmission. Borrowing from technology developed for radio communication, Harrington's device operated by modulating the amplitude of a sinusoidal carrier wave. The use of a carrier wave resulted in lower distortion of the pulses. Two additional carrier waves at the same 2 KHz frequency but different amplitudes were used for timing and synchronisation pulses. These allowed data to be transmitted reliably at speeds of up to 1,600 bps. Another device at the other end of the line demodulated the pulses received and converted them back into digital data. The device was originally called a dataset but later became known as a modem, an abbreviation of modulator-demodulator.

In September 1950, prototype modems were successfully used to transmit data from the Microwave Early Warning radar at Hanscom Field in Bedford over commercial telephone lines to MIT's new Whirlwind computer in Cambridge at a rate of 1,300 bps. This experiment caught the attention of MIT professor George Valley whose air defence committee was in the process of setting up the study which would lead to the creation of the MIT Lincoln Laboratory and Project SAGE. Valley recognised the importance of digital communications to the SAGE air defence system concept and Harrington's group were transferred from the AFCRL to the Lincoln Laboratory in 1953 to become part of the large team working on Project SAGE. Over the next two years Harrington and colleague Paul Rosen worked to improve the design of the modem for use in Project SAGE, culminating in the filing of a US patent application in June 1955 which was granted in September 1958.

The Western Electric Company was contracted to manufacture the production version of Harrington and Rosen's modem. In addition to linking the radar sites to the AN/FSQ-7 computers in each of the 24 SAGE Direction Centers, the modems were used to link these machines with AN/FSQ-8 computers in the 3 Combat Centers, creating the world's first computer network. Western Electric also introduced the first commercially available modem, the Bell 101 Dataset, in 1959, choosing Harrington's technology over a rival 650 bps technique developed by the company's sister organisation Bell Telephone Laboratories.

The Bell 101 was the first in a series of modems introduced by Western Electric over the next few years. Later models, beginning with the Bell 103 in 1962, employed an alternative modulation technique, a form of frequency modulation known as frequency-shift keying, which was more reliable than the amplitude modulation technique developed by Harrington. The Bell 103 also permitted two-way (duplex) communication using different carrier frequencies for transmitting and receiving.

12. Bringing it all Together – The Graphics Workstation

A US law which forbade the attachment to the public telephone network of any device not furnished by official network provider AT&T gave Western Electric, a wholly-owned subsidiary of AT&T, an almost total monopoly in the US modem market. The market was only opened up to competition in June 1968 following a legal ruling known as the Carterfone Decision which permitted non-AT&T equipment to be connected to the telephone network providing that the equipment passed a series of stringent Federal Communications Commission (FCC) tests. However, this did not include connecting electrically therefore an indirect method of connection had to be used. Soon other firms such as Codex Corporation were offering alternative products which connected to telephone lines by means of an acoustic coupler, a device that converted the electronic signals from the modem into sound and fed them through the telephone handset. Duplex communication was also possible by converting sound received through the handset into electronic signals.

Early modems employed analogue electronic components which limited the speed at which the signals could be processed. The emergence of digital signal processing (DSP) integrated circuit devices in the late 1960s would push modem transmission rates up to 9,600 bps and beyond. The move from analogue to digital electronics also facilitated the inclusion of a microcontroller chip which could automatically dial a number to make the connection and hang up again when the data transfer was completed. Sequences of commands were transmitted in order to establish and terminate the connection with the receiving modem. This method, which was pioneered by Hayes Microcomputer Products in their 'Smartmodem' model introduced in June 1981, removed the need for a telephone handset to dial numbers manually, thus allowing the modem to be electrically coupled to the telephone line, something that had only become possible following an amendment to the FCC rules in 1975 which permitted devices to be directly connected to the telephone network through standard plugs and jacks. A directly connected modem equipped with a microcontroller could also be set to 'listen' for an incoming call and automatically establish a connection with the modem at the other end of the line.

Integrated circuits also reduced the cost of modems, making them an affordable accessory for computer hobbyists. Modem manufacturers such as Hayes in the US and Pace Micro Technology in the UK began targeting the hobby and home computer market with a series of low-cost modems for microcomputers. This led to the creation of computerised bulletin board systems, microcomputers equipped with a modem and software which allowed remote users to log in using a dial-up connection and read or post messages. These were ideal platforms for computer hobbyists wishing to exchange information or software, virtual computer clubs where members could interact without leaving the comfort of their own home. As their popularity spread, bulletin board systems were also adopted by microcomputer manufacturers and software developers as an inexpensive means of providing customer support, and by other hobbies for sharing information. By the mid 1980s there were several thousand bulletin board systems operating in the US alone.

Distributed Networks and Packet Switching

Modems allow pairs of computers to be connected but they are not suitable for interconnecting large numbers of computers. This was most apparent with the early bulletin board systems which only permitted a single user to connect at a time. Some later systems were equipped with time division multiplexers which allowed multiple remote users to connect by giving each connection a fixed time slot on a rotating basis. However, users were still limited to a single point-to-point connection with one computer at a time. What was needed was a method for allowing computers on a network to share capacity so that each could communicate with any other computer on the network without an open connection locking out the other computers at the same time. A major breakthrough in the realisation of this goal came with the development of distributed networks and packet switching.

The twin concepts of distributed networks and packet switching were first proposed by Polish-born electrical engineer Paul Baran while working at the RAND Corporation in Santa Monica, California. Shortly after joining RAND in 1959, Baran began work on a project to investigate the survivability of communications networks for strategic weapons command and control in the event of a nuclear attack. Having created the SAGE air defence system, the US Air Force had become concerned that it might be vulnerable to attack. A first strike from the enemy on certain critical parts of the system could knock out the entire communications network, thereby impeding the Air Force's ability to launch a counter attack.

Baran's solution borrowed from an earlier RAND idea for survivable radio communications. He envisaged a distributed store-and-forward communications network, a mesh of interlinked 'nodes' that would each act as switches. When a node received a message, it would store it, determine the best route to its destination and then forward it to the next node on that route. As the network had no centralised control, it would continue to function even if some of the nodes and links were destroyed. Furthermore, the reliability of individual links would no longer be critical to maintaining connectivity, making it possible to use lower cost hardware than with a conventional network.

One problem with this concept was that the Air Force's communications requirements extended to both voice messages and digital data. Voice messages were routinely digitised but the transmission rates were different from those used for data. Transmission rates also varied widely throughout the network depending on the quality and purpose of the link. To address this problem, Baran proposed to divide messages into standard blocks of equal 1,024-bit length. These would then be sent out separately over the network

12. Bringing it all Together – The Graphics Workstation

and reassembled in the correct order at the destination. Any delays caused to voice transmissions as a result of the messages being divided up could be minimised by having a sufficiently high data rate. The use of standard message blocks, or packets, would permit messages with different transmission rates to share the same network. Moreover, it would also allow multiple computers to use the network at the same time.

Baran presented his work to the Air Force in 1961. Following further refinement and validation through software simulation, it was published by RAND as an 11-volume series of memoranda in August 1964. Unusually, these documents were not given a security classification as Baran and his superiors believed that the work belonged in the public domain. A formal recommendation to construct the network was sent to the Air Staff in August 1965. However, construction never took place due to a fatal combination of strong resistance from AT&T and a lack of understanding within the Defense Communications Agency, the US government agency tasked with implementing it.

A theoretical foundation for packet switching was developed by the electrical engineer Leonard Kleinrock in 1962. Kleinrock was employed as a research assistant in MIT's Research Laboratory of Electronics while working towards a PhD on the subject of the mathematical theory of communication networks. His doctoral research involved creating analytical models to simulate the flow of message traffic in a hypothetical computer network, using an analysis based on queuing theory and a software simulation running on the Lincoln Laboratory's TX-2 computer. Among other things, Kleinrock showed that network response time could be improved dramatically by dividing messages into smaller pieces. Although Kleinrock made no reference to standardising the length of these pieces, his work would provide a future framework for analysing the performance of packet-switched networks.

Kleinrock's work was published in book form in 1964, the same year RAND published the memoranda describing Baran's work. However, both had little impact initially despite the importance of the work. Computers and telecommunications were separate fields and few computer researchers were interested in networking at this point. Fortunately, all this was about to change.

ARPANET

In April 1963, timesharing evangelist Joseph Licklider sent a memorandum to colleagues at the Advanced Research Projects Agency which he humorously addressed to "Members and Affiliates of the Intergalactic Computer Network". The memo proposed various topics for discussion at a forthcoming meeting, one of which was the development of a network control language for linking together several timesharing computer systems. Licklider envisaged an integrated network that would allow a user sitting at a terminal on one timesharing system to seamlessly access hardware, software or data on another system. The use of a common language would allow communication between different types of computer, irrespective of their architectures and operating systems. Licklider's memo did not immediately lead to any development work but it did stimulate considerable interest in the subject. The ensuing discussions caught the imagination of two talented computer researchers, Lawrence Roberts at MIT Lincoln Laboratory and Donald Davies at the National Physical Laboratory in the UK. Both would go on to play leading roles in the development of networking technology.

Larry Roberts first encountered networking concepts while studying for a PhD alongside fellow MIT doctoral student Leonard Kleinrock in 1962. After meeting Licklider at a conference in November 1964, Roberts became more actively involved and in 1965 he led an ARPA-funded experiment to link two timesharing computer systems located on opposite sides of the country. The experiment had been proposed by Thomas Marill, a former student of Licklider's who had founded a local timesharing company. Roberts and Marill employed modems and a leased telephone line to connect the Lincoln Lab's TX-2 computer to an IBM AN/FSQ-32 computer at the System Development Corporation in Santa Monica, California. A message protocol was developed to allow messages to be exchanged between the computers. Recalling Kleinrock's work, Roberts decided to impose a maximum message length of 119 characters, thereby requiring longer messages to be divided up. This was not the same as Baran's fixed-length packets but Roberts was clearly thinking along similar lines.

Donald Davies was a veteran computer scientist, having been a member of the NPL team which developed the Pilot Model ACE in the late 1940s. In May 1965, he attended the annual congress of the International Federation for Information Processing (IFIP) in New York City where he heard a presentation on Project MAC, the MIT timesharing initiative funded by ARPA. Recognising the importance of this work, Davies arranged for a group of MIT researchers to visit the UK a few months later to discuss their work in more detail. One of these visitors was Larry Roberts.

During the discussions, Davies began thinking about the mismatch between the transmission capability of telephone networks and the characteristics of timesharing computer systems, which communicated with multiple terminals using short, intermittent bursts of data. He realised that an efficient computer network should treat all traffic as short messages in order to maximise the available capacity. Drawing on his experience of queuing systems, Davies devised a distributed store-and-forward scheme for transmitting messages in fixed-length blocks of 50 characters each. Despite having no knowledge of Paul Baran's earlier work, and approaching the problem from an entirely different direction, Davies had conceived an almost identical solution.

Davies described his ideas in a series of internal NPL documents beginning in

12. Bringing it all Together – The Graphics Workstation

November 1965. In March 1966, he presented his work at a public seminar which attracted over 100 attendees including representatives from the UK Ministry of Defence (MoD) and the GPO, the UK government agency responsible for telecommunications. During the seminar one of the MoD attendees drew his attention to Baran's work at RAND. Davies was not unduly discouraged by this, as it helped to validate his own ideas, but he later admitted that he might not have pursued them if he had encountered Baran's work at an earlier stage.

The success of the seminar prompted Davies to write up his ideas in more detail. In June 1966 he produced a comprehensive 33-page document which proposed the construction of an experimental system to serve as the prototype for a national digital communications network. The document also introduced the term 'packet' to describe fixed-length message blocks. It was widely circulated to interested parties both within the UK and internationally. However, like Baran with AT&T, Davies encountered resistance from the GPO and his proposal was not taken forward. Fortunately, Davies was promoted to Divisional Superintendent in August 1966 and this allowed him to set up his own research group within NPL to progress his ideas, albeit at a much slower pace than would have been the case had his proposal received national endorsement.

Back in the USA, the first project which would realise the twin concepts of distributed networks and packet switching was taking shape. In February 1966, IPTO deputy director Robert W (Bob) Taylor obtained an agreement in principle from ARPA for $1 million funding to build on Roberts and Marill's earlier work by interconnecting timesharing computer systems at a number of ARPA-sponsored institutions. Like his mentor Joseph Licklider, Taylor had originally trained as a psychologist but had moved into engineering, working at NASA for several years before arriving at the IPTO in 1965. Taylor's determination to secure this funding was partly motivated by a personal frustration at having 3 separate terminals on his desk in order to access the various timesharing systems under his jurisdiction. The project would be called ARPANET.

Taylor's first task was to hire an experienced programme manager to lead the project. Larry Roberts was the obvious choice but he was reluctant to leave MIT and required considerable persuasion before finally agreeing in January 1967. In April 1967, Roberts held a design meeting at the University of Michigan which included the Principal Investigators from each of the ARPA-sponsored institutions who had agreed to participate in the project. Several of the PIs expressed concern at the computational overhead required for networking and the detrimental effect this might have on the performance of their timesharing systems. One of the participants was LINC designer Wesley Clark, who addressed these concerns by proposing the development of Interface Message Processors (IMPs), dedicated minicomputers at each node to handle the storing and forwarding of messages.

In October 1967, Roberts outlined his initial plans for ARPANET at an Association for Computing Machinery (ACM) symposium in Gatlinburg, Tennessee. It would be a distributed network linking 35 computers at 16 locations across the country. A small number of IMPs would be used to create a packet-switched backbone but the majority of the links would still involve dial-up connections. Also presenting at the same meeting was Roger A Scantlebury, the leader of the NPL research group set up by Donald Davies. Scantlebury discussed Davies' ideas and also referred to Paul Baran's work at RAND, prompting Roberts to visit Baran at RAND shortly afterwards. Baran subsequently became an informal adviser to the ARPANET project.

Roberts used this new information on the NPL and RAND work to refine the ARPANET design specification and in August 1968 a Request for Quotation (RFQ) was issued by ARPA for the packet switching equipment and operation of the network. The $500,000 contract was awarded to the Massachusetts consultancy firm of Bolt, Beranek and Newman, the same company where Licklider had conducted his early timesharing activities. BBN's remit was to develop and deliver the backbone network of 4 IMPs within a year. AT&T was also contracted to provide the dedicated long distance telephone lines necessary to connect the IMPs together.

The IMPs were based on a Honeywell DDP-516 minicomputer fitted with specially designed input/output boards and modems which were capable of duplex communication at a rate of 50,000 bps. Hardware development was led by Severo M Ornstein who had been a member of Wesley Clark's team at MIT before moving to BBN in 1967. Each IMP could handle up to 4 host computers. As the network would have to operate continuously, the hardware was also ruggedized to increase reliability.

The first IMP was delivered to Leonard Kleinrock's group at UCLA in August 1969 and the second a few weeks later to Douglas Engelbart's group at the Stanford Research Institute. The installation of two IMPs allowed communication over the network to begin and the first communication between two computers over ARPANET took place on 29 October when a UCLA student on a terminal connected to the university's SDS Sigma 7 computer system remotely logged in to an SDS 940 computer at SRI on his second attempt. By the end of the year all 4 IMPs had been installed. A further 9 IMPs were added the following year.

BBN supplied the software for the operation of the packet-switched backbone but the company's contract did not extend to the development of network services. These were gradually added through the Request for Comments (RFC) process, which was initiated by UCLA postgraduate student Stephen D Crocker in March 1969 as a means of documenting and sharing protocol developments. RFCs were issued through the Network Working Group which began as an informal gathering of staff and students from each of the four original ARPANET sites but grew in size and importance as the project

12. Bringing it all Together – The Graphics Workstation

progressed. This process resulted in the development of protocols for three essential network services; logging in to a remote host, transferring files from one host to another and sending a message to another user on the network. To enable these services and standardise the ARPANET network interface, a host-to-host communications protocol called Network Control Program (NCP) was also developed.

With the protocols agreed, software applications were then developed to implement each of the three essential network services. The first was 'Telnet' in 1969, which provided the user with a virtual terminal connection to a remote host. Next was File Transfer Protocol (FTP) which was developed at MIT in 1971. However, the application which had the greatest impact was e-mail. In 1971 BBN programmer Raymond S Tomlinson adapted an existing electronic messaging system developed for timesharing computers, SNDMSG, to operate over ARPANET, thus creating the first e-mail system. The first e-mail message was a test message sent by Tomlinson to himself in October 1971. By 2001, 31 billion e-mail messages per day were being sent.

Internetworking

NCP only allowed communication between computers on the same network. In order to realise Licklider's vision of a truly universal computer network, a protocol was needed which would permit intercommunication between different packet-switched networks. This came in 1974 with the development of the Transmission Control Program.

The Transmission Control Program was conceived by IPTO scientist Robert E Kahn in collaboration with Vinton G Cerf, an Assistant Professor at Stanford University. It was conceived as a means of linking ARPANET with an experimental packet-switched wireless data communications network based on radio technology. In order to achieve this, the Transmission Control Program was designed to support both connection-oriented links, such as those used in ARPANET, and a connectionless communication mechanism in which each packet contains sufficient information to be routed from the source to the destination computer without relying on earlier exchanges between them or the transporting network. These self-contained packets are known as datagrams.

Kahn and Cerf were influenced by the work of the French engineer Louis Pouzin who led the CYCLADES Computer Network project which began in 1971 at the Institut de Recherche d'Informatique et d'Automatique (IRIA) near Paris. CYCLADES was based on an alternative packet switching design in which the hosts are responsible for the delivery of data, rather than the network itself, using datagrams and associated end-to-end communications protocol mechanisms. Pouzin also coined the term datagram, an amalgam of the words data and telegram.

Cerf's networking research group at Stanford was contracted by ARPA to implement the Transmission Control Program concept. Two new communications protocols were developed as part of this work, Transmission Control Protocol (TCP) for connection-oriented links and Internet Protocol (IP), an end-to-end protocol for datagrams. These are collectively known as TCP/IP. In March 1982 TCP/IP was adopted by the US Department of Defense as the standard for all US military networking and by 1985 it had become the standard for networking throughout the computer industry. The uptake of TCP/IP facilitated the proliferation of computer networks and led to the birth of the global system of interconnected computer networks that we now call the Internet.

The Xerox Alto

In early 1970, the giant US photocopier firm Xerox was in the throes of a major diversification programme. This had been triggered by serious threats to the company's dominance of the photocopier market from IBM and Kodak, both of which were known to be developing photocopier products. To add to the company's woes, a patent covering a critical part of the xerographic process was due to expire soon, exposing Xerox to the possibility of direct competition in the form of copycat products. Chief Executive C Peter McColough saw the burgeoning computer industry as the way forward for Xerox and set his sights on emulating IBM's successful transformation from office equipment manufacturer to computing behemoth. The company's bold new mission would be to establish leadership in "the architecture of information". The first step towards this goal was the controversial $920 million acquisition of minicomputer pioneer Scientific Data Systems in May 1969.

McColough's plan was for SDS to provide a solid platform upon which to build the new computer business. As we have already seen in Chapter 9, events were to take a different turn but the SDS acquisition would lead to the formation of one of the most inventive research groups in the history of the computer.

Xerox PARC

Following the SDS acquisition, McColough tasked his Chief Scientist Jacob E Goldman with establishing a new corporate facility to perform research in support of the company's new mission. Despite already having a sizeable research and engineering centre in Webster, New York, Xerox had lost its way as an innovator. Goldman realised that what was needed was a different approach to research, one which would look much further into the future by conducting research at a more fundamental level. Therefore, the new facility would undertake basic research in the physical sciences as well as applied research in support of SDS.

12. Bringing it all Together – The Graphics Workstation

Goldman began by hiring a top-notch research administrator to help create and direct the operation of the new facility. As Provost of Washington University, George E Pake had been instrumental in persuading Wesley Clark's LINC group to up sticks and move the 1,000 miles from Massachusetts to Missouri. He had also helped to secure the ARPA funding which allowed the LINC development to continue. Pake shared Goldman's views on the importance of a longer term research agenda. Soon after joining Xerox in December 1969, he began formulating an organisational structure for the new facility. It would comprise three separate laboratories; General Science, Systems Science and Computer Science. Pake himself would take charge of the General Science Laboratory. For the other two, experienced managers would be headhunted to lead them and help recruit suitable researchers.

Pake had begun his career as a physicist and had maintained strong links with the physics research community but he was less familiar with the field of computer science. Following a suggestion from Wesley Clark, he approached ARPANET sponsor Bob Taylor, whom he was acquainted with through ARPA's support of the LINC work at Washington University, for advice on suitable candidates for the job of managing the Computer Science Laboratory. Taylor, who was known for his outspoken views, was scathing of the choice of SDS as an acquisition. He had harboured a low opinion of SDS since having a disagreement with founder Max Palevsky over the merits of timesharing during his time at the IPTO. Taylor also questioned Xerox's plans to target the business data processing sector and suggested focusing instead on office automation, as this would be a more natural fit with the company's existing products and markets. Despite Taylor's bluntness, Pake was impressed by his ideas.

A few weeks after their meeting, Taylor was surprised when Pake telephoned to offer him the job. Taylor had recently left the IPTO to join David Evans's computer graphics research centre at the University of Utah but felt unfulfilled in his new role. He accepted Pake's offer on condition that his responsibilities would not include corporate matters. An associate manager would be appointed to deal with the bureaucratic aspects of the job, freeing Taylor to concentrate on managing the research projects and supporting the researchers. Pake agreed to this and also granted Taylor the privilege of recruiting his co-manager. He hired Jerome I Elkind, an MIT scientist who had also worked at BBN, where he had directed the company's computer research programme.

Goldman and Pake also had to decide on a suitable location for their new research facility. It would have to be far enough away from the company's headquarters in Connecticut to ensure a degree of independence but also within reach of a leading research university for access to suitable research talent. After considering a number of potential locations, they chose Palo Alto in California for its proximity to both Stanford University and to SDS in El Segundo. The location also gave the new facility its name, the Xerox Palo Alto

The Story of the Computer

Research Center (PARC).

Building a Team

Xerox PARC opened for business in June 1970. Using his unrivalled contacts in the field, Taylor was able to recruit some of the best computer scientists and engineers then working in the USA, taking advantage of a recent downturn in US government spending on R&D which saw many researchers facing the threat of redundancy. He began with a group from Berkeley Computer Corporation, a struggling start-up which had been established in 1969 to commercialise technology developed under Project Genie, the ARPA-funded timesharing initiative at the University of California Berkeley. The group of eight included two physicists with extensive computer design experience, Butler W Lampson and Charles P (Chuck) Thacker, and a software specialist, L Peter Deutsch. These became the core of the Computer Science Laboratory's development team.

Another key recruit was Alan C Kay. As a doctoral student at the University of Utah in 1969, Kay had created the conceptual design for a personal interactive computer system which he called the Reactive Engine. Kay was inspired by the work of computer graphics pioneer Ivan Sutherland, who was one of his lecturers at Utah. He was also influenced by Douglas Engelbart's work at the Stanford Research Institute. Kay's Reactive Engine envisaged a computer for personal use which incorporated a high-resolution vector graphics display. The computer would be equipped with a digitising tablet and windowing hardware to allow the user to select for display any 1,024 by 1,024 sized portion of a large addressable display area of 16,384 by 16,384 points. His massive 656-page doctoral thesis also proposed a new programming language called FLEX which included visualisation features and would support user interaction with the hardware by operating in an interpretive mode where statements typed are executed immediately.

Although some experimentation was conducted using an IBM 1130 minicomputer, the Reactive Engine remained a concept and was never built. After completing his PhD, Kay moved to Stanford University where he obtained a position as a research assistant in the Stanford Artificial Intelligence Laboratory (SAIL). This brought him into contact with the South African mathematician Seymour Papert, one of the leading proponents of the use of new technologies for learning and teaching who was based at MIT's Artificial Intelligence Lab. Through Papert, Kay also became familiar with the theories of Swiss developmental psychologist Jean Piaget, who is best known for his work in the field of children's education. This set Kay thinking about the combination of interactive computing and learning through technology. He conceived a hypothetical personal computer for children of all ages which he called the Dynabook.

12. Bringing it all Together – The Graphics Workstation

The Dynabook would be a notebook-sized portable machine with a flat panel display, integrated keyboard and removable storage. Kay was aware that many of the electronic components necessary to build such a machine did not exist but he also knew that the rate of technological progress was such that it was only a matter of time before they would. He had seen some of the earliest work on flat panel display technology during a conference tour of the University of Illinois in 1968 and envisaged the future development of a 1,024 by 1,024 resolution LCD panel. The keyboard would be either a slim membrane keyboard or implemented virtually by increasing the size of the screen so that it covered the entire front face of the machine and fitting it with strain gauges under each corner to create a rudimentary touchscreen. In this way the size and position of the keyboard could be changed to suit the application. An existing magnetic tape cartridge or floppy disk unit could be re-engineered for use as a removable storage device by reducing its power requirements to make it suitable for battery operation.

Kay lacked the resources to develop the Dynabook concept. Instead, he made a number of cardboard models which he filled with lead pellets to simulate the weight and feel of the machine. He also described the concept in a series of papers, anticipating that advances in integrated circuit technology would eventually allow such a machine to be manufactured and sold for only $500.

Taylor was aware of Kay from his time at the University of Utah. Both men shared similar views on the importance of interactive computing and Taylor was keen to get him on board. Kay was equally keen, as he saw PARC as an opportunity to progress his Dynabook concept, but he had recently taken up the post at SAIL and was reluctant to leave his new job so soon. Instead, he began providing occasional input in the form of consultancy beginning in September 1970 before finally becoming a Xerox PARC employee in early 1971 where he would be allowed to lead his own research group in simplifying programming. However, rather than joining Taylor in the Computer Science Laboratory, it was decided that he should be assigned to the Systems Science Laboratory under lab manager William F (Bill) Gunning. He set up the Learning Research Group and began work on developing an object-oriented programming language based on his earlier work at Utah.

Converging Themes

Personal interactive computing and learning through technology would be the two recurring themes throughout Alan Kay's illustrious career. Such was Kay's missionary zeal that these themes would also come to influence the work at Xerox PARC. Fortunately, both themes overlapped with Bob Taylor's interest in man-machine communication, which he had developed through his connections with fellow psychologist Joseph Licklider.

In 1968 Licklider and Taylor had co-written a paper entitled 'The Computer as a Communication Device' in which they emphasised the need for extensive networks of interconnected computers with sophisticated user interfaces. Shortly after arriving at Xerox PARC, Taylor obtained a small number of Imlac PDS-1 standalone graphics systems. With its high resolution graphics display, light pen input device and editing software, the PDS-1 was specifically designed to support interactive graphics. Taylor could see that this was the right approach to providing a rich user interface but the PDS-1 lacked the ability to produce the high quality text necessary for office applications, as the machine's vector graphics display could only form characters using straight lines of uniform thickness. A raster display would have more flexibility but would require significantly more memory, a commodity which remained prohibitively expensive until the introduction of semiconductor memory in 1970.

Licklider and Taylor's paper had also referred to the work of computer mouse co-inventor Douglas Engelbart at SRI and this was to be another major influence on PARC. At the Fall Joint Computer Conference held in San Francisco in December 1968, Engelbart and his colleague Bill English demonstrated a remarkable computer-based system for collaborative working which he called NLS (oN-Line System). NLS ran on a medium-scale SDS 940 system equipped with two vector graphics display generators, each of which was coupled to a set of six high precision 5-inch CRT screens which could be individually addressed using a timesharing arrangement to produce 12 independent displays. Video cameras were used to transmit these screen images to 12 video display terminals distributed at various locations around the Augmentation Research Center. Each terminal was equipped with an alphanumeric keyboard, mouse and another input device known as a chorded keyset. The latter device took the form of a small keypad with 5 keys which could be pressed simultaneously in various combinations in the same way as playing a musical chord on a piano keyboard. It was designed to be used in combination with the mouse, where one hand would be used to control the mouse and the other to control the chorded keyset.

The NLS software allowed documents composed of text and simple vector graphics to be created and shared with other users on the system, thus allowing groups of users to contribute to documents. A shared electronic workspace provided group access to common files, such as records of ARC experiments, and allowed users to leave messages for each other. Other advanced features included keyword searching and the first practical implementation of hypertext, where links could be embedded in files which, when the appropriate text was clicked on with the mouse cursor, automatically accessed related files. The use of video technology to connect the display terminals also facilitated the embedding of live video within a screen display to create a basic video teleconferencing capability.

Engelbart's 90-minute presentation took the form of a live demonstration of

12. Bringing it all Together – The Graphics Workstation

NLS using a microwave communications link which had been set up between the conference centre and SRI 30 miles away. A video projector borrowed from NASA allowed the 2,000-strong audience to see what was happening on Engelbart's terminal screen. The impact of the demonstration was such that it has since entered into computer folklore as "the mother of all demos".

In 1971, Xerox initiated a collaborative project with SRI to implement NLS at PARC. The project was called the PARC On Line Office System (POLOS) and was funded by ARPA. Xerox licensed the mouse from SRI and hired several people from ARC who had worked on NLS, including chief engineer Bill English. However, implementing a system based on timesharing went against the grain in an organisation which was becoming smitten with the idea of personal computing. Instead, POLOS would be based on a network of inexpensive Data General Nova minicomputers fitted with custom electronic character generators designed by Butler Lampson and Roger Bates, and the NLS software would be distributed over the network so that each individual machine was responsible for a specific process.

Another strong influence was the GRAIL (GRAphical Input Language) Project which finished in September 1969. GRAIL was one of a series of ARPA-funded man-machine communication projects conducted at the RAND Corporation in Santa Monica during the 1960s. The objective of the project was to perform computer programming graphically using a flowcharting metaphor. Flowchart symbols displayed on the system's CRT screen could be moved around and resized by picking and dragging 'handles' on their top left and bottom right corners. The system also provided visual cues such as small circles to indicate which handles were active and a change in the brightness of objects when selected.

GRAIL was implemented on an IBM System/360 Model 40 fitted with two additional disk storage units for refreshing the display screen and a Burroughs prototype display controller equipped with a RAND Tablet pointing device. Despite the use of a medium-scale mainframe computer, the compute-intensive nature of the software was such that the system could only support a single graphics terminal.

NLS and GRAIL had demonstrated the power of interactive graphics and hinted at what might be possible if this technology was applied to office automation but they were based on expensive mainframe systems which made them incompatible with the new vision of personal computing being promulgated by Taylor and Kay. Wesley Clark, who acted as an occasional consultant to PARC, had shown the potential of a dedicated machine for individual use with his much-admired LINC design. This resulted in a gradual shift in focus within PARC from timesharing to distributed personal computing.

The Interim Dynabook

By late 1972, sufficient resources were in place for the Computer Science Laboratory researchers to begin tackling larger and more ambitious projects. The impetus for the first such project came from Alan Kay, who needed 15 to 20 "interim Dynabooks" for his Learning Research Group and also had a budget of around $230,000 to fund their development. Kay had attempted to initiate a project to realise his Dynabook concept shortly after arriving at PARC but flat panel display technology was in its infancy and developing such technology to the required specification would have been an extremely costly endeavour. Instead, Pake suggested sacrificing portability for a larger interim machine which could prove the principle of Kay's ideas using existing display screen technology. This also fitted nicely with Taylor's vision of interactive computing and Lampson's wish to develop a high-performance minicomputer which would be cheap to manufacture.

The new machine would be called the Alto in a reference to the Californian city where PARC was located. A design specification was drawn up by Kay, Taylor and Lampson. The detailed design and construction was led by Chuck Thacker. The Alto was based on a custom 16-bit processor with 128 Kilobytes of semiconductor memory and a 14-inch CRT display in portrait orientation with 606 by 808 resolution raster graphics. The machine could also be fitted with one or two Diablo Model 31 removable disk cartridges, each of which had a storage capacity of 2.5 Megabytes, using a custom disk controller designed by Edward M McCreight.

The Alto's processor design was loosely based on the Data General Nova. It was a discrete processor implemented using a set of 4 Texas Instruments 74S181 4-bit arithmetic logic units in a 'bit slice' arrangement. The clock speed was a respectable 5.9 MHz. A multiplexing scheme allowed the Alto to manage input/output tasks more efficiently by sharing the processor between 16 fixed-priority tasks, using two additional registers and a small amount of memory to store the task program counters. The processor also employed microprogramming, which breaks down machine instructions into sequences of microinstructions. This allowed the team to experiment with different instruction sets and also made for a much simpler logical design.

The Alto had no dedicated video memory. Instead, a section of main memory was set aside for graphics use and configured as a frame buffer, where each bit of memory directly represents a pixel on screen. By setting a bit to 1, a dot is displayed at the location on screen represented by that bit. If the value is 0 then that location will be blank. This technique had been pioneered at Brookhaven National Laboratory in 1968 and further developed by Bell Labs over the next two years for use in video telephone research. Known as bitmapped graphics, it allows complete control over what is displayed on screen. Characters can be created directly in memory without the need for an electronic character

12. Bringing it all Together – The Graphics Workstation

generator, thereby allowing different fonts to be provided. However, more video memory is required than with other methods and roughly half of the Alto's main memory was given over to the frame buffer.

The Alto's input devices included a 44-key alphanumeric keyboard with programmable function keys, a 3-button mouse and a chorded keyset. The mouse buttons were colour-coded and arranged in a vertical column. However, the keyset proved difficult to learn to use properly and was unpopular with most of the researchers.

In addition to interactive graphics, the other prerequisite necessary to realise the PARC vision of distributed personal computing was for the Altos to be able to communicate with one another over a fast local area network. The task of developing this network was given to Robert M (Bob) Metcalfe from the Computer Science Laboratory and David R Boggs from the Systems Science Laboratory. Metcalfe had studied the University of Hawaii's ALOHAnet system while working on his doctoral dissertation at Harvard in 1970. ALOHAnet was a packet-switched data communications network which used commercial radio equipment to connect computers wirelessly. Unlike a store-and-forward scheme, every node was allowed to broadcast messages directly over the network at any time. When a collision was detected, the sender node waited a random interval before retransmitting the message. Metcalfe used this technique to develop a wired local area network which he called Ethernet.

Using a 1 kilometre length of coaxial cable, Ethernet operated at a speed of 2.94 million bps (Mbps) and was capable of supporting up to 256 nodes. Individual computers were connected to the network using a 'vampire tap' which pierced the cable in order to connect to the inner wire. The Ethernet interface comprised a transceiver, which generated the analogue signals used to broadcast messages in a similar way to the radio signals used in ALOHAnet, and a data link controller which encoded and decoded the analogue signals and handled the data packets.

Two wire-wrapped Alto prototypes were completed by April 1973 and were demonstrated by displaying a test image of the Cookie Monster from the children's television programme Sesame Street which had been created by Kay using a carefully digitised drawing. A further 8 machines were also produced by the end of that year. Despite its personal computer aspirations, the Alto was an imposing sight. A large desk-side cabinet housed the circuitry, the upper section of which contained an aperture for the Diablo disk drive. The CRT display, keyboard, mouse and chorded keyset were placed on top of a desk and connected to the cabinet by means of cables. As well as an Ethernet interface, each Alto was equipped with a printer interface and other special-purpose interfaces were also developed to support various research projects ranging from music synthesis to machine vision. The Ethernet network was operational by November 1973 and the interface hardware retrofitted to those machines which had already been built.

Prompted by Taylor's belief that the fruits of research should always be put to work, a decision was taken to build a further 80 Altos to allow a wider group of PARC researchers to make use of the machine. However, this number was beyond the limited in-house manufacturing capability of the group so the work was subcontracted to Clement Designlabs of Mountain View, California. Clement also improved the aesthetic appearance of the Alto by designing custom cases and adding a stand for the display which the keyboard could be fitted into to reduce the amount of desk space required.

Xerox Alto

Deliveries of the first production machines began in May 1974. The Alto proved to be an excellent development platform and as more software became available for it, interest in the machine began to increase, both within Xerox and amongst PARC's research collaborators. Around 1,500 production Altos were eventually built by Clement Designlabs for use within Xerox and limited distribution to various collaborators, including SAIL, SRI and several government agencies.

12. Bringing it all Together – The Graphics Workstation

WIMP and WYSIWYG

The Xerox Alto was notable for the way in which it tightly coupled a minicomputer and raster graphics terminal but this was by no means unique and the machine would have had little impact were it not for the groundbreaking software developed for it. The driving force behind these developments was Alan Kay's Learning Research Group.

Kay's work on object-oriented programming had resulted in the development of a new language and programming environment which he called Smalltalk, the first complete version of which, Smalltalk-72, was completed in 1972. Object-oriented programming treats all program entities as objects which interact with one another and can be grouped into classes. This facilitates a more structured approach, encouraging the programmer to create modularised software which is easier to maintain and better suited to team-based projects. The object-oriented approach also lends itself to the development of graphics software.

Smalltalk-72 was ported to the Alto by Daniel H H Ingalls and was the first software available on the new machine. Kay and another colleague, Diana Merry, then used this software to create multiple windows on screen using the convention of corner handles for moving, resizing, cloning and closing which they borrowed from the GRAIL project. However, unlike GRAIL, these windows could be overlapped in order to make the most of the Alto's limited screen resolution, a concept which had first appeared in Kay's 1969 doctoral thesis. A window could contain text or graphics so Kay and Merry also created proportional fonts for the display of high quality text. Text could be scrolled up or down using the mouse buttons when the cursor was positioned within a scroll bar area on the left hand side of a text window. The shape of the cursor automatically changed to a double-headed arrow when within the scroll bar area to indicate when scrolling was active. For graphics, Ingalls developed an object-oriented version of an existing system called Turtle Graphics which had been created by Seymour Papert at MIT as an extension to his Logo programming language. This allowed simple line drawings to be produced.

The first version of the Alto software was completed in March 1974 but the display performance was found to be poor when using overlapping windows. Merry addressed this by inventing a new type of graphics operation which allowed entire rectangular blocks of pixels to be manipulated as a single entity in the frame buffer. Known as a bit-boundary block transfer or 'BitBlt', this significantly improved the performance of overlapping windows, allowing windows to be moved fluidly around the screen and text to be scrolled smoothly. For the next version of the software, Smalltalk-74 which was completed in November 1974, Ingalls implemented the BitBlt operation in microcode to further increase performance. He also added pop-up menus for accessing commonly used commands which were displayed at the cursor position by pressing the middle mouse button.

Kay's concept of overlapping windows led to the adoption of a desktop metaphor, where the display screen is treated as the user's desktop, with windows representing open documents on the desk. This metaphor was further enhanced through the development of a number of innovative software applications, the first of which was a word processing program called Bravo which was completed in October 1974. Unlike earlier word processors, Bravo made clever use of the bitmapped graphics capability of the Alto to display the formatted text exactly as it would appear when printed. This paradigm became known as what-you-see-is-what-you-get (WYSIWYG).

Bravo was conceived by Lampson and Charles Simonyi, a Hungarian-born software engineer who had worked alongside Lampson at Berkeley Computer Corporation and had followed him to PARC in 1972. It employed a novel data structure known as a piece table to keep track of modifications to the document text in an efficient way. Rather than representing every attribute of every character in the document, this method stored text in blocks, or pieces, with a table of descriptors to represent the attributes of each piece. This reduced the requirement to rewrite the entire document file every time a change was made. The use of a piece table data structure also helped to maintain performance when editing large documents.

Bravo was a powerful, feature-packed word processor but was difficult to learn and use due to a clumsy user interface which employed a separate mode for entering commands. This kind of interface, which is known as a modal interface, tends to exhibit inconsistent behaviour. In Bravo's case, this was where letters typed in command mode invoked a different response than when typed in text entry mode. As Bravo commands were represented by their first letter, inadvertently typing the word 'edit' in command mode could result in the entire document being replaced by the letter 't'. This inconsistency jarred with PARC's ethos of simplifying man-machine communication, prompting efforts to develop modeless interfaces. The chief protagonist in these efforts was Lawrence G (Larry) Tesler, a computer scientist who had joined Kay's Learning Research Group in 1973 from Stanford University where he had been working as a research assistant at SAIL.

Tesler's earliest work at PARC was on the development of the NLS-derived POLOS system, which he was highly critical of due to the system's complexity and heavy reliance on the dreaded chorded keyset for user input. His opportunity to create a modeless interface came in October 1974 as a result of a project to develop a computer-assisted editing system for book publishing. The project was prompted by a request for assistance from Ginn & Company, a Xerox-owned school textbook publisher based in Lexington, Massachusetts. Ginn management wanted to explore the use of computer technology to reduce publishing costs and turned to PARC for help. However, initial efforts to adapt

12. Bringing it all Together – The Graphics Workstation

POLOS for use in word processing and page layout were unsuccessful so Taylor suggested developing a system based on the Alto. Timothy Mott, a British computer scientist, was recruited by Ginn to work with Tesler. The resulting system was named Gypsy.

Gypsy was based on Bravo but with a modeless interface which made much greater use of the mouse. It was developed by Tesler and Mott with the assistance of Daniel C Swinehart and contributions from Lampson, Simonyi and others. The design was also shaped by a study Mott had made of the working methods of the book editors at Ginn. Gypsy's screen layout comprised a main document window with a smaller command window across the top and a 'wastebasket' window across the foot of the screen which displayed deleted text. Text entry was possible at all times whenever the cursor was positioned within the document window and was enabled by simply clicking the top mouse button. The mouse was also used to select text for editing and function keys were used for commands such as cut, copy, paste and undo. Other commands were available from a menu in the command window at the top of the screen.

Installation and testing of Gypsy began at Ginn in February 1975. An evaluation conducted by PARC's User Sciences Group in April the following year found that adopting the system had led to an increase in efficiency amongst Ginn's secretarial staff but that it lacked many of the features necessary for serious book editing work. Nevertheless, the success of Gypsy's modeless interface inspired Simonyi and colleague Thomas Malloy to develop an improved version of Bravo called BravoX which borrowed the three-window layout from Gypsy and added menus at the top of each window to create a modeless interface. BravoX also introduced the concept of 'styles' to represent a set of properties which control the formatting of a document, such as headings, fonts and paragraph alignment. Development of BravoX began in 1976 and was completed by 1978. BravoX became the word processor of choice amongst the staff at Xerox PARC and as it grew in popularity, the name was later changed to the Xerox Document System Editor.

Smalltalk's windows and menus had enabled the development of sophisticated graphical user interfaces, as seen in Bravo with its WYSIWYG display and Gypsy with its modeless interface. A third software application developed for the Alto provided another important feature in the evolution of the graphical user interface. The application was 'Draw', an interactive program for creating line drawings which was inspired by Ivan Sutherland's Sketchpad.

Draw was developed by Patrick C Baudelaire and Robert F Sproull, who were both members of PARC's Graphics Systems Group. It was implemented using Pico, a library of graphics functions created by the Group in 1974. Draw was one of several interactive graphics applications developed for the Alto but what made it special was the use of small pictograms or 'icons' to represent commands. A rectangular area down the left hand side of the screen contained two columns of icons. These were designed to symbolise each action, so for example a drawing of a paintbrush was used to represent the command for selecting line thickness. Commands were selected by placing the cursor over the icon and clicking the top mouse button. The shape of the cursor would then change to indicate that the command was active. Later projects at Xerox PARC would take icons onto the desktop itself. This type of user interface acquired the acronym WIMP (Windows, Icons, Menus and Pointers).

The development of the WIMP user interface and WYSIWYG software applications was truly revolutionary. It also marked the beginning of a sea change in the computer industry, where the technology would move from being hardware driven to software driven. However, in order for this revolution to begin, the technology developed at Xerox PARC would have to emerge from the laboratory and into the commercial marketplace.

Early Commercialisation Efforts

Xerox was slow to recognise the commercial potential of the technology developed at PARC. The planned synergy between the research capability of PARC and the product development resources of SDS failed to materialise due to the tightly compartmentalised structure of the company. PARC's remoteness from the company's headquarters had also become more of a curse than a blessing, with Pake and Goldman finding that their opinions held little sway over the Corporation's senior executives in Connecticut. More than four years were to pass before Xerox took steps to create a commercialisation path for PARC technology.

In January 1975, Xerox finally remedied the situation with the establishment of the Systems Development Division (SDD). The new division combined a product development capability in Palo Alto with a manufacturing facility in El Segundo staffed by engineers from SDS, which was in the process of being wound up. Its remit was to commercialise the technologies from PARC for the office automation market. Several key PARC staff, including Bob Metcalfe, Charles Simonyi and Ron Rider, transferred to the new division, enticed by the prospect of seeing their inventions transformed into products.

One of the earliest projects undertaken by SDD was the re-engineering of the Alto. The project was led by John Ellenby, a British computer scientist who had joined PARC in 1974 from the University of Edinburgh. While at Edinburgh, Ellenby had also been consulting for Ferranti on the design of high reliability computers for industrial process control and he used this experience to make the Alto easier to manufacture and service. The specification remained the same except for the ability to expand main memory up to 256 Kilobytes, twice that of the original machine, and the instruction set was extended slightly to include support for double length words. The new model was designated the Alto II. However, it was only intended to address maintenance issues with the

12. Bringing it all Together – The Graphics Workstation

original model and to help satisfy the growing demand for the Alto from within Xerox and PARC's collaborators. The new WIMP interface and WYSIWYG applications had highlighted the potential of the Alto platform but had also severely tested the limits of the hardware and performance was poor. To make this technology an attractive proposition for the office automation market, a new computer with significantly higher performance would be needed.

In early 1976, SDD head David E Liddle invited Alto designer Chuck Thacker to join the Division to develop such a computer. Thacker was reluctant to leave PARC but agreed to a secondment and brought with him the design for a new machine called the Dorado which he had been working on with Butler Lampson. The Dorado was intended to be the successor to the Alto as the main software development platform for PARC researchers and was designed to surmount the Alto's performance limitations and lack of virtual memory. It featured a 16-bit processor with a 4-stage instruction pipeline, an 8 Kilobyte memory cache and virtual memory support. The Alto's task switching mechanism was again employed to multiplex the processor between 16 fixed-priority tasks. The discrete processor design was implemented using the latest emitter-coupled logic (ECL) semiconductor technology, which facilitated a blazingly high clock speed of 16.6 MHz. Main memory could be expanded up to 16 Megabytes and the machine was also fitted with a high-capacity 80 Megabyte removable hard disk unit. However, the advanced architecture of the Dorado meant that the new machine would not be binary compatible with the Alto so an emulation mode was included to allow it to execute Alto software. The raw power of the Dorado was such that Alto software running through the emulator could run much faster than on the Alto itself.

The Dorado was equipped with a 1024 by 808 resolution bitmapped display in landscape orientation and featured dedicated BitBlt hardware to improve windowing performance, with 6 additional registers to support BitBlt operations. Support for a 640 by 480 resolution colour display was also provided. Another interesting feature of the Dorado was that it was optimised for the execution of software written in Mesa. This was achieved by means of a microprogrammed emulator to support the Mesa instruction set. Mesa was a high-level programming language and programming environment which was loosely based on Pascal. Originally developed at PARC in 1971, it had been ported to the Alto in 1975 but ran slowly. Despite this, Mesa became popular amongst PARC's researchers and in 1976 it was chosen as the standard programming environment for all SDD projects.

The Dorado was physically much larger than the Alto, with significantly higher power and heat dissipation requirements as a result of its fast ECL circuitry. It was also very expensive to build, the components alone costing $50,000. It soon became obvious that it would not make a suitable vehicle for Xerox's entry into the office automation market. Instead, it was decided to keep the Dorado as an internal development platform for PARC researchers and also use it as the controller for a new high performance printer which was being developed using a novel laser printing mechanism invented by optical engineer Gary K Starkweather at Xerox's Webster Research Center in 1969. A lower cost version of the Dorado would be created for the commercial market using cheaper TTL semiconductor components. This new machine was codenamed the Do or Dolphin.

The Dolphin was designed by a team led by Thacker and Brian Rosen, who had recently joined SDD from Three Rivers Computer Corporation, a spin-out from Carnegie Mellon University in Pittsburgh. The only major changes from the Dorado specification were a reduced clock speed of 7.3 MHz, due to the use of slower TTL components for the processor, and a fixed hard disk drive with a 23 Megabyte capacity instead of the more costly removable hard disk units fitted to the Alto and Dorado. Development was completed in 1978. However, despite the use of lower cost components, the Dolphin remained a complex, high specification machine which was still too expensive for the office automation market. When it did eventually reach the market 4 years later in 1982, it was sold as the Xerox 1108 Scientific Information Processor at a price of almost $40,000 and marketed for demanding scientific and technical applications such as Computer-Aided Design. At the top of the 1100 series was the Dorado, now renamed the 1132 Scientific Information Processor and priced at a whopping $59,719.

The Star Information System

After almost a decade of intensive development effort at PARC, Xerox still had no suitable computer platform for the office automation market. Instead, the company had opted to introduce a series of dedicated word processing systems developed by its Office Products Division in Dallas, Texas, beginning with the Xerox 850 Display Typing System in October 1977. These systems were intended to build on the success of the Xerox 800 Electronic Typing System, an early word processing system based on an electromechanical typewriter with magnetic tape storage which was introduced in October 1974. The 850 followed the industry trend of adding a CRT display and software to permit the interactive manipulation of text on the screen. Despite having a highly modular design which allowed the hardware to be expanded, the 850 was incapable of running any software other than the fixed word processing program stored in the machine's read-only memory.

Xerox had passed up an earlier opportunity to introduce a word processing system based on the Alto when John Ellenby proposed the Alto III in 1976. The Alto III would be a lower cost version of Ellenby's re-engineered Alto II. It would be optimised to run an advanced word processing program such as Bravo but would also be able to run other office software, such as database programs, just like any other general-purpose computer. However, a task force set up to evaluate the options for the company's next word processing

12. Bringing it all Together – The Graphics Workstation

product favoured the 850 and plans to develop the Alto III were shelved.

At the same time as the Dorado and Dolphin were being developed, SDD was also engaged on a project to develop a graphical user interface for office systems. Although this began as a software development project, it would result in the creation of the company's first computer for the office automation market, the Xerox 8010 Star Information System.

The project was initiated in 1975 by David Liddle under the codename Janus. Liddle's plan was to "look forward to office information systems and backward to word processing". This would be achieved by refining and integrating PARC's WIMP user interface technology and WYSIWYG software applications. The project was given a boost following the rejection of Ellenby's Alto III proposal, as one of the recommendations of the task force had been that the Office Products Division should adopt the new user interface technology. This was followed by a decision in the summer of 1978 to proceed with a product.

The Star graphical user interface was developed by a team led by David Canfield Smith, who also coined the term 'icon' in his doctoral thesis at Stanford University. It was programmed exclusively in the Mesa language and based on a design methodology drawn up by a joint PARC/SDD committee which had been formed in 1976 to advise on the design of the interface. The desktop metaphor pioneered on the Alto was extended to include icons for representing files, folders and objects which perform functions, such as an e-mail inbox, a calculator, a printer and a floppy disk drive. A set of icons was specially created for this purpose by Smith and graphics designer Norman L Cox.

The Star software also featured a suite of office applications, including a WYSIWYG word processor with embedded graphics capability which was based on Bravo and a tabular calculator with similar functionality to a spreadsheet program. Various networking utilities, such as e-mail and Telnet, were also provided. These applications and utilities were all tightly integrated with the WIMP user interface, allowing text to be cut and pasted between them. However, unlike the windows in Smalltalk, Star windows could not be overlapped. A study had shown that users seldom wanted to overlap windows, preferring instead to place the windows alongside each other, so the Star system would be equipped with a high resolution 17-inch display in landscape orientation to accommodate two documents displayed side by side.

The original intention had been to use the Dolphin as the hardware platform for Star but when it became apparent that this would be too expensive for the target market, the project was extended to incorporate the development of a new computer which would maintain the Dolphin's high performance but at a much lower cost achieved at the expense of configurability. Development was scheduled to take 27 months. The computer, which was codenamed Dandelion, was designed by Robert L Belleville, Robert B Garner and Ronald C Crane. The use of a 16-bit microprocessor was initially considered but rejected due to the need to retain compatibility with the Dolphin and Mesa software environment. Instead, the new computer would be based on a paper design for a discrete processor called Wildflower by Butler Lampson and Roy Levin which was intended to address some of the shortcomings of the D-series (Dorado/Dolphin) architecture while maintaining backward compatibility.

The Dandelion followed the example of Wildflower by ditching the Alto's task switching mechanism in favour of synchronising all peripheral devices to the processor, thus reducing the requirement for costly input/output buffers. An Intel 8085 microprocessor handled input/output tasks, freeing up the 16-bit processor to cope with the heavier computational load of the Star software. It was equipped with 384 Kilobytes of memory which was expandable up to 1.5 Megabytes, a choice of 10, 29 or 42 Megabyte hard disks and an 8-inch floppy disk drive. The circuitry was housed in a slim desk-side cabinet, with the 1024 by 808 resolution 17-inch CRT display, 79-key keyboard and 2-button mouse placed on a desk. A fast 10 Mbps Ethernet interface was also included to facilitate distributed computing and allow the machine to connect with Xerox's new range of office laser printers.

Xerox 8010 'Star' Information System (Photo © Xerox Corporation)

The new machine was introduced as the Xerox 8010 'Star' Information System in April 1981 and showcased at the National Computer Conference held in Chicago the following month. The $16,595 price included the full suite of Star software. However, the all-important Ethernet interface was an optional extra. Without it, the advanced networking features built into the Star software environment were useless. To take full advantage of the new

12. Bringing it all Together – The Graphics Workstation

machine's capabilities, customers would have to purchase a minimum of two computers fitted with Ethernet interfaces plus a $12,000 laser printer, pushing the total cost up to nearer $50,000. This was prohibitively expensive for most businesses. Performance was also sluggish, as the Star software soaked up every drop of compute power that the Dandelion processor could supply.

Only four months after the launch of the Xerox 8010, IBM introduced the 5150 PC which cost a fraction of the price. Despite its higher specification and revolutionary graphical user interface which made the 8010 much more user-friendly than other personal computers, companies seeking to introduce computers into their business were unlikely to opt for the Xerox model when they could buy 5 of the IBM machines for the same money. The 8010 was an over-specified premium product in a marketplace which valued low cost over usability. The machine's huge performance advantage over the IBM 5150 was also masked by the ponderous Star software. Xerox had also decided not to release the Mesa language and programming environment, which made it virtually impossible for third-party software developers to create applications for the 8010. A handful of optional applications were available from Xerox, including database software and an advanced graphics package, but customers seeking to run industry standard business software such as VisiCalc on an 8010 would be sorely disappointed.

Despite these shortcomings, around 30,000 Xerox 8010 systems were sold, encouraged by a dramatic price cut in 1984 which brought the price of a basic model down to $8,995. However, this was only a tiny fraction of the sales totals for the IBM 5150 and Apple][over the same period. Had the 8010 design team based the machine on a microprocessor, it should have been possible to bring the price down further to a level where it could compete directly with the IBM and Apple models but the decision to stick with what was essentially a minicomputer architecture was a costly mistake. Xerox had failed to recognise the changes taking place elsewhere in the computer industry while the 8010 was under development. By the time it was introduced, the industry had moved away from minicomputer-based systems to microcomputers and the office automation market was already awash with low-cost, microprocessor-based machines.

Xerox persevered and in early 1985 introduced a new model, the 6085 Professional Computer System, which addressed the 8010's performance issues and also featured a revamped version of the Star software called ViewPoint. Third-party software applications were supported by means of an optional Intel 80186 co-processor which allowed the machine to run programs written for the IBM PC. However, the 6085 was again based on the D-series architecture. Consequently, prices remained high and sales remained poor. To cap it all, other manufacturers who had taken inspiration from PARC's revolutionary graphical user interface technology were now poised to release their own microprocessor-based products.

PARC's Legacy

There has been considerable criticism of the Xerox Corporation in recent years for its apparent failure to commercially exploit the groundbreaking technology developed at PARC. This is both unfair and inaccurate. The laser printer was enormously successful for Xerox, with total earnings from sales of products and technology licensing estimated to have more than covered the costs incurred by PARC until the division's incorporation as an independent, wholly-owned subsidiary in 2002. Ethernet became established as the industry standard for connecting computers over a local area network following its adoption by DEC and Intel in 1980. PARC technologies were successfully transferred into industry through the migration of many key staff, who left PARC for pastures new, taking their radical ideas with them or in some cases starting up their own companies. PARC also spawned numerous spin-outs directly, several of which, such as 3Com and Adobe Systems, became household names. Most importantly, PARC's revolutionary graphical user interface technology encouraged an industry trend towards this technology and no modern personal computing device is complete without a user interface that owes its existence to the work of Xerox PARC.

Defining the Market

Although the Xerox 8010 was a market failure as a personal computer for business, mainly as a result of its uncompetitive price and closed architecture, the notion of a high performance personal computer with bitmapped graphics and integrated networking capabilities was not without merit. The Alto had proved itself as a highly desirable platform for researchers within Xerox and amongst PARC's external research collaborators. This attraction was so powerful for one researcher, the Swiss computer scientist Niklaus E Wirth, that he built an almost identical system on his return to Switzerland following a sabbatical at PARC in 1976-77. Here was a clear sign of the commercial potential of such high-end personal machines in scientific and technical applications.

Three Rivers Computer Corporation

The earliest attempt to exploit this potential by introducing a high performance personal interactive computer for the scientific and technical market was made by the Three Rivers Computer Corporation. Three Rivers was formed in Pittsburgh, Pennsylvania in 1974 by five researchers from the Computer Science Department at Carnegie Mellon University led by Brian Rosen. The company's first products were based on digital audio equipment and high-performance graphics subsystems for computer science research which the group had developed while at CMU.

12. Bringing it all Together – The Graphics Workstation

In November 1976, Rosen left Three Rivers to join the Xerox Systems Development Division where he co-designed the Dolphin. Following his return to Three Rivers in September 1978, Rosen lobbied for the company to develop an Alto-like computer which he called the 'Pascalto'. This happened to coincide with a request for proposals from CMU, which wanted to replace a large timeshared computer facility for computer science research with a distributed network of high performance personal machines as part of a project called SPICE (Scientific Personal Integrated Computer Environment). Three Rivers submitted a proposal based on Rosen's new machine and secured the contract.

The new machine was designed by Rosen as his version of the Xerox Alto and Dolphin and he deviated very little from the D-series architecture. The processor was a discrete design with microprogramming capability, a 16-bit word length and a 5.9 MHz clock speed. The machine supported up to 1 Megabyte of RAM and 1024 by 768 resolution bitmapped graphics. It also incorporated dedicated BitBlt hardware to improve windowing performance. However, as the Mesa programming environment was unavailable outside of Xerox, the processor was optimised instead to run p-code, a type of portable intermediate or 'assembly' code generated using a Pascal compiler.

The machine was packaged into the familiar desk-side cabinet with a 60-key detachable alphanumeric keyboard, a 15-inch CRT display in portrait orientation and a fixed hard disk of 12 or 24 Megabyte capacity. One departure from the familiar Xerox specification was the use of a touch-sensitive tablet instead of a mouse. Another new feature was the inclusion of an IEEE-488 interface for laboratory instrument control in addition to the 3 Mbps Ethernet interface. The machine also included an integrated speech synthesiser and loudspeaker for synthesised speech output, reflecting the company's audio equipment origins.

The new machine was introduced as the Three Rivers PERQ (an abbreviation of perquisite) in May 1979, three years ahead of its Xerox equivalent, the 1108 Scientific Information Processor. The price was $19,500 for a basic specification model fitted with 256 Kilobytes of RAM and a 12 Megabyte hard disk. System software comprised the PERQ Operating System (POS), a command-based operating system with multiprogramming capability, plus a display window manager and a distributed file system for access to files on other computers on the network. To facilitate programming, each machine was also supplied with an integrated Pascal complier, a symbolic debugger for debugging programs and a WYSIWYG text editor. Options included the Ethernet interface, a 1 megabyte floppy disk drive and a 4 kilobyte writable control store which allowed programmers to perform their own microprogramming. This latter option also facilitated the development of other operating systems for the PERQ and it later became available with a choice of operating systems; POS, a Unix variant called PNX which was developed by ICL and Accent which was

The Story of the Computer

developed by CMU.

Three Rivers marketed the PERQ for scientific, engineering and other technical applications such as publishing. One of the earliest customers for the machine was the UK's Rutherford Appleton Laboratory (RAL) in Didcot near Oxford. Before making a final decision, RAL scientists had explored various options and had been in discussions with British computer giant ICL regarding the company's plans to develop a graphics workstation based on the Intel 8085 microprocessor. However, as the ICL machine was at least two years away from production, they suggested to the company that they consider going into partnership with Three Rivers instead. RAL was also the coordinating organisation for a UK government sponsored research programme in distributed personal computing which would need more than 100 machines of similar specification to the Xerox Alto as a platform for the software development. As a direct descendent of the Alto, the PERQ would be the ideal choice but the UK Science Research Council which held the purse strings for the research programme was uncomfortable with an overseas supplier for a programme of national importance and wanted to buy British. This led to the Council persuading ICL to take on the manufacturing and marketing of PERQ machines for the UK, Europe and other territories.

The negotiations between ICL and Three Rivers coincided with a period of financial instability for both companies and took some time to conclude. Three Rivers was also experiencing production problems which had delayed PERQ deliveries by more than a year. Nevertheless, the deal was concluded by August 1981 and the first UK-manufactured PERQ machines, complete with ICL badging, began rolling off the production line in February 1982.

With ICL's help Three Rivers introduced a revised model in March 1983, the PERQ 2, which had similar performance but was cheaper to manufacture. It also featured 1 Megabyte of RAM as standard, an optional 19-inch display with 1280 by 1024 resolution graphics and a 3-button mouse in place of the touch-sensitive tablet. However, others had also spotted the emerging market for graphics workstations and several companies had begun introducing rival machines based on microprocessors. ICL attempted to counter this by developing the PERQ 3, an all-new model with multiprocessor capability based around a 32-bit Motorola microprocessor, but it never reached the market and the partnership's days were numbered following a decision by ICL to reposition itself as a solution provider. Three Rivers changed its name to PERQ Systems Corporation and struggled on for a while but the PERQ 2 was becoming increasingly outdated and the company finally went out of business in October 1985, having delivered a total of around 5,000 workstations.

Apollo

Another US company which had also been under consideration by the

12. Bringing it all Together – The Graphics Workstation

Science Research Council as a potential manufacturing partner for ICL was Apollo Computer Incorporated. Apollo was established in Chelmsford, Massachusetts in February 1980 by electrical engineer John William (Bill) Poduska of US minicomputer manufacturer Prime Computer plus six of his colleagues. Poduska had also co-founded Prime in 1972 following a period spent working in Honeywell's Research Center. Having helped build Prime into a highly successful firm by targeting the scientific and technical market, Poduska wanted to do the same again with microcomputers.

Start-up financing came from a trio of venture capital firms, one of which, Greylock Partners, had also been a major investor in Prime. The initial business plan was to develop a microprocessor version of a Prime minicomputer but this was soon changed after Poduska and co-founder David L Nelson learned of CMU's SPICE project which had specified the development of a high performance personal interactive computer for scientific computing. Although they did not know it at the time, this was the same project that had prompted the development of the Three Rivers PERQ.

The computer was designed by Nelson and another co-founder, Paul J Leach. It was based around the new Motorola MC68000 microprocessor. Introduced in September 1979, the MC68000 was one of the earliest 32-bit microprocessors to reach the market. It quickly became established as the 32-bit microprocessor of choice amongst computer designers due to its 16-bit data bus and 24-bit external address bus which made the chip much less expensive to produce and easier to implement than a full 32-bit device. The use of general-purpose registers and a powerful instruction set also made the MC68000 relatively easy to program. Nelson and Leach's design employed dual MC68000 processors running at a clock speed of 8 MHz. Dual processors were necessary in order to implement a virtual memory scheme in the absence of any virtual memory support on the MC68000, with the second processor used to handle page faults, requests for pages which are not yet loaded into physical memory. The machine also featured up to 3 Megabytes of RAM and 1024 by 1024 resolution bitmapped graphics with dedicated BitBlt hardware. The actual display resolution was slightly lower at 800 by 1024, with the additional memory used as temporary storage for character font tables. Another unusual feature was the choice of a proprietary network interface based on an alternative to Ethernet called token ring which operated at 12 Mbps. An Intel Multibus interface was also included to accommodate expansion boards.

The new machine was introduced as the Apollo DOMAIN DN100 in October 1980 at a price of $40,000 for a basic specification model fitted with 1 Megabyte of RAM and a 14 Megabyte hard disk drive. A bulky desk-side cabinet housed a 19-inch rack for the circuitry, with a separate 17-inch CRT display monitor in portrait orientation and keyboard featuring an integrated touch-sensitive tablet making up the rest of the system. System software comprised Apollo's Aegis operating system, an object-oriented operating system with

The Story of the Computer

multiprogramming capability and strong support for distributed computing. Like the PRIMOS system used in Prime minicomputers, Aegis was based on the influential timesharing operating system Multics. However, unlike these other systems, Aegis also featured an interactive display manager and a library of graphics primitives to facilitate graphics programming.

The first two DN100 machines were delivered to Harvard University in March 1981. Thanks to a highly effective marketing strategy directed by Edward J Zander, who joined the company from Data General in 1982, the DN100 and its successor, the DN420, became popular with academic researchers. It also found a ready market as a hardware platform for high-end CAD systems, with several turnkey CAD system vendors choosing to abandon their own hardware developments in favour of a rebadged Apollo workstation. This OEM (Original Equipment Manufacturer) market would later account for more than half of all Apollo's sales.

Apollo DOMAIN DN100 (Photo © Hewlett-Packard Company)

A single-processor desktop model, the DN300 Desktop Computational Node, was introduced in 1983 at a price of $15,995. This could be operated either as a diskless node by booting from another machine over the network or as a self-contained machine via an optional 34 Mb hard disk and 8-inch 1.2 Mb floppy disk. The DN300 was also notable for its unusual L-shaped case, with the integrated 17-inch CRT display positioned above the horizontal leg and fixed to the side of the vertical leg by a horizontal shaft which allowed the monitor to be rotated in order to adjust the angle of the display screen. The optional disk drives were housed in a separate unit which connected to the DN300 using two ribbon cables.

12. Bringing it all Together – The Graphics Workstation

As the first manufacturer to introduce graphics workstations based on the new generation of 32-bit microprocessors, Apollo enjoyed spectacular growth. A stock market launch in the spring of 1983 broke the previous record for the highest valued initial public offering (IPO) in the US technology sector. However, within two years the company was struggling against newer competitors who were offering cheaper products equipped with the non-proprietary Ethernet network interface and running Unix, another Multics-derived operating system developed by Bell Labs which had become firmly established as the de facto standard for scientific and technical computing. This resulted in Apollo posting its first ever losses in the third quarter of 1985.

Poduska left Apollo in November 1985 to start his third computer company, Stellar Computer Incorporated, with the aim of developing powerful graphics supercomputers. His successor, Thomas A Vanderslice, a veteran corporate executive who had been brought into the company the previous year to take it to the next stage in its growth, attempted to salvage the worsening financial situation by introducing cost-cutting measures and laying off 12% of its 3,400 strong workforce. As a stopgap, the company began offering an older version of the Unix operating system as an option while a new operating system called Domain/OS was being developed. Domain/OS was finally released in July 1988. It combined a common kernel (the part of the operating system that manages input/output requests) with three operating environments to support Aegis plus the two most popular flavours of Unix.

In order to survive, Apollo needed to introduce new low-cost models that could compete effectively in an increasingly crowded marketplace which now included scientific computing heavyweights DEC and Hewlett-Packard. In April 1986, the company entered negotiations with leading microcomputer manufacturer Apple Computer on a project to port Aegis to Apple's forthcoming Macintosh II personal computer and market it as an entry-level graphics workstation. Apollo would buy 40,000 Macintosh II machines and rebadge them as an Apollo model. However, the deal was rejected by Apple CEO John Sculley in January 1987.

Apollo's deteriorating financial position meant that there was little money available for new product development. This was made worse in 1987 when an unauthorised foreign exchange transaction by one of the company's employees resulted in a further loss of $6.5 million. By the end of 1988 the situation had become critical and Vanderslice began looking for a buyer for the company. In April 1989, Apollo was acquired by Hewlett-Packard for $476 million and its products merged with HP's workstation range. The company which had defined the graphics workstation market had ceased to exist just as the market was experiencing unprecedented growth of over 50% per annum. Total sales of Apollo workstations are estimated to have been around 100,000 machines, an impressive figure but one that would soon be eclipsed by the new leaders in the graphics workstation market.

The Stanford University Network

The two companies that would come to dominate opposite ends of the graphics workstation market both had their origins in an initiative which began at Stanford University in the autumn of 1979 when Xerox donated 18 Altos plus a laser printer and file server to the University's Computer Science Department. The University also secured funding from the Defense Advanced Research Projects Agency (formerly ARPA) to install an experimental campus-wide Ethernet network as part of a DARPA-funded programme of research into the design of VLSI (very-large-scale integration) integrated circuits. The result was the Stanford University Network (SUN) which initially incorporated the Xerox equipment plus two DEC VAX 11/780 minicomputers. This network would provide the infrastructure for a series of landmark research projects on different aspects of VLSI design.

Sun Microsystems

One of the supporting projects at Stanford focused on the development of a graphics workstation which could be built relatively cheaply and deployed in large numbers to supplement the Xerox Altos and provide a suitable platform for the development of VLSI design software. To keep the costs low, the workstations would operate as diskless nodes and boot from the DEC minicomputers over the network. The project was led by Associate Professor Forest Baskett, who had recently returned to Stanford following 18 months working at Xerox PARC on VLSI research.

The SUN workstation was designed by German-born postgraduate student Andreas (Andy) von Bechtolsheim. Using a minicomputer-hosted CAD system from another project, Bechtolsheim designed a set of three circuit boards comprising a processor board, a graphics display board and a 3 Mbps Ethernet interface board. Each board was equipped with an Intel Multibus edge connector for interconnection via a Multibus backplane. The processor board was self-contained and could also be operated as a standalone computer. It featured a fast 10 MHz version of the Motorola MC68000 microprocessor with support for virtual memory via an onboard memory management unit. Physical memory comprised 256 Kilobytes of DRAM and 32 Kilobytes of ROM but up to 2 additional memory expansion boards of 768 Kilobytes capacity each could also be connected via a separate memory bus. The graphics board incorporated dedicated BitBlt hardware and a 1024 by 1024 pixel frame buffer which was implemented using dual-port video memory to provide bitmapped graphics with a display resolution of 1024 by 800, with the additional memory available for use as temporary storage for different fonts, cursors and special characters.

12. Bringing it all Together – The Graphics Workstation

Bechtolsheim and his colleagues were able to build a small number of workstations themselves using parts scrounged from around the Computer Science Department and local electronic components suppliers. However, a suitable manufacturer would need to be found in order to supply the workstations in the required quantities. With no established workstation manufacturers to turn to, the University looked to the minicomputer sector for help. The two most likely candidates were scientific minicomputer market leaders DEC and Prime Computer. Both firms were approached but neither company was interested in taking it on so, following an expression of interest from Bechtolsheim, the Intellectual Property Rights were assigned to Bechtolsheim as inventor.

In May 1980, Bechtolsheim set up a company called VLSI Systems to exploit his newly acquired IP. Using $25,000 of his own money, he built more prototypes which he then used to demonstrate his board designs to prospective manufacturers. This approach was reasonably successful and Bechtolsheim licensed his designs to 5 separate firms for a fee of $10,000 each. However, 3 of these were manufacturers of board-level products and another was using the processor board as a laser printer controller. Only one licensee had any intention of producing an entire workstation, with plans to develop a workstation-based turnkey CAD system, but this company was newly formed and not yet in a position to begin manufacturing. Despite having recouped his investment and made some extra cash to help pay his way through university, Bechtolsheim was no further forward in finding a company to manufacture the SUN workstation.

Towards the end of 1981 Bechtolsheim was contacted by Vinod Khosla, an Indian immigrant who had obtained a master's degree in business administration (MBA) at Stanford the previous year and was working for Daisy Systems, a CAD systems developer based in Mountain View, California. Khosla was on the lookout for a better hardware platform for his company's electronic design automation software and could see the potential of Bechtolsheim's SUN workstation. On hearing of Bechtolsheim's difficulties in finding a company willing to manufacture it, he suggested they go into partnership together to produce and sell graphics workstations based on Bechtolsheim's board designs. Khosla then approached his friend and former MBA classmate Scott G McNealy, who was working as Manufacturing Manager at microcomputer firm Onyx Systems in San Jose, hoping to obtain an introduction to the former CEO of Onyx, Douglas W Broyles, as a potential investor. McNealy made the introduction but also offered to join them in setting up the new venture.

Bechtolsheim, Khosla and McNealy then sat down together and drafted a short 8-page business plan which outlined an ambitious plan to bring the SUN workstation to market within only 4 months. The low-cost modular design of the workstation and its adherence to emerging industry standards, such as the MC68000 processor, Multibus, and Ethernet, were emphasised. Despite the brevity of this document, they succeeded in securing $300,000 equity investment from Broyles and Robert Sackman of US Venture Partners, a new venture capital firm based in Menlo Park, California. Armed with an upfront payment of $100,000, they formed a new company called Sun Microsystems in Santa Clara, California, in February 1982, the name chosen to reflect the company's roots in the Stanford University Network.

One of the earliest decisions was the choice of operating system for the workstation. They had decided from the outset that it would run Unix, which was fast becoming the industry standard for scientific and technical computing. However, Unix came in many flavours and the version chosen would need to have a distributed file system to support diskless nodes plus a WIMP user interface to make the most of the workstation's bitmapped graphics. One version, BSD (Berkeley Software Distribution) Unix, which was under development at the University of California Berkeley, fitted these requirements so Baskett's colleague Vaughan R Pratt arranged for Bechtolsheim, Khosla and McNealy to meet the principal developer, a postgraduate student and part-time teaching assistant by the name of William N (Bill) Joy. Joy's programming skills were legendary, making him an ideal person to lead the company's software development activities. Following a successful meeting at Berkeley, Joy accepted an offer to join Sun and was given full co-founder status in recognition of his importance to the company.

Sun-1 Workstation (Photo © Mark Richards)

In May 1982, Sun Microsystems introduced the production version of Bechtolsheim's workstation, the Sun-1. They had met the tight 4-month development schedule by subcontracting most of the manufacturing and

12. Bringing it all Together – The Graphics Workstation

assembly work and by using off-the-shelf parts wherever possible. The specification was largely unchanged from the original, using the same boards fitted into a standard Multibus card cage. Two versions were available, the Model 100 desktop machine with a 7-slot card cage and the Model 150 deskside machine with a larger 15-slot card cage. Both were supplied with a 17-inch CRT display, keyboard and a choice of hard disks which connected to the system via a third-party disk controller board. Early examples were supplied without a mouse due to an initial lack of software support for the device which was later remedied by changing the operating system from UniSoft Corporation's UniPlus Version 7 Unix to one based on Joy's latest version of BSD Unix. The starting price for a diskless Model 100 fitted with 256 Kilobytes of RAM was under $10,000, a fraction of the price of the cheapest Apollo model, although this rose to nearer $20,000 for a typical specification machine.

With very little marketing effort, Sun achieved a respectable first year sales total of $8 million, selling around 400 Sun-1 machines mainly to academic institutions. However, Khosla had his sights set on the more lucrative OEM market and in the summer of 1983, Sun stole a march over market leader Apollo by winning a $40 million, three-year OEM contract to supply workstations to leading CAD systems manufacturer Computervision. Sun's lower manufacturing costs and commitment to industry standards for hardware and software had paid off. The Computervision systems would be based on Sun's forthcoming new model, the Sun-2, but fitted with Computervision's own graphics hardware instead of Sun's graphics display board in order to boost graphics performance.

The Sun-2 was introduced in November 1983. It featured the new Motorola MC68010 microprocessor, a revised version of the MC68000 with integrated support for virtual memory and improved loop instruction execution. Other improvements over the Sun-1 included an increase in main memory capacity from 2 Megabytes to 8 Megabytes, a faster 10 Mbps Ethernet interface and higher resolution graphics of 1152 by 900 pixels. To make the most of the increased graphics resolution, a 19-inch CRT display and a 3-button mouse were supplied as standard. The Intel Multibus and modular design of the earlier model were retained, which allowed existing customers to upgrade their Sun-1 systems to a Sun-2 by simply replacing the circuit boards.

The new model was marketed to what Sun called "knowledge workers"; scientists, engineers and other professionals who required access to high performance interactive computing. As with Apollo, CAD and other design automation applications formed the mainstay of sales. However, a new type of application called Desktop Publishing (DTP) had recently emerged which, like CAD, relied on high resolution interactive graphics and had previously only been available on dedicated minicomputer-based systems. DTP combined WYSIWYG word processing and embedded graphics with advanced page layout capability to create the digital equivalent of the typography process used in the production of newspapers and magazines. When coupled with a laser printer, a DTP system allowed a relatively unskilled user to produce high quality documents in-house.

DTP manufacturers soon began ditching their expensive custom hardware and porting their software to a less costly graphics workstation platform instead. One of the first to do so was Interleaf Incorporated of Cambridge, Massachusetts, which introduced its Interleaf OPS-2000 system for the Sun-2 in May 1984 at a price of $38,000 for the workstation and software or $62,500 including a laser printer. The drop in the price of DTP systems as a result of moving to a lower cost hardware platform boosted sales of both the DTP products themselves and the graphics workstations they ran on. Sun and Apollo both benefitted from this effect and by 1987 DTP accounted for around 9% of all graphics workstation sales.

In June 1984, Sun commenced the development of its third generation graphics workstation, the Sun-3, which would be based on Motorola's powerful new MC68020 processor, the first full 32-bit implementation of the MC680x0 architecture. Other differences from the previous model included the addition of a Motorola MC68881 co-processor to boost floating-point arithmetic performance, a doubling of main memory capacity from 8 to 16 Megabytes and a change from the Intel Multibus to the newer VMEbus which had been created by Motorola specifically for the MC680x0 microprocessor family. The move to the slightly larger format VMEbus also prompted the Sun-3 designers to integrate the graphics circuitry and Ethernet controller onto the processor board, thus reducing the number of circuit boards in a standard system from three to a single board. This reduced manufacturing costs and facilitated the more compact 'pizza box' packaging for desktop models.

The Sun-3 was introduced in September 1985. Three versions were available at launch; the 3/75 desktop model with a 2-slot card cage, the 3/160 desk-side model and the 3/180 rack-mountable model, both with 12-slot card cages. All were equipped with a 16.7 MHz MC68020, floating-point co-processor, Sun's version of BSD Unix (known as SunOS) and a full suite of programming tools, including compilers for the FORTRAN, Pascal and C programming languages. Prices ranged from under $10,000 for a diskless desktop model to over $60,000 for a fully specified desk-side server. Other models were gradually added to the Sun-3 range over the next three years with the final model introduced in April 1989. By the time that the range had been discontinued, more than 160,000 Sun-3 systems and processor boards had been produced.

The success of the Sun-3 would propel its manufacturer into the big league. Between 1985 and 1989 Sun Microsystems was reputed to be the fastest growing company in the United States. This remarkable period of growth was presided over by McNealy, who had taken over from Khosla as CEO in the autumn of 1984 following Khosla's resignation after a clash with the Sun board

12. Bringing it all Together – The Graphics Workstation

of directors. An IPO in March 1986 provided a cash injection which allowed the company to invest heavily in new product development and the next generation of Sun workstations would feature Sun's own design of microprocessor, the SPARC (Scalable Processor Architecture). Sun also led the computer industry in the development of software protocols for distributed file systems, with the company's Network File System (NFS) becoming established through the RFC process as the industry standard for accessing files over a computer network. Sun's commitment to networking was reflected in the phrase "The Network is the Computer" which was coined by Sun chief researcher John B Gage in 1984 and subsequently adopted as the company slogan.

In April 1987, Sun slashed the price of its entry-level Sun 3/50 model to under $5,000, triggering a price war which arch rival Apollo was ill-equipped to endure. By the end of that year Sun had overtaken Apollo in sales revenues to become the largest player in the US workstation sector with a 27% market share. Recent entrant DEC had climbed into second place, mainly by selling low-cost graphics workstations to its existing customer base as an upgrade for graphics display terminals on minicomputers. In October 1987, Apollo marketing guru Ed Zander left the company and moved to Sun, exemplifying the changing fortunes of the two companies. The following year Sun achieved $1 billion in annual sales.

Silicon Graphics

The same DARPA-funded programme of research into the design of VLSI integrated circuits at Stanford University which had spawned Sun Microsystems also led to the creation of another leading graphics workstation manufacturer, Silicon Graphics Incorporated. Silicon Graphics would come to dominate the upper end of the workstation market with a range of powerful machines designed specifically for 3-D graphics.

One of the DARPA-funded projects initiated at Stanford in 1979 was concerned with the design of a custom co-processor for boosting the graphics performance of microprocessor-based computers. The small project team was led by Associate Professor James H Clark. As a postgraduate student Clark had studied under CAD pioneer Ivan Sutherland at the University of Utah, obtaining his doctorate in computer science there in 1974 with a thesis entitled '3-D Design of Free-Form B-Spline Surfaces'. Following three years spent at the University of California Santa Cruz and a short spell working as a consultant in the aerospace industry, he moved to Stanford in 1979 on the recommendation of Forest Baskett who also secured DARPA's support for the project. The idea for the project came when Clark recalled the clipping divider which Sutherland and Robert Sproull had developed while at Harvard University. The aim would be to implement Sutherland and Sproull's clipping divider as a single integrated circuit.

Clark and electrical engineering PhD student Marc R Hannah completed the initial design of the device, which they called the Geometry Engine, in 1980. It incorporated 4 identical floating-point function units for accomplishing three common geometric operations in computer graphics; coordinate transformations on matrices of up to 4 by 4 elements, window clipping of both 2-D and 3-D objects, and scaling to the coordinate system of the output device. A 5 kilobyte control store held the microcode for performing the geometric operations. Taking inspiration from the pipelined architecture pioneered in the Evans & Sutherland LDS-1 graphics display system, they designed the Geometry Engine so that multiple devices could be arranged in a pipeline, with the devices grouped into separate subsystems for each of the three geometric operations. A configuration register in each device was employed to set the operation type for the device when the system was first powered up. A typical pipeline would comprise 10 or 12 Geometry Engines, with 4 devices allocated to coordinate transformation, 4 or 6 to clipping and 2 to scaling.

The Geometry Engine was implemented as a VLSI integrated circuit in a standard 40-pin package. In order to create such a complex device, Clark and Hannah received advice and assistance from Lynn Conway who led the LSI Systems group at Xerox PARC. Conway was an acknowledged expert in VLSI design and co-author of the standard textbook on the subject, 'Introduction to VLSI Systems' along with Carver A Mead of Caltech. Clark and Hannah followed the book's design methodology closely, using the example given in the book for the design of the Geometry Engine's arithmetic logic unit. They also received support from Stanford colleague John L Hennessy, an Assistant Professor of Electrical Engineering, making use of Hennessy's microcode simulation language known as SLIM (Stanford Language for Implementing Microcode) to create the microcode for the device.

The first working prototypes of the Geometry Engine were fabricated and tested at Xerox PARC's Integrated Circuits Laboratory. Performance was in excess of 100,000 lines per second for a 12-device pipeline, a figure which was comparable with dedicated graphics display systems costing several hundred thousand dollars. The commercial potential of this technology was clear. In April 1981, Clark filed a US patent application for the Geometry Engine. He then set about the task of commercialising the device.

Clark initially attempted to license the Geometry Engine to existing manufacturers, approaching both IBM and DEC with the suggestion that they incorporate it into their graphics display terminals and Apollo for use in their workstations, but none of them were convinced of the need for such a device. He then realised that he would have to set up his own company to manufacture the entire system. In order to do so, he took a year's sabbatical from Stanford. The company, Silicon Graphics Incorporated, was founded in January 1982 in Mountain View, California by Clark, Hannah and six of their fellow Stanford researchers (Kurt Akeley, David J Brown, Tom Davis, Mark

12. Bringing it all Together – The Graphics Workstation

Grossman, Charles S Kuta and Charles C Rhodes) plus Abbey Silverstone, an experienced development manager from Xerox PARC. To finance it, Clark arranged a $25,000 personal loan from a friend and used this to leverage an equity investment of $800,000 from Mayfield Fund, one of Silicon Valley's oldest established venture capital firms.

The first Silicon Graphics product, a high-end graphics display terminal named the Integrated Raster Imaging System (IRIS), was introduced in 1983. It combined a Geometry Engine pipeline with a raster graphics subsystem and a Motorola MC68000 microprocessor. The processor board was based on Andy Bechtolsheim's design for the SUN workstation, which the company obtained through licensing, but with a slightly slower 8 MHz processor. The 256 Kilobytes of onboard DRAM was supplemented by an additional memory board to provide 768 Kilobytes of memory. The graphics pipeline and raster subsystem were implemented on 4 circuit boards to deliver full colour graphics at a resolution of 1024 by 1024. A third-party Ethernet board provided the network interface. The boards were connected through an Intel Multibus backplane and housed in a desk-side cabinet with a separate 19-inch colour CRT display, 83-key keyboard and 3-button mouse.

The IRIS Terminal was designed to be connected to a DEC VAX or similar high performance minicomputer. It was programmed using commands written in the IRIS Graphics Library (GL) which were sent to the terminal from its minicomputer host over the Ethernet interface. Two models were available, the IRIS 1000 with a 10-slot backplane and the IRIS 1200 with a 20-slot backplane and larger cabinet. An optional floppy disk drive was available for storing demonstration programs. Other hardware options included a Dial Box with 8 independently programmable dials and a Switch Box with 32 independently programmable switches and 32 independently programmable LED indicator lights. Prices started at $37,500 for an IRIS 1000 model fitted with a 10 Geometry Engine pipeline. This compared favourably with the popular Evans & Sutherland PS 300, a colour graphics terminal with 3-D graphics hardware which boasted a high-quality vector display but came without an Ethernet interface.

In 1984 Silicon Graphics introduced its first standalone system, the IRIS 1400 Workstation. The IRIS Workstation was based on the IRIS 1200 Terminal but with an uprated 10 MHz MC68010 processor, an additional 768 Kilobytes of memory and a choice of 72 or 474 Megabyte hard disk drives. An optional magnetic tape cartridge drive was also available. The operating system chosen was a version of AT&T's Unix System V from UniSoft Corporation but with various enhancements including the IRIS Graphics Library. Priced at $59,500, the IRIS 1400 Workstation was a fraction of the cost of a system based on a suitably specified minicomputer and 3-D graphics display terminal. The first production machine was purchased by Carnegie Mellon University for use in electronic imaging research. The machine also found favour in molecular graphics, a new discipline in which the structure of molecules is modelled using 3-D computer graphics in order to analyse their properties.

The IRIS 1400 Workstation was followed by the IRIS 2000 series in August 1985, which incorporated various incremental improvements, and the IRIS 3000 series in February 1986, which was based on the more powerful MC68020 processor. Encouraging sales totalling around 3,500 of these machines prompted the company to go public with an IPO in October 1986 led by Edward R McCracken, an experienced senior executive from Hewlett-Packard who had taken over from Clark as CEO two years earlier. Under McCracken's leadership, Silicon Graphics would grow from a minor player servicing a niche market to a billion dollar company which dominated the upper end of the graphics workstation market. A major factor in this success was the adoption of a new microprocessor architecture which, like the Geometry Engine, had also come out of one of the DARPA-funded VLSI design projects, the Stanford MIPS project.

The Stanford MIPS (Microprocessor without Interlocked Pipe Stages) project began in 1981 and was led by John Hennessy, the Assistant Professor who had helped Clark and Hannah with the microcode for the Geometry Engine. The aim of the project was to develop a new single-chip VLSI processor architecture which would achieve high performance through the use of a simplified instruction set. This approach, which is known as RISC (Reduced Instruction Set Computing), had been pioneered in the CDC 6600 supercomputer and further developed by John Cocke at IBM's Thomas J Watson Research Center in the late 1970s as part of the IBM 801 experimental minicomputer project. By adopting the RISC approach, Hennessy and his team designed a 32-bit microprocessor which executed instructions several times faster than a Motorola MC68000 running at the same clock speed. However, they were initially unable to explain exactly why this should happen. Consequently, the semiconductor industry took little notice of their work.

As an academic, Hennessy was expected to move on to a new research project following publication despite the obvious commercial potential of the MIPS design. Fortunately, he was encouraged by Gordon Bell of DEC not to abandon the technology but to set up a company to exploit it himself. The normally prohibitive start-up costs associated with semiconductor firms could be minimised by operating as a 'fabless' company, concentrating on the design of the devices and outsourcing fabrication to other manufacturers. Revenues would be generated by licensing the design to the device manufacturers and to the firms which used the devices in their products.

The company, MIPS Computer Systems, was founded in September 1984 by Hennessy plus Edward P (Skip) Stritter, a Stanford graduate who had worked on the design of the MC68000 at Motorola, and John Moussouris from IBM, who was also a visiting researcher at Stanford and had worked on the ground-

12. Bringing it all Together – The Graphics Workstation

breaking IBM 801 project. Hennessy followed in Jim Clark's footsteps by taking a year's sabbatical in order to establish his company but, unlike Clark, he did return to Stanford and would later rise to the position of President of Stanford University. To facilitate his intended exit, Hennessy wisely decided not to lead the company but instead recruited an experienced executive from Intel by the name of Vaemond H Crane to run it. After his return to academia, Hennessy continued to work with the company in a consultancy capacity.

In January 1986, MIPS Computer Systems announced the MIPS R2000 chip set which comprised the R2000, a 32-bit microprocessor with a 5-stage instruction pipeline and 32 general-purpose registers based on the Stanford prototype, the R2010 floating-point co-processor and R2020 write buffer. The chips would be manufactured by Sierra Semiconductor in San Jose, California. With a performance three times that of the industry leading Motorola MC68020, the R2000 created considerable attention. The first customer was struggling minicomputer manufacturer Prime Computer which also helped to bankroll the development. Silicon Graphics had initially planned to go with a rival RISC microprocessor, the Clipper from Fairchild Semiconductor, due to a delay in introducing the R2010 co-processor but Clark was won over by Hennessy who reminded him of the common origins of the two companies and the close links between them which included a shared board of directors.

Silicon Graphics introduced the first in a powerful new family of workstations based on the MIPS chip set in March 1987. The Professional IRIS 4D/60 featured an 8 MHz MIPS R2000 processor coupled with a Weitek floating-point accelerator and a high performance graphics pipeline which boasted no less than 38 individual chips. Taking advantage of the move to a new processor architecture, the Multibus backplane used in previous models was replaced by the newer VMEbus standard and the machine was also supplied with the new IRIX operating system, a Silicon Graphics enhanced version of Unix System V with BSD extensions plus the company's own file system. The new model was packaged as a twin tower desk-side cabinet with 12 VMEbus slots and a beefy 1,000-watt power supply. Prices started at $74,000 for a machine equipped with 4 Megabytes of RAM and a 170 Megabyte hard disk.

The performance of the new Professional IRIS models far exceeded any other personal computer system on the market. The machines were particularly suited to applications where high performance 3-D graphics was a prerequisite, such as the emerging market for digital visual effects in the motion picture industry. However, with the price of a typical system nudging towards six figures, few organisations could afford to buy them. Fortunately, those who could did buy them and sales of the Professional IRIS resulted in the company's revenues doubling to $116 million in 1987 but the graphics workstation market was also expanding rapidly and its market share actually dropped slightly to a fraction over 4%. Also, Silicon Graphics was no longer the only workstation manufacturer serving the high performance 3-D graphics market, as both Hewlett-Packard and Tektronix had introduced workstations with 3-D graphics capability. Silicon Graphics needed to dramatically lower the cost of its technology in order to attract new customers and increase its market share. Its next series of workstations, the Personal IRIS range, would do exactly that.

In October 1988, Silicon Graphics introduced the Personal IRIS 4D/20, its first product to be targeted at the mid-range segment of the workstation market. Development was funded using cash generated from selling a 20% stake in the company to Control Data Corporation for $68.9 million in March of that year. The new machine featured a 12.5 MHz MIPS R3000 processor, an improved version of the R2000 with virtual memory support, and supported up to 32 Megabytes of memory. Several graphics options were available, ranging from 8-bit colour with a single Geometry Engine to full 24-bit colour with four Geometry Engines and advanced 'z-buffering' hardware for hidden surface removal. The use of the latest generation GE5 Geometry Engine allowed 3-D graphics performance to be maintained with a much lower chip count which significantly reduced the number of circuit boards required and the size of the desk-side cabinet used to house them. The lower manufacturing costs allowed the Personal IRIS 4D/20 to be priced at under $20,000 for a basic specification system with 8 Megabytes of memory and 8-bit colour graphics. Glowing reviews in the technical press hailed the new machine as a breakthrough in affordable high performance graphics. With an IRIS 4D range for both the upper and mid-range segments of the market, Silicon Graphics had consolidated its position as the leading manufacturer of high performance graphics workstations.

Silicon Graphics was soon followed in its adoption of RISC technology by other graphics workstation manufacturers using new microprocessors with RISC architectures from companies such as Hewlett-Packard and Motorola. These second generation machines blurred the line between minicomputers and graphics workstations and further widened the performance gap with microcomputers. By 1989, the graphics workstation had established itself in many of the application areas traditionally associated with minicomputers, with the result that the previously dominant minicomputer market began suffering badly.

In October 1991, MIPS Computer Systems introduced the first commercially available 64-bit microprocessor, the MIPS R4000. With an 8-stage instruction pipeline, integrated floating-point co-processor and backward compatibility with the R3000, the R4000 would provide the ideal platform for Silicon Graphics' next generation of workstations and design work was already at an advanced stage. However, the high cost of development of the new device had led to serious financial problems for MIPS. To protect its investment in the R4000 technology, Silicon Graphics moved to acquire MIPS Computer Systems in a stock swap valued at $333 million and the deal was concluded by March 1992. The R4000 made its debut in the high-end Silicon Graphics Crimson

12. Bringing it all Together – The Graphics Workstation

which was introduced in January 1992 but it was also used in a revised version of the IRIS Indigo, the company's entry-level desktop model which was priced at less than $10,000. The following year Silicon Graphics achieved $1 billion in annual sales, leapfrogging both DEC and IBM to become the third largest player in the graphics workstation market.

Workstation Technology for the Masses – The Apple Mac

Despite the incredible success of companies like Silicon Graphics and Sun Microsystems, graphics workstation technology remained unaffordable for the majority of business or home computing applications. Only one microcomputer manufacturer showed any interest in adopting graphics workstation technology for use in mainstream models. That manufacturer was Apple Computer Company.

In December 1979, the Xerox Alto was seen by a team led by Steve Jobs from fledgling microcomputer manufacturer Apple Computer on a visit to Xerox PARC. The visit was arranged by Jef Raskin, a mathematician and computer scientist who had joined Apple in January 1978 and had recently begun work on the conceptual design for a new low-cost model codenamed Macintosh which would be easy to use, semi-portable and could be sold for under $1,000. Raskin was already familiar with the work of Xerox PARC, having toured the facility several times in 1973 while a visiting scholar at Stanford University, and he wanted to show his boss PARC's WIMP user interface technology as an example of what could be done to improve usability. Jobs was deeply impressed by the demonstration and expressed his amazement that Xerox had not yet brought this revolutionary technology to market.

The Apple Lisa

As well as Raskin's work on the new low-cost model, Apple had initiated a parallel project to develop a new business computer codenamed Lisa, the initial specification for which had been written by Jobs and marketing director William M (Trip) Hawkins in October 1978. The design was to be carried out by Apple co-founder Steve Wozniak but Wozniak lost interest in the project and in July 1979 Apple hired Ken Rothmuller, a former engineer at Hewlett-Packard, to manage it. Following the Xerox PARC visit, Hawkins rewrote the specification to include a WIMP user interface, WYSIWYG software applications and a local area network.

A bitmapped display was already included in the Lisa specification but the machine would need a suitable pointing device. However, as the target price was a tight $2,000, this would have to be significantly cheaper to manufacture than the precision engineered 3-button mouse used by Xerox. Jobs asked Dean Hovey of local industrial design firm Hovey-Kelley Design to come up with a suitable design that could be produced for less than $35. The resulting one-button mouse with its distinctive box shape has come to be regarded as a classic example of product design.

Development of Lisa's WIMP interface and WYSIWYG applications would require a mammoth development effort. To meet this challenge, Rothmuller assembled a software development team which boasted several former Xerox PARC researchers including BravoX co-creator Tom Malloy, who had joined Apple in 1978, and WYSIWYG pioneer Larry Tesler, who arrived in July 1980. A fundamental component of the software was the LisaGraf graphics library which was written by senior software engineer William D (Bill) Atkinson and systems programmer Andy Hertzfeld. LisaGraf was used to build a graphical user interface called LisaDesk which closely resembled the icon-based desktop interface created by Xerox for the 8010 Star but also included some novel features such as pull-down menus, a drag-and-drop capability and a clipboard. The LisaDesk interface ran on top of a proprietary operating system with multiprogramming capability which the user loaded from a floppy disk drive under the control of a monitor program stored in ROM. A full suite of WYSIWYG office applications was also developed, including the LisaWrite word processor (written by Malloy), LisaDraw presentation graphics package, LisaCalc spreadsheet and LisaProject project management tool.

Lisa's hardware was equally demanding, though far less innovative. Wozniak had originally planned to follow the example of Xerox and employ a discrete processor design which would be implemented using bit slice components. Fortunately, this was changed to a less costly design based on the ubiquitous Motorola MC68000 following the microprocessor's introduction in September 1979. Less fortuitous was the decision to develop an Apple floppy disk drive for Lisa rather than use a third-party drive as these were considered insufficiently reliable for a business computer. Development of the drive was beset with problems and when Rothmuller complained that the 1981 target date for release was no longer achievable, he was fired and replaced by another former Hewlett-Packard engineer by the name of John D Couch. Jobs, whose uninvited participation on the Lisa project had been increasing, asked Apple CEO Michael Scott if he could take control of the project from Couch but this was refused as his input was felt by the team to be detrimental to progress.

The Apple Lisa was finally unveiled in January 1983, more than a year late. Shipments began in June. It featured a 5 MHz MC68000 with virtual memory support, 1 Megabyte of RAM as standard, monochrome bitmapped graphics with a resolution of 720 by 364 and 3 expansion slots. A substantial desktop enclosure also housed the integrated 12-inch CRT monitor and twin 5.25-inch floppy disk drives. A high-quality 73-key keyboard, one-button mouse and 5 Megabyte external hard disk drive were separate and connected to the main unit via cables.

12. Bringing it all Together – The Graphics Workstation

The new model was initially offered in one configuration only at a price of $9,995. The price included the full suite of office applications but not software development tools or a printer, both of which were optional extras, and the planned network interface would not be available for another year. Development had taken an estimated 200 person-years and cost Apple more than $50 million, and the inflated price was apparently chosen in an effort to recoup this investment. However, sales were poor as a result and the machine also suffered from embarrassingly slow performance due to the heavy computational demands of the WIMP interface and the absence of any graphics acceleration hardware, which left the processor to do all the work. Performance was further compromised by a reduction in the clock speed of the processor from 8 to 5 MHz during development to accommodate the slower access time of the video memory. Apple's investment in graphics workstation technology had not been wasted, however, as Raskin's Macintosh project would deliver a wildly successful product which would lead to a quantum shift in the development of the personal computer.

The Macintosh

In contrast to the Lisa development, the Macintosh project had remained a relatively modest effort, with a full-time staff of four led by Jef Raskin. The specification of the machine was equally modest and was based around the Motorola MC6809, a second generation 8-bit microprocessor with a 2 MHz clock speed. In December 1980, Macintosh team member Burrell C Smith, a self-taught engineer who had begun his career at Apple as a service technician, redesigned the prototype to use the more powerful MC68000 processor which had now become a viable alternative due to falling prices of the device. Smith also used six Programmable Array Logic (PAL) devices for several of the functions in order to minimise the overall number of chips required. Smith's clever redesign caught the eye of Steve Jobs, who, having recently been barred from further involvement in the Lisa development, was on the lookout for another project to devote his creative energies to.

Jobs had previously criticised the Macintosh project but he now saw it as way of obtaining revenge on those who had ousted him from the Lisa development by making a better product than Lisa and getting it to market first. He decided to take control of the project and began expanding the project team using staff from the company's Apple][development group which was in the process of being wound down. Jobs himself would take charge of the hardware development, leaving Raskin in charge of software. However, both men had very strong personalities and it soon became apparent that they also had markedly different visions of what kind of computer the Macintosh should be. After a year of heated arguments which culminated in an attempt by Jobs to remove Raskin's responsibility for software development, Raskin resigned in February 1982. Two months later, Jobs hired Xerox 8010 co-designer Bob

The Story of the Computer

Belleville from Xerox PARC as head of engineering for the Macintosh project.

The change to the same microprocessor as Lisa facilitated the adoption of Lisa's WIMP technology for use in the Macintosh. The LisaGraf graphics library was simplified then rewritten in assembly language to optimise efficiency and renamed QuickDraw. This would be stored in ROM for immediate access along with an operating system which sacrificed multiprogramming capability in favour of higher performance. A WIMP interface was then developed which shared many of the same features as the Lisa interface but incorporated several improvements and stripped out unnecessary features. The hardware specification was also uprated to accommodate the demands of a graphical user interface, with RAM doubled from the planned 64 Kilobytes to 128 Kilobytes, an increase in both the screen size and resolution of the display, and a change in pointing device from a joystick to Lisa's one-button mouse. Unlike Lisa, the Macintosh had no dedicated video memory. Instead, 21 Kilobytes of main memory was set aside as a frame buffer for the 512 by 342 resolution bitmapped graphics. However, as memory access could be interleaved between the processor and video circuitry, this allowed the full 8 MHz clock speed of the processor to be maintained. The same interleaving technique was also employed to access the on-board sound generation hardware. A single floppy disk drive was included but this would be a high density 3.5-inch drive sourced from Japanese manufacturer Sony rather than the problematic 5.25-inch Apple drive developed for Lisa.

Unlike other Apple models, the Macintosh would be a closed system with no third-party expansion capability. This was partly due to Raskin's desire for semi-portability and partly to control the quality and engineering integrity of the system, as Apple support was always the first port of call for irate Apple][owners when third-party boards developed a fault. However, this attitude did not extend to software and considerable effort went into producing technical documentation which would provide third-party developers with all the information necessary to create software applications for the Macintosh.

As the machine would be aimed at computer novices, considerable effort also went into making the Macintosh appear as friendly and unintimidating as possible. With input from Jobs, Apple][case designer Jerry Manock and colleague Terry Oyama designed a compact desktop case which evoked the look of a domestic appliance and minimised the amount of desk space required, facilitated by Smith's low chip count circuitry which required only two small circuit boards. The case also housed the integrated 9-inch CRT monitor and floppy disk drive, which were stacked above the circuit boards so that they would not increase the machine's 10 by 10 inch footprint. More space was saved by mounting the internal loudspeaker directly onto one of the circuit boards. A compact 58-key keyboard and mouse connected to the main unit via cables, the sockets for which were labelled with universal symbols rather than text so that no changes would be necessary for international markets. In a nod

12. Bringing it all Together – The Graphics Workstation

to Raskin's original specification, the top of the case incorporated a recessed carrying handle. Jobs also arranged for the signatures of every member of the Macintosh development team to be moulded into the inside surface of the case in recognition of their work.

Apple Macintosh (Photo © Apple Computer, Inc.)

The Apple Macintosh was introduced in January 1984 at a price of $2,495. Only one configuration was available at launch but the price included two full-featured WYSIWYG applications, the MacWrite word processor and MacPaint graphics package. Jobs had been unable to beat Lisa to market, having underestimated the amount of development work required, but he did succeed in creating a better product. The Macintosh offered a similar user experience to Lisa but performance was significantly faster due to the higher clock speed and more efficient ROM-based system software. More than 70,000 were sold in the first 3 months, exceeding the total sales for the Apple Lisa, despite a selling price which was more than double the original target. One of the reasons for this price hike was to recoup the cost of a $15 million advertising campaign which included the famous '1984' television commercial first aired during the 1984 US Super Bowl broadcast. Apple also gifted machines to 50 selected high-profile individuals in the creative and media industries. The advertising campaign, which was the brainchild of new CEO John Sculley, was a major risk for Apple as the increased price would almost certainly frighten off potential customers in the hobby and home market. Fortunately, Sculley's gamble paid off and the Macintosh rapidly became established as the computer of choice for professionals who wanted interactive graphics but could not justify spending upwards of $10,000 for a full blown graphics workstation. With the Macintosh, Apple had effectively created a graphics workstation for the masses.

Apple encouraged third-party software developers to create new WYSIWYG applications for the Macintosh, as these would help to drive sales. One application in particular, Aldus PageMaker, led to a spectacular increase in Macintosh sales following its release in July 1985. PageMaker, which was developed by ex-newspaperman Paul Brainerd, was the first DTP software package to be aimed at a low-cost computing platform. Its development had been triggered by Apple's plans to introduce an affordable laser printer, the Apple LaserWriter.

The LaserWriter was based on a desktop printing engine developed by Japanese imaging firm Canon which Jobs had seen while on a trip to Japan for the negotiations with Sony on the supply of 3.5-inch floppy disk drives for the Macintosh. Apple would not be first to reach the market, as Hewlett-Packard had introduced the HP LaserJet printer in May 1984 which also used the Canon engine, but the LaserWriter would be the first to support the new PostScript page description language from Xerox PARC spin-out Adobe Systems. Unlike the simple printer control languages used in other printers, PostScript enabled the LaserWriter to replicate a bitmapped image exactly as it appeared on the screen. However, this was a computationally intensive process and the printer controller had to be equipped with a 12 MHz MC68000 processor, 512 Kilobytes of RAM and a 1 Megabyte frame buffer in order to cope, creating a bizarre situation in which the printer was more powerful than the computer it was connected to.

The Apple LaserWriter was introduced in January 1985 at a price of $6,995. The high specification was reflected in the high price of the LaserWriter, which was double that of an HP LaserJet, and initial sales were poor. However, this changed dramatically following the release of Aldus PageMaker six months later, as the combination of a Macintosh, LaserWriter and the PageMaker software provided a complete DTP solution for a fraction of the cost of a system based on an Apollo or Sun workstation. The desktop publishing revolution had begun.

Despite his leading role in the creation of a successful new product, Steve Jobs was rapidly falling out of favour with Apple's senior management as a result of his maverick ways. His relationship with John Sculley, whom Jobs had personally recruited from beverage giant PepsiCo in 1983, had broken down as a result of arguments over who was to blame for the company's lack of success in targeting the office automation market. The situation came to a head in May 1985 following an attempt by Jobs to depose Sculley as CEO and he was promptly relieved of his management duties by Apple's board of directors. Four months later Jobs quit Apple along with five colleagues

12. Bringing it all Together – The Graphics Workstation

and established a firm called NeXT to develop graphics workstations for the educational market. In February the following year, he also acquired a controlling interest in another company which was developing a special-purpose graphics workstation for motion picture visual effects. The company, Pixar, had started out as the computer graphics division of George Lucas's film production company Lucasfilm and was led by Edwin E Catmull, who had studied under Ivan Sutherland at the University of Utah, and Alvy Ray Smith, a visiting scientist with Xerox PARC.

Neither venture was a huge success for Jobs. After selling around 50,000 workstations NeXT ceased hardware production in 1993 to concentrate on software development and was acquired by Apple in December 1996. Pixar's workstation efforts were also unsuccessful. Having introduced the Pixar Image Computer in 1986 at a price of $125,000, the company only managed to sell around 100 machines despite efforts to target the larger medical imaging market. In 1990, Pixar sold its hardware division in order to concentrate on its burgeoning computer animation business. Jobs would eventually return to Apple in 1997 following the finalisation of the deal to acquire NeXT.

The Macintosh was the first in an extended family of models which would keep Apple at the forefront of personal computer technology for the next two decades. However, Apple remained unable to conquer the elusive office automation market. Instead, an alliance of the leading semiconductor company and an up-and-coming software firm from Washington State would come to dominate this market by also adopting the WIMP and WYSIWYG technology pioneered at Xerox PARC.

Further Reading

Bardini, T., Bootstrapping: Douglas Engelbart, Coevolution, and the Origins of Personal Computing, Stanford University Press, Stanford, California, 2000.

Barnes, S. B., Douglas Carl Engelbart: Developing the Underlying Concepts for Contemporary Computing, IEEE Annals of the History of Computing 19 (3), 1997, 16-26.

Bechtolsheim, A., Baskett, F. and Pratt, V., The SUN Workstation Architecture, Stanford University Computer Systems Laboratory Technical Report 229, March 1982.

Bell, C. G., Toward a History of (Personal) Workstations, Proceedings of the ACM Conference on the History of Personal Workstations, 1986, 1-17.

Bhide, A. V., Vinod Khosla and Sun Microsystems, Harvard Business School Case Study, September 1989.

Campbell-Kelly, M., Data Communications at the National Physical Laboratory, IEEE Annals of the History of Computing 9 (3/4), 1988, 221-247.

Chesbrough, H., Graceful Exits and Missed Opportunities: Xerox's Management of its Technology Spin-off Organizations, Business History Review 76 (4), 2002, 803-837.

Clark, J. H., The Geometry Engine: A VLSI Geometry System for Graphics, Computer Graphics 16 (3), 1982, 127-133.

Clark, W. A., The LINC was Early and Small, Proceedings of the ACM Conference on the History of Personal Workstations, 1986, 133-155.

Giannakis, G. B., Highlights of Signal Processing for Communications, IEEE Signal Processing Magazine 16 (2), 1999, 14-50.

Hafner, K. and Lyon, M., Where Wizards Stay Up Late: The Origins of the Internet, Simon and Schuster, New York, 1996.

Hall, M. and Barry, J. A., Sunburst: The Ascent of Sun Microsystems, Contemporary Books, Chicago, Illinois, 1990.

Harrington, J. V., Radar Data Transmission, IEEE Annals of the History of Computing 5 (4), 1983, 370-374.

Hennessy, J., Jouppi, N., Przybylski, S., Rowen, C., Gross, T., Baskett, F. and Gill, J., MIPS: A Microprocessor Architecture, IEEE Proceedings of the 15th Annual Workshop on Microprogramming, 1982, 17-22.

Hiltzik, M., Dealers of Lightning: Xerox PARC and the Dawn of the Computer Age, Orion Business Books, London, 2000.

Johnson, J., Roberts, T. L., Verplank, W., Smith, D. C., Irby, C. H. and Beard, M., The Xerox Star: A Retrospective, Computer 22 (9), 1989, 11-29.

Kessel-Hunter, K. L., Sun Microsystems and a Strategic Analysis of the Workstation Industry, MSc Thesis, Alfred P Sloan School of Management, MIT, May 1988.

Lampson, B. W., Personal Distributed Computing: The Alto and Ethernet Software, Proceedings of the ACM Conference on the History of Personal Workstations, 1986, 101-131.

Lewis, M., The New New Thing: A Silicon Valley Story, W. W. Norton & Company, New York, 1999.

Linzmayer, O. W., Apple Confidential: The Real Story of Apple Computer Inc., No Starch Press, San Francisco, California, 1999.

Nelson, D. L. and Leach, P. J., The Architecture and Applications of the Apollo Domain, IEEE Computer Graphics and Applications 4 (4), 1984, 58-66.

Noll, A. M., Scanned-Display Computer Graphics, Communications of the

ACM 14 (3), 1971, 143-150.

Pfiffner, P., Inside the Publishing Revolution: The Adobe Story, Peachpit Press, Berkeley, California, 2003.

Roberts, L. G., The Evolution of Packet Switching, Proceedings of the IEEE 66 (11), 1978, 1307-1313.

Smith, D. K. and Alexander, R. C., Fumbling the Future: How Xerox Invented then Ignored the First Personal Computer, iUniverse, Lincoln, Nebraska, 1999.

Thacker, C. P., Personal Distributed Computing: The Alto and Ethernet Hardware, Proceedings of the ACM Conference on the History of Personal Workstations, 1986, 87-100.

13. Getting Personal – The World According to Wintel

"In retrospect, committing to the graphics interface seems so obvious that now it's hard to keep a straight face", Bill Gates, 1994.

The Commoditisation of the Computer

During the PC era, a radical transformation took place in which computer technology changed from being hardware driven to software driven. Software had always been the computer industry's poor relation. In the early years of the industry computers were such an expensive purchase that manufacturers would willingly provide software with their machines at no extra cost in order to secure a sale. When other software was required, the customers, which tended to be large businesses, government agencies or universities, simply developed their own using in-house programming staff hired specifically for this purpose. The esoteric nature of early computers also encouraged the formation of user groups, such as SHARE and DECUS, whose members freely swapped their latest programs with other users around the world. Therefore, with customers either developing their own software or obtaining it free of charge, there was no reason for them to buy it.

Software's commercial prospects remained bleak until integrated circuits brought the price of computers down to the point where smaller organisations could afford them. As these customers had no in-house programmers, they looked to contract software developers to meet their needs. The increase in computer numbers also created a viable market for off-the-shelf software products. This trend accelerated with the introduction of microprocessor-based computers but the software industry remained relatively small until the PC era, when standardisation of components and economies of scale in manufacturing led to the commoditisation of the hardware. As hardware prices tumbled and profit margins dwindled, the industry turned its attention to software and the software companies would soon find themselves in the driving seat.

Genesis of a Giant

The company which would lead this transformation was founded by William H Gates III and Paul G Allen in April 1975. Gates and Allen first met in 1968 while pupils at Lakeside School, a private preparatory school in Seattle,

The Story of the Computer

Washington, and became firm friends through a shared interest in science and technology. Gates was born in October 1955, the privileged son of a lawyer and a schoolteacher, both of whom were prominent figures in Seattle society. Allen, who was two years older than Gates, came from a family of educators. His father was an academic and Associate Director of Libraries for the University of Washington and his mother was a primary school teacher with a passion for literature.

Gates possessed an unusual combination of social awkwardness and intellectual confidence bordering on arrogance. He excelled at mathematics and was fiercely competitive. Allen was more reserved but equally gifted intellectually. Both were also fortunate to attend Lakeside School, a prestigious single-sex institution with high tuition fees and stringent entry requirements which prided itself in its progressive attitude to education.

The Lakeside Programming Group

In early 1968 Lakeside School decided to introduce its privileged pupils to computer technology. A fundraising campaign by the School's Mother's Club raised $3,000. This was used to purchase a Teletype ASR-33 teletypewriter terminal and rent time on a large-scale DEC PDP-10 timesharing system operated by General Electric in Seattle. Gates and his classmate Kent H Evans immediately began to take an interest in the School's new facilities. After some lessons in the BASIC programming language from one of the mathematics teachers, they formed a club called the Lakeside Programming Group along with two older pupils, Richard W (Ric) Weiland and Paul Allen.

Bill Gates and Paul Allen working at the Teletype Terminal at Lakeside School

13. Getting Personal – The World According to Wintel

The Lakeside Programming Group's youthful enthusiasm for programming soon began to burn through the allocated budget for computer time and it was used up within a matter of weeks. The mother of another pupil then suggested that the School would obtain better value for money by switching to an alternative computer bureau service provider, the Computer Center Corporation (C-Cubed), a newly-formed company which she had recently helped to establish in Seattle. This was agreed and arrangements were made to allow the pupils to access the timesharing computer system at C-Cubed, which was another DEC PDP-10, but with the pupils' parents now footing the bill on an individual basis. First in the queue to use the new facilities were the four members of the Lakeside Programming Group.

As the Group's knowledge of the PDP-10 increased, they began to uncover bugs in the machine's timesharing operating system. C-Cubed was still within the acceptance period of its rental agreement with DEC for the machine so any reliability problems reported would allow the company to delay or reduce payments. A plan was hatched whereby the Group was encouraged to probe the system for bugs outside of normal working hours in return for as much free computer time as they wanted. All that was required of them was to carefully document any problems found so that C-Cubed could report these to DEC. The four boys were soon spending practically all of their spare time at C-Cubed's premises in downtown Seattle, often working through the night in an effort to feed their growing obsession with computing. However, this arrangement would only last a few months and, fearing that they would lose access to the computer, Gates and Allen began using their detailed knowledge of the system to hack into other user's accounts. This resulted in the Group being banned from using the C-Cubed system over the summer.

Allen's father was able to use his position at the University of Washington to arrange access to a number of computers on the Washington campus, including two PDP-10 machines, which allowed the Group to continue their computing activities until the C-Cubed ban was lifted. However, C-Cubed was now struggling financially following the establishment of an in-house bureau service by major client Boeing and the company went out of business a few months later in March 1970. After scrounging more computer time from the University of Washington, the Group was approached in early 1971 by another computer service provider, Information Sciences Incorporated (ISI) of Portland, Oregon. ISI was looking for experienced PDP-10 programmers to develop a payroll program and had been told about the Group by a former regional sales manager for DEC who had been involved with C-Cubed. A deal was then negotiated in which the Group would undertake the development in return for around $10,000 of free computer time. However, Weiland and Allen decided that there was insufficient work for four programmers so the two younger boys were initially left out of the project. They were allowed back in after a few months when it became clear that the project was far more

The Story of the Computer

challenging than originally anticipated but Gates only agreed to return on condition that he would be in charge of the project. This was a turning point for Gates and from now on he would be the undisputed leader of the Group.

Traf-O-Data

Following the completion of the ISI project, Gates and Allen began looking around for another software project. They spotted an opportunity to undercut the companies which provided data analysis services to municipal roads departments for the analysis of vehicle traffic. The raw data was collected by roadside traffic counter units that monitored vehicle traffic at key locations using a pneumatic tube laid across the road connected to a mechanical tape punch which used special 16-row tape. When a vehicle passed over the tube, the pulse of air triggered the punch unit and punched a 16-bit binary number representing the time recorded. This data was then analysed to produce vehicle traffic flow statistics. Gates and Allen decided to set up their own business, called Traf-O-Data, which would offer a cut-price analysis service for vehicle traffic flow data. After developing the analysis program on one of the University of Washington computers over the summer break, Allen left Seattle to begin his undergraduate studies in computer science at Washington State University 250 miles away in Pullman. Gates kept the business running in his absence by paying some fellow Lakeside School pupils to transcribe the punched tape data onto punched cards for input to the computer.

When Allen returned to Seattle the following summer, he was full of enthusiasm for Intel's new 8008 microprocessor which had been released in April. He suggested to Gates that they use the 8008 to develop a special-purpose microcomputer which would be capable of reading and analysing the traffic counter tapes directly without the need for manual transcription and the use of a timesharing computer system. Gates scraped together the funding necessary to buy the components. Allen then set about creating an 8008 emulator program for a PDP-10 which could be used by Gates to develop the software for the system in the absence of a suitable 8008 development platform. He also developed a macro assembler and modified the PDP-10's debugger to allow Gates to halt an 8008 program in mid execution so that any bugs in the code could be traced and eliminated. As neither of them had any hardware development experience, they recruited Paul Gilbert, an undergraduate electrical engineering student at the University of Washington who had been recommended by a mutual friend. Gilbert created a two-board design and assembled a wire-wrapped prototype using the University's facilities. A home-made tape reader was also built to read the non-standard punched tapes but this failed during a demonstration of the system to a potential customer and had to be replaced with a costly commercial unit manufactured by Enviro-Labs.

After a frustrating year spent tackling a problem with electrical noise affecting the reliability of the memory, the prototype system was fully operational

13. Getting Personal – The World According to Wintel

by August 1974 and was successfully used by Gilbert to analyse tapes from three clients. However, several US states including Washington soon began offering vehicle traffic analysis services to their municipal authorities for free. Not only would there be little point in developing a production version of the system but Traf-O-Data's analysis service was now also surplus to requirements. Nevertheless, the project had given Gates and Allen a taste for entrepreneurship, and the experience gained with microprocessors would prove invaluable in their next business venture.

During the protracted development of the Traf-O-Data system Gates and Allen had also been kept busy with other computer projects. In May 1972, the Lakeside Programming Group was approached by School administrators regarding the development of a scheduling program to help timetable classes in the wake of an impending merger with another school. With Allen and Weiland both having graduated from Lakeside the previous term, the project would be carried out by Gates and Evans but Evans was killed in a mountaineering accident before work could begin. He was only 17. Gates was devastated but Allen rallied round following his return to Seattle from first year at university and he and Gates worked together to complete the program over the summer break.

Over Christmas 1972, Gates received a telephone call from the manager who had hired the Group for the ISI payroll program project asking if they would be interested in paid work from a local aerospace company to assist with the development of a real-time operating system for the DEC PDP-10. The project had fallen seriously behind schedule and the company, TRW Incorporated, needed programmers with PDP-10 experience to help bring it back on track. Gates and Allen both jumped at the chance to further their knowledge of the mighty PDP-10 and earn some hard cash at the same time. As he was in his final year at Lakeside School, Gates was allowed to undertake the work as a final year project. Allen took leave of absence from his studies at Washington State University and the two teenagers moved to Vancouver, Washington in January 1973 where they would spend the next 3 months working as part of a large team of veteran programmers.

Micro-Soft

Bill Gates was now the last remaining member of the Lakeside Programming Group but he too would soon be moving on. In September 1973, Gates travelled 2,500 miles east to Cambridge, Massachusetts to take up a place studying law and mathematics at prestigious Harvard University. However, by the end of his first year life as an undergraduate student was beginning to lose its appeal so he decided to take a break from university and try for a programming job at one of the Massachusetts computer firms. He also persuaded Allen, who was similarly disaffected with his studies at Washington State University, to do the same. Both were offered jobs working as programmers for Honeywell's minicomputer division in Billerica near Boston. Allen accepted the offer and made the move to New England with his girlfriend at the start of the summer but Gates was under pressure from his parents to continue with his university studies. He declined the job offer and returned to Harvard at the end of the summer break.

With Gates and Allen now living in the same area, they were able to work together again in their spare time. Both were keenly aware of the huge potential of the new second generation microprocessors such as the Intel 8080 but they also knew that their strengths lay in developing software rather than hardware. Towards Christmas of 1974, the January 1975 issue of Popular Electronics announcing the MITS Altair 8800 reached the Harvard Square newsstands and was immediately spotted by Allen. After reading about the new microcomputer, Gates and Allen concluded that this was the opportunity they had been waiting for. Anticipating the demand for an easy-to-learn programming language that would allow hobbyists to program the computer, they sent a letter to MITS founder Ed Roberts on Traf-O-Data headed notepaper which Gates then followed up with a telephone call claiming that they were in the final stages of developing a BASIC interpreter for the Altair. Roberts was unimpressed, having already been approached by others with similar claims. He told them what he had told the others, that the first company to demonstrate a working version of a BASIC interpreter on an Altair fitted with 4 kilobytes of memory at MITS' premises in Albuquerque would win the contract. The machine was currently undergoing tests in an effort to eliminate problems with the troublesome 4,096 byte memory board but would be ready in a month or so.

The race was now on for Gates and Allen to design, develop and deliver a highly complex program within only a few weeks for a target platform that they had never seen. To provide a suitable software development environment, Allen used his experience from the Traf-O-Data project to create an Intel 8080 emulator program and macro assembler which Gates could run on a DEC PDP-10 system in Harvard's Aiken Computation Laboratory. Gates based the design of the interpreter on DEC's extended version of BASIC, BASIC-PLUS, but kept the feature set to a minimum so that it could run comfortably within the Altair's 4 kilobytes of memory.

An initial prototype of the BASIC interpreter was completed by late February and Allen made the journey to New Mexico in early March to demonstrate it to Roberts. After carefully loading the punched tape containing the 2,000 lines of assembly code into the Altair, the interpreter worked first time, much to Allen's relief. Roberts was delighted, as a BASIC interpreter would give the Altair a significant competitive advantage over rival microcomputer kits. He confirmed that Gates and Allen had won the race to deliver a working interpreter and asked them to draw up a suitable licence agreement which would allow MITS to sell the software.

13. Getting Personal – The World According to Wintel

Following Allen's triumphant return to Boston, he and Gates set about putting the legal framework in place that would allow them to commercialise their software. They registered a business partnership in Albuquerque on 4 April which Allen named 'Micro-Soft'. Gates insisted on a 60/40 split in his favour to reflect the greater effort he had put into the development. He also asked his lawyer father to help draft the licence agreement. Allen continued to liaise with Roberts, providing regular updates by telephone as work progressed on the release version of the interpreter. However, Roberts quickly realised that it would make more sense for Allen to be based at MITS. He persuaded Allen to quit his job at Honeywell and become a MITS employee, giving him the impressive title of Director of Software Development and a remit to assemble an in-house software development team. Allen moved to Albuquerque after Easter while Gates remained at Harvard for the time being.

As well as completing the minimalist version of the BASIC interpreter which Allen had demonstrated in prototype form to Roberts, Gates and Allen also planned to supply MITS with two further versions of the interpreter containing extra features for Altair models fitted with 8 or 12 kilobytes of memory. However, this would require a considerable amount of work so to help ease the load they employed Gates's fellow Harvard mathematics student Monte Davidoff to write the floating-point arithmetic routines for the interpreter over the summer break. One of the Lakeside School pupils who had helped transcribe the Traf-O-Data tapes, Christopher R Larson, was also recruited as a summer intern. Larson, Davidoff and Gates all moved to Albuquerque for the summer, sharing Allen's rented two-bedroom apartment which was only a short car journey from MIPS' premises near Albuquerque airport.

With Davidoff and Larson's help, the 4K and 8K versions of the BASIC interpreter were ready for release by July 1975 and the licence agreement could be concluded. The agreement gave MITS exclusive worldwide rights to the software for a period of 10 years in return for an upfront fee of $3,000 plus a royalty on each copy sold of between $30 and $60, depending on the version, with a cap of $180,000. Each party would receive half the selling price on copies of the software sold separately from the hardware. Revenues from the sublicensing of the source code to other manufacturers would also be split 50/50. To ensure MITS made the most of the deal, a clause was inserted requiring the company to employ its "best efforts" to licence, promote and commercialise the software.

Gates returned to Cambridge after the summer break to begin his third year at Harvard but took a leave of absence in November 1975 for another spell working alongside Allen in Albuquerque. However, Micro-Soft's income was not yet sufficient to support him and he went back to Harvard after a few weeks. The same month MITS introduced the new Altair 680 model. As this was based on a different microprocessor, the Motorola MC6800, it would require a new version of the BASIC interpreter to be developed but Gates had returned to his studies and Allen was now fully occupied with his MIPS duties. In April 1976 they hired Ric Weiland from the Lakeside Programming Group to work for them over the summer plus another former Lakeside classmate, Marc McDonald, who was recruited as a permanent member of staff. The first job for the new recruits was to adapt the BASIC interpreter for the MC6800 microprocessor.

Gates and Allen's choice of BASIC as the programming language for the Altair turned out to be a good one. With few off-the-shelf software applications available, BASIC became a must-have for hobbyists who wanted to put their new microcomputers to work. BASIC (Beginners All-purpose Symbolic Instruction Code) was designed in 1964 by two academics from Dartmouth College in New Hampshire, John G Kemeny and Thomas E Kurtz. Kemeny and Kurtz wanted students in fields other than engineering and mathematics to be able to use Dartmouth's new timesharing computer system so the language had to be very easy to learn. Implementing BASIC as an interpreter, which executes statements line by line as they are entered rather than having to compile a program in order to run it, further enhanced the ease of use and the language soon became popular as a means of introducing novices to the joys of computer programming. Micro-Soft's interpreter was not the first version of BASIC to be made available for microcomputers but it was the most capable and the only one to support floating-point arithmetic. Consequently, other computer manufacturers soon began taking an interest in acquiring a licence to Micro-Soft's code.

A flurry of interest in late 1976 resulted in MITS sublicensing the BASIC source code to several companies, including General Electric and NCR. The GE licence was in return for a one-time fee of $50,000 but NCR required modifications to the source code so that the software would run on the company's NCR 7200 intelligent terminal and this extra work was reflected in a higher fee of $175,000. The agreement with MITS gave Micro-Soft half the revenues from sublicensing and the extra cash meant that the company could now support Gates and Allen full-time. Allen resigned from his position at MITS and moved to Microsoft in November 1976, the hyphen in the company's name having been dropped when they registered it as a trade name. In January 1977 Gates followed suit and quit Harvard for good.

The first quarter of 1977 witnessed a string of announcements for new microcomputers including the so-called 1977 Trinity, the three models from Apple, Commodore and Radio Shack which together would set a new benchmark for home and hobby machines. Clearly, the Altair's days were numbered. Gates and Allen were keen to sublicense BASIC to all three manufacturers and immediately entered into negotiations with Commodore's Jack Tramiel but Roberts resisted, knowing that these companies would be direct competitors to MITS unlike the other licensees. As total royalties from end-user sales of the BASIC interpreter had now reached the $180,000 cap, Microsoft needed new

13. Getting Personal – The World According to Wintel

sublicenses to generate income but without Roberts' agreement these could not go ahead. Gates and Allen were also concerned about the forthcoming sale of MITS to Pertec Computer Corporation, as Pertec considered the Microsoft agreement to be one of MITS' most valuable assets and the transfer of ownership to Pertec would continue to tie them in to a poorly performing deal for a further 8 years until the agreement expired. In April 1977 they instructed their lawyers to issue MITS with notice of termination of the agreement on the grounds that the company was no longer employing its best efforts to licence, promote and commercialise the software. However, MITS and Pertec refused to give up without a fight and the legal dispute which followed took 7 months to resolve, during which time all royalty payments to Microsoft were frozen. Gates and Allen had no option but to borrow money to keep the company afloat until the dispute was settled.

Gates and Allen were now free to licence BASIC to any company that came calling and they would no longer have to split the revenues with MITS. As part of Microsoft's efforts to support other second generation microprocessors, Weiland and McDonald had produced a version of the 12K BASIC interpreter for the MOS Technology 6502, the same processor used in the Commodore PET and Apple][, so the source code for this version was licensed to both Commodore and Apple for one-time fees of $25,000 and $21,000 respectively. Both companies rebadged it as their own product, with Apple adding some extra features and using it to replace Steve Wozniak's Integer BASIC in later Apple][models. A version for the Zilog Z80 was also developed and licensed to Radio Shack for the TRS-80.

Another consequence of the split from MITS was that there was no longer any strong reason for Microsoft to be based in Albuquerque. New Mexico had little in the way of an indigenous software industry during this period which made it difficult to recruit suitable employees locally. Like Gates and Allen, many of the Microsoft staff had come from the Seattle area and were missing their families. By December 1978 the company had grown to 13 employees and would soon require larger premises in any case so when their lease ran out at the end of the year, they upped sticks and relocated to the city of Bellevue near Seattle.

The Deal with IBM

By 1980 Microsoft had cornered the market in programming languages for microcomputers, with sales of over 500,000 copies of BASIC plus a range of products which included compilers for FORTRAN, COBOL and Pascal, so when IBM began looking for software partners for the IBM 5150 Personal Computer, Microsoft was high on their list. The initial contact was made in July 1980 when Gates received a telephone call from Project Chess software team leader Jack Sams. A series of meetings were then held the following month under a strict non-disclosure agreement, during which the IBM representatives revealed their plans for the PC and confirmed Microsoft's status as the company's favoured development partner for programming languages. The negotiations were handled by Gates and Microsoft's recently appointed business manager, Steven A Ballmer, a former Harvard classmate of Gates with a degree in applied mathematics and previous experience as a product manager for consumer goods giant Procter & Gamble.

The discussions also touched upon IBM's requirements for an operating system, as it would be much simpler for IBM to deal with a single development partner for all its PC software needs. Microsoft did have some experience with operating systems, having licenced Unix Version 7 from Western Electric in February 1980 which it planned to release as an operating system for high-end microcomputers called 'Xenix', but had never actually developed its own code. In a rare moment of generosity, Gates suggested that IBM should approach California-based software company Digital Research Incorporated whom he knew to be developing a 16-bit version of its popular CP/M operating system, CP/M-86. Gates happened to be on friendly terms with the company's founder, Gary A Kildall, so he contacted him personally to arrange the visit. However, when Sams and his colleagues arrived at Digital Research's premises the following day for the meeting, Kildall was absent, having been unavoidably delayed on a business trip, and his wife and co-founder Dorothy refused to sign IBM's formidable non-disclosure agreement without her husband's approval. Accounts differ as to whether Kildall later met with Sams but it seems that his insistence on a royalty per copy rather than the one-time licence fee arrangement favoured by IBM may have scuppered any chance of a deal.

Frustrated by the failure to reach an agreement with Digital Research, Sams contacted Microsoft again and asked if they had any other suggestions as to where IBM might find a suitable operating system for the PC. Allen was aware of a local hardware manufacturer called Seattle Computer Products (SCP) which had developed its own 16-bit operating system as a stopgap measure when Digital Research failed to deliver CP/M-86 in time for the launch of the company's new 8086 microprocessor board for the S-100 bus. The operating system, which was codenamed QDOS (Quick and Dirty Operating System), was written by SCP engineer Tim Paterson to provide a functional equivalent of CP/M, with sufficient compatibility that programs developed for a CP/M system would run under QDOS with little or no modification. Allen had worked with Paterson on porting Microsoft BASIC to SCP's new 8086 board and had been impressed by Paterson's technical prowess. Although Paterson had developed QDOS hurriedly in only two months, Allen reckoned that it might have the potential to fit IBM's requirements. He proposed that Microsoft obtain a licence to the source code for QDOS and use it as the basis of an operating system for the IBM PC.

With the backing of Kazuhiko (Kay) Nishi, Microsoft's Far East agent, Allen persuaded a reluctant Gates to agree to the QDOS proposal. He then contacted SCP and negotiated a licence to the QDOS source code for an upfront fee of

13. Getting Personal – The World According to Wintel

$10,000, with an additional $15,000 royalty payment for each sublicense to third parties. In order to preserve the terms of IBM's non-disclosure agreement, a clause was inserted which gave Microsoft the right to withhold the identity of any third-party sub-licensees. This had the added bonus of preventing SCP from gauging the potential size of the IBM sublicense deal and putting the price up accordingly.

When Microsoft finalised the software partnership agreement with IBM in November 1980, it included a licence to the QDOS-based operating system along with the company's four programming languages and an assembler. Only $45,000 of the $430,000 IBM paid Microsoft was for the operating system itself but the licence terms were non-exclusive so Microsoft would also be permitted to licence it to other manufacturers. However, the operating system software would have to be in place before the other software could be completed and IBM expected a working prototype by mid January in order to meet the tight 12-month Project Chess development schedule. This would require a massive effort from Microsoft to turn QDOS into a finished product and get it running on the unfamiliar PC hardware.

A project team was hastily assembled under the command of veteran Microsoft employee Robert (Bob) O'Rear and IBM provided two prototype machines under the strictest security. The team gradually increased in size to 35 as various problems were encountered and deadlines began to slip, with Paterson himself moving from SCP to join the team in May. Fortunately, the additional resource provided the necessary push and the operating system was completed by June 1981 in time for the launch of the PC in August. IBM renamed it PC-DOS and badged it as an IBM product but Microsoft also planned to market a standalone version to other manufacturers under the name MS-DOS. However, as Microsoft's licence to QDOS was non-exclusive, SCP could also sell it which would muddy the waters for MS-DOS. With less than three weeks remaining before the IBM announcement which would divulge the identity of Microsoft's sub-licensee, Allen secured exclusive rights to QDOS from SCP for an additional $50,000. Under the terms of the agreement SCP would also have the right to use Microsoft's code for its own products and would receive generous OEM discounts on Microsoft's other products.

Shortly before the August launch IBM sent evaluation models of the PC to a small number of influential users, one of whom was Andrew Johnson-Laird, a contract software developer who had worked for Digital Research. Johnson-Laird immediately recognised a strong resemblance between PC-DOS and CP/M and informed Kildall, who contacted IBM to express his deep concerns over the similarity. In order to avoid the threat of legal action, IBM agreed to offer CP/M-86 as an alternative operating system for the PC alongside a third operating system, UCSD p-System. However, only PC-DOS was available at launch. It was also much cheaper at only $40 compared with $240 for CP/M-86 and $675 for UCSD p-System as, unlike the others, IBM did not have to pay

515

Microsoft a royalty on every copy sold. Consequently, the vast majority of PCs were shipped with the PC-DOS operating system, thus helping to ensure that Microsoft's product would become the industry standard.

During this period Microsoft had expanded rapidly in size and now numbered more than 100 employees. The company had clearly outgrown its status as a business partnership and needed to incorporate to ensure future stability. In June 1981, the board of directors was strengthened by the inclusion of Silicon Valley venture capitalist David F Marquardt and the company registered as a corporation in the State of Washington. Gates and Allen remained the principal shareholders, with 51% and 30% respectively. The remaining shares were divided up between four minor shareholders, one of whom was Steve Ballmer. Remarkably, Microsoft had remained self-funded for over 6 years but this changed in September 1981 when Marquardt's VC firm Technology Venture Investors acquired a 5.1% shareholding for $1 million. However, the deal was done for strategic reasons as cash-rich Microsoft did not need the money.

The deal with IBM propelled Microsoft into the big league and MS-DOS would soon become the company's most profitable product. In contrast, SCP struggled to survive in an expanding microcomputer market. In 1985, SCP owner Rod Brock decided to sell his ailing company along with its most valuable asset, the agreement from Microsoft which gave the company a royalty-free licence to MS-DOS in perpetuity. When Microsoft's legal advisers informed him that the agreement was non-transferrable, Brock filed a $60 million lawsuit against Microsoft claiming that the company had knowingly withheld the identity of its sub-licensee in order to obtain the rights to QDOS cheaply. SCP eventually dropped the case in January 1986 following an out-of-court settlement in which Microsoft paid SCP a further $925,000 to terminate the agreement. This brought the total bill for QDOS to $1 million but Microsoft would earn many times that amount in licensing income from MS-DOS.

The Battle for the Desktop

The spectacular success of the IBM 5150 Personal Computer following its introduction in August 1981 attracted considerable attention amongst smaller companies keen to break into the personal computer market. With its open architecture and premium build quality, the 5150 was an easy target for copycat products which could be built more cheaply and sold for a lower price. The ready availability of the MS-DOS operating system from a third-party supplier rather than IBM further eased this task and it was only a matter of time before 'PC compatible' models began to appear.

13. Getting Personal – The World According to Wintel

PC Compatibles

The first company to seize this opportunity was Columbia Data Products, a small computer manufacturer based in Maryland. In June 1982, Columbia Data Products announced the MPC (Multi Personal Computer) 1600 at the Comdex Spring trade show in Atlantic City. The MPC 1600 was an almost exact copy of the IBM 5150 except that it included double the amount of RAM as standard and an additional 3 ISA expansion slots. With a price tag of $2,995, the Columbia Data Products model was also around 30% cheaper than a similarly specified 5150.

Although the IBM PC possessed an open architecture, the BIOS firmware, which contained the bootstrap loader plus low-level routines for hardware initialisation and input/output, was proprietary code. Fortunately, IBM had published a detailed specification of the BIOS for third-party hardware and software developers which the Columbia Data Products team were able to use to reverse-engineer it, replicating the functionality without using any of the same code in order to avoid copyright infringement claims.

Columbia Data Products was swiftly followed by Eagle Computer of Los Gatos, California, which introduced the Eagle PC later in 1982 at an even lower price of $2,385. Sales of the Eagle PC were strong from the outset, helped by a rationing of IBM 5150 deliveries to retail stores as production had been unable to keep up with the voracious demand. However, Eagle had not taken the trouble to reverse-engineer the BIOS code but had merely copied the IBM firmware. In February 1984, IBM lawyers took legal action against Eagle for copyright violation along with fellow PC compatible manufacturers Corona Data Systems and Handwell Corporation. All three companies settled out of court and agreed to halt shipments while they re-implemented their BIOS firmware but Eagle failed to recover lost ground and went out of business in 1986.

The most successful of the early PC compatible manufacturers was Compaq Computer Corporation. Compaq was established in Houston, Texas in February 1982 by three senior managers from semiconductor firm Texas Instruments. Rather than produce a clone of the IBM 5150, Compaq followed in the footsteps of another PC compatible manufacturer, Dynalogic Corporation, which had introduced a portable model, the Hyperion, in June 1982. Despite a high price of $4,995 and poor BIOS compatibility, the Hyperion had attracted considerable attention due to its self-contained, semi-portable design which weighed less than 10 kilograms. Compaq's founders recognised an opportunity to go one better and the Compaq Portable PC was introduced in November 1982 at a price of $2,995 for a basic specification machine fitted with 128 Kilobytes of RAM and a single floppy disk drive, a full $2,000 cheaper than the Hyperion and undercutting a similarly configured IBM 5150 desktop model by over $700.

The Story of the Computer

The Compaq Portable PC was advertised as the first 100% IBM PC compatible. Besides portability, it also featured a video display board which combined features of IBM's MDA text-only display board and CGA colour high-resolution graphics board to display both text and graphics on the integrated monochrome 9-inch CRT screen. Glowing reviews in the computer press praised the machine's IBM-like build quality and confirmed its' near perfect compatibility. The careful reverse-engineering process employed for the BIOS had cost the company $1 million but this investment was rewarded with sales of over $111 million in the first year of trading.

Compaq Portable PC

Compaq's spectacular success set an example for other firms and by early 1984 a survey by PC Magazine had identified 31 manufacturers selling PC compatible machines in the US alone. However, those manufacturers which had chosen to stay legal and reverse-engineer the IBM BIOS found that it was a very difficult task and compatibility varied widely. Of the 31 manufacturers identified in the PC Magazine survey, 14 declined to participate in a group test due to concerns over compatibility.

Entry into the PC compatible market became much easier in May 1984 when Californian software firm Phoenix Software Associates began offering licenses to its own BIOS firmware which it claimed to be 100% compatible. The company had employed a 'clean room' development process, whereby a group of engineers produced a detailed specification for the firmware based on IBM's technical literature which was then implemented by a single programmer working in isolation who had never seen any of the IBM literature and was also unfamiliar with Intel microprocessors. The entire process was carefully documented to provide an audit trail in the event of a legal challenge. PC compatible manufacturers licensing the Phoenix BIOS firmware were also

13. Getting Personal – The World According to Wintel

protected against copyright infringement claims by a $2 million insurance policy.

In June 1984 IBM launched an offensive against the massed ranks of PC compatible manufacturers by announcing price cuts averaging 20% across its entire PC range, which now included the enhanced 5160 PC XT model, low-end 4860 PCjr and a portable model, the 5155 Portable PC. This brought the price of a typical specification 5150 down to around $2,500, but it was not enough. The same month Compaq introduced its first desktop PC, the Deskpro Model 1 which offered a faster processor and higher specification for the same price as a PC XT. IBM countered two months later in August 1984 with the release of the PC AT, a second generation design which featured Intel's new 80286 processor and a wider 16-bit version of the ISA expansion bus amongst other performance enhancements. However, this was soon eclipsed by another new model from Compaq, the Deskpro 286, which featured a faster version of the 80286 for a similar price. Stiff competition at the lower end of the market also came in November 1984 when electronics retail giant Radio Shack introduced a PC compatible model, the Tandy 1000, at the bargain basement price of $1,149 or $999 without the video monitor.

In September 1985, an article in Creative Computing magazine reported that Michigan-based Zenith Data Systems had won a $99 million order from the US military for its Z-151 PC compatible model. IBM was not only losing out to the PC compatible manufacturers in retail sales but had also begun to lose its iron grip on the lucrative government market.

Rival Platforms

The arrival of PC compatibles in 1982 did not immediately spell the end of the road for manufacturers of alternative personal computer platforms. The Apple][range continued to sell in large numbers, although Apple's attempt to create a successor with the Apple III in 1980 had been less successful. Computers based on the CP/M operating system also remained popular, particularly in the UK and Europe where IBM had delayed the introduction of the 5150. The most successful of these was the Victor 9000 from California-based Sirius Systems Technology, which was introduced in late 1981 and designed by a team led by Commodore PET designer Chuck Peddle.

The Victor 9000 was a high specification business model with a 5 MHz Intel 8088 processor, up to 896 Kilobytes of RAM and integrated 800 by 400 resolution graphics. It ran the 16-bit version of CP/M (CP/M-86) but was also capable of running MS-DOS due to its 8088 processor despite a lack of PC compatibility. The high specification was reflected in the pricing which started at $4,995 but this included two high density floppy disk drives, a 12-inch monochrome video monitor, audio hardware and both operating systems. Despite the high price, the Victor 9000 became the most popular 16-bit business computer in Europe and the machine's success in these markets is rumoured to have provoked IBM into introducing the enhanced specification PC XT model.

Offering both CP/M and MS-DOS on the same hardware platform allowed manufacturers to hedge their bets. This was also the approach taken by leading minicomputer manufacturer DEC for its first foray into the personal computer market. Prior to the introduction of the IBM PC in August 1981, DEC had largely ignored microcomputers, its office automation efforts confined to dedicated word processing systems, but the spectacular success of the IBM PC forced the company to consider the development of a business-oriented microcomputer. The result was the DEC Rainbow 100 which was introduced in May 1982.

The Rainbow 100 was an ingenious dual processor design which incorporated a Zilog Z80A and an Intel 8088 in order to run both the 8-bit and 16-bit versions of CP/M plus MS-DOS. The processors communicated with each other through a shared 62 Kilobyte block of DRAM. Other functions were split, with the Z80A used for controlling the dual floppy disk drives and the 8088 for controlling the video display, keyboard and printer interface. The Rainbow 100 could also operate as a VT102 alphanumeric display terminal for connection to a minicomputer host via a modem or directly through the RS-232 interface.

Despite having an innovative design and impressive build quality, the Rainbow 100 failed to make an impact on the PC market. It was relatively expensive at $2,500 for a basic specification machine with only 64 Kilobytes of RAM and no graphics board. PC compatibility was also an issue, as DEC had made no attempt to replicate the IBM BIOS so only certain MS-DOS programs would run, and DEC's proprietary floppy disk format was incompatible with the format used by IBM so disks written on one manufacturer's machine could not be read by another. DEC's commitment to emerging industry standards was also questionable. As part of an overall strategy for desktop computing, the company had introduced two other desktop models at the same time as the Rainbow 100, the DECmate II and Professional 350, but both were based on a microprocessor implementation of DEC's proprietary PDP-8 minicomputer architecture.

Fellow US minicomputer manufacturer Hewlett-Packard was more successful in its efforts to tackle the PC market, its success with a series of desktop programmable calculators developed by the company's Corvallis Division in Oregon, beginning with the HP 9100A in 1968, prompting an earlier entry into the personal computer market. The company's first microcomputer was the HP-85, a semi-portable model which was introduced more than a year before the IBM PC in January 1980. Although it featured a 5-inch CRT display with high resolution graphics capability, 32 Kilobytes of RAM and a BASIC interpreter stored in ROM, the HP-85 was much closer in both concept and styling to a

13. Getting Personal – The World According to Wintel

desktop calculator than a microcomputer. Also, the custom 8-bit 'Capricorn' microprocessor, which had been designed in-house, was incompatible with any of the industry standard processors so software compatibility was non-existent. Hewlett-Packard targeted the machine at the scientific and technical sector where it sold in reasonable numbers despite a hefty price tag of $3,250.

In 1981 Hewlett-Packard introduced the first in a series of models targeted at business users, the HP-125. Unlike the Corvallis-designed HP-85, the new model was developed by the company's General Systems Division in Cupertino, California, and was firmly based on emerging industry standards. Like the later DEC Rainbow, the HP-125 was a dual processor design, incorporating two Zilog Z80A microprocessors, but these were essentially separate systems, with one processor allowing the machine to operate as an intelligent terminal and the other as a standalone computer running the CP/M operating system. To emphasise its capabilities as an intelligent terminal, the machine was enclosed in a standard HP 2621 terminal case complete with integrated 11-inch CRT display screen. However, the designers had made no provision for expandability other than an IEEE-488 interface for connecting external devices such as a floppy disk drive. The HP-125 was introduced in July 1981 at a price of $5,250 including a single 5.25-inch floppy disk drive. This was very expensive for a CP/M machine, especially one with a fixed 64 Kilobytes memory size, no graphics option and no expansion bus. HP's timing was also unfortunate, as the launch of the IBM PC the following month diverted attention away from the machine.

A second model introduced in November 1982, the HP-120, was identical in specification to the HP-125 but housed in a smaller case with an integrated 9-inch CRT display. Pricing was standardised for both machines at $2,775 for the basic machine or $3,975 with a single 3.5-inch floppy disk drive. The list of third-party software applications available for the machines included VisiCalc and the popular WordStar word processing program. However, most sales were to existing HP minicomputer customers attracted by their dual functionality and the machines made little impact on the PC market.

In October 1983, Hewlett-Packard introduced its third model in the Series 100 range, the HP-150. The new model was developed by HP's recently formed Personal Office Computer Division in Sunnyvale, California and addressed the shortcomings of the two previous machines by including high resolution graphics, increased memory capacity and two expansion slots. HP also acknowledged the industry's move towards the IBM PC architecture by switching to the MS-DOS operating system and Intel 8088 processor, though it stopped short of full PC compatibility. The intelligent terminal functionality of the earlier models was retained but the emphasis was now firmly on the machine's capabilities as a standalone computer.

Ease of use was a major design objective for the HP-150 and the specification included the industry's earliest example of a touchscreen interface. The touchscreen concept was first introduced in 1965 in an article by Eric A Johnson of the UK's Royal Radar Establishment. It was implemented in the HP-150 using rows of infrared LEDs built into two sides of the bezel of the integrated 9-inch CRT display which shone invisible beams of infrared light across the screen in a grid pattern where they were detected by photocells on the opposite sides. By placing a finger on the screen, the user would block the light at a particular row and column, thus providing the location of the digit to within the 41 by 27 resolution of the system. The touchscreen operated in combination with a software application called Personal Applications Manager (PAM) which displayed a series of 'buttons' on the screen to provide access to applications and utilities without having to enter MS-DOS commands. Performance was assured through the use of a fast 8 MHz version of the 8088 and 512 by 390 resolution bitmapped graphics. Compactness was also a feature, with the case from the HP-120 employed again to provide a system which occupied only one square foot of desk space.

Hewlett-Packard had high hopes for the HP-150. A market study commissioned by the company suggested that the new model had the potential to capture up to 22% of the US PC market within a year. However, at $3,995 for a basic specification machine fitted with 256 Kilobytes of RAM, the HP-150 was an extravagant purchase. Furthermore, with its HP BIOS and proprietary expansion bus, it was not PC compatible. Consequently, HP sold only 40,000 units in the first year, much less than anticipated though considerably higher than the company's two previous models. Nevertheless, Hewlett-Packard was now a serious contender in the personal computer market and would go on to secure a top 5 position with a new range of PC compatible machines beginning with the HP Vectra in October 1985.

Despite the gradual adoption of the IBM PC architecture by the majority of the established computer manufacturers in the US, CP/M models continued in popularity for several years in Europe. One of the most successful manufacturers of CP/M computers was the UK-based consumer electronics firm Amstrad which was led by the celebrated entrepreneur Alan M Sugar, later Lord Sugar. In September 1985, Amstrad introduced the first in an enduring series of low-cost personal computers targeted at the SOHO (Small Office/Home Office) market, the PCW 8256. Sugar's initial concept was for a dedicated word processing system with an integrated printer which could be sold for less than the price of an office typewriter but the specification was later extended to include the CP/M operating system and a BASIC interpreter.

The PCW 8256 was based on a 4 MHz Zilog Z80A processor with 256 Kilobytes of RAM and integrated video hardware for displaying text at up to 27 rows of 90 characters. Bundled software included Locomotive Software's LocoScript word processing package, which was specially written for the machine, plus interpreters for the BASIC and LOGO programming languages. The

13. Getting Personal – The World According to Wintel

price for the entire package, including an integrated 12-inch CRT display, single Matsushita 3-inch floppy disk drive and dot matrix printer, was £459 (equivalent to around $600 at 1985 exchange rates). This was a remarkably low price which was only possible as a result of some clever design work by engineer Roland Perry to minimise the number of electronic components and house all the circuitry within the monitor case, including the electronics for the printer.

As well as small businesses, the low price of the PCW 8256 also attracted hobby and home computer customers seeking to upgrade from minimalist models such as the Sinclair ZX81, with the result that Amstrad sold an impressive total of 700,000 machines in the first two years. New models introduced at regular intervals kept sales strong and by the time the PCW series was withdrawn in 1998 over 8 million had been sold.

The Personal System/2

IBM's adoption of an open architecture to encourage the development of third-party hardware and software for the PC was a risky strategy which had also left the door ajar for the introduction of PC compatibles. IBM could have retained some control over the PC market by licensing its BIOS firmware to PC compatible manufacturers but it chose not to and control was effectively relinquished when the BIOS was successfully reverse-engineered by companies such as Compaq and Phoenix. By April 1986, Compaq's sales had exceeded 500,000 units, consolidating the Texan firm's position as the undisputed leader in PC compatibles. However, despite the erosion of its market share, IBM remained the overall market leader with a 26% share of the PC market.

As the originator of the PC architecture, IBM also continued to dictate the agenda for the evolution of the platform. However, a dramatic change occurred in September 1986 when Compaq introduced the Deskpro 386, the first PC to feature Intel's new 32-bit 80386 microprocessor. IBM's technical leadership was now under threat from a company which had only been in business for 4 years. IBM's answer to this highly embarrassing situation was to introduce a new generation of models aimed at reclaiming control of the PC market by moving to an architecture with several proprietary features, the most important of which was a new high performance expansion bus.

The new expansion bus was called Micro Channel Architecture (MCA). It was designed by IBM senior engineer Chet Heath who adopted some of the concepts from the input/output channels used in the company's System/360 mainframe family. MCA increased speed over the ISA bus from 8 to 10 MHz and width from 16 to 32 bits on systems fitted with 32-bit processors, giving a data transfer rate of up to four times that of the ISA bus. It also supported bus mastering, where expansion boards equipped with their own microprocessor may operate independently of the system processor, and included provision

523

for automatic configuration. MCA was physically and electrically incompatible with the ISA bus which suited IBM's intention to keep the new technology proprietary. Detailed specifications for the bus were published but IBM made it clear that other manufacturers would have to obtain a licence to the technology if they wanted to adopt it.

Other new features included integrated graphics, liberal use of custom logic devices to reduce the chip count, which would also make the machines more difficult to clone, and modular construction to increase reliability. Software compatibility with PC models was maintained through a second BIOS ROM which held the original firmware. For the integrated graphics two new colour graphics standards were introduced; MCGA with 256 colours at 320 by 200 resolution, and VGA with 16 colours at 640 by 480 resolution. Both were backward compatible with IBM's older PC graphics boards but employed a 15-pin connector for the monitor rather than the 9-pin connector used previously.

The new generation was introduced as the IBM Personal System/2 (PS/2) in April 1987. Four models were available at launch, with prices ranging from $2,065 to $8,495. The entry-level Model 30 was a halfway house machine which retained the 8086 processor and ISA bus to provide some hardware compatibility with its PC predecessors but the others were all firmly based on the new architecture, with the mid-range Models 50 and 60 featuring an 80286 processor and the high-end Model 80 boasting a powerful 16 MHz 80386. All four models standardised on Sony's 3.5-inch floppy disk drive and were designed to run OS/2, a new multiprogramming operating system which IBM was jointly developing with Microsoft. However, this would not be ready for another 8 months so early deliveries were shipped with MS-DOS instead.

Sales of the PS/2 range started well, with 1.5 million units sold by January 1988. However, despite the growing need for a higher performance expansion bus, MCA did not become the new industry standard as IBM had hoped. Few manufacturers of PC compatibles chose to adopt it due to prohibitive licence fees which also required the licensee to compensate IBM for any previous PC compatible machines sold. Instead, Compaq and eight other leading PC compatible manufacturers including Hewlett-Packard formed an alliance known as the 'Gang of Nine' and developed an alternative standard called EISA (Extended Industry Standard Architecture) which was introduced in September 1988. Although not as fast as MCA, EISA was effectively a superset of ISA and could accept ISA circuit boards.

As new PC compatibles based on the EISA bus became available, PS/2 sales went into a sharp decline with the result that IBM's market share fell to around 13% by 1990. IBM's attempt to reclaim control of the PC market had failed. IBM executives could take some comfort from the knowledge that the PC standard which they had created would remain the dominant desktop platform for the foreseeable future but it would no longer be under IBM control. The

13. Getting Personal – The World According to Wintel

future of the PC would now be in the hands of two of IBM's suppliers, Microsoft and Intel.

Development of the Wintel Platform

Having cornered the market in programming languages for microcomputers and moved into operating systems with Xenix and the soon to be released MS-DOS, Bill Gates now decided to focus on business software applications. Gates knew that business applications were where the real money was to be made but in order to take full advantage Microsoft would also have to expand its modest sales operation and create retail distribution channels. A new Applications Development Group was set up in May 1981 under the command of WYSIWYG pioneer Charles Simonyi, who had recently moved to Microsoft from Xerox PARC on the recommendation of colleague Bob Metcalfe. The first member of the group was Richard R Brodie, another former Xerox PARC researcher who had worked on the development of the BravoX word processor. The company's sales and marketing operation was also strengthened in order to target retail sales, with veteran employee Vern L Raburn appointed as President of Microsoft Consumer Products.

Multi-Tool

Microsoft's first application product was an electronic spreadsheet program intended to compete with VisiCorp's VisiCalc, which had sold almost 400,000 copies since its release in October 1979. The spreadsheet was released as Multiplan for the Apple][in August 1982. Keenly priced at $239, Multiplan improved upon VisiCalc by having commands in plain English and supporting multiple sheets within the same workbook. Under Simonyi's guidance, considerable attention was paid to the user interface design, although this remained character-based due to the lack of bitmapped graphics on the target platform. Multiplan was written using a p-code compiler to make it easier to port to different operating systems. This allowed Microsoft to release versions for CP/M and MS-DOS simultaneously two months later in October 1982. However, it also meant that the software was not optimised to run on any one platform and performance was sluggish. Despite receiving favourable reviews in the computer press Multiplan failed to make an impact on the market and it was soon eclipsed by Lotus Development Corporation's Lotus 1-2-3 which was released to universal acclaim in January 1983.

The market failure of Multiplan was a serious blow to Microsoft's plans for a suite of 'Multi-Tool' applications which would include a word processor, business graphics package and a database. Applications would share a common user interface to make them easier to learn and use. The graphics package and database were shelved but Brodie had already begun working on the word processor so this was allowed to continue. Unlike Multiplan, which had been under development before Simonyi arrived at Microsoft, Simonyi had the opportunity to oversee the creation of the word processor so the user interface design was extended to include support for a mouse and multiple tiled (non-overlapping) windows. However, a full graphical user interface was not yet possible due to the lack of support for bitmapped graphics across all target platforms so the interface would remain character-based for the time being.

Fortunately, 1982 had been a good year for sales of Microsoft's other products. MS-DOS was now licensed to 50 separate manufacturers of PC compatibles and annual revenues had increased by over 50% to $24 million. Gates decided to plough extra resources into applications development and the word processor was released as Multi-Tool Word for the Xenix operating system in September 1983 at a price of $395 or $495 with a mouse included. The MS-DOS version was released two months later under the more familiar name of Microsoft Word. It was launched with a $3.5 million advertising campaign which included giving away 100,000 copies of the demonstration version free in a special edition of the November 1983 issue of PC World magazine. An additional 350,000 copies were also given away as part of a subscription offer.

Reviewers praised Word's windowing abilities and extensive collection of powerful features but were critical of the Multi-Tool user interface which exhibited inconsistent behaviour by employing a separate mode for entering commands. Consequently, Word was difficult to learn, even when used with a mouse, and it also suffered from the same performance issues as Multiplan. Nevertheless, Microsoft stuck with it, releasing improved versions every year or so until it eventually became the market leader.

Interface Manager

Ironically, Microsoft's poorly received efforts to create a consistent user interface would result in the development of its most successful product, Windows. In September 1981 the company had initiated a project called Interface Manager to look at new ways of making computers easier to use. A concept was developed by computer scientist Chase Bishop which involved creating a software 'wrapper' around applications to give them a consistent look and feel through standardising the user interface. The project carried a low priority until November 1982 when Gates saw a demo of VisiCorp's VisiOn at the Comdex Fall trade show in Las Vegas.

VisiOn was a mouse-driven graphical user interface and integrated applications suite for the PC platform which employed a desktop metaphor and supported multiple overlapping windows. Other than icons, which were apparently omitted due to insufficient graphics resolution on the target platform, VisiOn provided a full implementation of a WIMP interface. It also supported a form

13. Getting Personal – The World According to Wintel

of multiprogramming called co-operative multitasking which allowed the user to have several applications running on the desktop at the same time. VisiOn was not an operating system, as it required MS-DOS in order to run, but VisiCorp was moving dangerously close to Microsoft's core business in operating systems and Gates viewed this as a threat.

Earlier that year in January 1982, Microsoft had entered into a third-party developer agreement with Apple Computer to produce software for the forthcoming Macintosh model, including BASIC and Multiplan. Steve Jobs was concerned that Microsoft might gain a competitive advantage from having insider knowledge of Apple's new WIMP interface technology and so the agreement included a clause which prevented Microsoft from releasing any mouse-driven software applications for 12 months following the introduction of the Macintosh. However, this did not cover operating systems.

For Gates, the demo of VisiOn and Apple's commitment to the development of a WIMP user interface for its own products were confirmation that WIMP technology was the future for microcomputers. He upgraded the specification of Interface Manager to a mouse-driven graphical shell running on top of MS-DOS which would feature multiple overlapping windows and support co-operative multitasking. It would provide an 'application environment' for new WYSIWYG applications and also allow legacy MS-DOS programs to run in full-screen mode. A Graphics Device Interface (GDI) would be developed to provide device-independent graphics and allow the software to run on a range of different graphics hardware. Two experienced programmers, Dan McCabe and Rao V Remala, were then assigned to the project and a Xerox 8010 Star computer system was purchased so that the project team could study a fully-fledged WIMP interface.

McCabe and Remala worked hard over the next few months to develop a proof-of-concept demonstrator of Interface Manager which was ready by April 1983. However, this was only a mock-up with no underlying functionality and it was apparent that additional staff resources would be needed to tackle what was shaping up to be a major development. As the originator of WIMP technology, Xerox PARC remained the best place to find the expertise required so Simonyi approached his former colleague Scott A McGregor, a senior developer on the Xerox Star project, with an offer to become leader of the Interface Manager team. McGregor accepted the offer and was soon joined by two other members of the Star development group, Daniel E Lipkie and Leo Nikora. Marlin Eller, a mathematician who had joined Microsoft the previous year, was also assigned to the team which would now be called the Interactive Systems Group.

The introduction of the Apple Lisa in January 1983 generated an explosion of interest in WIMP technology for microcomputers. VisiCorp was now working hard to capitalise on this and was planning to release VisiOn before the end of the year. Rumours were also circulating of another product in development,

Quarterdeck Desq, which would have similar functionality to VisiOn but would also support third-party applications. With an enlarged team busy on the development of Interface Manager, Gates decided to take a risk and pre-announce the new product in an effort to divert attention away from the competition.

The announcement took place at the Plaza Hotel in New York City on 10 November 1983. Renamed Windows by Microsoft's recently appointed Vice President of Corporate Communications, Rowland Hanson, it would be released in April 1984 at a selling price of between $100 and $250. Also in attendance were 18 leading manufacturers of PC compatibles who were at the event in order to pledge their commitment to Windows. However, IBM was not among them. Microsoft's efforts to persuade IBM to licence Windows had failed, with IBM choosing instead to press ahead with its own character-based interface, TopView. Another notable absentee from the event was Microsoft co-founder Paul Allen who had left the company in February 1983 following a period of ill health, having been diagnosed with Hodgkin's lymphoma a few months earlier.

Windows

Following the November 1983 announcement, the race was on for the Interactive Systems Group to complete the development of Windows in time for the release date which was less than 6 months away. However, Microsoft managers had seriously underestimated the amount of development work remaining and the project schedule began to slip. Unlike VisiOn, Windows would also support third-party applications and Microsoft had worked hard to get developers to commit to the new platform. A developer conference was held in February 1984 but Microsoft was unable to provide the necessary development tools and the company was forced to admit that the April release date would not be met.

As the largest project Microsoft had ever undertaken, the Windows development was also making huge demands on the company's management structure, which had remained largely unaltered since the early days in Albuquerque. In August 1984 the company was restructured and additional staff drafted into the Interactive Systems Group from Simonyi's Applications Development Group. However, a constantly changing specification led to further delays and the relentless pressure began to take its toll on the Group. The catalogue of missed release dates also prompted press accusations of 'vaporware', a derogatory term used to describe the premature announcing of non-existent or partially developed software in order to gain competitive advantage. In January 1985, McGregor resigned as Group leader and was replaced by Steve Ballmer. Ballmer's tough, bombastic management style gave the project the final push needed and Windows was released in November 1985, 19 months later than originally planned. Development had taken a team of 30 programmers and

13. Getting Personal – The World According to Wintel

more than 80 person-years of effort.

Although much anticipated, the development delays meant that Windows had to compete against a quartet of similar products which had already reached the market. Besides VisiOn, which had been released more than two years earlier in October 1983, the market now also included Digital Research GEM, Quarterdeck Desq and IBM TopView. Windows was competitively priced at only $99 including two mini applications, Windows Write and Windows Paint. This compared with a price of $990 for the VisiOn suite. Windows was also designed to operate with a minimal hardware configuration of an 8088 processor, 256 Kilobytes of RAM and two floppy disks. However, it ran painfully slowly with this configuration and performance only became acceptable when running on a much higher specification machine. Furthermore, the overlapping windows functionality had been dropped by McGregor during development and the tiled appearance made Windows look inferior to VisiOn despite supporting colour graphics. Consequently, reviewers were unimpressed by Microsoft's initial efforts and further development would be needed if Windows was to succeed in the marketplace.

Fortunately for Microsoft, none of the rival products set the market alight either. IBM's TopView fared particularly badly due to a lack of support for DOS batch files and during development of the new OS/2 operating system in 1986, it was decided to drop TopView in favour of a graphics-based WIMP interface for OS/2. Ballmer negotiated an agreement in which Microsoft and IBM's Hursley Research Laboratories in the UK would work together on the development of the new interface which would be called Presentation Manager. The graphical front-end from Windows would be utilised but Microsoft's GDI device-independent graphics code would be replaced by software based on Hursley's Graphical Data Display Manager (GDDM) from the IBM System/370 in order to provide compatibility with mainframe applications.

The Presentation Manager deal was concluded in December 1986 but the project soon began to have an adverse effect on the development of the next version of Windows which had been progressing steadily under the direction of product manager Tandy Trower. Ballmer argued that there was no point in devoting valuable development resources to two similar projects and lobbied for Windows to be terminated on the assumption that OS/2 would become the dominant PC operating system but Gates was less sure and decided to keep both projects running, as Windows also allowed the company to sell more copies of MS-DOS. However, the Presentation Manager project was given priority so most of the Interactive Systems Group programmers were reassigned onto Presentation Manager with only a handful left to complete the Windows development. Despite this setback, Windows 2.0 was released in December 1987, 10 months ahead of Presentation Manager which had become a management nightmare as a result of the clash of cultures between Microsoft's free-spirited approach to software development and IBM's locked-down methodology.

Windows 2.0 was a significant improvement over the original version. The most obvious change was the visual appearance which now featured a full WIMP user interface with icons and overlapping, resizable windows. Performance was also improved through support for expanded memory but the minimum memory requirement had risen to 512 Kilobytes and a hard disk was now a recommended accessory. Since the release of Windows 1.0, a growing number of third-party software products had become available for the Windows platform including several major applications such as Aldus PageMaker. Microsoft had also released a Windows version of Excel, a powerful WYSIWYG spreadsheet program originally created for the Apple Macintosh, to coincide with the release of Windows 2.0. These products helped boost sales of Windows but the market was not yet convinced of the need for a graphical user interface for the PC and sales totals remained unimpressive.

Microsoft's adoption of a full WIMP user interface for Windows 2.0 soon attracted the attention of Apple. In March 1988, Apple filed a copyright infringement lawsuit accusing both Microsoft and Hewlett-Packard of copying the "look and feel" of the Macintosh user interface in Windows 2.0 and HP's NewWave, an object-oriented graphical desktop environment based on Windows. Apple had copyrighted the Lisa Desktop and Macintosh Finder program in May 1987. However, the situation was not as straightforward as it might seem, as Microsoft had been granted permission to use certain Apple visual copyrights in September 1985 following an earlier legal tussle between the two companies. Microsoft's lawyers also pointed out that Apple's user interface technology had actually originated at Xerox PARC, a move which roused slumbering giant Xerox into filing its own lawsuit against Apple in December 1989. Xerox accused Apple of unfair competition for having infringed the copyright on the Xerox 8010 Star software and was seeking $150 million in royalties and damages.

Following a protracted legal battle, Apple's case against Microsoft was eventually thrown out in May 1993. The judge ruled that all except a tiny handful of the contested visual features in Windows were indeed covered by the September 1985 agreement and the rest could not be protected under copyright law. Gates estimated Microsoft's legal costs at $10 million. Fortunately for Apple, the Xerox case was also dismissed on the grounds that Apple's actions were not considered to be unfair competition, the judge admonishing Xerox for not having taken action on the more straightforward claim of copyright infringement instead. Meanwhile, Microsoft had introduced a new version of Windows aimed at unleashing the full performance of the PC architecture by breaking through a barrier imposed by the BIOS which limited memory to 640 Kilobytes.

Released in May 1990 at the slightly higher price of $149, or $50 to upgrade

13. Getting Personal – The World According to Wintel

from a previous version, Windows 3.0 offered three different memory modes depending on which processor in the 80x86 family it was running on, supported virtual memory on machines equipped with an 80386 processor and included further improvements to the visual appearance. It was also the first version of Windows to be supplied pre-installed on new computers through an OEM distribution arrangement with PC manufacturers. The launch event held in New York on 22 May was broadcast by CCTV to venues across the US and transmitted live by satellite to 7 other countries. Gates was joined on stage by Microsoft co-founder Paul Allen, who had rejoined the company's board of directors a few weeks earlier after recovering from his health issues. The launch was also reinforced by a $10 million advertising campaign which included giving away 400,000 free demonstration copies in PC World magazine.

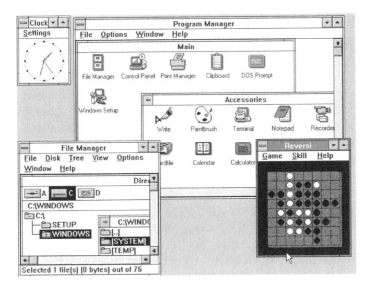

Microsoft Windows 3.0

With rave reviews appearing in the computer press and strong OEM demand, Windows 3.0 sold 2 million copies within the first 6 months of release. By the end of 1990 Microsoft had achieved $1 billion in annual revenues. Multimedia support was added in April 1992 with the release of Windows 3.1. This was followed six months later by the release of Windows for Workgroups 3.1 which added integrated networking and workgroup functionality. By April 1993 total sales of Windows 3.x had exceeded 25 million copies. In contrast, sales of OS/2 were languishing below the 1 million mark, hamstrung by a $325 price tag, poor MS-DOS compatibility and a confusing name which suggested it was only compatible with the PS/2 family of computers.

The huge success of Windows 3.x also helped to fuel demand for Intel's highly profitable 80386 microprocessor. Microsoft and Intel had become inextricably linked, with each new iteration of Windows seemingly coinciding with the introduction of a new Intel processor packing more power to satisfy the increasing performance requirements of the software. However, in reality the relationship between the two companies was not close, each viewing the other as having an unhealthy monopoly in their respective areas of the PC business. There had also been some friction between them over the inefficiency of Microsoft's programming tools. Nevertheless, from now on the PC's destiny would be controlled not by IBM but by the reluctant alliance of Microsoft and Intel. A new term soon appeared in the computer press, 'Wintel', a portmanteau of the words Windows and Intel, to describe what had previously been referred to as the PC platform.

The Battle Won

By 1991, more than 85% of all computers sold conformed to the Wintel standard. Windows 3.0 had captured the business desktop but other software platforms still dominated in scientific and technical applications, and in the growing network server market. Unix remained the most popular choice of operating system for graphics workstations and was now available with a range of sophisticated WIMP interfaces based on the X Window System, a powerful client/server windowing system for bitmapped displays which originated at MIT in 1984. Unix was also a popular choice for servers due to its robustness and scalability but shared this segment of the market with a number of network operating systems aimed exclusively at the PC platform.

These were the markets IBM and Microsoft were hoping to target with OS/2 which was available in two versions, a standard edition for desktop machines and an extended edition for servers. However, OS/2 only ran on Intel 80x86 machines, which limited its use to relatively low-end applications. In order to compete with Unix, OS/2 would have to run on a wider range of hardware platforms including those based on the new generation of RISC microprocessors.

Project Psycho

During the development of Windows 3.0 Microsoft had remained fully committed to OS/2 and Presentation Manager, with the new version of Windows regarded by the company as a stopgap product until the next version of OS/2 became available with improved support for MS-DOS applications. A third generation version which could run on non-80x86 microprocessors was also in the pipeline. Work started in early 1988 under Director of Special Projects Nathan Myhrvold, a mathematical physicist who had joined Microsoft in 1986 when the company acquired his software business, Dynamical Systems Research. Codenamed Project Psycho, the aim was to create a future operating

13. Getting Personal – The World According to Wintel

system architecture which would be scalable for implementation on a broad range of hardware platforms and written in a high level programming language in order to make it portable. Backward compatibility would be maintained by preserving OS/2's 16-bit Application Programming Interface (API) in addition to developing a new API for 32-bit processors.

Project Psycho received a boost towards the end of 1988 when David N Cutler arrived at Microsoft. Cutler was the principal architect of DEC's much admired minicomputer operating system VMS (Virtual Memory System) and had acquired a reputation as an exceptional software engineer with a ruthless work ethic. Based at DEC's DECwest facility in Bellevue, Washington, which was located only a few miles away from Microsoft's new headquarters in Redmond, Cutler had been working on the design of a new state-of-the-art operating system called Mica for use with DEC's next generation RISC processor codenamed Prism. However, the project was abruptly cancelled in June 1988 and word soon reached Microsoft that Cutler was looking for another job. Gates and Myhrvold lost no time in approaching him with an offer to take over the leadership of Project Psycho but Cutler's disdain towards microcomputers made persuading him a difficult task. He was eventually won over through a combination of Gates's compelling vision for the new operating system, a monetary incentive in the form of generous share options and an agreement that he could bring some of his DEC development team with him.

Cutler arrived on 31 October 1988 and was followed within a week by a group of 7 colleagues from DEC, including senior programmers Mark Lucovsky and Lou Perazzoli. The core of the development team was completed by one Microsoft employee, veteran Windows programmer Steven Wood. Cutler decided to dump the existing Project Psycho code, which he found to be inadequately documented, and the first few months were spent creating a detailed requirements specification for the new operating system with the design goals of portability, reliability, security and extensibility.

By February 1989, the specification was complete and coding could begin. Initial efforts were targeted at Intel's recently announced RISC microprocessor, the i860. Myhrvold had chosen the i860 over competing RISC designs despite Microsoft's wariness of Intel, as he felt both companies shared a commitment to reducing their dependence on the ageing 80x86 architecture and remaining leaders in their respective markets. However, as the i860 had not been released yet, there were no computers available which incorporated the new device so a group of hardware engineers which Cutler had brought with him from DEC were tasked with designing an i860-based computer to serve as the development platform. A software simulation would be employed until the new machines became available. The project also acquired a new name, NT, a shortened version of N-Ten which was Intel's codename for the i860.

Cutler had estimated an ambitious 18-month timescale for the NT development but this soon began to slip due to the unforeseen complexity of the networking and security elements of the system. Following delivery of the i860 development machines in late 1989, it also became apparent that the performance of Intel's new RISC processor did not live up to expectations. In December 1989 the target processor was changed to the more capable MIPS R3000, a relatively straightforward task due to the inbuilt portability of the NT code. By the spring of 1990, the team had a minimal version of NT running on an R3000 system and were also working on an implementation for the 80x86 architecture. However, a product release was still some way off, as the integration of the graphics software was proving more time-consuming than expected. The team had also taken the opportunity to make some improvements to the Presentation Manager interface which further slowed progress.

The spectacular success of Windows 3.0 following its release in May 1990 gave the NT team pause for thought. If Windows was outselling OS/2, perhaps they should drop compatibility with OS/2 and give NT a Windows personality instead. This would allow existing Windows customers to migrate seamlessly to NT as their computing needs grew. Additional work would be required but NT's modular design would make this a relatively straightforward task. To test the idea Lucovsky and Wood took the 16-bit Windows API functions and expanded them out to 32 bits, thus creating a new API that would make NT behave like Windows. They then presented their work to a design review group in August 1990 which wholeheartedly approved the change. Despite the potential ramifications of such a major shift, Microsoft's future operating system would now be based on Windows rather than OS/2. The project was given full product development status with a planned release date of summer 1992 and the team was increased substantially from around 25 members to an eventual total of more than 220 by transferring staff from the OS/2 development.

When IBM learned of Microsoft's plans for NT in January 1991, it worsened an increasingly difficult relationship between the two companies which had began to turn sour following changes in senior personnel at IBM's Entry Level Systems unit in Boca Raton, Florida. The inevitable divorce came in June 1992 in the form of a source code exchange agreement which freed both companies from the pain of further co-development by giving them access to each other's code for the next 15 months, thereby allowing them to work independently on their respective operating systems. The agreement would protect Microsoft from any claim IBM may have had over jointly developed technology used in NT and would also permit IBM to continue to use a Windows-based graphical front-end in future versions of OS/2. Microsoft also paid IBM around $25 million to licence certain patents but would receive a royalty on every copy of OS/2 sold in return.

With the divorce from IBM finalised Microsoft was now free to introduce the new Windows-based operating system, but not without further delays while

13. Getting Personal – The World According to Wintel

various stability and performance issues highlighted during beta testing were addressed. It was eventually released as Windows NT 3.1 on 26 July 1993, two years later than originally planned. Microsoft's marketing department had decided to keep the NT codename, as it also happened to be an abbreviation for 'New Technology', and the version number was chosen to match the current version of Windows. Two variants were available, a desktop edition priced at $495 and an 'Advanced Server' edition at $1,495. Features included pre-emptive multitasking, support for multiple users and an advanced multiprocessing capability known as symmetric multiprocessing. NT was also designed as a network operating system with powerful features such as remote administration and support for multiple communications protocols to allow a Windows NT system to be integrated into other networks. The initial release included versions for the MIPS R3000 and Intel 80x86 processors, and a version for DEC's new 64-bit Alpha processor was released two months later in September 1993.

The development of Windows NT is estimated to have cost Microsoft $150 million, having taken a project team of more than 200 people almost 5 years to develop. The resulting operating system comprised over 4 million lines of source code. However, it seems that Cutler and his DEC colleagues may have been tempted to bring some of the code from their cancelled operating system project with them when they moved to Microsoft. Cutler's former bosses at DEC certainly suspected as much. Following NT's release, they hired an operating systems expert from MIT to investigate and an embarrassing similarity between portions of NT code and Mica was indeed discovered. DEC used the threat of legal action against Microsoft to strike a deal in 1995 which involved Microsoft maintaining NT support for DEC's unpopular Alpha processor. Microsoft is also reported to have paid DEC around $105 million, most of which was provided in order to bolster DEC's NT support operation.

The Impact of Windows NT

The uptake of Windows NT as a desktop operating system began slowly due to an initial lack of 32-bit applications and most early sales came from businesses requiring a network operating system for their local area network. The market leader in this sector was Novell with a product called NetWare which ran on PC compatibles, turning the machine into a file and print server for other machines running MS-DOS or CP/M. However, NetWare was expensive at $2,495 for a 10-user licence and a whopping $6,995 for up to 100 users. It also lacked NT's multiprocessing capability and support for RISC processors. Consequently, many businesses looking to install or upgrade their network servers began choosing Windows NT instead.

NT's ability to run on more powerful hardware also made it a viable alternative to Unix for organisations seeking to downsize from minicomputer-based servers. The minicomputer market had been in decline since the mid 1980s, mainly as a result of the downsizing trend which saw many organisations replace their centralised computer systems with networks of desktop PCs. High-end microcomputers running Unix had become popular for use as network servers but the Unix market was becoming badly fragmented, with dozens of different versions of Unix available and poor compatibility between them. Unix was also somewhat intimidating, with a command-line interface and long list of obscure commands that was in stark contrast to the familiar Windows interface of NT.

Windows NT's release in 1993 coincided with a step change in the performance of the PC platform. Intel's new Pentium processor, which brought instruction-level parallelism to the 80x86 architecture, had been introduced in March and a sixth generation 80x86 processor with a 14-stage instruction pipeline was at an advanced stage of development. A high specification PC fitted with one or more of these new processors and running Windows NT was now a credible alternative to a low-end minicomputer for most server applications. Leading PC compatible manufacturer Compaq was once again ahead of the game with the introduction of the Compaq Systempro range of PC-based servers in 1990. Early Systempro models had to make do with dual 80386 processors and a custom version of Unix but the Texan firm was perfectly placed to take advantage of the new Wintel technologies as soon as they became available. As other manufacturers followed suit and began introducing PC-based servers, the writing was on the wall for the minicomputer. The minicomputer's slow demise was complete by January 1998 when long-time minicomputer market leader DEC was acquired by Compaq for $9.6 billion.

By the mid 1990s, as the number of 32-bit applications available for Windows NT increased and the hardware specification of PC compatibles continued to rise, the lowly PC also emerged as a credible alternative to a graphics workstation in all but the most demanding of applications. Intel had targeted its sixth generation 80x86 processor, which was introduced as the Pentium Pro in November 1995, at precisely this market. With an integrated floating-point unit, out-of-order execution engine and clock speed of up to 200 MHz, the Pentium Pro offered equivalent computational performance to the latest RISC microprocessors on the PC platform. Graphics performance could also be boosted to near workstation levels by using graphics boards with 2-D and 3-D accelerated graphics from specialist firms such as Matrox Graphics and Diamond Multimedia.

Microsoft worked hard with manufacturers to ensure that most Pentium Pro models were supplied with Windows NT pre-installed. The importance of this new market trend was also recognised by some of the established players in computer graphics, notably Evans & Sutherland which designed the 'REALimage' graphics controller chip set used in many of the graphics accelerator boards and Silicon Graphics which heavily promoted the adoption of the open standard version of its IRIS Graphics Library, OpenGL, for use on

13. Getting Personal – The World According to Wintel

the Wintel platform.

Just as graphics workstations had gradually edged out minicomputers from scientific and technical applications, high-end Wintel machines were now threatening to do the same to graphics workstations. Those workstation manufacturers who also had PC-based product lines, such as Hewlett-Packard and IBM, were in a position to address this threat by simply introducing their own high-end Wintel machines based on the Pentium Pro. However, others such as Sun Microsystems and Silicon Graphics had little option but to remain wedded to their RISC processors and Unix operating systems rather than entering unfamiliar territory but revamped their low-end models in an effort to make them more attractive to Wintel customers. Sun had already failed in an earlier attempt to introduce an Intel 80x86-based model, the Sun386i, in 1988 which served to reinforce the company's commitment to its own RISC microprocessor architecture, SPARC. As the owner of leading RISC processor firm MIPS, Silicon Graphics was also heavily committed to RISC technology but the company did eventually introduce a series of Wintel models, the SGI Visual Workstation range, in January 1999. However, this move came too late and the switch to a non-MIPS processor also unsettled the company's existing customers.

As the performance of the PC platform continued to increase and economies of scale pushed down the cost of components, manufacturers of RISC/Unix graphics workstations were no longer able to compete on price/performance and by 2003 almost 90% of graphics workstations sold were based on the PC platform. Sun and Silicon Graphics both retreated to the high end of the market, with Silicon Graphics concentrating on its lucrative supercomputer business and Sun on high-end servers. The other manufacturers quietly dropped RISC/Unix workstations from their product ranges, with only Hewlett-Packard holding out for a few more years with a series of machines based on its PA-RISC architecture.

Microsoft also dropped support for non-80x86 processors from its next version of Windows NT which was released in February 2000. Named Windows 2000, the new release also merged the Windows and NT product families into a single unified product. With exactly the same operating system now available on every class of machine from the lowest specification PC to a high performance multiprocessor system, the line between graphics workstations and PCs had been completely erased.

Making Enemies

Despite persistent criticisms over security and stability, Microsoft Windows has remained the dominant desktop operating system for over 20 years. During that time Microsoft has grown rich and powerful. An IPO in March 1986 raised $61 million and the subsequent rise in share price created three billionaires and several thousand millionaires amongst the Microsoft employees. In 1990, when Windows 3.0 was released, Microsoft had just become the first software firm to exceed $1 billion in annual sales. By 2010 annual sales revenues had risen to $62 billion, making Microsoft the largest technology company in the world. However, this success is not entirely due to Windows. Microsoft's business software applications have also sold exceedingly well and this has made the company a regular target for accusations of unfair competition.

In April 1991, an investigation was initiated by the US Federal Trade Commission following allegations from competitors that Microsoft gained an unfair advantage as a result of the company's application developers having insider knowledge of its operating system code. Microsoft was already under investigation by the FTC over accusations of collusion with IBM to control the PC operating system market but the focus of the investigation shifted when it became apparent that Microsoft's relationship with IBM was deteriorating. After a two-year investigation, in which some evidence of undocumented system calls being used by Microsoft application developers was discovered, the FTC commissioners failed to agree on whether to initiate legal action or not.

In August 1993 the case was handed over to the US Department of Justice, which had taken an interest in Microsoft following complaints from UK-based PC compatible manufacturers that their Microsoft OEM agreement required them to pay a royalty for MS-DOS on every computer sold irrespective of whether it was installed or not. The case was eventually settled in July 1994 when Microsoft signed a consent decree which prohibited the company from tying in the licensing of its software applications to its operating system products. However, this would not be the end of Microsoft's legal woes. Even more difficult challenges lay ahead as a result of a new technology for accessing information that would present both an opportunity and a threat to Microsoft's dominance.

The World Wide Web

By the early 1990s the PC had become firmly established as the standard tool for business, with tens of millions sold each year to satisfy a huge worldwide market. The advent of low-cost graphics accelerator boards from firms such as ATI Technologies and S3 Graphics also helped to position the PC as a viable alternative to dedicated games consoles for the burgeoning computer games market. However, the computer industry had not yet cracked the elusive home market. What was needed was a 'killer app', a must-have application which would fuel the demand for home PCs in the same way that electronic spreadsheets and word processing programs had done for business computing.

In a 1985 interview for Playboy magazine, Apple co-founder Steve Jobs made

13. Getting Personal – The World According to Wintel

a bold prediction that "The most compelling reason for most people to buy a computer for the home will be to link it to a nationwide communications network. We're just in the beginning stages of what will be a truly remarkable breakthrough for most people – as remarkable as the telephone.". Four years later in 1989, a British research physicist working at CERN in Geneva, Switzerland, proposed the development of an information management system based on the notion of distributed hypertext. The subsequent implementation of this system would help to make Jobs' prediction a reality. It would also spark a revolution in information technology which has changed the world.

Hypertext

The concept of hypertext, a method of interconnecting related documents using embedded links which allow the reader to move non-sequentially from one to another, was first introduced by the American computer visionary Theodor H (Ted) Nelson in a paper presented at the ACM National Conference in New York in August 1965. Nelson's paper described his efforts to create a computer-based personal file system and proposed an information structure based on what he called 'zippered lists', where elements in pairs of lists are linked to indicate entries in one list that correspond to entries in the other, in combination with an 'Evolutionary List File' structure which stored files by means of the lists. In this way, the user could link a section of text in one file to identical or related elements in other files irrespective of the order in which the files were stored.

Nelson took his inspiration from an essay by analogue computing pioneer Vannevar Bush entitled 'As We May Think' which was published in the July 1945 issue of the Atlantic Monthly magazine. In his essay Bush envisages a 'memex', a desk-shaped document storage device which is capable of joining related items together in an "associative trail". Bush wrote the essay in 1939 prior to the introduction of electronic digital computers, so there is no mention of the electronic technology from his Rapid Arithmetical Machine project. Instead, he proposed employing microfilm as the principal storage medium and an electromechanical mechanism for retrieving stored material. Nevertheless, the concepts he described not only provided the inspiration for Nelson's work but they are also regarded as the source of the branch of computing science known as Information Retrieval.

Nelson was primarily a sociologist and he lacked the technical knowledge necessary to implement his ideas. In April 1967, he met computer graphics researcher Andries (Andy) van Dam at the Spring Joint Computer Conference in Atlantic City. Both had studied together as undergraduates at Swarthmore College in Pennsylvania but had not seen each other for several years so Nelson took the opportunity to tell van Dam of his work on hypertext. Nelson's enthusiasm for his subject captured van Dam's imagination and led to Nelson collaborating with van Dam, who was based at Brown University in Providence,

Rhode Island, on the development of system to explore the use of hypertext as a tool for editing text files. The Hypertext Editing System (HES) was a single-user system which ran on an IBM System/360 Model 50 computer equipped with a timesharing operating system and an IBM 2250 Display Unit. The use of a graphics terminal and timesharing computer allowed the user to interact with the system to create, edit and print documents composed of alphanumeric characters. A hypertext facility also allowed sections of text to be connected between documents using links and branches in order to simplify navigating and browsing.

Van Dam's main research in computer graphics was funded through an industrial research contract from IBM and HES was initially run as a side project, with most of the work carried out by his students. However, this changed following a demonstration of the initial prototype to Sam Matsa, the senior IBM scientist who had been assigned to monitor the progress of van Dam's research. Following Matsa's endorsement, IBM began funding the work as an official project. The additional resources allowed the project team to develop HES to a high standard of completion and one system was sold by IBM to NASA in 1969 for use by the Houston Manned Spacecraft Center to create the documentation for the Apollo space programme. IBM also hawked it around a number of established customers in the publishing industry but none were convinced of the need for such advanced technology.

A follow-on project known as FRESS (File Retrieval and Editing System) was initiated in 1969 following van Dam's witnessing of Douglas Engelbart's fabled NLS demonstration at the Fall Joint Computer Conference in December 1968. The NLS demonstration served to highlight the limitations of HES and what might be possible with a device-independent multi-user system that would facilitate on-line collaboration. As a result FRESS embodied many of the features of NLS but without the rigid structure employed by Engelbart to control content. It also included several powerful new features, such as 'autosave' and undo, plus a link visualisation facility and an improved hypertext implementation which permitted links to be made down to the level of individual characters.

FRESS ran on a powerful IBM System/370 mainframe equipped with the VM/370 timesharing operating system. It supported multiple users through a range of devices from simple text terminals to the Imlac PDS-1 Programmable Display System. Van Dam struggled to convince senior colleagues at Brown University to take the project seriously so resources were scarce but the system was eventually completed and used for humanities teaching. It was also adopted by Katholieke Universiteit Nijmegen and Philips Research Eindhoven in the Netherlands.

The impact of Engelbart's NLS demo and van Dam's work at Brown University generated further research activity on hypertext and the concept was gradually

13. Getting Personal – The World According to Wintel

extended to include multimedia through landmark projects such as MIT's Aspen Movie Map in 1978 and Xerox PARC's NoteCards in 1984. However, hypertext remained a research tool until the introduction of Apple HyperCard in August 1987. HyperCard was developed for the Macintosh by Apple senior software engineer Bill Atkinson. Atkinson utilised hypertext concepts to create a programming tool and application program which allowed the user to produce multimedia content in the form of a stack of virtual cards. Cards could contain a number of interactive objects such as text fields, check boxes and buttons. The viewer browsed the stack by navigating from card to card using built-in navigation features, a powerful search mechanism, or through user-created scripts written in the accompanying HyperTalk scripting language.

HyperCard was designed for use by non-programmers and was reasonably priced at only $49.95. With nothing else like it on the market, it was an immediate success. However, Atkinson's implementation of hypertext was incomplete as it did not support embedded links within text. It also lacked the ability to operate over a network. Nevertheless, as the first commercially successful application of hypertext, HyperCard demonstrated the remarkable increase in readability that could be achieved when content was linked in this way. It also provided the inspiration for later developments.

Online Information Services

In the early 1970s a number of systems were developed to transmit pages of text-based information through the medium of broadcast television. The idea had first surfaced in 1965 when William D Houghton of RCA Laboratories developed a technique for transmitting auxiliary information using the vertical blanking interval (VBI) which is inserted between each frame of the television signal to allow time for the scanning beam to return to the start position. As the beam is switched off during this period, any extra information transmitted would not affect the normal television picture. RCA employed this technique in a television-based facsimile system for home use called 'Homefax' which was completed in 1967. RCA formed a partnership with US television network NBC to introduce Homefax as a service but the technology was deemed too impractical and the service was never marketed.

The VBI transmission technique then resurfaced in the UK a few years later as a means of providing subtitles for the deaf and hard of hearing. Research conducted in 1971 at the Mullard Application Laboratory in Mitcham, Surrey showed that an entire page of text could be transmitted in less than a second if the text was digitally encoded. Moreover, the advent of LSI integrated circuit technology made it feasible to manufacture inexpensive VBI decoder chips which could be built into domestic television sets. This opened up the possibility of a television-based information system which could be used to transmit information such as news reports, weather updates and programme listings in addition to subtitles.

The Story of the Computer

The Mullard researchers built a working prototype in 1972 which was demonstrated to UK broadcasters, resulting in the adoption of what became known as 'Teletext' by both the British Broadcasting Corporation and the Independent Broadcasting Authority. The BBC Teletext service, which was called CEEFAX (see facts), was launched in September 1974 and the IBA service, ORACLE (Optional Reception of Announcements by Coded Line Electronics), appeared a few months later. Both initially employed different formats which were incompatible and required separate decoders. In 1976 both broadcasters agreed to standardise on a 24 by 40 character page layout with 8 colours and a character set that included some semigraphics characters for creating simple block graphics. Up to 800 pages were available which the viewer selected by keying in the appropriate three-digit number using the television set's remote control. The selected page would then be displayed after a short delay.

BBC CEEFAX

Teletext was designed to be accessed through a domestic television set but it could also be accessed using a microcomputer fitted with a suitable adapter. One such product was created for use with the BBC Microcomputer, a popular home and hobby model manufactured by Acorn Computers as part of the BBC's Computer Literacy Project. The adapter operated in combination with a special graphics mode which emulated the appearance of a Teletext display. By using this equipment, information transmitted via Teletext could be captured and stored for printing or further processing. The BBC also used CEEFAX to distribute free software for the BBC Microcomputer, a service which it referred to as 'Telesoftware'.

Teletext was a one-way system, with interaction limited to viewers browsing

13. Getting Personal – The World According to Wintel

information by selecting which pages to view, but it was also used as the basis for a two-way system known as 'Viewdata'. Viewdata was developed by the UK Post Office to provide an interactive information service for businesses and home subscribers. It grew out of research into videophone technology by telecommunications engineer Samuel Fedida at the Post Office Research Station in Dollis Hill, London in the early 1970s. Fedida quickly established that a domestic videophone service would be impractical due to the requirement to install a second telephone line but a text-based information service which gave users direct access to a computerised information system through their existing telephone line would be feasible. Moreover, the cost of the equipment could be minimised by using a television set as the display device.

The parallel development of Teletext and Viewdata led to the Post Office adopting Teletext transmission and display formats for Viewdata, thereby making it easier for television set manufacturers to support both systems. However, as Viewdata communicated via the user's telephone line, an additional unit known as a Viewdata terminal was also required which incorporated a 1,200 bps duplex modem and alphanumeric keyboard. Alternatively, a microcomputer equipped with a suitable modem and the appropriate terminal emulation software could also be used to access Viewdata.

The Post Office launched its Viewdata service in March 1979 under the name 'Prestel', an abbreviation of press telephone. A huge number of pages were available, with the content supplied by hundreds of information providers who were each charged fees according to the number of pages they published on the system. The use of modems permitted two-way communication so subscribers could use the Prestel system for interactive tasks such as making airline reservations or managing their bank account. A simple electronic mail facility known as Prestel Mailbox was also added in 1983. However, unlike Teletext which was a free service, Prestel subscribers had to pay a monthly subscription for the use of the service plus telephone charges for the connection time. An additional expense was the cost of the terminal. Consequently, the service only attracted around 90,000 home subscribers, far fewer than the Post Office had expected. Fortunately, it was reasonably popular with business users, particularly those in the travel industry where it remained the main booking system of high street travel agents until well into the 1990s.

Despite the poor uptake in the UK, the Post Office succeeded in selling the Prestel service to telecommunications agencies in several other countries including Australia and Italy. Viewdata technology was also adopted by France Telecom for its highly successful Teletel/Minitel system, a government subsidised online information service which boasted almost 9 million terminals and 25 million users at its peak. Teletel/Minitel was hailed as a technological marvel and was a great source of pride for the French nation. It also had a long and productive life, operating for over 30 years before finally being switched off in June 2012.

The Story of the Computer

Teletext and Viewdata paved the way for Internet-based information systems by showing what was possible with an online service and whetting the public's appetite for instant information. The Internet had expanded rapidly throughout the 1980s and had matured sufficiently to support a global information system. Apple HyperCard and the pioneering hypertext research projects conducted at SRI and Brown University had demonstrated the power of interconnecting related material using embedded links. Now all that was required was for someone to put them together.

Weaving the Web

The breakthrough in combining the Internet with hypertext to create a global information system was made by the British research physicist and computer scientist Sir Timothy J Berners-Lee. Tim Berners-Lee was born in London in June 1955. His parents were both mathematics graduates and had worked as programmers for Ferranti during the company's heyday as a leading computer manufacturer in the 1950s. However, their son chose physics for his undergraduate degree which he received from the University of Oxford in 1976.

Berners-Lee had gained some experience with microprocessors during his time at Oxford and his first job after graduation was as an engineer for UK-based electronics multinational Plessey, working at the company's telecommunications division in Dorset on a variety of computer-related projects including distributed transaction systems, message relays and bar code technology. After two years at Plessey he and colleague Kevin Rogers moved to D G Nash Limited, a small Dorset company developing computer interface equipment for typesetting machines, where they worked on the software for a microprocessor-based printer for proof checking. From mid 1979 to early 1981 Berners-Lee worked as an independent software consultant and it was during this period that he first worked at CERN, the European Organisation for Nuclear Research in Geneva, Switzerland. He spent 6 months at CERN working with other software contractors, including Rogers, on a new control system for a particle accelerator.

In 1981 Berners-Lee returned to Dorset to take up an enticing offer from John Poole, one of the founders of D G Nash, to work for Poole's new venture, Image Computer Systems Limited. This was a more senior position than his previous role and he was given overall responsibility for the design and development of software for the company's entire range of specialist printing products. However, his earlier experience at CERN had left a lasting impression so towards the end of 1983 he applied for one of CERN's prestigious fellowships in applied science and engineering. The application was successful and in September 1984 he returned to CERN, this time as a research fellow, to work on distributed real-time systems for scientific data acquisition and system control.

13. Getting Personal – The World According to Wintel

At CERN, Berners-Lee became increasingly aware of the difficulty of managing information within such a large and amorphous organisation. Projects typically involved large numbers of researchers and contractors, many of whom were based in other locations, and generated vast amounts of documentation. CERN's computer facilities were similarly unstructured, comprising a wide range of incompatible systems loosely tied together by a network running the TCP/IP protocols. In March 1989, Berners-Lee drafted a document entitled 'Information Management: A Proposal' which sought to address this issue by proposing the development of a system that linked these hierarchical information systems together using distributed hypertext to provide a unified information management system. During his earlier stint at CERN, Berners-Lee had created a program for his own use called Enquire which employed hypertext to keep track of all the dependencies between people, programs and hardware. His proposal built on this experience and also drew upon more recent work to implement a special type of network protocol known as a remote procedure call (RPC) for data acquisition and control applications.

Berners-Lee's proposal was well received by his immediate superiors but failed to attract the attention of CERN's senior management. One colleague, the Belgian systems engineer Robert Cailliau, was particularly enthusiastic, helping Berners-Lee to redraft the proposal and lobbying management for resources for the project. When no action was taken, Cailliau also helped look for a commercial solution but, while hypertext had now caught the computer industry's attention, nobody seemed interested in developing it for use over the Internet.

In March 1990, Berners-Lee was given permission by his manager to purchase a NeXT graphics workstation and use it to build a prototype of his information management system. Launched in October 1988, NeXT workstations had created quite a stir amongst the CERN researchers due to the company's 'NeXTStep' software which combined a Unix-based real-time operating system with an object-oriented software development environment and a suite of powerful programming tools including a graphical user interface builder. The new computer arrived in September and by the end of the year Berners-Lee had created a working prototype using the NeXTStep development tools supplied with the machine.

Berners-Lee's prototype, which he called 'WorldWideWeb', employed a client-server architecture, with a server program servicing requests for 'web' pages and a client application known as a browser which ran on the user's computer for displaying the requested pages. The browser made full use of the NeXTStep graphical user interface, supporting WYSIWYG text and graphics, and also functioned as an editor for creating and modifying web pages. The layout and content of pages were specified in a document formatting language which he called HTML (HyperText Mark-up Language). The design of HTML was closely based on an established formatting language, SGML (Standard

Generalised Mark-up Language), which had been created a few years earlier to enable the sharing of electronic documents between different systems. A new network protocol, HTTP (HyperText Transfer Protocol), was also devised to make it easier for the browser to request pages from the server over a network.

The combination of a prototype under development and a redrafted proposal convinced CERN management to allow the project to proceed, but with fewer resources than Berners-Lee and Cailliau had requested. Both would be permitted to work full-time on the project for 6 months but they would have to make do with limited assistance from colleagues for the additional development work necessary to implement the system. As Berners-Lee's prototype browser was specific to NeXT graphics workstations, one of the earliest tasks was to create a generic web browser which could run on a wider range of computers. This task was carried out by Nicola Pellow, an undergraduate mathematics student from Leicester Polytechnic who was spending a year at CERN as part of her sandwich course. Pellow created a simple line mode browser with a text-only interface so that it could run on most types of display equipment from alphanumeric terminals upwards. Another colleague, Bernd Pollermann, wrote a gateway program which facilitated hypertext access to information stored on CERN's IBM mainframe computer system. Pollermann also made the CERN internal telephone directory available as web pages in an effort to encourage others to use the new technology.

The first web site, a collection of pages describing the WorldWideWeb project and the resources available, went live in August 1991. Berners-Lee announced it by posting a description of the project on the 'alt.hypertext' Internet newsgroup along with an invitation for others to make use of the library code he had written. However, an accompanying copyright notice made it clear that the code belonged to CERN and was only available for non-commercial use in collaborating non-military academic institutes. Those organisations wishing to use the code for commercial purposes would have to apply to CERN to obtain the specific conditions under which this would be permissible. Following persistent lobbying from Berners-Lee and Cailliau, this restriction was lifted in April 1993 when CERN issued a public statement declaring that the three components of web software (the line-mode browser, the basic server and the library code) would be placed in the public domain.

With the basic components in the public domain, the door was now wide open for commercial implementations of web technology. As interest grew and other software implementations began to appear, it soon became apparent that industry standards would have to be established and maintained in order to ensure compatibility between products. Berners-Lee took the lead in this activity, leaving his job at CERN in October 1994 to found the World Wide Web Consortium (W3C) at the Massachusetts Institute of Technology. W3C is the main international standards organisation for what has become known as the World Wide Web.

13. Getting Personal – The World According to Wintel

Browser Wars

Berners-Lee's online invitation to make use of his library code soon led to a number of web development projects at academic institutions across Europe and the USA. As NeXT graphics workstations were relatively uncommon, much of this effort focused on developing graphics-enabled web browsers for more popular machines. Two early examples were 'Erwise', which was written by a group of four Master's students at Helsinki University of Technology in Finland and released in April 1992, and 'ViolaWWW' which was written by Pei-Yuan Wei, an undergraduate student at the University of California Berkeley, and released the following month. Both browsers were developed for Unix workstations running the X Window System and featured WIMP interfaces but ViolaWWW also supported advanced features such as embedded objects and tables plus a rudimentary stylesheet mechanism for attaching styling information to a document. Consequently, it became the recommended browser at CERN.

In late 1992, the ViolaWWW browser caught the attention of Marc L Andreessen, a computer science undergraduate who was working as a student intern in the Software Design Group of the National Center for Supercomputing Applications (NCSA) at the University of Illinois at Urbana-Champaign. Andreessen was acutely aware of the potential of the World Wide Web as a means of facilitating access to the Internet for a wider audience. He could also see that the key to making this happen was having a multi-platform web browser which was simple to install and use. ViolaWWW clearly had potential but could be improved upon by making it more user-friendly and developing versions for a range of platforms. Andreessen was also beginning to lose interest in the 3D visualisation work he was doing at NCSA and needed a new challenge so the development of a user-friendly browser was an attractive prospect.

Andreessen spent a weekend creating a rough mock-up of his browser which he showed to colleague Eric J Bina, an NCSA employee with a reputation as a gifted Unix programmer. Bina was sufficiently intrigued that he offered his services as co-developer and obtained permission from his manager to work on the project. Andreessen and Bina then devoted the next two months to the project. The result was X Mosaic, a new browser for the X Window System which was released as an alpha prototype in January 1993. When their Software Design Group colleagues saw it, they were keen to lend their programming expertise in other software environments to the project. By September 1993 versions of the Mosaic browser for Microsoft Windows and the Apple Macintosh had also been released along with a Unix-based web server.

Part of NCSA's remit was to assist the scientific research community by producing widely available, non-commercial software, so Mosaic was made freely available for non-commercial use. Mosaic's ease of installation and multi-platform support helped to ensure its popularity and by mid-1994

The Story of the Computer

downloads from NCSA's file transfer facility had reached 50,000 copies per month. Some industry observers have also cited Mosaic as a major factor in the explosive growth of the World Wide Web during this same period. In August 1994, NCSA assigned the Mosaic commercial rights to Spyglass Incorporated, a company established by the University of Illinois in 1990 to commercialise NCSA technology. Spyglass rewrote the Mosaic source code to improve consistency across the three platforms and began licensing it to other companies for commercial use. However, Mosaic's dominance was about to be challenged by the very person who had created it.

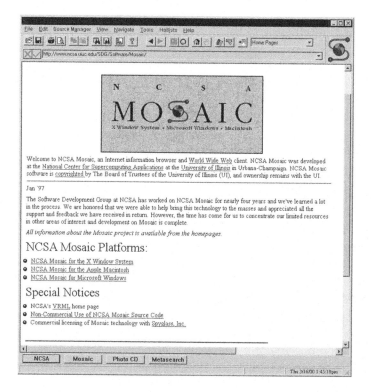

NCSA Mosaic

Following graduation in December 1993 Andreessen moved to California for a job at a small company in Palo Alto developing security products for the Web. In nearby Mountain View, Silicon Graphics founder Jim Clark had just announced his resignation as Chairman of the company and was looking for a new opportunity which he could devote his entrepreneurial energies to. Clark's resignation had been triggered by a clash with the board of directors over the future of Silicon Graphics which he felt should be directed towards consumer products. He had identified interactive television as having considerable promise but it was not yet clear how this immature technology could be

13. Getting Personal – The World According to Wintel

developed for the consumer. On his last day at Silicon Graphics, one of Clark's employees showed him the Mosaic browser. Clark immediately spotted the potential of web browsers for accessing video over the Internet and could see how this might provide a future delivery mechanism for interactive television. When he learned that Mosaic's creator was now based only a few miles away, he immediately contacted Andreessen to arrange a meeting. After meeting Clark, Andreessen jumped at Clark's offer of joining him in this new venture.

In April 1994 Clark and Andreessen established a company called Mosaic Communications Corporation in Mountain View to develop and market an enhanced version of the Mosaic web browser which would be called Mosaic NetScape Network Navigator. Clark financed the venture using $4 million of his own personal wealth in an effort to retain more control over the company than he had with Silicon Graphics. A development team was quickly assembled by recruiting 5 of the 6 original Mosaic developers from NCSA plus several engineers from Silicon Graphics. The name of the company was changed to Netscape Communications Corporation in November 1994, as Spyglass owned the Mosaic trademark, and the web browser was released the following month under the shorter name of Netscape Navigator. Features included support for the JPEG image format and secure communications using encryption and server authentication to facilitate electronic commerce applications. Performance was also optimised for use with relatively slow dial-up modem connections, as most users would not have access to a fast Internet connection.

Netscape's business model was to give the browser away free for non-commercial use but charge $99 per copy for commercial use. The company also charged handsomely for its web server software which was available as a basic version, NetSite Communications Server, for $1,495 and an e-commerce version, NetSite Commerce Server, for $5,000. This business model initially worked well and Netscape products quickly cornered the market, with the browser peaking at almost 80% market share in 1996. An IPO in August 1995 valued the company at $2.9 billion and sparked a trend for investing in Internet businesses that would later be called the dot-com boom. However, the growing popularity of the Web and the rapidly expanding market for web browsers had caught the attention of Microsoft and it was only a matter of time before the software giant muscled in.

In May 1995, Microsoft Chairman Bill Gates sent a lengthy memo entitled 'The Internet Tidal Wave' to all executive staff declaring the Internet to be the most important development since the IBM PC and setting out detailed plans to refocus the entire business on Web-enabled products. Microsoft had been uncharacteristically slow in identifying the potential of the Web, focusing instead on online dial-up services, but would now use its vast resources to make up for lost time. First up would be a web browser that could be integrated into the Windows environment and bundled with the operating system in order to achieve deep market penetration. To reduce the development timescale, the code for the Windows version of Mosaic would be licensed from Spyglass, reworked slightly to make it more consistent with the Windows user interface and rebadged as Internet Explorer.

Internet Explorer was introduced in August 1995 but arrived too late to be included with Windows 95 so it was initially released as part of the Microsoft Plus! add-on package which sold for $50. A version for Windows NT was also released in January 1996. Later versions were included in Windows service packs or delivered pre-installed with new PCs until the release of the next version of the operating system, Windows 98, in June 1998 when Internet Explorer 4 was finally bundled with the operating system. Sales of Windows 98 ensured that Internet Explorer 4 was widely distributed. It also outperformed the latest version of Netscape Navigator so Windows users had no reason to install another browser. By the end of 1998 Internet Explorer had replaced Netscape Navigator as the world's most popular web browser.

Legal Woes

Microsoft's strategy of bundling the browser with the operating system did not go down well with Spyglass, as the licence agreement for the Mosaic code was based on a royalty fee estimated at around 50 cents for each copy of Internet Explorer sold, with a minimum quarterly payment of $400,000 and an annual cap of $5 million. As Microsoft was not selling Internet Explorer as a separate product, it was only willing to pay the minimum quarterly amount. The matter came to a head in January 1997 when Spyglass lawyers demanded an audit to determine exactly how many copies of Internet Explorer had been distributed. To avoid legal action, Microsoft agreed to make a one-time payment of $7.5 million in cash and $500,000 in software and other considerations in return for the freedom to use the Mosaic technology with no further payments.

By bundling Internet Explorer with Windows, Microsoft had won the browser wars. The company also pursued a similar strategy for its web server software, Internet Information Server, which was bundled with the advanced server edition of Windows NT, beginning with version 4.0 in July 1996. Sales of Netscape's NetSite server products were badly affected but, unlike Microsoft, Netscape had no other products to rely on for sales revenues. A series of redundancies in January 1998 heralded Netscape's demise and the company was acquired by multinational Internet services corporation AOL in March 1999. Co-founders Clark and Andreessen moved on to new entrepreneurial opportunities, albeit separately, but would wisely avoid any future ventures which came into competition with Microsoft.

Microsoft's bundling strategy also triggered further legal action from the US Department of Justice which filed an anti-trust lawsuit against the company in May 1998 for unlawfully monopolising computer software markets. Government lawyers called for Microsoft to be split into two separate

13. Getting Personal – The World According to Wintel

companies as a punishment and for a while this looked likely but in November 2001 both parties agreed a settlement which avoided a company breakup. The consent decree, which was extended twice before finally expiring in May 2011, forced Microsoft to make Windows interoperable with non-Microsoft software and barred the company from entering into any OEM agreements which restricted manufacturers from working with Microsoft's competitors. A similar action by the European Commission in 2009 forced Microsoft to offer alternative browsers during installation of Windows for copies of the operating system sold in European territories. Microsoft's failure to maintain this option following the release of a service pack in 2011 resulted in the company being fined €561 million in March 2013.

Another lawsuit in December 2000 threatened to upset the entire industry. In July 1976, the UK Post Office filed a US patent application covering certain aspects of the Viewdata system. The patent described the data format for Viewdata pages which had been developed by engineer Desmond J Sargent. This featured an additional block for storing information on how the page should be processed by the system, such as the address of another block for linking related pages. The Sargent patent gathered dust amongst the Post Office's portfolio of 15,000 patents until it was rediscovered in a review conducted by British Telecom in 1997 following the transfer of Post Office intellectual assets to BT in 1981. Intellectual property specialists hired by BT concluded that the patent was sufficient to give the company a valid claim on the invention of the hyperlink and in June 2000 they contacted 16 leading Internet Service Providers in the US requesting the payment of a licence fee to BT for hyperlink usage. Unsurprisingly, this request was refused by all 16 ISPs so BT lawyers singled out one, Prodigy Communications Corporation, as a test case and filed a lawsuit for damages in December 2000. The case was thrown out in August 2002, with the judge ruling BT's claim invalid due to several obvious technical differences between the system described in the Sargent patent and the way in which hyperlinks are implemented on the Internet.

These high profile legal cases are an indication of how important the World Wide Web has become. When Mosaic was introduced in 1993, the number of Internet users was estimated at just over 14 million. By July 2014 this had grown to nearly 3 billion, which is around 40% of the world's population. The amount of material published on the Web has also risen dramatically, prompting the development of sophisticated search engine technology to help users find information from the 1 billion web sites and countless pages available. As the Web has grown in importance, Tim Berners-Lee has become something of a reluctant hero for his invention and in 2004 he received a knighthood from Queen Elizabeth II in recognition of his services to the global development of the Internet through the invention of the Web.

The Information Age

Nowadays, the World Wide Web is accessible on a huge range of devices as a result of the proliferation of portable computing devices such as smartphones and tablet computers which took place in the early years of the 21st century. These devices make use of advances in flat panel displays, miniaturised electronics, electric batteries and wireless communications technology to create highly portable devices with sufficient power to run a web browser. Their ubiquity has led to an explosion in the use of the Web.

There has been significant effort, even from the early days, to make computing equipment small enough to be carried, beginning with Sir Samuel Morland's adding machine in 1666. Microprocessors allowed manufacturers to reduce the size and weight of computers, from the boxy luggable models of IBM and Compaq in the early 1980s to the sleek laptops that became an essential part of most companies' product ranges in the 1990s. However, there has been very little innovation shown until recently, as manufacturers have tended to create smaller and lighter versions of existing products. It is really only since the convergence of portable computing and mobile telecommunications technologies, heralded by the introduction of the Nokia 9000 Communicator in 1996, that the industry has begun to view portable computing devices as a separate species which could be developed along different lines. As a result, the future of the industry is now back in the hands of the hardware manufacturers.

This trend has not been lost on Microsoft, which in recent years has diversified into hardware manufacture in an effort to preserve its diminishing power. The software giant's first foray into hardware was the Xbox video game console in November 2001 but it has also targeted the growing market for portable computing devices with the launch of the Surface range of tablet computers in June 2012 and the acquisition of the mobile phone business of Finnish telecommunications corporation Nokia in April 2014.

The industry has moved full circle, returning to its original aim of providing convenient tools to aid everyday tasks, except that the tasks are now different. The advent of portable computing devices with an always-on Internet connection has radically altered the way in which society uses and considers computers, enabling new methods of social interaction through a myriad of social networking services. We have entered the Information Age and the future is very bright indeed.

Further Reading

Allan, R. A., A History of the Personal Computer: The People and the Technology, Allan Publishing, London, Ontario, 2001.

Allen, P., Idea Man: A Memoir by the Co-founder of Microsoft, Penguin

13. Getting Personal – The World According to Wintel

Books, London, England, 2011.

Barnet, B., Crafting the User-Centered Document Interface: The Hypertext Editing System (HES) and the File Retrieval and Editing System (FRESS), Digital Humanities Quarterly 4 (1), 2010, 44-57.

Bush, V., As We May Think, The Atlantic Monthly, July 1945.

Campbell-Kelly, M., From Airline Reservations to Sonic the Hedgehog: A History of the Software Industry, MIT Press, Cambridge, Massachusetts, 2003.

Edstrom, J. and Eller, M., Barbarians Led by Bill Gates: Microsoft from the Inside, Henry Holt & Company, New York, 1998.

Gillies, J. and Cailliau, R., How the Web Was Born: The Story of the World Wide Web, Oxford University Press, Oxford, 2000.

Grehan, R., Lemmons, P., Malloy, R., Thompson, T., White, E., Williams, G. and Wszola, S., First Impressions: The IBM PS/2 Computers, Byte Magazine 12 (6), 1987, 100-114.

Hamm, S. and Greene, J., The Man Who Could Have Been Bill Gates, BusinessWeek Magazine, 24 October 2004.

Naughton, J., A Brief History of the Future: The Origins of the Internet, Weidenfeld & Nicolson, London, England, 1999.

Nelson, T. H., A File Structure for the Complex, the Changing and the Indeterminate, Proceedings of the ACM 20th National Conference, 1965, 84-100.

Paterson, T., An Inside Look at MS-DOS, Byte Magazine 8 (6), 1983, 230-252.

Rivkin, J. W., Porter, M. E., Bruin, C. E., Chappel, M., Galizia, T. M. and Worrell, L. J., Matching Dell (A), Harvard Business School Case 799-158, 1999.

Rumelt, R. P., VisiCorp 1978-1984 (Revised), UCLA Anderson School of Management Case POL-2002-01, 2002.

Slater, M., Intel Boosts Pentium Pro to 200 MHz, Microprocessor Report 9 (15), 1995, 1-5.

Stephens, K., Xerox Finally Wakes Up, But Is It Too Late?, Santa Clara High Technology Law Journal 6 (2), 1990, 407-420.

Sukumar, S., Touchscreen Personal Computer Offers Ease of Use and Flexibility, Hewlett-Packard Journal 35 (8), 1984, 4-6.

Wallace, J. and Erickson, J., Hard Drive: Bill Gates and the Making of the Microsoft Empire, John Wiley & Sons, New York, 1992.

Wolf, G., The Curse of Xanadu, Wired 3 (6), 1995.

Zachary, G. P., Showstopper!: The Breakneck Race to Create Windows NT and the Next Generation at Microsoft, The Free Press, New York, 1994.

Index

Symbols
3Com 478
3M Company 217
1977 Trinity 431–435, 512

A
abacus 7
Bruno Abdank-Abakanowicz 60
Académie des Sciences 26, 59
Accademia di Belle Arti 58
accumulator 97, 146, 181
ACE (Automatic Computing Engine) 176–180
ACM (Association for Computing Machinery) 458
ACM National Conference 539
Acorn Computers 542
AC Spark Plug Company 279
Adage 391
Adage Ambilog 200 391
Adage Graphics Terminal 391
Charles W Adams 356, 373
Adams Associates 373, 375
address space 101, 251, 328, 333
Admiralty Fire Control Table 76, 230
Adobe PostScript 500
Adobe Systems 478, 500
Aegis 481
AEI 1010 271
Aerodynamischer Versuchsanstalt (AVA) 105
AES-90 CRT Text Editor 440
Howard H Aiken 85, 115–120, 123, 133, 155, 168, 184, 189, 192, 194, 212, 240
Al-Bīrūnī 11–12
Aldus PageMaker 500, 530
Paul G Allen 505–516, 528, 531
ALOHAnet 467
ALWAC 361
Gene M Amdahl 222, 246, 297
American Airlines 292
American Arithmometer Company 224
American Mathematical Society 110, 128
American Research and Development Corporation (ARDC) 331
American Telephone & Telegraph Company (AT&T) 108, 126, 285, 453, 455
American Totalizator Company 204
Ampex 290, 303, 387
Jakob Amsler 59
Amstrad 522
Amstrad PCW 8256 522
Analytical Engine 47–51, 54–55, 85, 94, 115, 156
AN/ASQ-38 359
Harlan E Anderson 330–331
Marc L Andreessen 547–550
AN/FSQ-7 358, 360, 364, 452
AN/FSQ-8 360, 452
AN/FSQ-32 456
AN/FST-2 Coordination Data Transmitting Set 228
Antikythera Mechanism 10–11
AN/USQ-17 254, 256
AOL 550
APE(X)C 232

555

API (Application Programming Interface) 533
APL (A Programming Language) 438
Apollo Computer Incorporated 481–483, 487, 489, 490, 500
Apollo DOMAIN DN100 481–482
Apollo DOMAIN DN300 Desktop Computational Node 482
Apollo DOMAIN DN420 482
Apple][428–431, 432, 435, 436, 445, 447, 477, 497, 498, 513, 519, 525
Apple-1 426–427, 431, 437
Apple Computer Company 424–431, 432, 441, 445, 483, 495–501, 512, 527, 530
Apple Corps 426
Apple HyperCard 541, 544
Apple III 445, 519
Apple LaserWriter 500
Apple Lisa 495–497, 499, 527, 530
Apple LisaCalc 496
Apple LisaDesk 496
Apple LisaDraw 496
Apple LisaGraf 496, 498
Apple LisaProject 496
Apple LisaWrite 496
Apple Macintosh 495–501, 527, 530, 547
Apple Macintosh II 483
Apple MacPaint 499
Apple MacWrite 499
Apple QuickDraw 498
Archimedes of Syracuse 10
ARDS (Advanced Remote Display Station) 381, 384
Argo Clock 71–74
Arithmometer 36–39, 52, 86
Arma Corporation 278
Arma Engineering 76
ARPA (Advanced Research Projects Agency) 367, 379, 392, 455, 457, 461, 465
ARPANET 455–459, 461
ASCA (Airplane Stability and Control Analyzer) 353
ASCC (Automatic Sequence Controlled Calculator) 116–119, 146, 213
Astrarium 14, 17
John V Atanasoff 132–137, 144, 150, 168
Atanasoff-Berry Computer 132–137, 227
Atari 425, 427, 429, 430
ATI Technologies 538
William D (Bill) Atkinson 496, 541
Atlantic City Personal Computer Show 423, 427
Atlantic Monthly magazine 539
Atlas Airborne Digital Computer 278
Atlas Computer Laboratory 252
Atlas Guidance Computer (AN/GSQ-33) 277, 280
Atlas missile 276, 291
Atomic Energy of Canada Limited (AECL) 333
Atomic Energy Research Establishment 249, 252
Atomic Weapons Research Establishment 248, 249, 252
Audion 125
Isaac L Auerbach 277–278
Auerbach Associates 278
Augmentation Research Center (ARC) 380, 464
Autarith 85
auto-index register 333
Automated Engineering Design (AED) 369, 376

Index

Automatically Programmed Tool (APT) 369
Automatic Control magazine 364
Automatic Electronic Systems 440
Automatic Relay Computer 192

B

Charles Babbage 39–55, 85, 94, 115, 156, 176, 230
Henry P Babbage 53
John Backus 222, 246
BAC TSR-2 307
Bailey Meter Company 336
Ballistic Computer 112
Ballistic Missile Early Warning System (BMEWS) 291
Steven A Ballmer 514, 516, 528–529
Bank of America 288, 309
Paul Baran 454–455
John Bardeen 265, 267
Barr & Stroud 69, 71–72
BASIC 421, 427, 429, 433, 434, 435, 439, 441, 444, 450, 506, 510, 512, 527
Forest Baskett 484, 489
batch processing 364
Allen J Baum 426, 430
BBC (British Broadcasting Corporation) 542
BBC CEEFAX 542
BBC Microcomputer 542
BBN (Bolt, Beranek and Newman) 332, 366, 393, 458
Robert M Beck 317, 335, 348
Norman Bel Geddes 119
Alexander Graham Bell 87, 267
C Gordon Bell 333, 342, 345, 347, 415, 492

Bell 101 Dataset 452
Bell 103 452
Robert L Belleville 475, 497
Bell & Howell 314
Bell Laboratories Command Guidance System 280
Bell Labs Model V Relay Calculator 113–114, 149
Bell Punch Company 398
Bell Telephone Laboratories (BTL) 92, 108–114, 123, 147, 150, 161, 168, 262, 263–267, 271, 278, 280, 281, 287, 301, 353, 378, 386, 452, 466, 483
Robert W Bemer 364
Bendix Aviation Corporation 288, 317–319, 323, 334
Bendix D-12 317, 335
Bendix G-15 316–319, 335, 364, 392
Bendix G-20 319
Benson-Lehner Corporation 362
Berkeley Computer Corporation 462, 470
Sir Timothy J Berners-Lee 544–547, 551
Clifford E Berry 134–137, 227
Julian H Bigelow 171–173
Eric J Bina 547
BINAC 202–205, 277
binary arithmetic 88, 101, 158, 331
binary-coded decimal (BCD) 109, 204, 210, 211, 214, 218, 225, 226, 227, 228, 239, 285, 290, 293, 295, 296, 304, 314, 322, 323, 406
BIOS (Basic Input/Output System) 443, 517, 523
biquinary-coded decimal 112, 140, 219, 241, 282
Vannoccio Biringuccio 28

557

The Story of the Computer

Birkbeck College 191, 398
James W Birkenstock 216–217
Birmingham Small Arms Company (BSA) 74
BitBlt 469, 473, 479, 481, 484
BIZMAC 286, 289, 290
Honoré Blanc 34
Bletchley Park 137–142, 155, 165, 175, 183
Felix Bloch 263
Boeing 507
George Boole 89
Boolean Algebra 89, 91–92, 100, 134
Andrew D Booth 163, 191, 194, 232, 398
bootstrap loader 409, 443
Sir Charles Vernon Boys 61, 72
Lee L Boysel 400
John G Brainerd 145, 354
Walter H Brattain 264–265, 267
Karl Ferdinand Braun 263
Bravo 470, 474, 475
BravoX 471, 496, 525
Breakout 429
Daniel S Bricklin 435–436
Bristol Bloodhound 329
British Aircraft Corporation (BAC) 307
British Tabulating Machine Company (BTM) 107, 138, 232, 303
British Telecom (BT) 551
Rod Brock 516
Richard R Brodie 525
Brookhaven National Laboratory 466
Frederick P Brooks 245, 297
Brown & Sharpe 33
Brown University 539, 544

Douglas W Broyles 485–486
Marc Isambard Brunel 43
James W Bryce 116
BSD (Berkeley Software Distribution) Unix 486, 487
BTM 1200 232, 303
Werner Buchholz 223, 245, 298
Compagnie des Machines Bull 308, 321, 412
bulletin board system 453
Bull Gamma 3 321
BUNCH 308–311, 348
John P Bunt 327
Henry Burkhardt III 345
Arthur W Burks 144, 146, 171
William Seward Burroughs 224
Burroughs 220 228, 294
Burroughs B5000 293
Burroughs Corporation 86, 200, 224–229, 232, 288, 293, 294, 302, 314, 330, 465
Burroughs E101 Desk Size Electronic Computer 225, 314, 450
Burroughs Electronic Accounting Machine (BEAM) 225, 228
Burroughs Laboratory Computer 225
Burroughs Research Center 225, 277
Burroughs Research Division 277
Vannevar Bush 77–82, 91, 128–129, 168, 539
Busicom 401, 405
Busicom Junior 402
Busicom LE-120A 402
Business Week magazine 259, 440
byte 298
Byte magazine 422, 427, 435, 436
Byte Shop 426, 434

Index

C

Stan Cagle 416
Robert Cailliau 545–546
Calculating Clock 17–20
CALDIC 192, 221, 227
Samuel H Caldwell 81, 128, 184
California Computer Products Company (Calcomp) 362, 374
California Institute of Technology (Caltech) 319, 375
Cambridge Air Force Computer 282
Cambridge University Mathematical Laboratory 253
Canon 401, 500
Canon Pocketronic 402
Carnegie Mellon University (CMU) 345, 347, 375, 474, 478, 491
Cartesian 400
CCC DDP-24 344
CCC DDP-116 309, 344
CCC DDP-416 344
CCC DDP-516 344, 458
CDC 160 256, 324, 333
CDC 160-A 325, 330, 380
CDC 160G 326
CDC 1604 254, 256, 259, 324
CDC 1604A 386
CDC 3200 375
CDC 6600 253–259, 284, 391, 492
CDC 7600 259
CDC Digigraphic System 270 375
CEC Model 30-201 227
Cedar Engineering Incorporated 254
Vinton G Cerf 459
CERN (European Organisation for Nuclear Research 259, 390, 539, 544

CGA (Color Graphics Adapter) 443, 518
Charlie Chaplin 444
chart recorder 361
Thomas B Cheek 384
chorded keyset 464, 467, 470
Wen Tsing Chow 279
Marcus Tullius Cicero 11
City & Guilds College 127
James H Clark 489–490, 492, 548–550
Wesley A Clark 273, 375, 448–449, 457, 461, 465
Richard Clayton 449
Joseph Clement 43–47, 53
Clement Designlabs 468
client-server architecture 545
clipping divider 393, 489
COBOL 513
Sir John Cockcroft 252
John Cocke 247, 492
Codex Corporation 453
Arnold A Cohen 207–208
John S Coleman 225
Colossus 137–142, 148, 150, 155, 161, 165, 177, 183
Colossus Mark 2 141–142
Columbia Data Products 517
Columbia Data Products MPC (Multi Personal Computer) 1600 517
Columbia University 93, 116, 126, 213, 215, 238
Comdex Fall 526
Comdex Spring 517
Commodore Business Machines 428, 431–433, 437, 441, 445, 512
Commodore C64 434
Commodore CBM 8032 445

The Story of the Computer

Commodore PET 2001 431–433, 435, 440, 447, 451, 513, 519

Commodore VIC-20 434

Communications Supplementary Activities - Washington (CSAW) 206–207

Compaq Computer Corporation 517–519, 523–524, 536, 552

Compaq Deskpro 286 519

Compaq Deskpro 386 523

Compaq Deskpro Model 1 519

Compaq Portable PC 517

Compaq Systempro 536

Compatible Time-Sharing System (CTSS) 365, 371, 383

Complementary Diode-Transistor Logic (CDTL) 322

Complex Number Calculator 109, 144, 451

Computek 385

Computek Model 400/20 terminal 385

Computer-Aided Design (CAD) 368–377, 397, 474, 482, 485, 487

Computer Aided Design Centre 253

Computer Center Corporation (C-Cubed) 507

Computer Control Company (3C) 309, 344, 405

Computer Design magazine 400

Computer Displays Incorporated 381, 384

Computer History Museum 54

ComputerLand 444

computer mouse 380–382, 384, 464, 467, 495, 526

computer network 451, 455, 489

Computer Research Corporation (CRC) 362

Computer Terminal Corporation (CTC) 407

Computervision 487

Computing-Tabulating-Recording Company 98, 212

Leslie J Comrie 183, 215

James B Conant 119

Consolidated Engineering Corporation 136, 227

Consolidated Vultee Aircraft Corporation (Convair) 276, 279, 359

Consumer Electronics Show (CES) 432

Continuous Integraph 77

Control Data Chippewa Laboratory 256

Control Data Corporation (CDC) 254–259, 302, 319, 324–326, 330, 375, 382, 494

Control Instrument Corporation 225

Lynn Conway 490

John M Coombs 191, 358

co-operative multitasking 527

coordinate transformation 391

Fernando J Corbató 365

Ralph J Cordiner 287, 289

Cornell University 108

Corning Glass Works 332

Corona Data Systems 517

Cosmic Engine 13

John D Couch 496

counting board 7

Courant Institute of Mathematical Sciences 259

Norman L Cox 475

CPC (Card-Programmed Electronic Calculator) 215–216

C programming language 488

Perry O Crawford 190, 354

Index

Seymour R Cray 237, 254–259, 280, 284, 324
CRC 105 362
Creative Computing magazine 451, 519
Arthur J Critchlow 221
Cromemco Dazzler 429
Loring P Crosman 200
Charactron CRT 359
Typotron CRT 359
CRT (Cathode Ray Tube) 163, 258, 315, 354
CTC Datapoint 2200 407
David N Cutler 533, 535
CYCLADES 459
cycle stealing 388

D

Daisy Systems 485
Sidney Darlington 280
DARPA (Defense Advanced Research Projects Agency) 484, 489
Dartmouth College 110, 144, 365, 451, 512
Data Equipment Company 380
Data General Corporation 344–347, 482
Data General Nova 345, 397, 424, 465, 466
Data General Nova 800 347
Data General Nova 1200 346
Data General Supernova 347
datagram 459
Datamatic Corporation 309
Dataram Corporation 390
DATAR (Digital Automated Tracking and Resolving) 382
Monte Davidoff 511

David Taylor Model Basin 244, 284
Donald W Davies 177, 456
Lester T Davis 256
Malcolm R Davis 379
Davis & Rock 336
Sir Humphry Davy 42
DEC 4000 Series 448
DEC Alpha 535
Edson D de Castro 334, 342, 344
DEC DECmate II 520
DEC (Digital Equipment Corporation) 313, 330–334, 337, 341–343, 347–348, 357, 366, 368, 374, 382–383, 386–388, 415, 435, 449, 478, 483, 485, 489, 490, 495, 507, 520, 535, 536
DEC General Purpose Experimental Display System 386
DEC Microprocessor Series (MPS) modules 438
DEC PDP-4 332–333, 343
DEC PDP-4C 386
DEC PDP-5 333–334, 338, 341, 382, 387, 449
DEC PDP-6 348, 368
DEC PDP-8 341–343, 347, 387, 389, 390, 406, 411, 438, 449, 520
DEC PDP-8/E 343
DEC PDP-8/I 343, 450
DEC PDP-8/S 343
DEC PDP-10 393, 436, 506, 508, 510
DEC PDP-11 347, 397
DEC PDP-11/20 348
DEC PDP-12 450
DEC Professional 350 520
DEC Programmed Data Processor Model 1 (PDP-1) 331, 337, 366, 373, 382

561

DEC Rainbow 100 520
DEC Type 30 Precision CRT Display 382
DEC Type 33 Digital Symbol Generator 382
DEC Type 330 383
DEC Type 338 Programmed Buffered Display 387, 390
DEC Type 340 Precision Incremental Display 382, 386
DEC UNIBUS 347, 421
DECUS 505
DEC VAX 11/780 484
DEC VMS (Virtual Memory System) 533
DEC VT78 Video Data Processor 438
DEC VT102 520
DECwest 533
Giovanni de' Dondi 14, 17
Defense Calculator 215–217, 239
Lee de Forest 125
Dekatron 398
Robert H Dennard 404
Department of Scientific and Industrial Research (DSIR) 178, 249, 252
Baron Gaspard de Prony 41
depth cueing 392
Design Augmented by Computers 370–373
Desktop Publishing (DTP) 487, 500
Deutsche Versuchsanstalt für Luftfahrt (DVL) 103, 131
D G Nash Limited 544
Diablo Model 31 466
Diamond Multimedia 536
Difference Engine 42–47, 146
Difference Engine No. 2 51–52, 53
Differential Analyzer 79–81, 143, 145, 168, 271, 361

Digigraphic Laboratories 375
Digital Research CP/M 444, 514, 519, 521, 522
Digital Research GEM 529
Digital Research Incorporated 514
digital signal processing (DSP) 453
digital-to-analogue converter 356, 357
digitising tablet 378–380, 462
Thomas L Dimond 378
Direct-Coupled Transistor Logic (DCTL) 257, 284
direct memory access (DMA) 331, 409
Direct-View Bistable Storage Tube (DVBST) 383, 450
diskless node 482
display buffer 372, 387
display console 359, 364
display list 363, 387, 390
display processor 391
Benjamin Disraeli 52
L Edwin Donegan 310
Georges F Doriot 331, 341
dot matrix printer 523
Willis K Drake 253
Draw 471
Dr Dobb's Journal 427
Lieutenant Frederic C Dreyer 73–74
Dreyer Fire Control Table 73
dual in-line package (DIP) 400, 406
Dumaresq 69, 73, 230
Geoffrey W A Dummer 298, 338
Stephen W Dunwell 218, 245, 246, 248
Benjamin M Durfee 116, 119
Dynabook 462, 466
Dynalogic Corporation 517

Index

Dynalogic Hyperion 517
Dynamical Systems Research 532
DYSEAC 357

E

Eagle Computer 517
Eagle PC 517
EAI PACE 361
Eastern Joint Computer Conference 332, 379
EasyWriter 444
J Presper Eckert 123, 143–151, 157, 160, 161, 164, 167–169, 190, 194, 201–206, 262, 293, 354
Wallace J Eckert 116, 213, 215, 238
Eckert-Mauchly Computer Corporation 201, 209, 228, 269, 277
Edinburgh Review magazine 52
Thomas Alva Edison 123, 190
EDSAC (Electronic Delay Storage Automatic Calculator) 184–187, 198
EDVAC (Electronic Discrete Variable Automatic Calculator) 158–160, 167, 169, 174, 176, 184, 269, 354
Dai B G Edwards 181, 250
EELM System 4 307
Lewis C Eggebrecht 442
Eidgenössische Technische Hochschule 106
Albert Einstein 158
EISA (Extended Industry Standard Architecture) bus 524
Electric & Musical Industries Limited (EMI) 178, 304
ElectroData Cardatron 228, 313
ElectroData Corporation 227–228, 288, 293, 313, 315
ElectroData Datatron 223, 227, 293, 294, 313
electromagnetic relay 87, 92, 100
Electronic Arrays Incorporated 417
Electronic Associates Incorporated (EAI) 361
electronic calculator 130–131, 158, 226, 303, 398
electronic character generator 372, 385, 422, 429, 465
Electronic Control Company 201
Electronic Drafting Machine (EDM) 373–375
Electronic News 407
Electronic Recording Machine Accounting (ERMA) 289
electronic spreadsheet 436, 496, 525, 530, 538
electrostatic storage 163, 239
John Ellenby 472, 474
William S (Bill) Elliott 230–231, 253
Elliott 401 222, 230
Elliott 402 231
Elliott 405 231
Elliott 801 327
Elliott 802 327
Elliott 803 327, 332, 338
Elliott 803B 328
Elliott-Automation 307, 327, 402
Elliott Brothers 39, 69, 73, 76, 222, 230–231, 253, 307
Elliott Type 152 Naval Gunnery Control Computer 230
Thomas O Ellis 379
e-mail 459
EMI EMIDEC 1100 304
EMI EMIDEC 2400 304
emitter-coupled logic (ECL) 247, 473

Douglas C Engelbart 380–381, 384, 458, 462, 464, 540

William K (Bill) English 381, 464

English Electric Company 179, 196, 231, 287, 304, 306

English Electric Computers Limited 307

English Electric DEUCE 231, 398

English Electric KDP10 287

English Electric Leo Computers Limited 306

English Electric Leo Marconi Computers (EELM) 306

Howard T Engstrom 206

ENIAC (Electronic Numerical Integrator and Computer) 123, 143–151, 155, 157, 161, 164, 167, 177, 228, 262, 286, 313, 317, 319, 354

Enigma machine 137

Enviro-Labs 508

ERA 1101 207

ERA 1103 208, 218, 222, 223, 237, 240, 255, 284

ERA (Engineering Research Associates) 191, 201, 206–210, 219, 225, 228, 253

Ergon Research Laboratories 264

ESL Display Console 383

Ethernet 467, 476, 478, 479, 483, 484, 485, 491

European Commission 551

Bob O Evans 297, 363

David C Evans 392, 461

Kent H Evans 506, 509

Evans & Sutherland Computer Corporation 393, 536

Evans & Sutherland Line Drawing System Model 1 (LDS-1) 393, 490

Evans & Sutherland PS 300 491

Robert R Everett 354–355

F

Facit 402

Federico Faggin 406, 434

Fairchild 3800 400

Fairchild Clipper 493

Fairchild Semiconductor 257, 298, 339, 345, 400, 403, 406, 428, 493

Fall Joint Computer Conference 346, 373, 464, 540

Robert M Fano 367, 384

FASTRAND 244

Samuel Fedida 543

Harold Feeney 408

Jean H Felker 272

Lee Felsenstein 422–423

Sidney Fernbach 255

Ferractor 282

Ferranti Argus 328

Ferranti Argus 100 329

Ferranti Atlas 250, 257, 329, 367

Ferranti Atlas 2 253

Ferranti Canada 382

Ferranti ICT 305

Ferranti ICT 1900 Series 305

Ferranti Limited 196, 230, 250, 304–305, 329, 336, 472, 544

Ferranti Mark 1 196–198, 223, 250, 270

Ferranti Mercury 250, 251, 328

Ferranti Orion 368

Ferranti Packaged Computer 231

Ferranti-Packard 305

Ferranti-Packard FP-6000 305, 368

Ferranti Pegasus 231, 241, 347

Index

File Transfer Protocol (FTP) 459
flat panel display 463, 552
John Ambrose Fleming 124, 264
FLEX 462
Flexowriter 320, 336, 365
Flip Chip 341, 345, 348, 449
flip-flop 127, 128, 146, 161, 340, 378, 397, 404
F L Moseley Company 361
floppy disk drive 192, 436, 443
Thomas H Flowers 139–141, 177
Hannibal C Ford 75–76
Ford Instrument Company 75–76, 209
Ford Motor Company 302
Ford Rangekeeper 74–76
Jay W Forrester 195–196, 273, 330, 354
FORTRAN 223, 334, 371, 447, 488, 513
Fortune magazine 297, 370
Four-Phase Systems 400, 405
Four-Phase Systems AL1 400
Four-Phase Systems Model IV/70 400
Foxboro Corporation 332
frame buffer 466, 484, 498, 500
France Telecom 543
France Telecom Teletel/Minitel 543
Stanley P Frankel 319, 335
Franklin Institute 79
Robert M Frankston 436
Edward Fredkin 366
Donald French 433
Gordon French 421, 423
FRESS (File Retrieval and Editing System) 540
Friden EC-130 398
Paul J Friedl 438–439

Thornton C Fry 108
functional parallelism 256
Daniel H Fylstra 436

G

Gang of Nine 524
GARDE (Gathers Alarms Records Displays Evaluates) 327
William Gascoigne 32
William H (Bill) Gates III 505–516, 525–531, 549
GE 312 Digital Control Computer System 326
GE-415 322
GE-425 295
GE-435 295
GE-455 295
GE-465 295
GE-600 series 295, 308
geared astrolabe 12
GE Compatibles/400 295
GE Computer Department 289, 295, 337
GE Model 210 287
General DataComm Industries 412
General Electric Company (GEC) 182, 196, 271, 303
General Electric (GE) 76, 77, 268, 277, 281, 285, 287–291, 294–296, 302, 308, 326, 506, 512
General Electric Radio Tracking System (GERTS) 277
General Electric Research Laboratory 126
General Instrument 402, 408
General Motors 370
General Motors Research Laboratories (GMR) 370

565

General Post Office (GPO) 139, 155, 457
General Precision Equipment Corporation 320
Geometry Engine 490–492
Georgia Institute of Technology 207
François Gernelle 411
Paul Gilbert 508
John T (Jack) Gilmore 356, 373
Ginn & Company 470
Glenn L Martin Aircraft Company 280
Globe-Union Incorporated 339
Jacob E Goldman 460, 472
Herman H Goldstine 145–148, 158, 170, 184, 198
Tito Gonnella 58
I Jack Good 141, 180
Göttingen University 26
Government Code and Cypher School 137
Government Communications Headquarters 141
Grafacon 380, 386
GRAIL (GRAphical Input Language) 465, 469
GRAPHIC 1 remote graphical display console 386
Graphical Data Display Manager (GDDM) 529
graphical user interface 471, 475, 526, 545
3-D graphics 489
bitmapped graphics 258, 466, 478, 481, 484, 495–496, 525
molecular graphics 491
raster graphics 355, 373, 376
vector graphics 356, 359, 363, 371, 387, 390, 462, 464

Graphics Device Interface (GDI) 527
Gresham College 9
Greylock Partners 481
Jean-Baptiste Vaquette de Gribeauval 33–34
Richard L Grimsdale 270
Roberto A Guatelli 16
Benjamin M Gurley 331, 366, 374, 376
Gypsy 471

H

John W Haanstra 297, 309
Philipp Matthäus Hahn 27, 37, 42
Francis E Hamilton 116, 119, 214
Handwell Corporation 517
Marc R Hannah 490
Harmonic Analyser 64–67
John V (Jack) Harrington 451–452
James R Harris 272
Douglas R Hartree 168, 177, 184, 192, 198
Harvard Aiken Computation Laboratory 510
Harvard Architecture 120, 160, 277, 279, 280, 329
Harvard Business School 435
Harvard University 89, 115, 128, 168, 212, 392, 467, 482, 489, 509
Byron L Havens 238, 315
William M (Trip) Hawkins 495
Hayes Microcomputer Products 453
Hayes Smartmodem 453
Munro K Haynes 268
Hazeltine Corporation 359
Heath Robinson 138
Heinrich Hertz Institut für Schwingungs-

Index

forschung 132
Helsinki University of Technology 547
John L Hennessy 490, 492
Olaus Henrici 66
Joseph Henry 87
Henschel Aircraft Company 100, 103
Johann M Hermann 57–58
Hero of Alexandria 10
John F W Herschel 40–41, 44, 53
Hewlett-Packard 424, 430, 433, 437, 483, 492, 494–495, 500, 520–522, 524, 530, 537
hidden line removal 377
Jack Hill 208, 210
Helmut Hoelzer 276
Jean A Hoerni 340
Marcian E Hoff 406–408
Herman Hollerith 92–98, 189, 232
Hollerith Electric Tabulating System 95
Hollerith Electronic Computer 232
Rod Holt 430
Homebrew Computer Club 421, 422, 424
Homestead High School 424
Honeywell 150, 298, 302, 309, 326, 343, 398, 405, 481, 509, 511
Honeywell H-200 298, 322, 344
Honeywell H-290 327
Honeywell H-800 368
Honeywell Kitchen Computer 415
Honeywell Series 16 309, 344, 415
Honeywell Series 6000 309
Robert Hooke 15
Hovey-Kelley Design 495
HP-65 programmable calculator 425
HP-85 425, 520

HP-120 521
HP-125 521
HP-150 521
HP 2116A 344
HP 2621 521
HP 9100A 445, 520
HP Capricorn 521
HP Corvallis Division 520
HP General Systems Division 521
HP LaserJet 500
HP NewWave 530
HP PA-RISC 537
HP Personal Applications Manager (PAM) 522
HP Personal Office Computer Division 521
HP Vectra 522
HTML (HyperText Mark-up Language) 545
HTTP (HyperText Transfer Protocol) 546
Hughes Aircraft Company 319, 359, 399
Cuthbert C Hurd 216, 217, 219, 245
Harry D Huskey 177, 185, 316–318
Christiaan Huygens 24
Gilbert P Hyatt 399
hyperlink 551
hypertext 464, 539–541
Hypertext Editing System (HES) 540

I

IAS Electronic Computer 170–175, 187, 269, 284, 353, 357
IBM 305 RAMAC (Random-Access Memory Accounting Machine) 221
IBM 701 Electronic Data Processing Machine 217, 237, 240, 241, 363

567

The Story of the Computer

IBM 702 Electronic Data Processing Machine 218, 315

IBM 704 Electronic Data Processing Machine 223, 227, 240, 246, 247, 291, 363, 365, 370

IBM 705 294, 323

IBM 709 291, 292, 294, 363, 365

IBM 740 CRT Output Recorder 362

IBM 780 CRT Display 363, 370

IBM 801 492

IBM 1130 343, 388, 439, 462

IBM 1130/2250 Graphic Data Processing System 388

IBM 1401 Data Processing System 283, 295, 299, 304, 321

IBM 1410 296

IBM 1620 Data Processing System 296, 321, 330, 343

IBM 1710 Industrial Control System 330

IBM 2250 Display Unit 372, 383, 388, 390, 540

IBM 4860 PCjr 519

IBM 5100 Portable Computer 439

IBM 5110 Computing System 440

IBM 5120 Computing System 440, 441

IBM 5150 Personal Computer 437–444, 477, 513, 516

IBM 5155 Portable PC 519

IBM 5160 PC XT 519

IBM 7030 Data Processing System 247, 252, 259

IBM 7070 294, 296, 299

IBM 7080 294, 299

IBM 7090 Data Processing System 248, 255, 291, 294, 296, 299, 322, 371

IBM 7094 253, 303, 367, 387

IBM 7950 Harvest 248

IBM 7960 Special Image Processing System 371

IBM and the Seven Dwarves 302

IBM Corporate Management Committee (CMC) 442

IBM Endicott Laboratories 213, 219, 239, 294, 296, 322, 359, 363

IBM Entry Level Systems (ELS) 439, 534

IBM General Systems Division (GSD) 438

IBM Hursley Laboratory 297, 529

IBM (International Business Machines Corporation) 16–17, 99, 116, 133, 135, 150, 167, 189, 195, 200, 212–224, 231, 232, 238, 245, 256, 262, 268, 281, 287, 291–292, 296–300, 314, 319, 321–323, 330, 343, 358–361, 367, 382, 388, 460, 490, 495, 513–516, 523, 532, 534, 537, 538, 540, 552

IBM Los Gatos Advanced Systems Development Laboratory 439

IBM OS/360 299

IBM PALM processor 438

IBM Palo Alto Scientific Center 438

IBM PC AT 519

IBM PC-DOS 444, 515

IBM Personal System/2 523–524

IBM PL/1 299

IBM Poughkeepsie Laboratory 216, 219, 239, 241, 245, 281, 292, 296, 322, 358

IBM SCAMP 438

IBM SPREAD Task Group 297, 309

IBM System/23 Datamaster 440

IBM System/34 441

IBM System/360 296–300, 305, 309, 343, 368, 388, 400

IBM System/360 Model 30 299

Index

IBM System/360 Model 40 465
IBM System/360 Model 50 540
IBM System/360 Model 67 368
IBM System/370 310, 439, 529, 540
IBM System/370 Model 145 310
IBM TopView 529
IBM Type 350 Disk File 221
IBM Type 405 Accounting Machine 216
IBM Type 407 363
IBM Type 603 Electronic Multiplying Punch 213, 216, 261
IBM Type 604 Electronic Calculating Punch 213, 219, 281, 321
IBM Type 607 321
IBM Type 608 Transistor Calculator 281
IBM Type 610 Auto-Point Computer 316, 322, 450
IBM Type 650 Magnetic Drum Data Processing Machine 220, 222, 229, 282, 294, 315, 321, 323
IBM VM/370 540
ICI (Imperial Chemical Industries) 329
ICL (International Computers Limited) 308, 479
icon 471, 475
Iconoscope 164, 170
ICT 1301 304
ICT 1501 304
ICT (International Computers & Tabulators Limited) 303
IEEE-488 interface 433, 447, 450, 479, 521
ILLIAC 357
Image Computer Systems Limited 544
Imlac Corporation 389–390
Imlac PDS-1 Programmable Display System 389–391, 464, 540
Imlac PDS-4 391
Imperial College 177
Independent Broadcasting Authority (IBA) 542
Industrial Revolution 27, 29, 36, 41
Industria Macchine Elettroniche (IME) 398
inertial guidance 275
Information International Incorporated 332
Information Processing Techniques Office (IPTO) 380, 392, 449, 457
information retrieval 539
Information Sciences Incorporated (ISI) 507, 509
Daniel H H Ingalls 469
Gary Ingram 421–422
initial public offering (IPO) 483, 489, 492, 537, 549
Institut de Recherche d'Informatique et d'Automatique (IRIA) 459
Institute for Advanced Study (IAS) 158, 168, 170, 187, 192, 198, 201, 216
Institut National de la Recherche Agronomique (INRA) 411
instruction lookahead 246
instruction pipelining 241, 246, 250, 536
Integraph 60–61
integrated circuit 298, 301, 338–340, 397, 399
Integrating Tabulator 97
Intel 1101 404
Intel 1102 405
Intel 1103 405
Intel 3101 404
Intel 4004 407, 411
Intel 8008 408, 411, 418, 438, 508

Intel 8080 409, 417, 425, 430, 434, 510
Intel 8080A 420, 423, 440
Intel 8085 440, 476, 480
Intel 8088 442, 520, 521
Intel 80286 519
Intel 80386 523, 531
Intel Corporation 403–410, 415, 434, 478, 525, 531, 536
Intel i860 533
Intel Multibus 481, 484, 485, 487, 491
Intel Pentium 536
Intel Pentium Pro 536
Intel SIM4-01 409
Intel SIM8-01 409, 413, 418
interactive computer games 382, 538
interactive computer graphics 355, 368, 435, 465
Interactive Data Corporation 436
interchangeability 33
intercontinental ballistic missile (ICBM) 275, 280, 360
Interface Age magazine 427
Interface Message Processor (IMP) 457
Interleaf Incorporated 488
Interleaf OPS-2000 488
International Federation for Information Processing (IFIP) 456
Internet Protocol (IP) 460
Intersil IMS 6100 438
Iowa State College 132, 144
IRIS Graphics Library (GL) 491, 536
IRIX 493
ISA (Industry Standard Architecture) bus 443, 519, 523
Harold Isherwood 70–71, 76
Itek Corporation 373

ITT 7300 ADX 332
ITT (International Telephone & Telegraph Corporation) 332

J

Rob Janoff 430
Charles G Jarvis 49, 51
Thomas Jefferson 34
W Stanley Jevons 89
Steven P Jobs 424–428, 432, 495–501, 527, 538
Johns Hopkins University 90, 143
Eric A Johnson 522
Timothy E Johnson 377
Andrew Johnson-Laird 515
William N (Bill) Joy 486

K

Robert E Kahn 459
William M Kahn 398
Katholieke Universiteit Nijmegen 540
Alan C Kay 447, 462–463, 466
John G Kemeny 512
Johannes Kepler 17, 20, 21
Vinod Khosla 485–486, 488
Tom Kilburn 164–167, 180, 196, 197, 249, 270
Jack S Kilby 339–340, 401
Gary A Kildall 514–515
King Henry VIII 14
Kirtland Air Force Base 416
Leonard Kleinrock 455–456, 458
Kodak 460
Alan Kotok 333
Daniel Kottke 426
Herbert Kroemer 282

Index

Thomas E Kurtz 512

L

Clair D Lake 116, 119
Lakeside Programming Group 506–507, 512
Lakeside School 505–506, 511
Butler W Lampson 462, 465, 466, 470, 473, 476
Laplaciometer 133
LARC (Livermore Automatic Research Computer) 241–244, 245, 246, 248, 249, 252
large-scale integration (LSI) 399, 541
Christopher R Larson 511
laser printer 474, 478, 488
Lawrence Radiation Laboratory 255, 259
T Vincent Learson 222, 296
Gottfried Wilhelm Leibniz 24–26, 88, 97
Leicester Polytechnic 546
Steven W Leininger 434
Ernest H Lenaerts 185, 198
John Leng 313
John J Lentz 315
LEO Computers Limited 199, 306
LEO II 306
LEO III 368
LEO (Lyons Electronic Office) 198–199
Leonardo da Vinci 16–17, 30
Leprechaun 272
Librascope Equation Solver 319
Librascope Incorporated 317, 323
Joseph C R Licklider 366, 380, 392, 455, 457, 463
David E Liddle 473, 475
light gun 355, 359, 372

light pen 372, 376, 378, 387, 392, 464
Julius E Lilienfeld 264
LINC (Laboratory Instrument Computer) 448–450, 457, 461, 465
line buffer 385
Linotype & Machinery Company 69, 72, 74
Lockheed 375
Sir Ben Lockspeiser 197
Locomotive Software LocoScript 522
logarithms 8
Logical Piano 89
logic machine 18, 89, 138
Logistics Research Company 361
LOGO 522
LOGRINC Digital Graph Plotter 361
Lord Kelvin 57, 61–67, 68, 70, 72, 79, 82, 361
Lorenz machine 138
Los Alamos Scientific Laboratory (LASL) 175, 245, 247, 259, 319
Lotus 1-2-3 525
Lotus Development Corporation 525
Ada Lovelace 55, 156
William C Lowe 441
George Lucas 501
Lucasfilm 501
Mark Lucovsky 533
Herman Lukoff 202, 241
J Lyons & Company 185, 198, 205, 230, 306

M

machine tool 27–31
numerically controlled (NC) machine tool 369

Walter H MacWilliams 271
MADDIDA 192, 203, 317, 361
magnetic amplifier 268
magnetic core storage 189, 193–196, 228, 242, 246, 255, 273, 278, 280, 286, 290, 322, 323, 354, 358, 372, 387
magnetic disk storage 221, 247, 257, 320, 374
magnetic drum storage 189, 242, 251, 270, 280, 282, 313, 315, 317, 354, 358, 366, 375, 380
magnetic ink character recognition (MICR) 289–290
Magnetic Switching Computer (Magstec) 280
magnetostrictive delay line 163, 230, 327, 335
mainframe computer 313, 436
Thomas Malloy 471, 496
Manchester Mark 1 180–182, 187, 192, 222, 270
Manchester University 82, 164, 168, 177, 187, 192, 217, 249, 252, 270, 272
Manchester University Transistor Computer 270–271
Manhattan Project 158, 319
Jerrold C Manock 430, 498
Guglielmo Marconi 91, 124
Marconi Company 306
marginal checking 359
Thomas Marill 456
Johann Maritz 29
Mark-8 417, 433
Armas C (Mike) Markkula 428, 435
Allan Marquand 90
Robert M Marsh 421
Martin Marietta 375

Massachusetts Institute of Technology (MIT) 67, 77–82, 91, 93, 128, 190, 195, 262, 273, 278, 353, 383, 393, 532, 546
mass production 27, 33, 36
Mathatronics Incorporated 398
Mathatronics Mathatron 398
Matrox Graphics 536
Matsushita 523
John W Mauchly 143–151, 157, 160, 167–169, 184, 201–206
Henry Maudslay 31, 32, 35, 43
James Clerk Maxwell 62
MAYBE 372
Mayfield Fund 491
Maze War 390
Stanley Mazor 406–407
MCA (Micro Channel Architecture) bus 523
Dan McCabe 527
John McCarthy 365–367
C Peter McColough 460
Marc McDonald 512–513
McDonnell Douglas Corporation 390
W Wallace McDowell 261, 281
Harold McFarland 347
MCGA (Multi-Color Graphics Array) 524
Scott A McGregor 527–528
MCI Contourama IV 399
Scott G McNealy 485–486, 488
John C McPherson 215, 238
MDA (Monochrome Display Adapter) 443, 518
Carver A Mead 490
Cecil Mead 304, 308
Ralph I Meader 206

Index

Mean Time Between Failure (MTBF) 261, 279, 280, 326
measuring instrument 31
mechanical integrator 57
medium-scale integrated (MSI) circuit 346
memex 539
Luigi Federico Menabrea 55
mercury delay line 161
Diana Merry 469
Mesa 473, 475
metal-oxide-silicon (MOS) 400
Robert M (Bob) Metcalfe 467, 472, 525
Meteorological Analyser 65
Metropolitan-Vickers Electrical Company 82, 164, 271
Metropolitan-Vickers MV950 271
Mica 533
microcomputer 415, 422, 425, 428, 431, 433, 435, 438
Micro Computer Incorporated (MCI) 399
microcontroller 403, 453
micrometer 32–33
microprocessor 397–410
microprogramming 298, 466, 479
Microsecond Engine (MUSE) 249–253
Microsoft 54, 512–516, 525–538, 549, 552
Micro-Soft 433, 509, 511
Microsoft Applications Development Group 525
Microsoft Excel 530
Microsoft Interactive Systems Group 527, 528
Microsoft Interface Manager 526
Microsoft Internet Explorer 550
Microsoft Internet Information Server 550
Microsoft MS-DOS 515, 519, 524, 526, 538
Microsoft Multiplan 525
Microsoft Surface 552
Microsoft Windows 528–538, 547
Microsoft Windows 2.0 529
Microsoft Windows 3.0 531, 534, 538
Microsoft Windows 3.1 531
Microsoft Windows 95 550
Microsoft Windows 98 550
Microsoft Windows 2000 537
Microsoft Windows for Workgroups 3.1 531
Microsoft Windows NT 533–537, 550
Microsoft Windows Paint 529
Microsoft Windows Write 529
Microsoft Word 526
Microsoft Xbox 552
Forrest M Mims III 416
MINAC 319
minicomputer 313, 334, 338, 340, 535
Ministry of Defence 457
Ministry of Supply 339
Ministry of Technology 307
MIPS Computer Systems 492–494, 537
MIPS R2000 493
MIPS R2010 493
MIPS R2020 493
MIPS R3000 494, 534
MIPS R4000 494
Mischgerät 276
MIT Artificial Intelligence Lab 462
MIT Aspen Movie Map 541

Mitchell Camera Corporation 363

MIT Communications Biophysics Laboratory (CBL) 448

MIT Computation Center 365

MIT Electronic Systems Laboratory (ESL) 383

MIT Lincoln Laboratory 273, 330, 358, 373, 375, 448, 452, 456

MIT Radiation Laboratory 145, 161, 164, 238, 315, 354

MIT Research Laboratory of Electronics 273, 366, 455

MITS 300/25 Business System 420

MITS 680bt 420

MITS 816 417

MITS 8800bt 420

MITS Altair 680 420, 511

MITS Altair 680b 420

MITS Altair 8800 418–421, 425, 429, 433, 451, 510

MITS Altair 8800b 420, 423

MIT Servomechanisms Laboratory 354, 369

MITS (Micro Instrumentation and Telemetry Systems) 416–420, 510–513

modal interface 470

modeless interface 470

Model Rocketry magazine 416

Modem 451–453

Monroe Calculating Machine Company 86, 115, 402

Monroe Royal Digital III 402

Gordon E Moore 403–404, 406

Moore School of Electrical Engineering 82, 143–151, 155, 157, 167–169, 187, 190, 225, 262, 284, 354

Sir Samuel Morland 23, 552

Samuel Morse 87

Mosaic Communications Corporation 549

MOS Technology 6502 410, 425, 431, 513

MOS Technology Incorporated 410, 415, 431

MOS Technology KIM-1 431

Mostek Corporation 401

Mostek MK6010 401

Motorola 401, 410, 415, 494

Motorola MC6800 410, 420, 425, 434, 450, 511

Motorola MC6809 497

Motorola MC6845 443

Motorola MC68000 481, 484, 485, 491, 492, 496

Motorola MC68010 487, 491

Motorola MC68020 488, 492

Motorola MC68881 488

Timothy Mott 471

Frank C Mullaney 208, 253

Mullard 251

Mullard Application Laboratory 541

Multics 482, 483

multiprogramming 252, 367, 479, 482, 496, 524, 527

Museum of the History of Science 12

Nathan Myhrvold 54, 532

N

John Napier 8–10

Napier's Bones 8–10, 19

NASA Ames Research Center 390

NASA Goddard Space Flight Center 337

NASA (National Aeronautics & Space Administration) 215, 325, 332, 337, 349, 381, 390, 449, 457, 465, 540

Index

James Nasmyth 51, 53
National Advisory Committee on Aeronautics 114
National Center for Atmospheric Research (NCAR) 259
National Computer Conference 476
National Defense Research Committee 129, 148
National-Elliott 328
National Institute for Research in Nuclear Science (NIRNS) 252
National Physical Laboratory (NPL) 175, 183, 184, 187, 196, 231, 232, 317, 456
National Research Development Corporation (NRDC) 230, 249, 304, 367
National Security Agency (NSA) 208, 245, 248, 281, 283, 376, 386
National Semiconductor 428, 434
National Semiconductor SC/MP 434
Nautical Almanac Office 64, 215
NBC 541
NCR 310 325
NCR 7200 512
NCR (National Cash Register Company) 129, 130, 200, 206, 231, 288, 290, 302, 325, 328, 512
NCSA Mosaic 547–550, 551
NCSA (National Center for Supercomputing Applications) 547
Neiman-Marcus 415
David L Nelson 481
Theodor H (Ted) Nelson 539
Netscape Communications Corporation 549–550
Netscape Navigator 549–550
NetSite Communications Server 549–550

Network Control Program (NCP) 459
Network File System (NFS) 489
Max H A Newman 138, 155, 165, 169, 175, 180
Newsweek magazine 150
Sir Isaac Newton 26, 40, 53, 67, 426
NeXT 501, 545, 547
NeXTStep 545
NICHOLAS 230
Nippon Calculating Machine Company 401, 405
Kazuhiko (Kay) Nishi 514
NLS (oN-Line System) 464–465, 470, 540
NM Electronics 403
Nobel Prize for Physics 267, 340
Nokia 552
Nokia 9000 Communicator 552
NOMAD 372
NORC (Naval Ordnance Research Calculator) 238–241, 245, 315
William C Norris 206, 210, 253
North American Aviation 400
Northrop Aircraft Company 192, 202, 215, 317, 361
Northwestern Aeronautical Corporation 206
Novell NetWare 535
Robert N Noyce 339–340, 403, 406

O

OARAC 287
object-oriented programming 469
OEM (Original Equipment Manufacturer) 482, 487, 531, 538, 551
Office of Aerospace Research 367
Office of Air Research 171

Office of Naval Research 168, 171, 192, 207, 238, 357
Russell S Ohl 263
Homer R Oldfield 289, 291, 295
Ing. C Olivetti & Co. SpA 308
Olivetti-General Electric 308
Kenneth H Olsen 273, 330, 341, 366, 397, 415
Onyx Systems 485
John R Opel 439, 441
OpenGL 536
ORACLE (Optional Reception of Announcements by Coded Line Electronics) 542
OS/2 524, 529, 531, 532, 534
Oscillation Valve 124
William Oughtred 9
out-of-order execution 257, 536
Charles E Owen 230–231

P

Pace Micro Technology 453
Packard Bell 335
Packard Bell PB250 335
packet switching 454
page fault 481
George E Pake 461, 466, 472
Max Palevsky 334, 341, 348, 461
Jean Laurent Palmer 33
Ralph L Palmer 213, 217, 245, 291
Seymour Papert 462, 469
William N Papian 195, 273
parametric modelling 377
PARC On Line Office System (POLOS) 465, 470
John E Parker 206

John T Parsons 368
Parsons Corporation 368
Blaise Pascal 20–23
Pascaline 20–23
Pascal programming language 444, 447, 473, 479, 488, 513
Tim Paterson 514
PC compatible 516
PC Magazine 518
p-code 479, 525
PC World magazine 526, 531
Charles I (Chuck) Peddle 410, 431, 519
Charles Sanders Peirce 89
Nicola Pellow 546
pen plotter 361–362, 379, 451
PepsiCo 500
PERQ Operating System (POS) 479
PERQ Systems Corporation 480
Personal Automatic Calculator (PAC) 315
personal computer (PC) 445, 447
personality module 423
Personal Software 436
Pertec Computer Corporation 420, 513
Peter the Great 26
Byron E Phelps 213–214
Philco 2000 Model 212 303
Philco Corporation 242, 273, 278, 280, 281, 283–285, 291, 302
Philco-Ford Corporation 302
Philco TRANSAC S-1000 285
Philco TRANSAC S-2000 285, 291, 302
Philco Western Development Laboratories 285
Philips Research Eindhoven 540
Philon of Byzantium 10

Index

Phoenix Software Associates 518, 523
Jules Piccus 16
Greenleaf Whittier Pickard 263
PICO1 402
Pico Electronics Limited 402
piece table 470
John R Pierce 265
Pilot Model ACE 178, 187, 231, 232, 456
John M M Pinkerton 199
Pitney Bowes 290
Pixar 501
Pixar Image Computer 501
Planimeter 58–60, 62
plated-wire memory 301
Playboy magazine 538
Plessey Company 251, 339, 544
plugboard 98, 313
PNX 479
John William (Bill) Poduska 481, 483
Polar Planimeter 59–60
Polish Cipher Bureau 138
Arthur Joseph Hungerford Pollen 69–74
Bernd Pollermann 546
James H Pomerene 171, 245, 248
Popular Electronics magazine 416, 418, 422, 427, 510
Popular Mechanics magazine 313
Popular Science magazine 433
Post Office Research Station 139, 177, 543
Valdemar Poulsen 190
Louis Pouzin 459
Powers Accounting Machine Company 98, 200
Powers-Samas Accounting Machines 303
Powers-Samas Programme Controlled Computer (PCC) 303
Pratt & Whitney Machine Tool Company 96
Presentation Manager 529, 532, 534
Prestel 543
George R Price 370
Prime Computer 481, 485, 493
PRIMOS 482
Princeton University 90, 158, 171, 393
Processor Technology Corporation 421
Processor Technology VDM-1 Video Display Module 422
Procter & Gamble 514
Prodigy Communications Corporation 551
Product Integraph 79
Program for Numerical Tooling (PRONTO) 369
programmable array logic (PAL) 497
programmable read-only memory (PROM) 279
Project Alpine 372
Project Athena 279–281
Project Chess 441–445, 513, 515
Project GEM (Graphic Expression Machine) 371
Project Genie 392, 462
Project MAC (Project on Mathematics and Computation) 367, 369, 384, 456
Project Mercury 438–440
Project Psycho 532–535
Project PX 146–149
Project SAGE 222, 357–361, 371, 452, 454
Project Stretch 245–249, 282, 284, 292, 296, 322, 367

Project Whirlwind 262, 273, 330, 353–357, 365, 369, 373, 383, 452
Prudential Insurance Company 204, 206
Earle W Pughe 373
pull-down menu 496
punched card tabulating machine 92–98

Q

QDOS (Quick and Dirty Operating System) 514–515
QST magazine 413
Quarterdeck Desq 528
Queen Christina of Sweden 23
Queen Elizabeth II 551
Queen Ludwika Maria Gonzaga 23
Queen Victoria 67

R

R2E Micral N 411
Jacob Rabinow 221
Radio-Electronics magazine 417, 434
Radio Receptor Company 281
Radio Shack 431, 432, 433–435, 437, 512, 519
Radio Shack TRS-80 433–435, 513
Jan A Rajchman 172, 195, 286
dynamic RAM (DRAM) 404
static RAM (SRAM) 404
Ramo-Wooldridge Corporation 246, 326
Ramo-Wooldridge RW-300 Digital Control Computer 326
Jesse Ramsden 30–31
RAND Corporation 175, 276, 360, 362–363, 379, 454–455, 465
random-access memory (RAM) 404
RAND Tablet 379, 465

Rapid Arithmetical Machine 128–129, 539
Jef Raskin 495, 497
Raytheon 135, 167, 201, 266, 291, 309, 358, 398
RCA 301 287, 300, 304
RCA 501 Electronic Data Processing System 286, 300
RCA 601 287
RCA Homefax 541
RCA Laboratories 170, 195, 286, 541
RCA (Radio Corporation of America) 126, 149, 167, 207, 281, 285–287, 290, 300, 302, 304, 309, 339, 363
RCA Spectra 70 300, 305, 307, 310
Reactive Engine 462
read-only memory (ROM) 251, 329, 406
Réalisation d'Études Électroniques (R2E) 411
Redstone Arsenal 335
David Rees 169, 180
Regis McKenna 430
William M Regitz 405
Relay Interpolator 111
Reliance Engineering 416
Rao V Remala 527
Remington Rand Athena 280
Remington Rand Incorporated 146, 200–212, 217, 223, 228, 241, 245, 280, 288, 314, 358
remote procedure call (RPC) 545
Research Corporation 135, 137
Richard of Wallingford 14
Ronald E Rider 382, 472
ring counter 128, 140, 146, 161, 398
ring structure 376
RISC (Reduced Instruction Set Comput-

Index

ing) 257, 492, 532
RLE timesharing system 367, 382
John V Roach 432, 434
H Edward Roberts 416–421, 422, 510–513
Lawrence G Roberts 377, 456–458
Alec A Robinson 181, 197
Arthur J Rock 336, 341, 348, 404
Rockefeller Differential Analyzer 91, 129, 170, 184
Rockefeller Foundation 91, 170, 192
Kevin Rogers 544
Brian Rosen 474, 478–479
Douglas T Ross 369, 376, 383
Ken Rothmuller 495, 496
Royal Air Force 164, 307
Royal Artillery 73
Royal Astronomical Society 40
Royal Institution 40
Royal McBee Corporation 320
Royal Navy 68, 70, 76, 77, 230
Royal Precision LGP-21 320
Royal Precision LGP-30 319–320, 323, 335, 366
Royal Radar Establishment 298, 338, 522
Royal School of Mines 61
Royal Scottish Society of Arts 62
Royal Society 24, 40, 42, 45, 49, 51, 52, 64, 65, 67, 125, 165
Royal Society Computing Machine Laboratory 178, 180
RS-232 interface 520
Morris Rubinoff 284
Stephen R Russell 382
Rutherford Appleton Laboratory (RAL) 480
Arnold J Ryden 253

S

S3 Graphics 538
S-100 bus 421, 514
SABRE (Semi-Automated Business Research Environment) 292, 361
Jack Sams 442, 513
Arthur L Samuel 217, 262, 281
Desmond J Sargent 551
David Sarnoff 286
Jack H Scaff 264
scale-of-two counter 128, 133, 139
SCELBAL 413
Scelbi-8B 414
Scelbi-8H 413
Scelbi Computer Consulting Incorporated 413–414
Georg Scheutz 52
Wilhelm Schickard 17–20
Helmut T Schreyer 101–102, 130–132
Science 445
Science Museum 53, 65
Scientific American 150, 326, 417
Scientific Data Systems (SDS) 334–338, 341, 344, 348, 368, 404, 460, 472
Scope Writer 373–374
Michael M Scott 428, 496
John Sculley 483, 499
SDS 92 341
SDS 900 Series 336
SDS 910 337, 338, 341
SDS 920 337
SDS 930 338
SDS 940 458, 464

579

SDS Sigma series 344, 458
SEAC (Standards Eastern Automatic Computer) 268, 272, 357
Sears Roebuck & Co 336, 444
Seattle Computer Products (SCP) 514
Robert R Seeber 214
Seiko S-500 408
Charles L Seitz 393
Selectron 172, 195
semiconductor 262–267
semiconductor diode 263, 264
semiconductor memory 397, 464
Sequoia Capital 428
L J Sevin 401
SGML (Standard Generalised Mark-up Language) 545
Claude E Shannon 91, 99, 102, 108, 169, 376
SHARE 505
Louay E Sharif 401
Sharp Compet CS-31A 399
Sharp Corporation 398
T Kite Sharpless 146, 161, 169
Robert F Shaw 146, 171, 202
C Bradford Sheppard 162, 164, 202
shift register 131, 406, 407, 426
Masatoshi Shima 406, 409, 434
William B Shockley 161, 264, 340
Sierra Semiconductor 493
Signetics Corporation 341
silicon chip 340, 397
silicon gate technology 406
Silicon Graphics Crimson 494
Silicon Graphics Incorporated (SGI) 489–495, 536–537, 548
Silicon Graphics IRIS 1000 491
Silicon Graphics IRIS 1200 491
Silicon Graphics IRIS 1400 Workstation 491
Silicon Graphics IRIS 2000 series 492
Silicon Graphics IRIS 3000 series 492
Silicon Graphics IRIS Indigo 495
Silicon Graphics IRIS (Integrated Raster Imaging System) 491
Silicon Graphics Personal IRIS 4D/20 494
Silicon Graphics Professional IRIS 4D/60 493
Silicon Graphics SGI Visual Workstation 537
Charles Simonyi 470, 472, 525, 527
Sinclair ZX81 523
Sirius Systems Technology 519
Sirius Victor 9000 519
Sketchpad 375–377, 471
slide rule 9
SLIM (Stanford Language for Implementing Microcode) 490
Ralph J Slutz 171, 269
Small-Scale Experimental Machine (SSEM) 180, 197
Smalltalk 471
smartphone 552
Burrell C Smith 497
David Canfield Smith 475
E E 'Doc' Smith 382
Software Arts 436
Richard G Sogge 345
SOHO (Small Office/Home Office) 522
Solid Logic Technology (SLT) 298, 343
solid-state 242, 245, 261–267
Leslie Solomon 418, 422
Sol Terminal Computer 421–424, 427,

Index

429, 431, 433
Sony 498, 500
Southern California Computer Society 427
Southwest Technical Products Corporation 416
Spacewar! 382
SPARC (Scalable Processor Architecture) 489, 537
Sperry Corporation 209
Sperry Gyroscope Company 75, 209
Sperry Rand Corporation 150, 224, 244, 249, 253, 282, 293, 296, 301, 302, 310, 319
SPICE (Scientific Personal Integrated Computer Environment) 479
Sprague Electric Company 340, 401
Spring Joint Computer Conference 376, 539
Robert F Sproull 393, 471, 489
Spyglass Incorporated 548, 550
SSEC (Selective Sequence Electronic Calculator) 213–215, 217, 219, 238–239, 315
Standard Modular System (SMS) 247, 322, 338
Standard Telephones & Cables 199, 270
Oliver Standingford 198
Stanford Artificial Intelligence Laboratory (SAIL) 462, 468, 470
Stanford MIPS (Microprocessor without Interlocked Pipe Stages) project 492
Stanford Research Institute (SRI) 288, 380, 384, 458, 462, 468, 544
Stanford University 459, 461, 462, 470, 475, 484, 489, 495
Stanford University Network (SUN) 484
Gary K Starkweather 474

Stellar Computer Incorporated 483
Stepped Reckoner 24–27, 37, 47
stepped wheel 25
George R Stibitz 108–114, 123, 144, 155, 168, 201, 451
stored-program concept 156–157
Robert H Stotz 383–384
Christopher Strachey 231, 367
Straza Symbol Generator 384
Almon B Strowger 87
Frank L Stulen 368
Stylator 378
Alan M Sugar 522
Sumlock ANITA 1000 402
Sumlock ANITA Mk VIII 398
Frank H Sumner 251
Sun-1 486
Sun-2 487
Sun-3 488
Sun 3/50 489
Sun 3/75 488
Sun 3/160 488
Sun 3/180 488
Sun386i 537
Sun Microsystems 484–489, 495, 500, 537
SunOS 488
supercomputer 238, 259, 324
Su Sung 13
Ivan E Sutherland 375–377, 386, 392, 462, 471, 489, 501
SWAC (Standards Western Automatic Computer) 270, 317
Swarthmore College 539
Sycor 409
William L Sydnes 441, 442

581

Sylvania Electric Products 262, 291

symbolic logic 88, 100

System Development Corporation (SDC) 361, 456

T

tablet computer 552

Tabulating Machine Company 97

Tandon Corporation 443

Tandy 1000 519

Tape Processing Machine 217–218

Norman H Taylor 373

Robert W (Bob) Taylor 457, 461–465, 466

TCP/IP 460, 545

Herbert M Teager 365

Techniques Incorporated 418

Technische Hochschule 99, 130

Technitrol Engineering Company 169, 222

Technology Venture Investors 516

Tektronix 383–385, 450, 494

Tektronix 4051 Graphics Display Computer 450

Tektronix Type 564 Storage Oscilloscope 384

Tektronix Type 601 Storage Display Unit 384

Tektronix Type 611 Storage Display Unit 385

Tektronix Type T4002 Graphic Computer Terminal 385, 390

Telecommunications Research Establishment (TRE) 139, 164, 177, 181, 183, 184

Teledyne 399

Telegraphone 190

Teletext 542

Teletype 109, 384, 385, 409, 414, 420, 422, 451

Teletype ASR-33 334, 342, 346, 450, 506

Telex Corporation 374

Edward Teller 240–241

Telnet 459

Paul Terrell 426–427

Nikola Tesla 90

Lawrence G (Larry) Tesler 470, 496

Test Assembly 239

Test-Experimental Computer Model 0 (TX-0) 273, 281, 330, 333, 373, 375, 448

Test-Experimental Computer Model 1 (TX-1) 273

Test-Experimental Computer Model 2 (TX-2) 274, 330, 375, 448, 455

Texaco 326

Texas Instruments 282, 289, 298, 339, 390, 401, 403, 406, 417, 431, 517

Charles P (Chuck) Thacker 462, 466, 473

thermionic valve 124–126, 131, 134, 139, 161, 261–262, 266, 281, 286, 288, 398

Charles Xavier Thomas de Colmar 36–39

Thomas J Watson Research Center 492

Thomas J Watson Scientific Computing Laboratory 213, 238, 315

Alan Thompson 329

Raymond Thompson 198

Thompson Ramo Wooldridge TRW-230 332

James Thomson 61, 66, 78

Joseph J Thomson 124

James E Thornton 256, 259, 280, 281

Three Rivers Computer Corporation 474, 478–479

Index

Three Rivers PERQ 479–480
Three Rivers PERQ 2 480
Thyratron 128, 139, 140, 146, 161
Tide Predictor 62–63
Time magazine 445
timesharing 364–368, 482, 506, 540
Titan missile 280
Jonathan A Titus 417, 433
token ring 481
Raymond S Tomlinson 459
Geoff C Tootill 165, 197
Leonardo Torres y Quevedo 86
Gianello Torriano 15
touchscreen 463, 522
trackball 387
TRADIC (Transistorized Airborne Digital Computer) 271–273, 274
traffic counter 508
Traf-O-Data 508–510
Jack Tramiel 431, 512
transfer trapping 365
drift transistor 247, 282, 286
junction transistor 267, 281, 286, 408
MADT transistor 285
point-contact transistor 266, 270
surface barrier transistor 242, 251, 273, 278, 280, 281, 284
transistor 242, 264–267, 397
Transistor Test Computer (Transtec) 280
transistor-transistor logic (TTL) 348, 389, 408, 474
Transmission Control Program (TCP) 459
Irven Travis 167, 168, 225, 277

TRICE (Transistorised Real-Time Incremental Computer Expandable) 335
André Thi Truong 411
TRW Incorporated 509
Tufts University 77
Alan M Turing 138, 156, 159, 175–180, 181
Turing's Bombe 138
Turtle Graphics 469

U

UCSD p-System 444, 515
UDEC 225
UK Atomic Energy Authority (UKAEA) 249, 252
UK Post Office 543, 551
UK Science Research Council 480
Francis O Underwood 321
UniSoft Corporation 487, 491
UNIVAC 90/70 310
UNIVAC 490 Real-Time System 294
UNIVAC 1004 301
UNIVAC 1005 301
UNIVAC 1103AF 255
UNIVAC 1107 Thin-Film Memory Computer 294
UNIVAC 9000 Series 301, 310
UNIVAC File-Computer 210, 283
UNIVAC I 202–206, 209, 219, 222, 224, 241, 243, 261, 269
UNIVAC II 210, 294
UNIVAC III 244, 293
UNIVAC Solid-State 282, 285, 294
University College London 124
University of Altdorf 24

University of California Berkeley 192, 221, 227, 317, 393, 424, 462, 486, 547
University of California Los Angeles (UCLA) 334, 458
University of California Radiation Laboratory 240, 241, 244
University of California Santa Cruz 489
University of Cambridge 39, 53, 62, 82, 127, 138, 156, 169, 177, 183, 298
University of Chicago 170, 336
University of Edinburgh 472
University of Glasgow 62
University of Hawaii 467
University of Helmstedt 26
University of Illinois 175, 262, 268, 330, 357, 463, 547, 548
University of Leipzig 24
University of Massachusetts Amherst 16
University of Miami 416
University of Michigan 457
University of Minnesota 254
University of Munich 58
University of Oxford 12, 164, 544
University of Padua 14
University of Rochester 314
University of Tübingen 17, 18, 20
University of Utah 392, 461, 462, 489, 501
University of Washington 506
University of Wisconsin 222
Unix 479, 483, 486, 491, 493, 514, 532, 536, 545
Ursinus College 144
US Air Defense Command 363
US Air Force 202, 228, 269, 272, 273, 274, 276, 278, 280, 287, 291, 339, 357, 363, 368, 373, 416, 454
US Air Force Cambridge Research Center (AFCRC) 282, 374, 451
US Air Force Office of Scientific Research 380
US Air Reserve Records Center 287
US Armed Forces Security Agency 208
US Army 228, 286, 291, 319, 335, 376
US Army Air Forces (USAAF) 276
US Army Ballistic Research Laboratory 114, 143, 158, 174
US Army Corps of Engineers 314
US Army Map Service 202, 269
US Army Ordnance Department 143, 168, 171
US Army Ordnance Tank and Automotive Command 286
US Army Signal Corps 145, 272, 339
US Atomic Energy Commission 210, 241, 245, 256
US Census Bureau 98, 167, 201, 218, 269
US Census Office 93, 96
US Civil Aeronautics Administration 254
US Defense Communications Agency 455
US Department of Defense 460
US Department of Justice 267, 441, 538, 550
US Federal Communications Commission (FCC) 453
US Federal Trade Commission (FTC) 538
US National Academy of Sciences 449
US National Bureau of Standards 169, 178, 201, 221, 238, 268, 317, 357
US National Defense Research Committee 111
US National Institutes of Health (NIH)

Index

449
US National Science Foundation 365
US Naval Computing Machine Laboratory 206
US Naval Ordnance Laboratory 136, 168, 238, 240
US Naval Postgraduate School 255
US Naval Proving Ground 240
US Naval Research Laboratory 112, 113, 383
US Naval Reserve 118, 206
US Navy 74, 76, 191, 200, 225, 238, 244, 254, 276, 278, 284, 286, 319, 353, 393
US Navy Bureau of Aeronautics 354
US Navy Bureau of Ordnance 74–75, 119, 238
US Patent Office 93
US Supreme Court 126, 127
US Venture Partners 486

V

V-2 rocket 275
Donald T Valentine 428
George E Valley 358, 452
Andries (Andy) van Dam 539
Thomas A Vanderslice 483
vaporware 528
Vermont Research Corporation 366
Versuchsmodell-1 100
vertical blanking interval (VBI) 541
VGA 524
Vickers 68, 82
Vickers Range Clock 68–69
video display board 422
video teleconferencing 464

Viewdata 543, 551
ViolaWWW web browser 547
virtual memory addressing 250, 481
VisiCalc 435–437, 444, 477, 521, 525
VisiCorp VisiOn 526
VLSI Systems 485
VLSI (very-large-scale integration) 484, 489
VMEbus 488, 493
Andreas (Andy) von Bechtolsheim 484–486, 491
John von Neumann 158–160, 167–168, 170, 183, 201, 216, 227, 239, 277, 353, 357
von Neumann Architecture 160

W

Nat Wadsworth 412–414, 415
An Wang 194
Willis H Ware 171
Washington State University 508
Washington University 449, 461
wastebasket 471
Thomas J Watson 99, 116, 189, 201, 212, 214, 216
Thomas J Watson Jr. 216, 248, 259, 299
James Watt 29, 32
Ronald G Wayne 425–428
Wayne State University 225, 317
web browser 545–550
Richard W (Ric) Weiland 506–507, 512, 513
Weitek 493
Western Electric Company 96, 280, 452, 514
Kaspar Wetli 59
Sir Charles Wheatstone 63

585

wheel cutting engine 15
David J Wheeler 186–187
Eli Whitney 34
Sir Joseph Whitworth 35, 46, 51, 53
Norbert Wiener 171
Maurice V Wilkes 169, 183–187, 196, 198, 253, 298
James H Wilkinson 176
John Wilkinson 29
Frederic C Williams 164–167, 177, 180–183, 196
Williams Tube 164–167, 172, 180, 197, 217, 218, 230, 239, 354, 357
Alan H Wilson 263
WIMP (Windows, Icons, Menus and Pointers) 469–472, 473, 475, 495, 526, 530, 547
Wintel 525, 532, 536
Niklaus E Wirth 478
John R Womersley 175, 184, 232
Steven Wood 533
word length 100
word processing system 373, 435, 440, 470, 474, 487, 522, 538
WordStar 521
World Wide Web 538–551
World Wide Web Consortium (W3C) 546
Stephen G Wozniak 424–430, 432, 436, 495–496, 513
Charles E Wynn-Williams 127, 139
WYSIWYG 469–472, 473, 475, 479, 487, 495

Xerox 1108 Scientific Information Processor 474, 479
Xerox 1132 Scientific Information Processor 474
Xerox 6085 Professional Computer System 477
Xerox 8010 Star Information System 475–478, 496, 497, 527, 530
Xerox Alto 460–468, 480, 484, 495
Xerox Alto II 472, 474
Xerox Alto III 474
Xerox Computer Science Laboratory 462, 466
Xerox Corporation 348, 382, 460–478, 484, 530
Xerox Dandelion 475
Xerox Data Systems (XDS) 348
Xerox Document System Editor 471
Xerox Dolphin 474, 479
Xerox Dorado 473
Xerox General Science Laboratory 461
Xerox Office Products Division 474
Xerox PARC (Palo Alto Research Center) 460–461, 484, 490, 495, 501, 525, 527, 530
Xerox Systems Development Division (SDD) 472, 479
Xerox Systems Science Laboratory 463, 467
Xerox ViewPoint 477
Xerox Webster Research Center 474
Xerox Wildflower 476
X Window System 532, 547

X

Xenix 514, 525, 526
Xerox 800 Electronic Typing System 474
Xerox 850 Display Typing System 474

Y

Yale University 125
William Yates 417–418

Z

Edward J Zander 482, 489

z-buffering 494

Zeiss 74

Zenith Data Systems 519

Zenith Z-151 519

Ziff Davis 418

Zilog Z80 434, 513

Zilog Z80A 520, 521, 522

zippered list 539

Konrad Zuse 99–107, 110, 115, 123, 130, 155, 157, 223, 241, 246, 257

Vladimir K Zworykin 164, 170

Made in the USA
Monee, IL
03 December 2020